This Book
Belongs To:
BOB
GUSTAFSON

CCNP Cisco Networking Academy Program: Multilayer Switching Companion Guide

Cisco Systems, Inc.

Cisco Networking Academy Program

Wayne Lewis, Ph.D.

Cisco Press

201 West 103rd Street

Indianapolis, Indiana 46290 USA

CCNP Cisco Networking Academy Program: Multilayer Switching Companion Guide

Cisco Systems, Inc.

Cisco Networking Academy Program

Wayne Lewis

Published by:
Cisco Press
201 West 103rd Street
Indianapolis, IN 46290 USA

Printed in the United States of America 1 2 3 4 5 6 7 8 9 0

First Printing August 2002

Library of Congress Catalog-in-Publication Number: 2001090451

ISBN: 1-58713-033-5

Trademark Acknowledgments

All terms mentioned in this book that are known to be trademarks or service marks have been appropriately capitalized. Cisco Press and Cisco Systems, Inc. cannot attest to the accuracy of this information. Use of a term in this book should not be regarded as affecting the validity of any trademark or service mark.

Warning and Disclaimer

This book is designed to provide information about multilayer switching. Every effort has been made to make this book as complete and accurate as possible, but no warranty or fitness is implied.

The information is provided on an "as is" basis. The author, Cisco Press, and Cisco Systems, Inc. shall have neither liability nor responsibility to any person or entity with respect to any loss or damages arising from the information contained in this book or from the use of the discs or programs that may accompany it.

The opinions expressed in this book belong to the author and are not necessarily those of Cisco Systems, Inc.

Feedback Information

At Cisco Press, our goal is to create in-depth technical books of the highest quality and value. Each book is crafted with care and precision, undergoing rigorous development that involves the unique expertise of members of the professional technical community.

Reader feedback is a natural continuation of this process. If you have any comments regarding how we could improve the quality of this book, or otherwise alter it to better suit your needs, you can contact us through e-mail at networkingacademy@ciscopress.com. Please be sure to include the book title and ISBN in your message.

We greatly appreciate your assistance.

Publisher — *John Wait*
Editor-in-Chief — *John Kane*
Executive Editor — *Carl Lindholm*
Cisco Systems Management — *Michael Hakkert, Tom Geitner*
Production Manager — *Patrick Kanouse*
Acquisitions Editor — *Shannon Gross*
Project Manager — *Tracy Hughes*
Development Editor — *Ginny Bess Munroe*
Senior Project Editor — *Sheri Cain*
Copy Editor — *Gayle Johnson*
Technical Editors — *Barbara J. Nolley, Howard Rahmlow, Todd White, Stanford M. Wong*
Team Coordinator — *Sarah Kimberly*
Cover Designer — *Louisa Klucznik*
Composition — *Scan Communications Group, Inc.*
Indexer — *Tim Wright*

CISCO SYSTEMS

Corporate Headquarters
Cisco Systems, Inc.
170 West Tasman Drive
San Jose, CA 95134-1706
USA
http://www.cisco.com
Tel: 408 526-4000
800 553-NETS (6387)
Fax: 408 526-4100

European Headquarters
Cisco Systems Europe
11 Rue Camille Desmoulins
92782 Issy-les-Moulineaux
Cedex 9
France
http://www-europe.cisco.com
Tel: 33 1 58 04 60 00
Fax: 33 1 58 04 61 00

Americas Headquarters
Cisco Systems, Inc.
170 West Tasman Drive
San Jose, CA 95134-1706
USA
http://www.cisco.com
Tel: 408 526-7660
Fax: 408 527-0883

Asia Pacific Headquarters
Cisco Systems Australia,
Pty., Ltd
Level 17, 99 Walker Street
North Sydney
NSW 2059 Australia
http://www.cisco.com
Tel: +61 2 8448 7100
Fax: +61 2 9957 4350

Cisco Systems has more than 200 offices in the following countries. Addresses, phone numbers, and fax numbers are listed on the Cisco Web site at www.cisco.com/go/offices

Argentina • Australia • Austria • Belgium • Brazil • Bulgaria • Canada • Chile • China • Colombia • Costa Rica • Croatia • Czech Republic • Denmark • Dubai, UAE • Finland • France • Germany • Greece • Hong Kong • Hungary • India • Indonesia • Ireland Israel • Italy • Japan • Korea • Luxembourg • Malaysia • Mexico • The Netherlands • New Zealand • Norway • Peru • Philippines Poland • Portugal • Puerto Rico • Romania • Russia • Saudi Arabia • Scotland • Singapore • Slovakia • Slovenia • South Africa • Spain Sweden • Switzerland • Taiwan • Thailand • Turkey • Ukraine • United Kingdom • United States • Venezuela • Vietnam • Zimbabwe

About the Author

Wayne Lewis is the Cisco Academy Manager for the Pacific Center for Advanced Technology Training, based at Honolulu Community College (HCC). Since 1998, Wayne has taught Networking Academy instructors from universities, colleges, and high schools in Australia, Canada, Central America, China, Hong Kong, Indonesia, Mexico, Singapore, South America, Taiwan, and the U.S., both onsite and at HCC. Prior to teaching computer networking, Wayne began teaching math at age 20 at Wichita State University, followed by the University of Hawaii and HCC. Wayne received a Ph.D. in math from the University of Hawaii in 1992. Wayne works as a contractor for Cisco Systems Worldwide Education, developing curriculum for the Networking Academy program. Wayne enjoys surfing the North Shore of Oahu when he's not distracted by work.

About the Technical Reviewers

Barb Nolley is the president and principal consultant for BJ Consulting, Inc., a small consulting firm that specializes in networking education. Since starting BJ Consulting, Barb has developed and taught training courses for Novell's Master CNE certification, as well as several courses for Cisco Systems' Engineering Education group and a CCNA track for the University of California-Riverside Extension. Her certifications include the CCNA, CNE, and CNI. She lives in and works out of an RV with her husband, Joe.

Howard Rahmlow, CCAI CCNP MCSE MCNE, has worked in the computer field for more than 20 years. Currently, he is a senior systems analyst for Unisys Corp. He also is a part-time Networking Academy instructor for Montgomery County Community College. He would like to thank his wife, Pat, and daughter, Caroline.

Todd White, CCIE #7072, taught for five years at Mainland High School in Daytona Beach, Florida. He also ran the Cisco Academy Training Center (CATC) for the southeast while there. In addition, he was the CCNP Development Lead for version 1 of the CCNP curriculum.

Stanford M. Wong, CCAI CCIE #8038 MCSE PE, has worked in the networking and computer technology area for more than 18 years. Currently, he is an independent consultant who provides the U.S. Government with design and implementation services for enterprise routing architectures. He also teaches computer operating systems/networking courses at the University of Phoenix and Honolulu Community College in Honolulu, Hawaii. He holds a BSEE degree from the University of Hawaii.

Dedication

To my beautiful wife, Leslie, and daughters, Christina and Lenora.

Leslie, you are such an inspiration to me. Our meeting was true serendipity. Thank you for putting up with your silly husband and for being the best mom in the world. And thank you so much for managing all the business stuff.

Christina, thank you being such a good person, for doing so well in school, and for being such a great piano player.

Lenora, thank you for being such a good person, for always saying that math is your favorite subject, and for being such an expert at Kenpo Karate.

Putting together a book is really a selfish effort, but the support of you three has enabled me to write a book that will hopefully be useful to the readers.

Acknowledgments

This is the part where I list all the folks who made writing this book possible. Well, there's obviously not enough room, but here's my best attempt.

Mom, thanks for showing me how to realize my goals and keep a positive attitude. Dad, thanks for instilling in me a love of learning and scientific endeavor. Thanks to my brother, Richard, for keeping me out of worse trouble than I managed to find. Thanks to my sisters, Nancy, Barbara, Debra, and Sandra, for showing me the things I want to see come true for my daughters. Okaasan, thanks for always being there for our family in Hawaii.

Ginny Bess Munroe, Carl Lindholm, and Tracy Hughes at Cisco Press and Shannon Gross of Cisco Systems, thanks for keeping me on track and providing the professional guidance that made this effort possible.

Thanks to Kevin Johnston and Todd White of Cisco Systems for being the founders of the CCNP Academy Program. This book is built on their efforts.

Thanks to George Ward of Cisco Systems for creating the Cisco Networking Academy Program. Were it not for his vision, I would not have had the opportunity to start a career in computer-networking education.

Thanks to Dennis Frezzo of Cisco Systems for his guidance in all things Cisco Academy-related. Without Dennis, there would be no voice for the students and instructors in the Networking Academy Program.

I am especially thankful to Vito Amato of Cisco Systems for giving me the opportunity to write this book. Vito has been 100 percent supportive of my work since I met him. I am eternally indebted to him for the many opportunities he has made available to me.

Alex Belous of Cisco Systems doesn't get the acclaim he deserves because he's always working behind the scenes, making things happen. Thanks, Alex, for providing support when no one is aware of it.

I want to thank my CATC partner, Dallas Shiroma, for showing me what a good teacher is. Dallas serves as the ideal role model for teachers. Year after year, Dallas selflessly and diligently continues to provide the best in science and technology education.

Don Bourassa and Ramsey Pedersen are my bosses at Honolulu Community College. I cannot say enough about how grateful I am to them for paving the career path for me that began in 1993. Ramsey has been absolutely supportive of me from the beginning. Don is a great boss who leads in a way that all want to follow.

Mark McGregor blazed the trail for me to follow in writing this book. Thanks, Mark, for setting a standard for curriculum development that we can all strive for. (Sorry about ending that sentence in a preposition.)

Special thanks go to this book's technical reviewers: Barb Nolley, Stanford Wong, Todd White, and Howard Rahmlow. They have done an incredible job in revamping the original draft.

Thanks to the Hawaii CCIE Group Study: Stanford (the godfather), Michael Jordan, Rob Rummel, Frank Buffington, Nick Pandya, Robert Yee, Torrey Suzuki, and Errol Gorospe. Stanford and Rob passed, and the rest of us just keep trying.

Finally, I'd like to thank the Hawaii currdev team. I feel privileged to work with you: Torrey Suzuki, Errol Gorospe, Jim Yoshida, and Nick Pandya.

Contents at a Glance

Contents

Introduction

CCNP Cisco Networking Academy Program: Multilayer Switching Companion Guide is designed to supplement your study of multilayer switching concepts in the Cisco Networking Academy Program. This book is useful in its own right as a Cisco multilayer switching reference. Through examples, technical notes, and Check Your Understanding questions, this book provides intermediate- and advanced-level students with the knowledge and skills needed to pass the CCNP switching exam (BCMSN) and to further their career opportunities in computer networking.

Concepts covered in this book include campus network design, VLANs, Spanning-Tree Protocol (STP), inter-VLAN routing, multilayer switching (MLS), Cisco Express Forwarding (CEF), Hot Standby Router Protocol (HSRP), security with Catalyst switches, and an in-depth look at the role of switches in multicasting.

As with all advanced networking topics, students will find that their studies are best complemented by hands-on lab exercises. To that end, Cisco Press offers the *CCNP Cisco Networking Academy Program: Multilayer Switching Lab Companion*, which includes comprehensive lab exercises that can be completed individually or in small groups.

Who Should Read This Book

This book's audience includes students who are seeking advanced Cisco switching configuration skills and certification. In particular, this book is targeted toward students in the CCNP Cisco Networking Academy Program, which is offered in schools around the world. In the classroom, this book can serve as a supplement to the online curriculum.

Another audience for this book includes network engineers presently working in the industry and individuals striving to become network engineers. This book was designed to have a broad appeal and is useful both as a reference and as an introductory text on multilayer switching. For corporations and academic institutions to take advantage of modern networking technologies, a large number of individuals need to be trained in the design and operation of networks.

This Book's Organization

The book has ten chapters, three appendixes, and a glossary.

Chapter 1, "Campus Networks and Design Models," introduces you to campus network design.

Chapter 2, "Gigabit Ethernet," describes the vital role of Gigabit Ethernet in modern campus networks.

Chapter 3, "Switch Administration," introduces you to the command line and the basic management of Catalyst switches.

Chapter 4, "Introduction to VLANs," explains the various types of VLANs. It explains how to best deploy VLANs in a campus network and how to configure VLANs on Catalyst switches.

Chapter 5, "Spanning-Tree Protocol," describes in detail the operation of STP in campus networks. It also details STP configuration and STP design implications.

Chapter 6, "Inter-VLAN Routing," demonstrates the configuration of routing between VLANs using a Cisco router or a route processor specific to a particular Catalyst switch platform.

Chapter 7, "MLS and CEF," explores the two primary technologies used to deliver wire-speed routing with Catalyst switches.

Chapter 8, "Hot Standby Router Protocol," details the use of HSRP to provide Layer 3 redundancy in the context of the core-distribution-access campus network design model.

Chapter 9, "Multicasting," provides an exhaustive survey of the multicast protocols used today to enable the propagation of rich multimedia content over campus networks.

Chapter 10, "Security," describes how to create a security policy and secure Catalyst switches in a campus network.

Appendix A, "Check Your Understanding Answer Key," provides the answers to the Check Your Understanding questions found at the end of each chapter.

Appendix B, "Transparent Bridging with Routers," explains how to configure inter-VLAN bridging with Cisco routers and Catalyst switches.

Appendix C, "Miscellaneous Campus LAN Switch Technologies," introduces a number of emerging campus LAN switch technologies.

The glossary includes the key terms used throughout this book.

This Book's Features

This book contains several elements that help you learn about advanced networking concepts and Cisco IOS technologies:

- **Figures, examples, and tables**—This book contains figures, examples, and tables that help explain concepts, commands, and procedures. Figures illustrate network layouts and processes, and examples provide sample IOS configurations. In addition, tables provide command summaries and comparisons of features and characteristics.
- **Notes and tech notes**—Notes highlight important information about a subject. This book also includes tech notes that offer background information on related topics and real-world implementation issues.

- **Chapter summaries**—At the end of each chapter is a summary of the concepts covered in that chapter. It provides a synopsis of the chapter and can serve as a study aid.
- **Check Your Understanding questions**—After the chapter summary are ten review questions that serve as an end-of-chapter assessment. These questions reinforce the concepts introduced in the chapter and help you evaluate your understanding before moving on to another topic.

The conventions used to present command syntax in this book are the same conventions used in the *Cisco IOS Command Reference*:

- **Bold** indicates commands and keywords that are entered literally as shown. In examples (not syntax), bold indicates user input (for example, a **show** command).
- *Italic* indicates arguments for which you supply values.
- Braces ({ }) indicate a required element.
- Square brackets ([ ]) indicate an optional element.
- Vertical bars (|) separate alternative, mutually exclusive elements.
- Braces and vertical bars within square brackets (such as [x {y | z}]) indicate a required choice within an optional element. You do not need to enter what is in the brackets, but if you do, you have some required choices in the braces.

After completing this chapter, you will be able to perform tasks related to the following:

- Hierarchical network design model for a campus network
- Switch blocks in a campus network
- Key switching technologies in a campus network
- Small, medium, and large campus network designs
- Catalyst switch products

Chapter 1

Campus Networks and Design Models

"The only constant in life is change." Whoever coined this phrase might as well have been a network engineer. The design and implementation of modern networks are in a constant state of flux. However, the fundamental design principles of switching at OSI Layers 2 and 3 have not changed appreciably. The addition of new product lines of Cisco Catalyst switches affords a range of possibilities that allows network planners to optimize their existing designs to promote scalability and increased application availability. Adding to the picture are enhanced features and functionality, such as quality of service and multilayer services, as well as increased trunking bandwidths, increased Gigabit Ethernet port densities, and increased switching capacities.

This book explores the state of the modern campus network. You'll become familiar with the technologies common in such a network. You also learn how to configure Catalyst OS (CatOS)-based and Cisco IOS based switches to implement these technologies.

This chapter provides an overview of the campus network and associated design models. In particular, you learn about the Cisco-defined core-distribution-access layer model; the core-distribution-access layer model is a general switching model that provides a framework for design methodology, much as the OSI model provides a framework for networking in general. The various types of switching, such as Layer 3 switching, are also introduced in this chapter. The Cisco Catalyst switch lines that are appropriate for various switched network designs are also presented.

TECH NOTE: CAMPUS NETWORKS

We have already used the term *campus network* a few times. This term needs some clarification. For many people, the first thing they think of when they hear the term *campus network* is a network on a university campus. It used to be that research universities and the U.S. military were the only game in town with respect to networking. That's how the term *campus network* came about (referring to university campuses). However, you should dissociate in your mind the terms *campus network* and *university network*. The term *campus network* refers to a network that connects devices within and between a collection of buildings. LAN switching and high-speed routing provide connectivity among buildings. Typically, the term *campus network* refers to a set of buildings comprising a corporate presence. It is common to focus on the role of switching in a network when referring to the campus environment.

Noncampus environment is a term used here to describe networks or portions of networks in which LAN switching does not play a major role. For example, Frame Relay, xDSL, and ISDN networks are technologies that might appear in noncampus environments.

Hierarchical Design Model for Campus Networks

If you have been keeping current with the role of switches in campus design over the last ten to 15 years, you have witnessed some dramatic changes. Switches' capabilities have evolved considerably during this time, and as a result, campus network design has changed accordingly. Here, you look at some of the hierarchical models used in campus network design.

NOTE

The term *multiservice application* generally refers to applications that support voice, video, or broadband communication.

A hierarchical network design includes the following three layers:

- Core layer—Provides optimal transport between sites
- Distribution layer—Provides policy-based connectivity
- Access layer—Provides user access to the network for multiservice applications and other network applications

Figure 1-1 shows a high-level view of a hierarchical network design.

The following sections discuss the functions of each layer, starting with the core layer.

Function of the Core Layer

The *core layer* is a high-speed switching backbone. It should be designed to switch packets with minimum latency. Just as with Open Shortest Past First (OSPF) area 0, the term core is synonymous with backbone. This layer of the campus network should not perform any packet/frame manipulation, such as processing access lists and filtering, because this would slow down the switching of packets. It is now common for the core layer to be a pure Layer 3 switched environment, which means that VLANs and VLAN trunks (discussed in Chapter 4) are not present in the core. It also means that spanning-tree loops (discussed in Chapter 5) are typically avoided in the core. But the key function of the core layer is to provide high-speed connectivity between all distribution layer devices in the campus network.

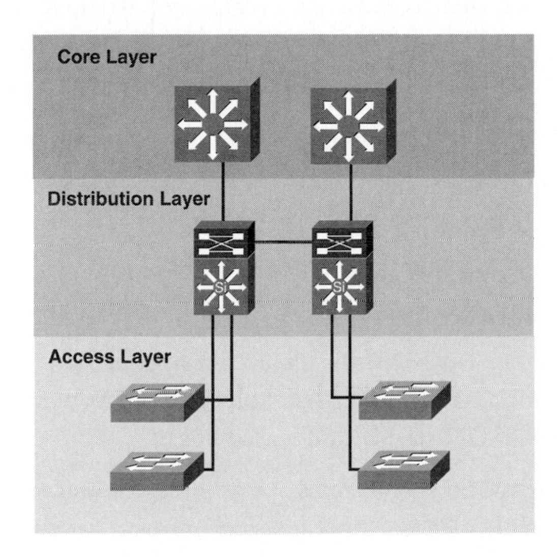

Figure 1-1 A hierarchical network design presents three layers—core, distribution, and access—with each layer providing different functionality.

Functions of the Distribution Layer

The *distribution layer* of the campus network is the demarcation point between the access and core layers. It helps define and differentiate the core. The purpose of this layer is to provide network boundary definition. It is the place where packet/frame manipulation can take place. In the campus environment, the distribution layer can include several functions:

- Address or area aggregation
- Departmental or workgroup access connectivity to the backbone
- Broadcast/multicast domain definition
- Inter-VLAN routing
- Media transitions
- Security

In a noncampus environment, the distribution layer handles redistribution between routing domains and typically is the demarcation between static and dynamic routing protocols. It can also be the point at which remote sites access the corporate network. The distribution layer can be summarized as the layer that provides policy-based connectivity.

Packet manipulation, filtering, route summarization, route filtering, route redistribution, inter-VLAN routing, policy routing, and security are some of the primary roles of the distribution layer.

Functions of the Access Layer

The *access layer* is the point at which local end users are allowed to attach to the network. This layer might also use access lists or filters to optimize the needs of a particular set of users, such as those who regularly participate in videoconferencing. Frequently, Layer 2 switches play a significant role at the access layer. In this context, switches are called *edge devices* because they reside on the network's boundary.

In the campus environment, access layer functions can include the following:

- Shared bandwidth
- Switched bandwidth
- MAC layer filtering
- Microsegmentation

In a noncampus environment, the access layer can give remote sites access to the corporate network via wide-area technologies such as plain old telephone system (POTS), Frame Relay, ISDN, xDSL, and leased lines.

Some people mistakenly think that the three layers (core, distribution, and access) must exist in clear and distinct physical entities, but this does not have to be the case. The layers are defined to aid successful network design and to represent functionality that must exist in a network. The instantiation of each layer can be in distinct routers or switches, can be represented by physical media, can be combined in a single device, or can be omitted altogether. How the layers are implemented depends on the needs of the network being designed. However, for a network to function optimally, hierarchy must be maintained.

Switch Blocks in Network Design

Another concept that aids in hierarchical network design is switch blocks. A *switch block* is a set of logically grouped switches and associated network devices that provide a balance of Layer 2 and Layer 3 services. A switch block is a relatively self-contained, independent collection of devices, typically binding a collection of Layer 3 networks. Switch blocks are used as a tool to facilitate communication about switched network design.

A switch block contains broadcasts by providing a Layer 3 boundary in the form of routers or Layer 3 switches, as shown in Figure 1-2. In addition, a switch block, somewhat like an OSPF area, prevents a network failure from affecting other switch blocks.

Within a switch block, Spanning-Tree Protocol (STP) enables redundant links between switches and prevents bridge loops. STP does not scale well in a campus environment, so it indirectly limits the size of a switch block. Other limiting factors include the type of traffic, the geographic span of the switch block, and the number of workgroups. A switch block contains multiple VLANs, multiple networks, and multiple instances of spanning tree (one per VLAN with the standard Cisco implementation). Typically, a switch block supports fewer than 2000 users.

Figure 1-2 Switch blocks allow for hierarchical network design. Switch blocks include access layer and distribution layer devices, usually including redundant links.

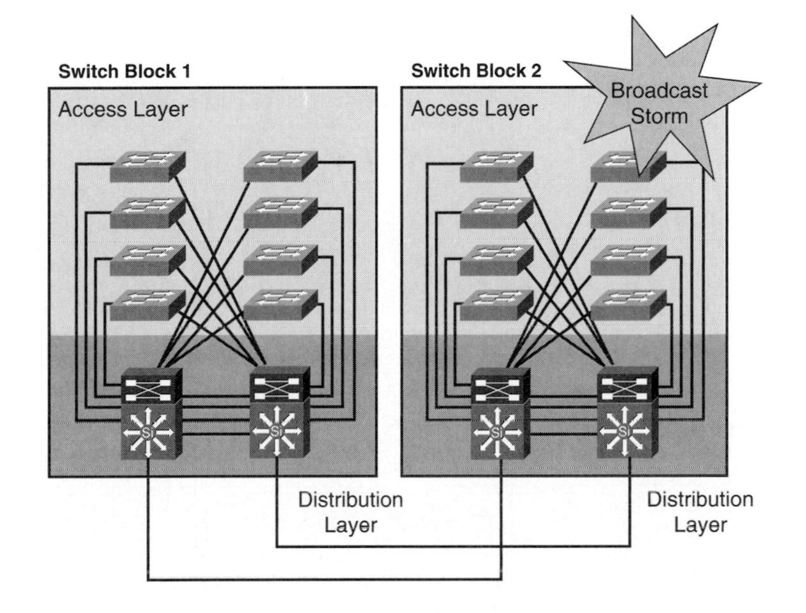

A *core block* is a switch or set of switches that interconnect multiple switch blocks. A switch block consists of access layer and distribution layer devices in the hierarchical model. The core block's duty is to pass data with minimum latency between switch blocks. The core block is isolated by the Layer 3 devices within the respective switch blocks, and routing protocols handle topological changes in the network. Normally the core block has two or more switches to provide redundancy and load balancing. Two designs for the core block are the collapsed core, shown in Figure 1-3, and the dual core, shown in Figure 1-4. As you can see, the dual core has the advantage of providing redundant connections between switch blocks.

The switch block and the core block are the two primary elements used in what is called the building block approach to campus network design. The building block design allows a network design to scale to the required size. Other blocks—such as the mainframe block, the server block, and the WAN block—are common in this approach. The mainframe block normally includes an IBM mainframe and a Token Ring infrastructure. The server block includes enterprise servers, providing Web services, DNS, and e-mail. The WAN block includes devices that provide a portal to the outside world or to other campus networks within the enterprise.

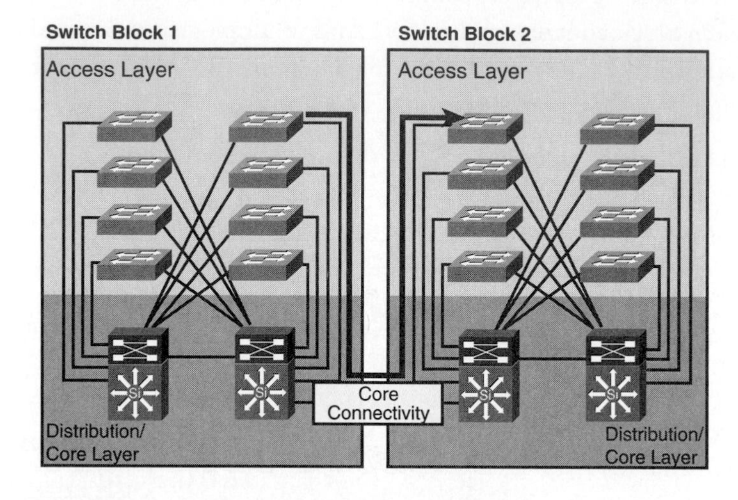

Figure 1-3
The collapsed core design is more prevalent in smaller campus networks.

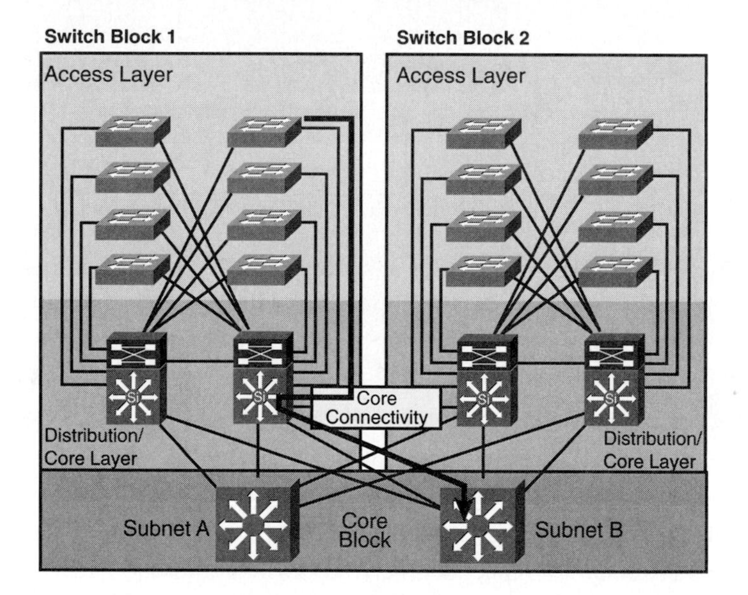

Figure 1-4
With the dual core design, each distribution layer switch can maintain two equal-cost paths to each destination network.

Governing Factors in Switched Network Design

The success of a switched network is determined by several governing factors. The two primary factors are availability and performance. Other factors include reliability, scalability, and cost.

Before we get into more detail on switched network design and campus networks, let's take a look at some of the factors that weigh in on the network architect's decision-making process.

The array of services now available on modern switches can be overwhelming. It's important to keep in mind that these services are driven by customer needs and that they will continue to evolve over time. So it helps to focus on the core concepts involved with these technologies and not expect to master each and every one of them, especially considering the fact that many will be replaced by completely independent solutions in the years to come.

Modern networks should support data, voice, and video convergence. This means that the devices must support quality of service (QoS) and multicast technologies. QoS is needed to provide acceptable levels of service to the end user. The network should also have a high-speed backbone. This means that you need one or more switches that support large capacities on the switch backplane and that offer significant Gigabit Ethernet port densities.

TECH NOTE: ARCHITECTURE FOR VOICE, VIDEO, AND INTEGRATED DATA (AVVID)

The Cisco Architecture for Voice, Video, and Integrated Data (AVVID) provides the framework for today's Internet business solutions. Cisco AVVID provides the road map for combining business and technology strategies into one cohesive model. AVVID provides the baseline infrastructure that allows an enterprise to design networks that scale to meet e-business demands. The plethora of services provided by Cisco's switching products are an integral part of the AVVID solution.

Networks should be scalable, which means that they need devices that can be upgraded without the so-called "forklift upgrade." (This means upgrading by using a forklift to move out the old equipment and replace it with the new equipment. Of course, using a forklift is normally not necessary, but at one time this term did have a direct connotation.) So each switch purchased for the backbone must have a chassis containing sufficient slots to reasonably support module upgrades over time that are appropriate for your network. You should also have a switch that supports high-speed switching at Layer 3+ (Layer 3 to Layer 7). This means that your switches should support technologies such as MLS (multilayer switching) or CEF (Cisco Express Forwarding).

Finally, your network should have redundancy. This means that your switches need to support such features as HSRP, IRDP, VTP, trunking, channeling, IGMP, CGMP, and, of course, spanning tree.

NOTE

If you don't understand terms and acronyms such as HSRP, IRDP, channeling, and spanning tree, you've found the right book! In this book, you learn what these terms mean and, more importantly, how to implement these technologies.

> **NOTE**
> This book focuses primarily on the following series of switches, especially with respect to configuration: Catalyst 2900, 2950, 3500, 3550, 4000, 5000/5500, 6000/6500, and 8500.

When you purchase a switch, the features just listed are some of the considerations you need to keep in mind. A variety of other features are also supported by modern switches. A good example of a popular series of switches that meets the criteria listed is Cisco's Catalyst 6000 family. The Catalyst 4000, 5000, and 6000 series switches also support modules allowing for a long list of WAN connectivity options.

Three key technologies that play a role in the design of modern switched networks are OSI Layer 3+ switching, QoS, and multicast. Each of these is discussed briefly in the following sections, and in more detail throughout this book.

Switching and the OSI Model

The OSI model provides a convenient way to describe processes that take place in a networking environment. Each layer in the OSI model serves a set of specific functions. This book focuses on Layers 2, 3, and 4 of the OSI model.

Protocols normally can be categorized as fitting into one or more of the specific OSI layers. Each protocol exchanges information called protocol data units (PDUs) between peer layers of network devices. A given layer has specific names for each of its PDUs. Table 1-1 provides examples of specific PDUs for Layers 2, 3, and 4 in addition to the device types that traditionally process these PDUs.

Table 1-1 Protocol Data Units

Layer	PDU Type	Device Type
Data link (Layer 2)	Frames	Layer 2 switch or bridge
Network (Layer 3)	Packets	Router
Transport (Layer 4)	Segments	Router or server (TCP or UDP port)

> **NOTE**
> The term *datagram* is often used to describe a Layer 3 PDU. Also, you might occasionally see the term *packet* used in the context of Layer 2, almost as a generic term for *data*. In general, we stick to the frame/packet/segment terminology and the associations detailed in Table 1-1.

The underlying layer services each protocol peer layer. For example, Transmission Control Protocol (TCP) segments are encapsulated in Layer 3 packets, and Layer 3 packets are encapsulated in Layer 2 frames. A layer-specific device processes only those PDUs for which it is responsible. This is done by inspecting the PDU header.

Modern switches are available that can make decisions based on PDU header information at Layers 2, 3, and 4. Content service switches, such as the Cisco CSS 11800, provide switching decisions based on information up through Layer 7.

Multilayer switching combines Layer 2 switching, Layer 3 routing functionality, and the caching of Layer 4 port information. Multilayer switching provides wire-speed (hardware-based) switching enabled by Application-Specific Integrated Circuits (ASICs) within a switch, which share in the duties normally assumed by the CPU.

ASICs are specialized hardware that handle frame forwarding in a switch. The point is to speed up the internal process of switching packets between ingress and egress ports. Layer 3 switching can be described as hardware-based routing enabled by ASICs; however, as you'll see in Chapter 7, there's a little more to it than that.

With multilayer switching, the switch caches information about a given *flow* of data through the switch, inspecting the PDU header information of the first frame in the flow and switching the remaining data at *wire speed* based on the information gleaned from the first frame. This explains the phrase *route once, switch many* used in reference to multilayer switching. Flow is a term used loosely to describe a stream of data between two endpoints across a network. The flow can be based on Layer 2 and Layer 3 information or on a combination of Layer 2, 3, and 4 information.

The term *Layer 4 switching* is sometimes used to describe switching frames based on flows characterized by a combination of source/destination MAC addresses, IP addresses, and port numbers. One advantage of Layer 4 switching is that it allows QoS to be applied to the traffic. QoS is typically applied to classes of traffic defined by port numbers. In this sense, Layer 4 switching can be said to provide application-based prioritization.

The role of multilayer switching can sometimes be clouded by the marketing efforts of the vendors selling these switches, but the importance of this technology in a modern network cannot be overstated. It's important to have a good handle on these technologies when making decisions about purchases and deployments. Multilayer switching is discussed in detail in Chapter 7.

Quality of Service

Quality of service (QoS) for voice or video over IP consists of providing minimal packet loss and minimal delay so that voice/video quality is not affected by conditions in the network. The brute-force solution is to simply provide sufficient bandwidth at all points in the network so that packet loss and queuing delay are small. A better alternative is to apply congestion management and congestion avoidance at oversubscribed points in the network. Most modern Layer 3+ switches support configuration options for congestion management and congestion avoidance.

A reasonable design goal for end-to-end network delay for Voice over IP (VoIP) is 150 milliseconds. At this level, delay is not noticeable to the speakers. To achieve guaranteed low delay for voice at campus speeds, it is sufficient to provide a separate outbound queue for real-time traffic. Bursty data traffic, such as file transfers, is placed in a different queue from real-time traffic. Low-latency queuing (LLQ) is the preferred method for queuing in this context, giving priority to voice traffic while data traffic is typically governed by weighted fair queuing. Auxiliary VLANs (see Chapter 4) also provide an effective means for optimizing voice traffic in a campus LAN.

If low delay is guaranteed by providing a separate queue for voice, packet loss will never be an issue. Another option, Weighted Random Early Detection (WRED), achieves low packet loss and high throughput in any queue that experiences bursty data traffic flow.

QoS maps well to the multilayer campus design. Packet classification is a multilayer service that applies at the wiring-closet switch, the ingress point to the network. VoIP traffic flows are recognized by a characteristic port number. The QoS features in Cisco Catalyst switches select network traffic, prioritize it according to its relative importance, and use various techniques to provide priority-based handling. QoS features base their handling decisions on the class of service (CoS) values that are carried in (Layer 2) frames or in the type of service (ToS) or Differentiated Services Code Point (DSCP) values carried in (Layer 3) IP packets. 3 bits in the IP header govern the ToS values, ranging from 0 for the lowest priority to 7 for the highest priority. 6 bits in the IP header govern the possible DSCP values: 0, 8, 10, 16, 18, 24, 26, 32, 34, 40, 46, 48, and 56.

Improper CoS settings are common, often stemming from a misunderstanding of the basic operations of CoS on the Catalyst switches. CoS is implemented at Layer 2 and, like ToS, ranges in value from 0 to 7 (lowest to highest). Classifying frames for priority involves marking 3 bits in an interswitch link (ISL) frame header or 3 bits in the tag of a frame based on the IEEE 802.1Q standard. (ISL and 802.1Q are explored in Chapter 4.)

Wherever VoIP packets encounter congestion in the network, switches and routers apply the configured congestion management and congestion avoidance techniques based on the CoS, ToS, and DSCP values.

QoS plays a significant role in designing a modern switched network. With VoIP and streaming video now commonplace, and videoconferencing on the rise, it's important to understand the particular demands of this type of traffic on the network and how to design your switched network to optimally support it. Some QoS techniques are discussed in Appendix C.

Multicast

Applications such as Cisco's IP/TV depend on multicast functionality in the campus network. Within Cisco Systems' own worldwide network, multicast is configured throughout its worldwide network so that one-to-many and many-to-many communication with voice and video are realized on a day-to-day basis. Multicast is becoming an integral part of network design, although it poses one of the greatest challenges to network engineers. A quote from a respected network sage expresses this sentiment: "When you see a bunch of engineers standing around congratulating themselves for solving some particularly ugly problem in networking, go up to them, whisper 'Multicast,' jump back, and watch the fun begin . . . "

Because end users are the recipients of multicast traffic in a campus network, a good multicast network design is fundamentally important for any enterprise that wants to enjoy the many benefits of multicast technology. If the devices in your network are not specifically configured for multicast, you are not only precluding the possibility of clients receiving multicast streams from multicast servers, but you are also leaving the door open to the indiscriminate flooding of multicast traffic in portions of your network. (In this case, it is effectively equivalent to broadcast traffic.)

The multilayer campus design is ideal for the control and distribution of IP multicast traffic. Layer 3 multicast control is provided primarily by the Protocol Independent Multicast (PIM) protocol. For more information on PIM, see Chapter 9. Multicast control at the wiring closet is provided by Internet Group Management Protocol (IGMP) and IGMP Snooping or Cisco Group Multicast Protocol (CGMP). Multicast control is extremely important because of the large amount of traffic involved when several high-bandwidth multicast streams are provided.

At the wiring closet, IGMP Snooping and CGMP are multilayer services that prune multicast traffic between switch ports and directly connected routers. Without IGMP Snooping or CGMP, multicast traffic would flood all the switch ports. IGMP itself prunes unnecessary multicast traffic on router interfaces connected to client workstations. Without IGMP, multicast traffic would flood all switch ports and all multicast-configured router interfaces.

In PIM sparse mode (which is discussed in more detail in Chapter 9), a *rendezvous point* is used as a broker or proxy for multicast traffic. A design goal of multicast networks is to place the rendezvous point and backup rendezvous point(s) to affect the shortest path. If this is accomplished, there is no potential for suboptimal routing of multicast traffic. Ideally, the rendezvous point and the optional backup rendezvous point(s) should be placed on the Layer 3 distribution switches in the server farm close to the multicast sources. If redundant rendezvous points are configured on routers (instead of switches) in your network, recovery is much slower and more CPU-intensive.

In any case, it's critical that the network design take multicast functionality into account. The switches should support IGMP Snooping or CGMP, as well as the various PIM modes and Multicast Source Discovery Protocol (MSDP), a relatively new protocol that can, among other things, enhance dynamic rendezvous point selection.

Campus Network Design

Not long ago, a campus network consisted of a single LAN connected by hubs and repeaters. By now you're aware of the major limitation of this type of network—distance! This type of campus network was restricted to a small geographic area, such as a building or part of a building. In the case of Ethernet, the number of collisions on this type of network was proportional to the number of computers (one collision domain). Depending on how far back you go in time, hosts on a campus network such as this were connected via coaxial cable (10Base2 or 10Base5) or, more recently, by twisted-pair cable supporting 10BaseT; the connections were half-duplex and ran at 10 Mbps.

Present-day campus networks are, of course, much more complex and robust. To better understand modern campus networks, it helps to categorize them according to size. The next sections describe small, medium, and large campus networks as commonly deployed today.

Small Campus Networks

Small campus networks, as shown in Figure 1-5, are typically contained within one building. In most cases, network redundancy is not the top priority—cost-effectiveness is. Requirements for these designs typically include

- High performance for applications such as voice, video, and IP multicast
- Support for applications based on Novell IPX, DECnet, AppleTalk, and SNA

NOTE

DECnet is group of communications products (including a protocol suite) developed and supported by Digital Equipment Corporation. Systems Network Architecture (SNA) is a large, complex, feature-rich network architecture developed in the 1970s by IBM. SNA is commonly used by financial institutions.

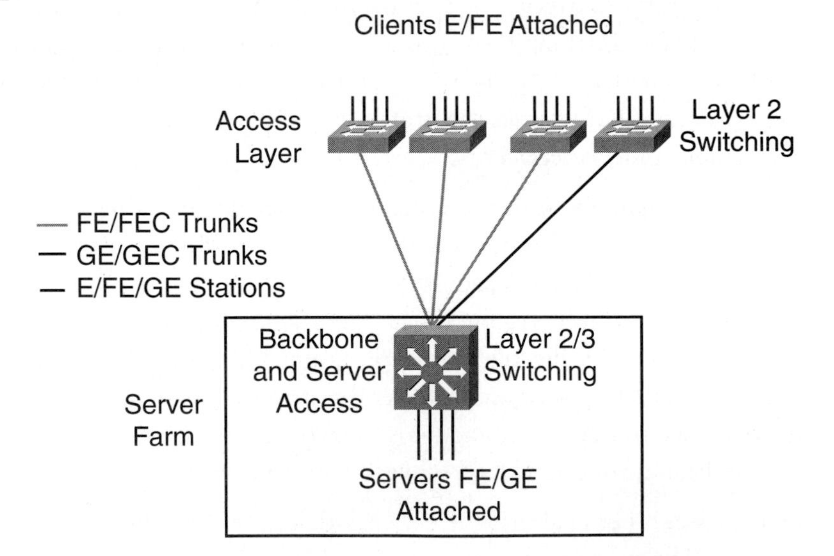

Figure 1-5 Small campus networks are the most common. In this design, it is not unusual to have a single multilayer switch servicing the entire network.

A design solution that meets these requirements provides a high-performance, switched infrastructure for a building-sized intranet with hundreds of networked devices. The network backbone consists of a Layer 3+ switch. Access layer switches provide connectivity to clients and servers. Cisco IOS Software supports QoS, security, troubleshooting, and common management features from end to end, making the multilayer Catalyst switches ideal for such a deployment.

TECH NOTE: SERVER FARMS

Whether your network is small, medium, or large, you need services to get business done. These services are run on servers with network operating systems such as Microsoft .NET or a particular flavor of UNIX.

A *server farm,* shown in Figure 1-6, is the industry term used to describe a set of enterprise servers housed in a common location, with high-speed links to the intranet, typically deployed in concert with a secure firewall and Web caching servers. The services provided by server farms can include database access, e-mail, DNS, and many others. The servers in a server farm are frequently used as sources of multicast traffic as well. Often a server farm is implemented within the so-called demilitarized zone (DMZ), allowing "outsiders" access to the services run on the servers while maintaining a secure corporate intranet. The DMZ is a portion of the network, normally housing a set of servers, that separates internal users (such as company employees) from outsiders (who typically access Web sites hosted by the company).

The highly resilient server farm represents a building block in a multilayer campus design that allows for scalable, fault-tolerant, centralized connectivity for enterprise servers. The server farm design addresses multiple connectivity options in both speed and redundancy to provide a flexible network deployment. As shown in Figure 1-6, the servers replace the access layer switches seen in a typical switch block.

A network engineer will likely build in deterministic paths for users to access the server farm. This allows preferred paths to be used for certain applications or users.

To build modular, resilient enterprise server farms, you need switches that offer the following features:

- Highly fault-tolerant hardware.
- A variety of connectivity options.
- Highly optimized software features. In Cisco parlance, this includes UplinkFast, PortFast, Per-VLAN Spanning Tree (PVST), Unidirectional Link Detection (UDLD), OSPF, EIGRP, Hot Standby Routing Protocol (HSRP), and high-speed integrated server load balancing.

UplinkFast, PortFast, PVST, and UDLD are discussed in Chapter 5.

Primary advantages of deploying a server farm include enhanced security, ease of management, and reduced costs.

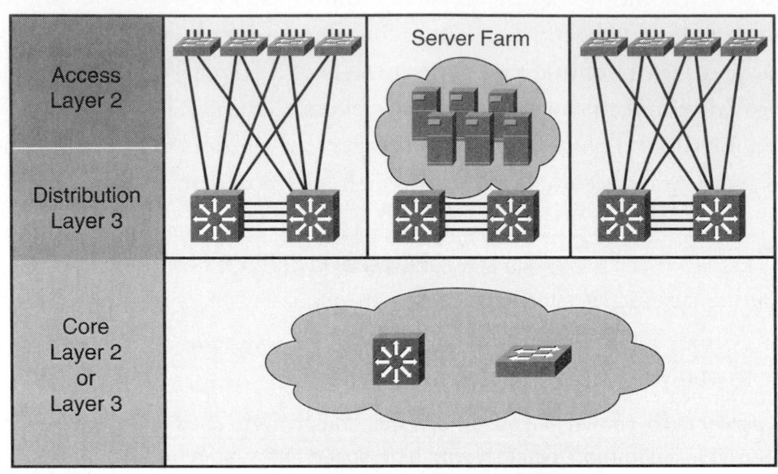

Figure 1-6 Server farms form a building block in campus network design. A server farm consists of high-speed servers used to host web sites, provide DNS services, manage e-mail accounts, and deliver other services required throughout the network.

Medium Campus Networks

Medium campus networks, as shown in Figure 1-7, consist of one large building or several buildings. Networking for a medium campus is designed for high availability, performance, and manageability. This type of design is also called a *collapsed backbone* design. Additional requirements of these designs typically include

- High performance and availability for applications such as voice, video, and IP multicast
- Support for applications based on Novell IPX, DECnet, AppleTalk, and SNA

These additional requirements parallel those of small campus networks. The major difference between small and medium campus networks is the number of users. Building in redundancy is also more important in medium campus networks.

This design solution provides a manageable switched infrastructure for a campus intranet with more than a thousand networked devices. The high-performance collapsed backbone uses Layer 3 switching. Network redundancy is provided for clients and servers. HSRP provides fast recovery of switch-to-router link failures. HSRP is discussed in more detail in Chapter 8.

TECH NOTE: THE 80/20 RULE (OR IS IT THE 20/80 RULE?)

Whether your campus network is small, medium, or large, the traffic patterns are likely to have one thing in common: the amount of traffic local to a given VLAN versus nonlocal traffic.

The conventional wisdom of the 80/20 rule underlies the traditional design models of years past. Even with a campus model using VLANs that span the enterprise, the legacy campus network was organized under the assumption that 80 percent of traffic was to be contained within the VLAN. The remaining 20 percent of traffic was to leave the network or subnet through a router.

The traditional 80/20 traffic model arose because each department or workgroup had a local server on the LAN. The local server was used as file server, logon server, and application server for the workgroup. The 80/20 traffic pattern has been changing rapidly with the rise of corporate intranets and applications that rely on distributed IP services. Many new and existing applications are moving to distributed Web-based data storage and retrieval. The traffic pattern is moving toward what is now referred to as the 20/80 model or 20/80 rule. In the 20/80 model, only 20 percent of traffic is local to the VLAN and 80 percent of the traffic is destined for other VLANs, a server farm on the network, or locations outside the campus network. The 20/80 rule is much closer to representing the reality of modern campus networks.

With the 20/80 rule comes an implicit categorization of campus services: local, remote, and enterprise. Local services reside in the same VLAN, remote services reside in other VLANs, and enterprise services (which all users access) are located in a server farm.

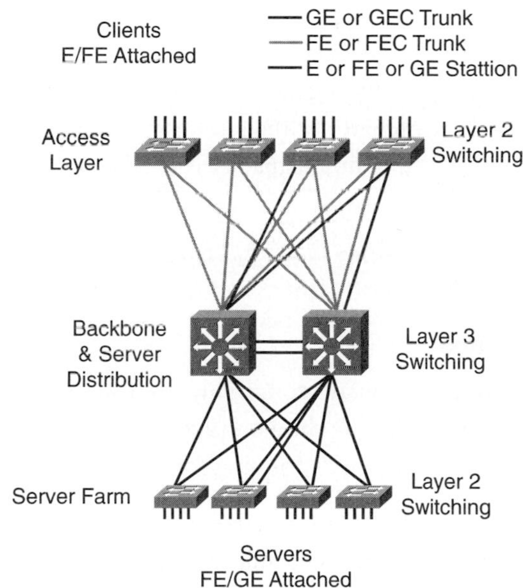

Figure 1-7
A medium campus network is characterized by a collapsed backbone, where the distribution layer and core layer merge. Unlike a small campus network, redundancy is built in to a medium campus network.

Large Campus Networks

Businesses operating large campus networks, as shown in Figure 1-8, are increasingly looking for infrastructure upgrades to

- Support high-bandwidth applications such as voice, video, and IP multicast
- Support applications based on Novell IPX, DECnet, AppleTalk, and SNA
- Offer high availability, performance, and manageability for the company's intranet

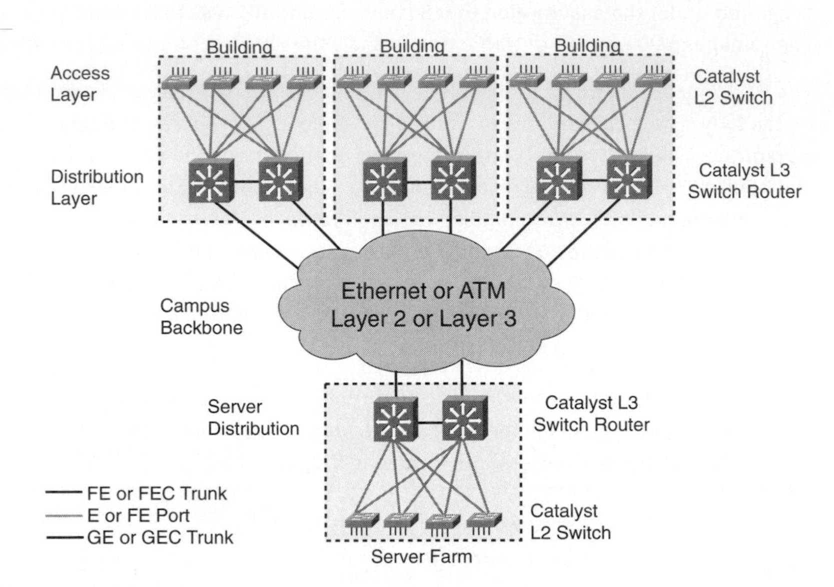

Figure 1-8 Large campus networks have it all—redundancy, high bandwidth, and well-defined core, distribution, and access layers.

In typical designs, the buildings or different parts of the campus are connected across a high-performance switched backbone. Network redundancy and high availability are provided at each layer. A high-capacity centralized server farm provides resources to the campus.

Large campus networks use Layer 2, Layer 3, or ATM backbone solutions to expand the network. Each of these is discussed in the following sections.

Layer 2 Backbone

Layer 2 backbone designs are commonly used when cost-effectiveness and high availability are high priorities, as shown in Figure 1-9. Businesses operating large campus networks are also increasingly looking for infrastructure upgrades to

- Support high-bandwidth applications based on IP
- Provide high performance and fast recovery from failures
- Support applications based on Novell IPX, DECnet, AppleTalk, and SNA

Campus building blocks and server farms connect to the backbone with redundant Gigabit Ethernet or Gigabit EtherChannel trunks. See Chapter 5 for more details on EtherChannel trunks. Layer 3 switching across the backbone uses routing based on EIGRP, OSPF, or IS-IS for load balancing and fast recovery from failures.

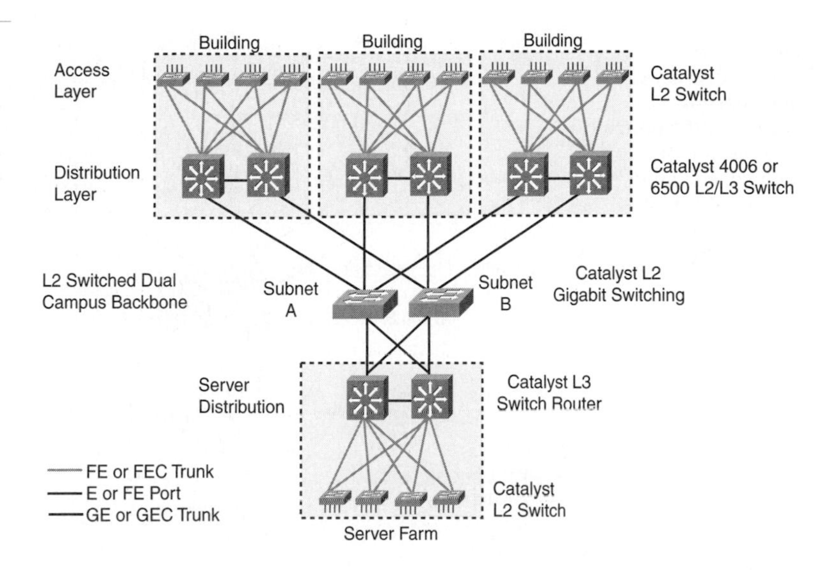

Figure 1-9
Layer 2 backbones now are characterized by Gigabit and 10 Gigabit Ethernet interfaces and links.

Layer 3 Backbone

Layer 3 backbone designs, shown in Figure 1-10, are commonly used when very high performance is desired for supporting multimedia applications based on IP unicast and multicast. Additional requirements of these designs typically include

- Nonblocking campus backbone that scales to many Gbps of throughput
- Broadcast containment
- Very fast deterministic failure recovery, campus-wide
- Support for applications based on Novell IPX, DECnet, AppleTalk, and SNA

This solution provides a manageable switched infrastructure that scales to a huge campus with many buildings and tens of thousands of networked devices.

Buildings connect to a very high-performance, nonblocking Layer 3 switched backbone. Fast failure recovery is achieved campus-wide using a Layer 3 routing protocol such as EIGRP, OSPF, or IS-IS. One or several high-capacity server farms provide application resources to the campus.

The Layer 3 campus backbone is ideal for multicast traffic, because PIM runs on the Layer 3 switches in the backbone. PIM routes multicasts efficiently to their destinations along a shortest path tree, as opposed to the multicast flooding inherent in a Layer 2 switched backbone design.

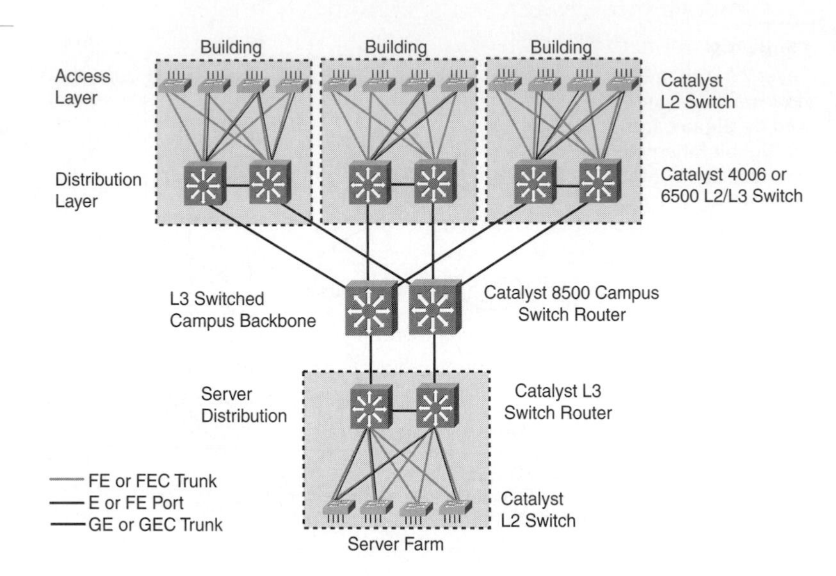

Figure 1-10 The Layer 3 backbone design is becoming the norm as a result of the high-speed Layer 3 switching now common in core and distribution layer switches.

ATM Backbone

An Asynchronous Transfer Mode (ATM) backbone design, shown in Figure 1-11, is typically used for a very large switched campus intranet that demands high performance and availability. Requirements of these designs typically include the following:

- High performance and very high availability for IP applications
- Trunking for native real-time voice and video applications
- Fast deterministic failure recovery for high application availability
- Support for applications based on Novell IPX, DECnet, AppleTalk, and SNA

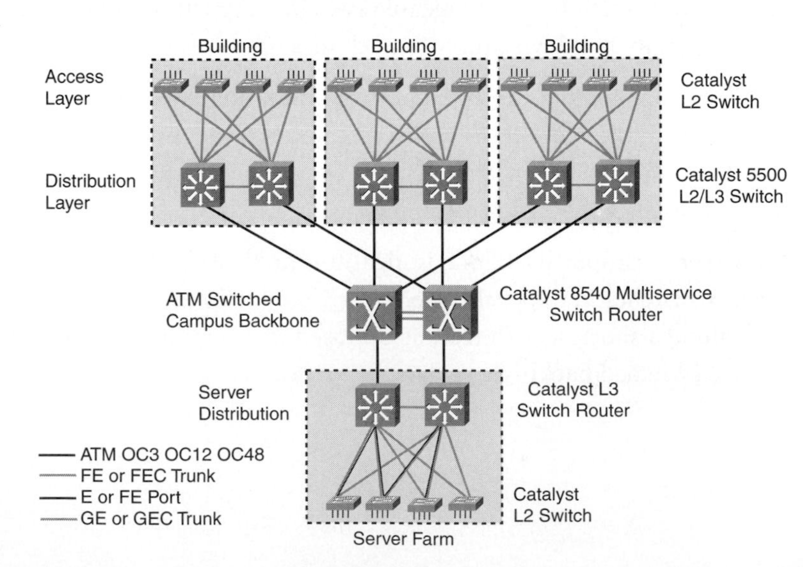

Figure 1-11 The ATM backbone is still fairly common, but it appears to be losing ground to Gigabit and 10 Gigabit Ethernet. Ethernet solutions are generally cheaper and less complex than ATM solutions.

This solution provides a manageable switched infrastructure for a large campus with thousands of networked devices.

Buildings are connected across a high-performance ATM switched backbone. IP routing protocols as well as ATM routing with Private Network-to-Network Interface (PNNI) provide network redundancy and high availability. One or more high-capacity server farms provide resources to the campus.

TECH NOTE: SONET AND ATM

SONET is the ANSI standard for synchronous data transmission on optical media. The international equivalent of SONET is Synchronous Digital Hierarchy (SDH). Together, they ensure interoperability so that digital networks can interconnect internationally and so that existing conventional transmission systems can take advantage of optical media through tributary attachments.

SONET provides standards for a number of line rates up to the maximum line rate of 9.953 Gbps. Actual line rates approaching 20 Gbps are possible. SONET is considered the foundation of the physical layer of broadband ISDN (BISDN).

SONET defines a base rate of 51.84 Mbps and a set of multiples of the base rate known as Optical Carrier (OC) levels. The common OC rates, delivered by line cards in networking devices, are OC-1 (~52 Mbps), OC-3 (~155 Mbps), OC-12 (~622 Mbps), OC-48 (~2.5 Gbps), and OC-192 (~10 Gbps).

ATM runs as a layer on top of SONET as well as on top of other Layer 1 technologies, typically implemented over OC-3, OC-12, or OC-48 modules on a switch or router.

Although ATM is still popular in the world of telecommunications, its use as a campus backbone solution appears to be on the decline.

Catalyst Switching Solutions for Campus Networks

Cisco provides the most comprehensive set of campus switching solutions in the industry, meeting the requirements of the smallest wiring closet to the most routing-intensive network cores.

Campus networks continue to evolve to meet the demands of new network-intensive applications, the bandwidth they consume, and the policies to enforce fairness and security throughout the network. Network managers are faced with the task of efficiently managing traffic loads, containing equipment and management costs, and planning for future growth.

To manage costs and increase productivity, corporations are increasingly leveraging their intranets to run business-critical applications. Technologies such as multicast are being employed to increase the sharing of information without compromising productivity. Manufacturing facilities now share resource-planning databases directly with suppliers to better plan for future capacity and to decrease product lead times. IP-based telephony continues to gain acceptance as corporations realize the huge financial savings of converged voice and data networks. Corporate intranets can also be used by employees for non-business-critical applications, such as networked games and Web surfing, which can consume precious network resources to the point of having harmful effects on business-critical applications.

Traffic must be segmented based on security, QoS, and traffic management. Resource contention at multigigabit rates is an expensive problem to solve with raw bandwidth. As corporations continue to leverage their intranets with an increased set of network applications, traffic must be classified and policies enforced to optimize resource utilization. Thus, multilayer services are being driven into all tiers of the network.

Desktop connectivity has migrated from 10 Mbps to 100 Mbps, putting more stress on the backbone to accommodate increased densities at much higher uplink rates. Backbones must also be capable of managing traffic in a way that does not require the continuous capital outlay associated with meeting bandwidth needs bit for bit, thus driving the need for intelligent networking—the ability to grant network resources based on application, user, time and day, or predetermined policy.

Increased corporate leverage of networked applications also demands that availability be maintained end to end. This requires not only hardware redundancy options, but also the software architecture to manage traffic and recognize and overcome failure conditions.

From access layer to network core, different design parameters must be used to select the most appropriate solution. For example, whereas low price/port and high port density might rule in wiring-closet environments, wire-rate multilayer-switching performance and policy-enforcement (such as QoS and security) features might drive the design of the core and distribution layers. No single equation addresses the requirements for all tiers of the network. For this reason, the campus switching market has begun to fragment into market segments that include wiring-closet switches, multilayer switches, and switch routers. Each of these is discussed in the following sections.

Wiring-Closet Switches

Wiring-closet switches are devices that provide client connectivity at the edge of the network (close to the users). For small wiring closets, the Catalyst 2900 and 3500 series provide a number of cost-effective solutions for small to medium wiring closets. For midrange and large closets, the Catalyst 4000 and 5000 families provide excellent modularity, scaling 10/100/1000 densities up to 240 ports for the Catalyst 4000 family and more than 500 10/100 ports for the Catalyst 5000 family.

Switches such as the Catalyst 3524-PWR-XL are becoming more popular due to their support for inline-power Ethernet ports, which can be used to provide both power and connectivity to IP phones attached to the switch ports. (A PC can be connected to the IP phone so that a single switch port can be utilized for both the IP phone and the PC.) The Catalyst 3550 series comprise relatively high-end wiring-closet switches that provide multilayer switching and up to ten 10/100/1000 Ethernet ports. Because of their support for QoS and multiservice technologies, the 2950s, the modular 2900 XLs and 3500 XLs, the 3550s, and the 4000s are now the Catalyst switches of choice in new wiring-closet deployments.

TECH NOTE: CATALYST 2950 AND 3550 SERIES SWITCHES

Cisco has announced the end-of-sales (EOS) of *fixed* configuration Catalyst 2900 XL switches (such as the 2912 and 2924). The last order date was 10/31/2001. These EOS fixed-configuration switches are being replaced by the Catalyst 2950 series switches (listed in Table 1-2), which provide superior functionality at the same price. Modular configuration Catalyst 2900 XL switches will continue to be sold and supported by Cisco. Cisco has also announced the EOS of Catalyst 3512 XL, 3524 XL, and 3548 XL switches. The last order date was 7/27/2002. The replacement products are the Catalyst 3550 series switches.

Table 1-2 Catalyst 2950 Series Switches

Product Name/Part Number	Product Description
Catalyst 2950G-12 WS-C2950G-12-EI	12 10/100 ports and two fixed GBIC-based 1000BaseX uplink ports One rack unit (RU) stackable switch Delivers intelligent services to the network edge Enhanced Software Image (EI) installed
Catalyst 2950G-24 WS-C2950G-24-EI	24 10/100 ports and two fixed GBIC-based 1000BaseX uplink ports One RU stackable switch Delivers intelligent services to the network edge Enhanced Software Image installed
Catalyst 2950G-48 WS-C2950G-48-EI	48 10/100 ports and two fixed GBIC-based 1000BaseX uplink ports One RU stackable switch Delivers intelligent services to the network edge Enhanced Software Image installed
Catalyst 2950G-24-DC WS-C2950G-24-EI-DC	24 10/100 ports and two fixed 1000BaseX uplink ports One RU stackable, DC-powered switch Delivers intelligent services to the network edge Enhanced Software Image installed

continues

Table 1-2 Catalyst 2950 Series Switches (Continued)

Product Name/Part Number	Product Description
Catalyst 2950T-24 WS-C2950T-24	24 10/100 ports and two fixed 10/100/1000BaseT uplink ports One RU switch Delivers intelligent services to the network edge Enhanced Software Image installed
Catalyst 2950C-24 WS-C2950C-24	24 10/100 ports and two fixed 100BaseFX uplink ports One RU switch Delivers intelligent services to the network edge Enhanced Software Image installed
Catalyst 2950-24 WS-C2950-24	24 10/100 ports One RU switch
Catalyst 2950-12 WS-C2950-12	12 10/100 ports One RU switch

Wiring-closet switches provide more and more features as time goes on. It is becoming more common for these switches to support multilayer switching and various QoS technologies. With IP telephony and multicast applications increasing in popularity, these switches will continue to increase their 10/100/1000 port densities over time. The demands of the end users and the operating systems they use are pushing the envelope on the requirements for wiring-closet switches. Multilayer switching functionality is essentially the last frontier to cross for wiring-closet switches. After that, it becomes harder to distinguish between these switches and the multilayer switches discussed in the next section.

Multilayer Switches

Multilayer switches (Layers 2, 3, and 4) provide diverse interfaces, high port densities, and extensible functionality suitable for high-function network access or backbone applications. Depending on the network design, multilayer switches can fulfill very high-end wiring closet, distribution, server aggregation, and backbone environments (where distribution and core are collapsed). They provide multiprotocol routing, policy networking, and services such as multicast, security, and mobility (allowing users to move devices around the network with intervention by a network administrator).

High-availability options are also key due to the number of clients supported per platform. Backbone switches must have redundancy options and also support network resiliency features such as Layer 2 and 3 load balancing and traffic-management techniques to guarantee immediate transmission of mission-critical traffic.

Cisco provides multilayer switching on the Catalyst 5000 and 6000 families for low- and high-density environments with a wide range of performance requirements. See Figure 1-12 for a comparison.

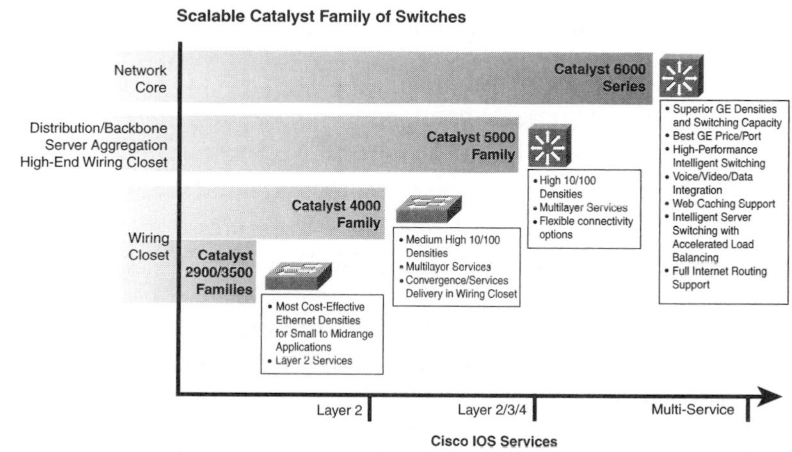

Figure 1-12
The Catalyst 2950, 3550, 4000, and 6000 family switches offer multiservice features. This is the status quo for Catalyst switches from here on out.

> **NOTE**
> Throughout www.cisco.com, you will see the Catalyst switch product lines described alternately as *family* and *series,* such as the Catalyst 6000 family of switches or the Catalyst 6000 series switches. The term *family* tends to be used to indicate all Catalyst switches beginning with the same number. The term *series* is normally used to indicate a subfamily, such as the S500 series (as opposed to the 5000 family).

The Catalyst 6000 family provides Fast Ethernet and Gigabit Ethernet densities, scaling up to 1152 10/100 Fast Ethernet, 388 Gigabit Ethernet, and 12 10 Gigabit Ethernet ports. You can also aggregate up to eight physical Fast Ethernet, Gigabit Ethernet, or 10 Gigabit Ethernet links using either Fast EtherChannel (FEC), Gigabit EtherChannel (GEC), or 10 Gigabit EtherChannel (10GEC), respectively, for logical connections up to 16 Gbps. Used in this configuration, the Catalyst 6000 family creates a robust Gigabit Ethernet backbone solution. The Catalyst 6000 family also supports high-performance ATM connectivity via a single-port OC-12 module, with up to 12 ports per chassis.

The architecture of the Catalyst 6500 series supports scalable switching bandwidth up to 256 Gbps and scalable multilayer switching up to 210 Mpps (million packets per second). If this level of performance is not required, the Catalyst 6000 series provides a more cost-effective solution, delivering 32 Gbps of backplane bandwidth and multilayer switching up to 30 Mpps.

Switch Routers

Switch routers are Layer 2, 3, and 4 switches focused primarily at the core of the network, providing multiservice wire-speed routing and services on all interfaces. Positioned in the core of the network, switching routers are not differentiated by extremely high densities, but rather the level of performance achieved per port. Like high-end/backbone routers, such as the Cisco 7600 series, switch routers have a broad suite of interfaces, including both campus backbone ports and MAN/WAN interfaces, and support for multiprotocol environments. However, switch routers scale switching bandwidths and multilayer performance to the gigabit speeds that are necessary for today's highest-performance network backbones.

The Catalyst 8500 series product line has four flavors of switch routers: the 8510 Campus Switch Router (CSR), the 8540 Campus Switch Router, the 8510 Multiservice Switch Router (MSR), and the 8540 Multiservice Switch Router.

The Catalyst 8500 series supports wire-rate multilayer switching across flexible densities of Fast Ethernet, Gigabit Ethernet, ATM, SONET, and Frame Relay. With support for ATM circuit emulation service (CES), the Catalyst 8500 series can also provide PBX and video CODEC (coder/decoder or compressor/decompressor) termination for a converged data, voice, and video network.

Campus Positioning of Catalyst Switches

For reference, here is a quick synopsis of where, how, and why you deploy a given Catalyst switch. We'll refer back to this list as we discuss particular technologies.

- Catalyst 2900 XL and 2950 series
  - Pure Ethernet environment
  - Cost-effective wiring-closet densities of less than 50 ports (10/100)
  - Layer 2 switching only
  - Primary drivers are low price per port, low densities, and easy management
- Catalyst 3500 XL series
  - Pure Ethernet environment
  - Gigabit uplinks
  - Cost-effective in small/medium wiring-closet densities up to 100 ports
  - Layer 2 switching only
  - Scales up to 384 ports
- Catalyst 2900 (2948G)
  - Pure Ethernet environment
  - Cost-effective 48 ports
  - Layer 2 switching only
  - Same command environment as the Catalyst 4000
- Catalyst 2948G-L3/4908G-L3
  - Cost-effective Layer 2 and 3 switching for small backbones
  - Cisco IOS Software routing support

- Catalyst 4840G
  - Cost-effective medium-density Layer 2, 3, and 4 switching
  - Cisco IOS Software server load balancing
  - Wire-speed Network Address Translation (NAT)
  - Firewall load balancing
- Catalyst 4000 family
  - Pure Ethernet environment
  - Cost-effective wiring-closet densities of up to 240 10/100/1000 ports
  - Small server switch for densities of fewer than 20 servers (reserving two ports for uplinks)
  - Layer 2/3 switching
  - Layer 2/3 packet classification
  - Primary reasons for purchase include low price per port, medium/high Ethernet densities, gigabit scalability, easy management, multicast support, traffic classification, and Layer 3 QoS
  - Access Gateway Module (not supported with Supervisor Engine III) provides IP WAN routing, VoIP gateway, and IP telephony services
- Catalyst 5000 family
  - Cost-effective densities for large wiring closets from 300 to 500+ ports
  - Interface flexibility of access media supporting 10BaseFL, 10/100TX, 100FX, and 10BaseTX
  - Interface flexibility of backbone connectivity supporting Token Ring, ATM, Packet over SONET (PoS), Fast Ethernet, Gigabit Ethernet, FEC, GEC, and even FDDI
  - Layer 2-, 3-, and 4-based packet classification
  - Extensive Layer 3 services such as security, QoS, and traffic management
  - Multiprotocol environments
  - High-availability environments
  - Primary drivers are high wiring-closet densities; support for broad connectivity options; uplink scalability; multiprotocol support; Layer 2, 3, and 4 traffic classification; and redundancy applications
- Catalyst 6000 family
  - Ethernet backbones and intelligent server switching environments requiring high-performance, cost-effective Web scaling technologies and very high densities of Fast or Gigabit Ethernet—up to 1152 10/100, 576 100FX, and 388 Gigabit Ethernet ports

  - High-performance ATM OC-12 connectivity
  - High-performance switching and services for large wiring closets
  - High-performance, scalable switching from 32 Gbps to 256 Gbps
  - High-performance, multilayer switching from 30 Mpps to 210 Mpps for Layer 2, 3, and 4 switching and policy enforcement
  - Extensive Layer 3 services such as security, QoS, and traffic management
  - Multiprotocol environments
  - Large network support via BGP4 (supporting full Internet routes) and IS-IS
  - Multicast support
  - High-availability environments
  - Primary drivers are very high 100/1000 Mbps Ethernet densities, scalable multilayer switching and services (security, routing, policy enforcement, and so on), extensive QoS support, multiprotocol support, and high availability
  - WAN integrated delivery, extended connectivity to the large enterprise
  - Convergence ready with integrated voice capabilities
- Catalyst 8500 series
  - Multiservice core backbones requiring a mix of high-capacity routed interface types, especially for Gigabit Ethernet, ATM, and SONET
  - High-density ATM and SONET support, scaling densities up to 64 T1/E1, 128 OC-3, and 32 OC-12
  - Scalable wire-rate switching from 10 Gbps to 40 Gbps
  - High-performance ATM switching up to OC-48
  - Wire-rate Layer 2, 3, and 4 switching from 6 Mpps to 24 Mpps
  - Extensive Layer 3 services such as security, QoS, and traffic management
  - Multiprotocol environments
  - Multicast support
  - Voice and video support via ATM CES (direct PBX or CODEC connectivity)
  - High-availability environments

- — Primary drivers are wire-speed switching in multiservice network cores, extensive multilayer services (such as security, routing, and policy enforcement), a broad range of LAN/WAN connectivity options, extensive QoS and multiprotocol support, and high availability
- — WAN interfaces including ATM, SONET, and Frame Relay

You might have noticed that no reference has been made to Token Ring switches thus far. Cisco does in fact have a line of switches dedicated to Token Ring—the Catalyst 3900 and 3920 switches. These switches allow for VLAN management and several other options, such as ISL trunking and source-route bridging. You might be interested to know that the 3920 switch is one of the many devices with which a CCIE candidate must become familiar.

Our focus in this book is on Ethernet switching theory and the configuration of Ethernet-specific Catalyst switch features.

Summary

In this chapter, you learned the basics of campus network design. The core-distribution-access layer model provides a convenient framework within which to discuss the various design models used for switched networks. Campus network designs are broken into small, medium, and large.

QoS, multicast, and multilayer switching functionality are now the primary factors governing decisions in campus network design and purchasing. A large campus has three possible backbone infrastructures: Layer 2, Layer 3, and ATM.

To support the various network designs, Cisco has the most comprehensive set of campus switching solutions in the industry, from the smallest wiring closet to the most routing-intensive network cores. When purchasing and deploying Catalyst switches, you can use the last section of this chapter as a reference.

Check Your Understanding

Test your understanding of the concepts covered in this chapter by answering these review questions. Answers are listed in Appendix A, "Check Your Understanding Answer Key."

1. At which layer of the hierarchical design model is shared access most likely to occur?
   A. Core
   B. Distribution
   C. Access
2. At which layer of the hierarchical design model is routing between virtual LANs most likely?
   A. Core
   B. Distribution
   C. Access
3. True or false: The core, distribution, and access layers should remain independent in all campus networks.
4. At which layer of the hierarchical design model is microsegmentation most likely?
   A. Core
   B. Distribution
   C. Access
   D. Server
5. Which Catalyst switch is best suited for deployment as a core switch in a new deployment of a medium campus network?
   A. 2900
   B. 4000
   C. 5000
   D. 6000
6. At which layer of the hierarchical design model is broadcast domain definition most likely?
   A. Core
   B. Distribution
   C. Access

7. The modern 20/80 rule states that _____% of the traffic leaves the local VLAN and _____% is contained within the local VLAN.

   A. 20/80

   B. 80/20

8. Which of the following Catalyst switches support ATM?

   A. 8540, 6506, 4006

   B. 8540, 5509, 2950

   C. 6506, 5509

   D. 8540, 2924

9. What are the three most common options for backbone infrastructure in a large campus network?

   A. ATM, SONET, Layer 3

   B. ATM, Layer 2, Layer 3

   C. ATM, SONET, Layer 2

   D. ATM, Layer 3, DPT

10. In a campus network, the placement of what device is critical for the optimal flow of multicast traffic when using PIM space mode?

    A. Server farm

    B. Route server

    C. Rendezvous point

    D. Distribution switch

Key Terms

20/80 rule Only 20 percent of traffic is local to the VLAN, and 80 percent of traffic is destined for other VLANs, a server farm on the network, or locations outside the campus network.

access layer The point at which local end users are allowed to attach to the network. Frequently, Layer 2 switches play a significant role at the access layer.

campus network A network that connects devices within and between a collection of buildings. LAN switching and high-speed routing provide connectivity among the buildings. It is common to focus on the role of switching when referring to a campus network.

core block A switch or set of switches that interconnect multiple switch blocks. A switch block consists of access layer and distribution layer devices in the hierarchical model. The main function of the core block is to pass data with minimum latency between switch blocks.

core layer A high-speed switching backbone in a campus network designed to switch packets with minimum latency.

distribution layer The demarcation point between the access and core layers. The distribution layer helps define and differentiate the core. The purpose of this layer is to provide a network boundary definition. This is where packet/frame manipulation takes place.

multilayer switching Combines Layer 2 switching, Layer 3 routing functionality, and the caching of Layer 4 port information. Multilayer switching provides wire-speed switching enabled by Application-Specific Integrated Circuits (ASICs).

server farm An industry term used to describe a set of enterprise servers housed in a common location, with high-speed links to the intranet, typically deployed in concert with a secure firewall and Web caching servers. The services provided by server farms can include database access, e-mail, and DNS.

switch block A set of logically grouped switches and associated network devices that provide a balance of Layer 2 and Layer 3 services. A switch block is a relatively self-contained, independent collection of devices, typically binding a collection of Layer 3 networks. Switch blocks are used as a tool to facilitate communication about switched network design.

After completing this chapter, you will be able to perform tasks related to the following:

- Gigabit Ethernet standards
- Media options for Gigabit Ethernet
- Gigabit Ethernet protocol architecture
- IEEE 802.1p quality of service
- Network design with Gigabit Ethernet

Chapter 2

Gigabit Ethernet

The demand for bandwidth continues to grow as new applications and services are added to existing networks. More than 85 percent of workstations are presently running Fast Ethernet, driving the need for servers with Gigabit connections. It is expected that autonegotiating 10/100/1000 Mbps NICs will enjoy almost complete market share by 2005. With users simultaneously editing a remote SQL database, downloading MP3s, sending an e-mail with a 10 MB attachment, and conducting a NetMeeting conference call over a multicast-enabled IP network, it's not surprising that Gigabit Ethernet is making its way to the desktop.

In the late 1990s, Gigabit Ethernet standards were approved for copper cable (IEEE 802.3ab) and fiber media (IEEE 802.3z), carrying on a 25-year tradition. Gigabit Ethernet options are now common for campus network devices, including switches, routers, servers, and desktops.

With more than 85 percent of the installed base of network ports utilizing Ethernet, the job of the network engineer is in some sense simplified with the increased availability of Gigabit Ethernet products. As the standards for Ethernet evolve, backward compatibility with 10 Mbps and 100 Mbps Ethernet installations continues to be maintained. More than 90 percent of existing Category 5 installations already meet the minimum requirements for deploying Gigabit Ethernet—so upgrading to Gigabit Ethernet is often just a matter of adding or swapping modules in a switch chassis.

The familiarity of Ethernet also serves as a psychological advantage for deploying Gigabit Ethernet. Most network engineers are much more comfortable with Ethernet deployments compared to other LAN/MAN technologies. Ethernet is easy to use and easy to upgrade. The upgrade process typically proceeds from the campus backbone to the server farm to the desktop.

Another advantage of Gigabit Ethernet is its support of quality of service (QoS) features, such as IEEE 802.1p traffic prioritization, which enhances audio and video communication. QoS is now pervasive in modern networks as voice, video, and data networks converge. This book does not delve into the huge array of QoS options for optimizing network traffic flow, but you are strongly encouraged to begin (or continue) your study of QoS to stay current in the field. Some QoS configuration options are explored in Appendix C.

More than increased bandwidth, ease of upgrades, or QoS options, the small cost-versus-bandwidth ratio of Gigabit Ethernet is the primary reason for its success. The economic bottom line generally serves as the deciding factor in major network upgrades.

This chapter explores Gigabit Ethernet in detail: the standards, the media, the protocol architecture, the encoding, and the pertinent QoS options. This is followed by another look at campus network design, with an emphasis on Gigabit Ethernet's role. This chapter prepares you for the tasks involved in implementing a switched network design, which is the subject of the upcoming chapters.

NOTE

This chapter contains quite a bit of highly technical content. If you prefer to jump right into the portion of the book dealing with Catalyst switch configuration, you might want to skim through this chapter and proceed to Chapter 3.

Gigabit Ethernet Standards

The original 10 Mbps Ethernet standards were devised more than 25 years ago. These include the industry standard 10BaseT. 10BaseT is supported by Category 3, 4, and 5 cable for up to 100 m (meters).

The 10BaseT standards were followed by the IEEE802.3u Fast Ethernet standards. Fast Ethernet includes specifications for 100BaseT, 100BaseT4, and 100BaseFX. 100BaseT utilizes Category 5 cable with pins 1 and 2 for transmit and pins 3 and 6 for receive and collision detection; the maximum cable run for 100BaseT is 100 m. 100BaseT4 utilizes all four wire pairs with Category 3 cable, and cable runs can extend to 100 m. 100BaseFX utilizes multimode or single-mode fiber, with maximum cable runs of 412 m and 10 km (kilometers), respectively. 100BaseFX is often used between buildings or in areas with high levels of radiated electrical noise.

At the other end of the spectrum, the first draft of the 10 Gigabit Ethernet standard, *IEEE 802.3ae*, was ratified on June 13, 2002. 10 Gigabit Ethernet is a full-duplex technology that targets the LAN, MAN, and WAN application spaces. Prestandard 10 Gigabit Ethernet products entered the market in 2001. 10 Gigabit Ethernet supports multimode and single-mode fiber-optic installations (note the fiber-only options). Coupled with devices supporting wave division multiplexing, 10 Gigabit Ethernet extends Ethernet to the WAN, enabling a common technology to run over both private and public networks. The technology supports distances of up to 40 km with single-mode fiber, connecting multiple campus locations within a 40 km range, as shown in Figure 2-1.

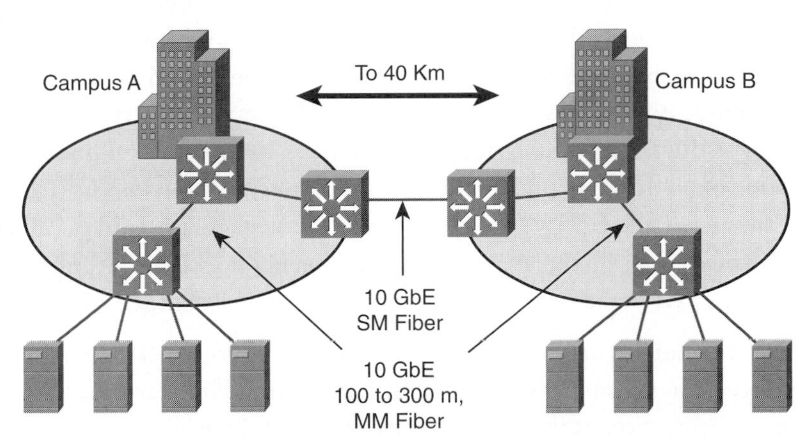

Figure 2-1
10 Gigabit Ethernet is ideal for intercampus single-mode fiber-optic links up to 40 or 50 km. These links can be bundled into 10 Gigabit EtherChannels to aggregate bandwidth.

Gigabit Ethernet standards have been in place since the late 1990s. Gigabit Ethernet ports are becoming as common as 10 Mbps switch ports were in the early 1990s. *IEEE 802.3z* specifies 1000 Mbps over fiber, and the IEEE 802.3ab specifies 1000 Mbps over Category 5 cable (1000BaseT). The *802.3ab* standard also provides support for 1000 Mbps over the EIA/TIA Category 5e specification. Figure 2-2 summarizes the Gigabit Ethernet standards.

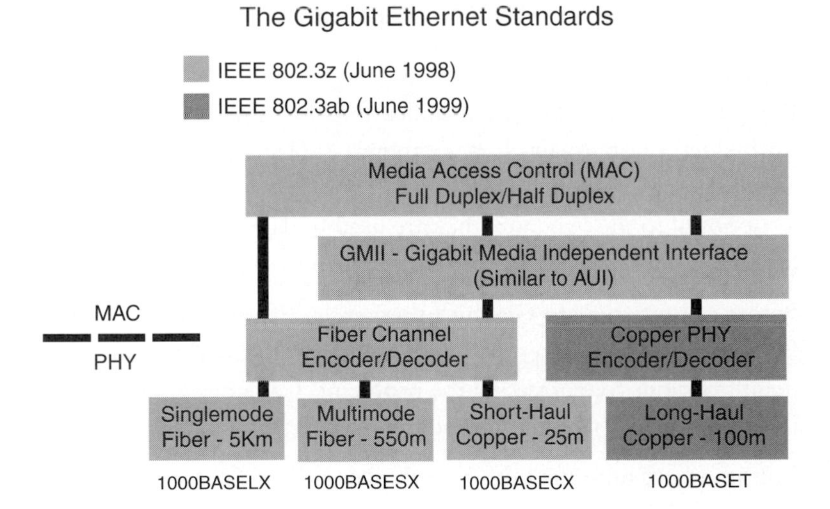

Figure 2-2
The IEEE 802.3z and IEEE 802.3ab standards provide the Gigabit Ethernet specifications for fiber-optic cable and Category 5 cable, respectively.

History of the Gigabit Ethernet Standards Process

More than a decade ago, 10BaseT hubs, allowing for greater manageability of the network and the cable plant, replaced the old 10Base5 and 10Base2 Ethernet networks. As applications increased the demand on the network, newer, high-speed protocols, such as FDDI and ATM, became available. However, Fast Ethernet became the backbone of choice over time because of its simplicity and its backward compatibility with Ethernet. The primary goal of Gigabit Ethernet was to build on that topology and knowledge base to create a higher-speed technology without forcing customers to replace existing networking equipment and infrastructure.

The standards body that worked on Gigabit Ethernet was called the IEEE 802.3z Task Force. The possibility of a Gigabit Ethernet standard was raised in mid-1995 after the final ratification of the Fast Ethernet standard. By November 1995 there was enough interest to form a group to study Gigabit Ethernet. This group met at the end of 1995 and several times during early 1996 to study the feasibility of Gigabit Ethernet. The meetings grew in attendance, reaching 150 to 200 individuals. Numerous technical contributions were offered and evaluated.

In July 1996, the 802.3z Task Force was established with a charter to develop a standard for Gigabit Ethernet. At the November 1996 IEEE meeting, the task force reached basic concept agreement on technical contributions for the standard. The first draft of the standard was produced and reviewed in January 1997; the final standard was approved in June 1998. The IEEE 802.3ab standard for Gigabit Ethernet over copper cable was approved one year later, in June 1999. The following sections discuss the standards for Gigabit Ethernet, starting with 802.3ab.

IEEE 802.3ab

IEEE 802.3ab describes the specifications for running Gigabit Ethernet over unshielded twisted-pair copper cabling. *1000BaseT* provides 1 Gbps bandwidth via four pairs of Category 5 UTP cable (250 Mbps per wire pair). All eight wires are used, as opposed to the two pairs that are used by 10BaseT and 100BaseT. Cabling distances for such an installation cannot exceed 100 meters.

Before you upgrade to Gigabit Ethernet, you should test existing Category 5 cable for far-end crosstalk and return loss. Should it fail, the ANSI/TIA/EIA TSB-95 standard details options for correcting the problems. Category 5e cable is recommended for new installations. These cables are terminated with RJ-45 connectors.

The standard supports both half-duplex and full-duplex operation, but full-duplex implementations are far more common. The CSMA/CD access method is used when Gigabit Ethernet is operating in half-duplex mode. In full-duplex mode, CSMA/CD is not utilized, but flow-control mechanisms handle buffering. Catalyst 4000 and 6000 family switches permit the configuration of Gigabit Ethernet flow-control parameters and the oversubscription of service-connected Gigabit Ethernet ports.

TECH NOTE: CSMA/CD AND COLLISIONS

With CSMA/CD, when stations detect that a collision has occurred, the participants generate a *collision enforcement signal* that lasts as long as it takes to propagate the smallest Ethernet frame size, or 64 bytes. This ensures that all the stations know about the collision and that no other station will attempt to transmit during the collision event. In addition, if an Ethernet device's counter exceeds a threshold value of 15 retries when trying to transmit a frame, the frame is discarded.

Gigabit Ethernet over Category 5 cable is presently the most common medium for horizontal cabling in ceilings and floors. However, as with Fast Ethernet, fiber-optic cabling is the medium of choice between buildings and between floors of a building.

1000BaseT can be deployed in three parts of a network:

- **Switch uplinks**—1000BaseT provides high-bandwidth connectivity from desktop switches to the next point of aggregation. With such uplinks, the switches can be linked to servers and other resources at gigabit speeds. These connections can substantially relieve network congestion, thus improving access to high-bandwidth applications and data.
- **Server connectivity**—1000BaseT links can be used to connect high-performance servers to the switch. This use dramatically improves traffic flow. Moreover, the price of 1000BaseT network interface cards has been falling as availability has been increasing.
- **Desktop connectivity**—As desktop network interface cards become available, users will begin implementing 1000BaseT at the desktop. At first, only power users will require such performance on the desktop. However, over time, 1000BaseT will migrate more to the desktop as prices continue to decrease.

IEEE 802.3z

The 802.3z specification calls for media support of multimode fiber-optic cable, single-mode fiber-optic cable, and a special balanced shielded 150-ohm copper cable. The current connector for Gigabit Ethernet is the SC connector for both single-mode and multimode fiber.

The signaling rate for Gigabit Ethernet is 1.25 Gbps. With the 8B/10B encoding scheme, data transmission equates to 1.0 Gbps.

The *Gigabit Interface Converter (GBIC)* was developed to allow network engineers to configure each gigabit port on a port-by-port basis for short-wavelength (SX), long-wavelength (LX), or long-haul (LH) interfaces. Figure 2-2 shows the various distances allowed for the 1000BaseSX and 1000BaseLX Gigabit Ethernet standards; these two standards are discussed in the next section. Vendor-specific LH GBICs extend the single-mode fiber distance from the standard 5 km up to 70 km. The LH option is value-added. Although it's not part of the 802.3z standard, LH allows switch vendors to build a single physical switch or switch module that the customer can configure for the required laser/fiber topology.

1000BaseSX and 1000BaseLX

Two laser standards for Gigabit Ethernet are supported over fiber: *1000BaseSX* (short-wavelength laser—850 nanometers [nm]) and *1000BaseLX* (long-wavelength laser—1300 nm). Short- and long-wavelength lasers are supported over multimode fiber. Only long-wavelength lasers are supported over single-mode fiber.

Two types of multimode fiber are available: 62.5 and 50 micron diameter fibers. (We are referring to the diameter of the core of the fiber-optic cable.)

NOTE

The Greek letter μ (pronounced "mew") is used in physics to denote microns. One micron is one micrometer, or 10^{-6} meters (.000001 m): one-millionth of a meter.

Single-mode fiber has a core diameter of 9 microns and is optimized for long-wavelength laser transmission.

The key differences between the use of short- and long-wavelength laser technologies are cost and distance. Lasers over fiber-optic cable take advantage of variations in cable attenuation. At different wavelengths, "dips" in attenuation are found over the cable. Short- and long-wavelength lasers take advantage of these dips and illuminate the cable at different wavelengths. Short-wavelength lasers are readily available because variations on these lasers are used in compact-disc technology. Long-wavelength lasers take advantage of attenuation dips at longer wavelengths in the cable. The net result is that although short-wavelength lasers cost less, they traverse a shorter distance. In contrast, long-wavelength lasers are more expensive but traverse longer distances.

Single-mode fiber has traditionally been used in cable plants to span long distances. With Gigabit Ethernet, for example, single-mode cable ranges can reach up to 10 km. Single-mode fiber, using a 9-micron core and a 1300-nanometer laser, makes possible the longest-distance technology. The small core and lower-energy laser elongates the laser's wavelength and allows it to traverse greater distances. This setup lets single-mode fiber reach the greatest distances of all media with the least reduction in noise.

NOTE

The Greek letter λ (pronounced "lam-duh") is used in physics to denote wavelength. Wavelength is frequently measured in nanometers (nm). A nanometer is 10^{-9} meters (.000000001 m): one-billionth of a meter.

Again, Gigabit Ethernet is supported over both 62.5-micron and 50-micron multimode fiber. 62.5-micron multimode fiber is typically seen in vertical campus and building cable plants and has been used for Ethernet, Fast Ethernet, and FDDI backbone traffic. This type of fiber, however, has a lower modal bandwidth (the cable's ability to transmit light), especially with short-wavelength lasers (1000BaseSX). In other words, short-wavelength lasers over 62.5-micron fiber can traverse shorter distances than long-wavelength lasers. Relative to 62.5-micron fiber, 50-micron multimode fiber has significantly better modal bandwidth characteristics and can traverse longer distances with short-wavelength lasers. The 62.5-micron option supports distances of up to 275 m. The 50-micron option supports distances of up to 550 m.

Table 2-1 summarizes the basic facts about 1000BaseSX and 1000BaseLX.

Table 2-1 **Multimode and Single-Mode Support for the 1000BaseSX and 1000BaseLX Standards**

Standard/Mode	Multimode Core diameter of 50 μ or 62.5 μ; λ = 850 nm	Single-Mode Core diameter of 10 μ; λ = 1300 nm
1000BaseSX (short wavelength)	Supported	Not supported
1000BaseLX (long wavelength)	Supported	Supported

TECH NOTE: DIFFERENTIAL MODE DELAY (DMD)

One of the problems that was solved in the process of establishing the IEEE 802.3z standards was that of differential mode delay (DMD). DMD affects only multimode fiber when using SX lasers. The problem is that when one mode of light experiences jitter (line distortion), this can cause a single mode to be divided into two or more modes of light, as shown in Figure 2-3. This results in lost data. Multimode fiber was originally designed for short-distance light-emitting diodes (LEDs), not lasers.

The fix is called a *conditioned launch* (see Figure 2-4). In other words, if the light that travels through the center of the core in a straight line is directed at a slight angle (or directed just off the core's center), the modal delay is corrected. To achieve a conditioned launch, a special mode-conditioning patch cable must be installed.

Figure 2-3 Differential mode delay affects multimode fiber when using SX lasers.

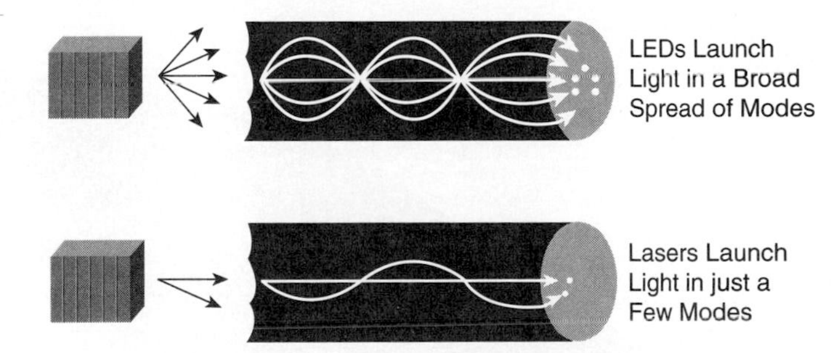

Modes traveling through the center (where the refractive index may dip) propagate too fast in some fibers, which causes the optical signal to be smeared.

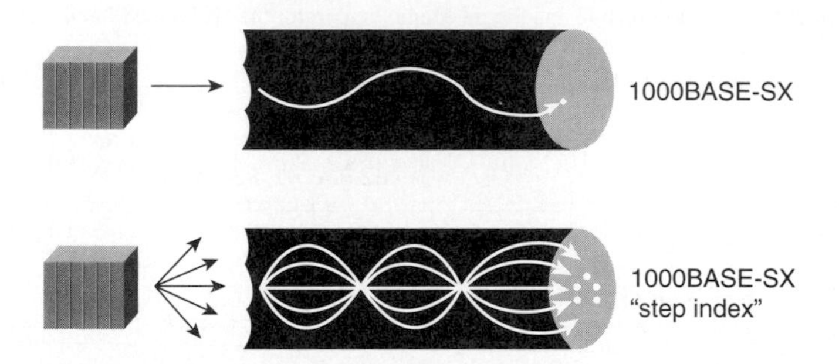

Figure 2-4
A conditioned launch solves the differential mode delay problem.

1000BaseCX

For shorter cable runs (25 meters or less), the *1000BaseCX* standard specifies Gigabit Ethernet transmission over a special balanced 150-ohm cable. This is a relatively new type of shielded cable; it is not unshielded twisted-pair cable or IBM Type I or II cable. In order to minimize safety and interference concerns caused by voltage differences, transmitters and receivers share a common ground. The return loss for each connector is limited to 20 decibels (dB) to minimize transmission distortions. Connector types for 1000BaseCX include a DB-9 connector and the High-Speed Serial Data Connector (HSSDC), which is shown in Figure 2-5.

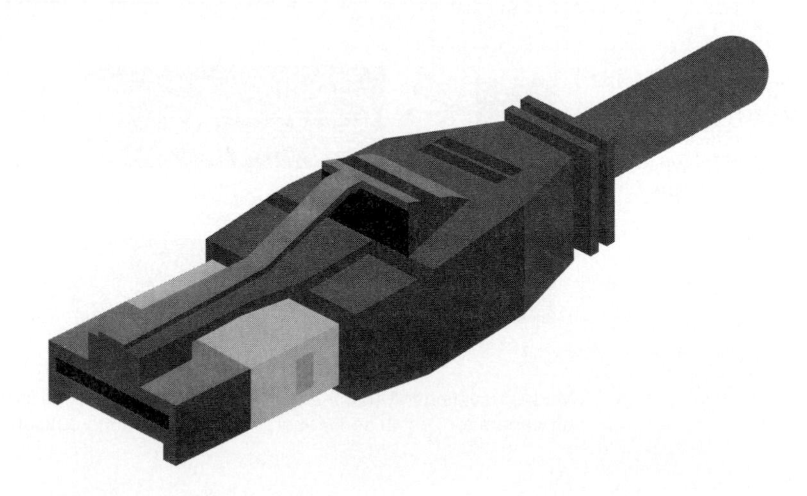

Figure 2-5
HSSDC is one option for the 1000BaseCX standard.

Applications of this type of cabling include short-haul data-center interconnections and inter- or intrarack connections. Because of the distance limitation of 25 meters, this cable does not work for interconnecting data centers to riser closets.

Figure 2-6 summarizes the possible distances for the various Gigabit Ethernet standards.

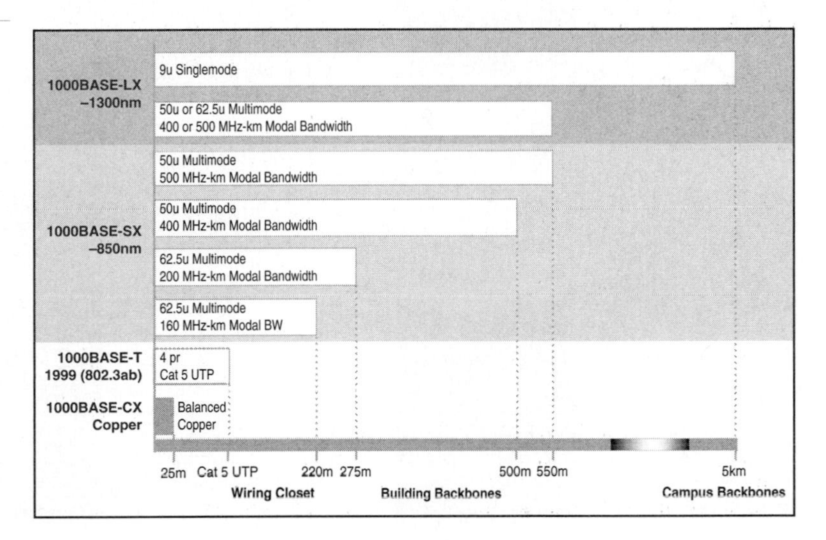

Figure 2-6 Various distance limitations are imposed by the media type used for Gigabit Ethernet transmission.

Gigabit Ethernet Protocol Architecture

In order to accelerate speeds from 100 Mbps Fast Ethernet to 1 gigabit per second (Gbps), several changes had to be made to the physical interface. The IEEE 802.3z Task Force decided that Gigabit Ethernet would look identical to Ethernet at the data link layer in terms of frame composition. The challenges involved in accelerating to 1 Gbps were resolved by merging two technologies: IEEE 802.3 Ethernet and ANSI X3T11 Fibre Channel. Figure 2-7 shows how the key components of each technology have been leveraged to form Gigabit Ethernet.

Leveraging these two technologies means that the designers of the standard can take advantage of Fibre Channel's existing high-speed physical interface technology while maintaining the IEEE 802.3 Ethernet frame format, backward compatibility for installed media, and the option of full duplex or half duplex with CSMA/CD.

The backward compatibility provided by the IEEE 802.3 Ethernet frame format helps minimize Gigabit Ethernet's technological complexity, resulting in a stable technology that can be quickly incorporated into existing networks.

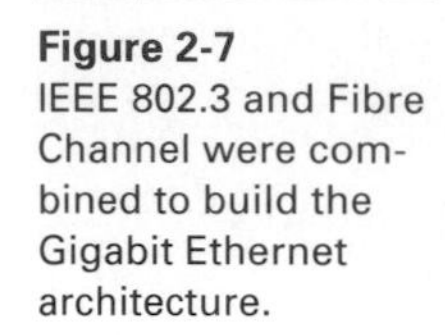

Figure 2-7
IEEE 802.3 and Fibre Channel were combined to build the Gigabit Ethernet architecture.

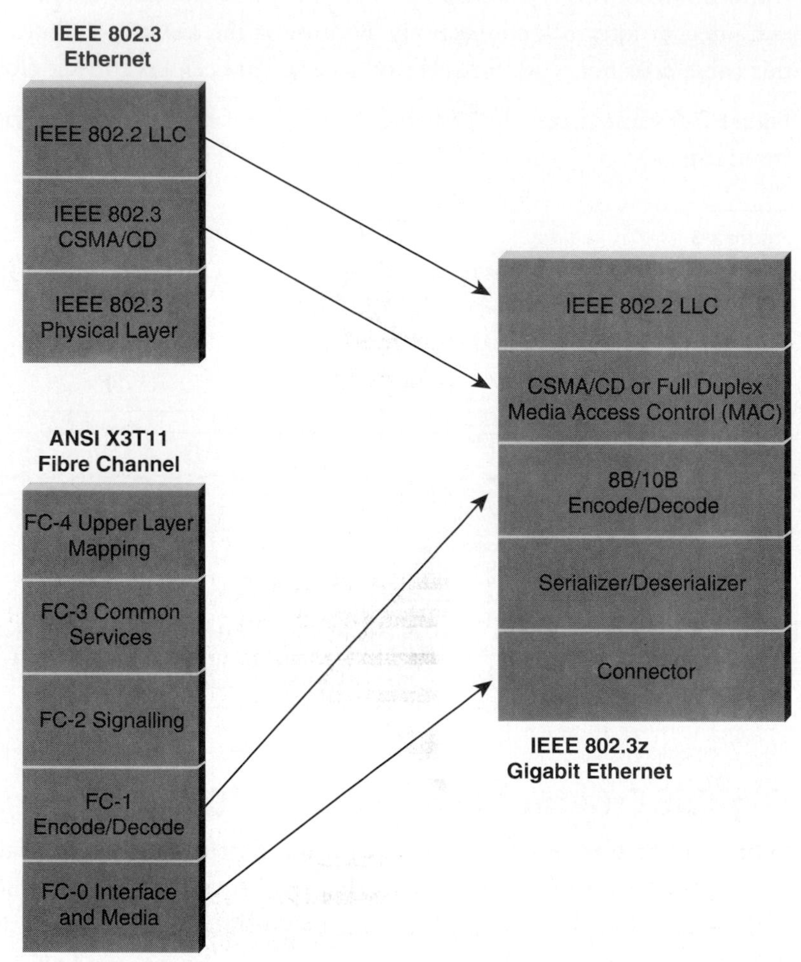

Recall that IEEE 802.3z and IEEE 802.3ab together comprise the standards for Gigabit Ethernet implementations. The architectural model for IEEE 802.3z is shown in Figure 2-8.

The reconciliation sublayer and the optional media-independent interface provide the logical connection between the MAC and the different sets of media-dependent layers.

The media-dependent physical coding sublayer (PCS) provides the logic for encoding, multiplexing, and synchronizing the outgoing symbol streams, as well as symbol code alignment, demultiplexing, and decoding of the incoming data.

The physical medium attachment (PMA) sublayer contains the signal transmitters and receivers (transceivers), as well as the clock recovery logic for the received data streams.

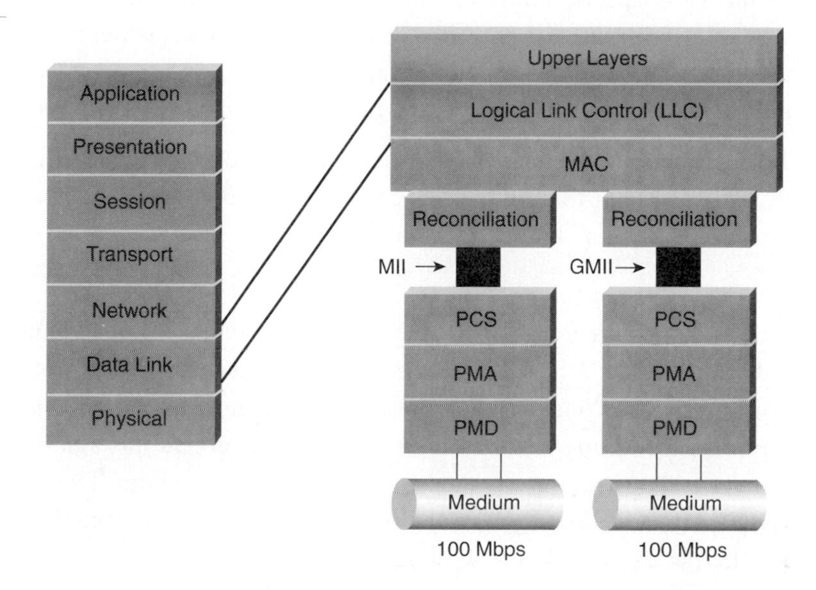

Figure 2-8
The architectural model for IEEE 802.3z maps to OSI Layers 1 and 2.

The physical medium-dependent (PMD) sublayer specifies the particular physical medium in use.

The physical interface for IEEE 802.3z and IEEE 802.3ab is detailed in Figure 2-9.

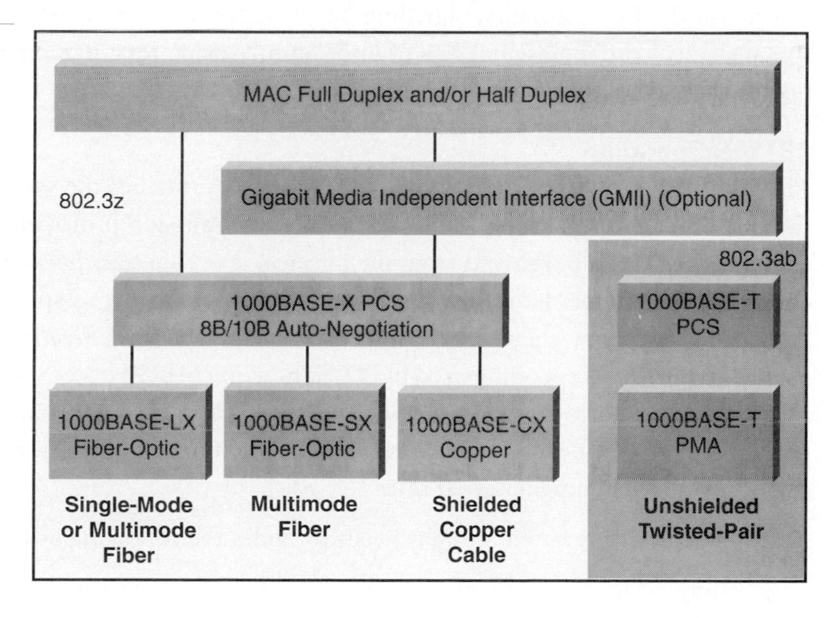

Figure 2-9
The physical interface for IEEE 802.3z and IEEE 802.3ab includes four media types.

The media-independent interface (MII) and Gigabit MII (GMII) are defined with separate transmit and receive data paths that are byte-serial (8 bits wide) for 1000-Mbps implementations. The media-independent interfaces and the reconciliation sublayer are common for their respective transmission rates and are configured for full-duplex operation.

The autonegotiation sublayer allows the NICs at each end of the link to exchange information about their individual capabilities and then negotiate and select the most favorable operational mode that both can support.

Serializer/Deserializer

The PMA sublayer for Gigabit Ethernet (refer to Figure 2-8) is identical to the PMA for Fibre Channel. The serializer/deserializer is responsible for supporting multiple encoding schemes and allowing presentation of those encoding schemes to the upper layers. Data entering the physical sublayer (PHY) enters through the PMD and needs to support the encoding scheme appropriate to that medium. The encoding scheme for Fibre Channel is 8B/10B, designed specifically for fiber-optic cable transmission. Gigabit Ethernet uses a similar encoding scheme. The difference between Fibre Channel and Gigabit Ethernet, however, is that Fibre Channel utilizes 1.062-gigabaud signaling, whereas Gigabit Ethernet utilizes 1.25-gigabaud signaling.

A different encoding scheme called PAM-5 is required for transmission over Category 5 cabling and is performed by the 1000BaseT PHY. PAM-5 provides better bandwidth utilization than simple binary signaling by using five different signaling levels. Each signal element can represent 2 bits of information (using four signaling levels). In addition, a fifth signal level is used for error correction.

8B/10B Encoding

The Fibre Channel FC-1 layer, shown in Figure 2-7, describes the synchronization and the 8B/10B encoding scheme. FC-1 defines the transmission protocol, including serial encoding and decoding to and from the physical layer, special characters, and error control. Gigabit Ethernet utilizes the same encoding/decoding as specified in the FC-1 layer of Fibre Channel. The scheme utilized is called *8B/10B encoding*. This scheme is similar to the 4B/5B encoding used in FDDI; however, 4B/5B encoding was rejected for Fibre Channel because of its lack of DC balance. The lack of DC balance can potentially result in data-dependent heating of lasers because a transmitter sends more 1s than 0s, resulting in higher error rates.

Encoding data transmitted at high speeds provides the following advantages:

- It limits the effective transmission characteristics, such as the ratio of 1s to 0s, on the error rate
- Bit-level clock recovery of the receiver can be greatly improved

- It increases the possibility that the receiving station can detect and correct transmission or reception errors
- It helps distinguish data bits from control bits
- It provides for octet and word synchronization

All these features have been incorporated into the Gigabit Ethernet specification.

In Gigabit Ethernet, the FC-1 layer takes decoded data from the FC-2 layer 8 bits at a time. The reconciliation sublayer (shown in Figure 2-8) "bridges" the Fibre Channel physical interface to the IEEE 802.3 Ethernet upper layers.

The 8B/10B code actually combines two other codes—a 5B/6B code and a 3B/4B code. The use of these two codes is simply for convenience; the mapping could have been defined directly as an 8B/10B code. In any case, a mapping is defined that maps each of the possible 8-bit source blocks into a 10-bit code block. There is also a function called *disparity control,* shown in Figure 2-10, that keeps track of the excess of 0s over 1s and vice versa. An excess amount in either direction is called a *disparity.* If there is a disparity, and if the current code block would add to that disparity, the disparity control block *complements* the 10-bit code block in binary. This complement has the effect of either eliminating the disparity or at least moving it in the opposite direction of the current disparity.

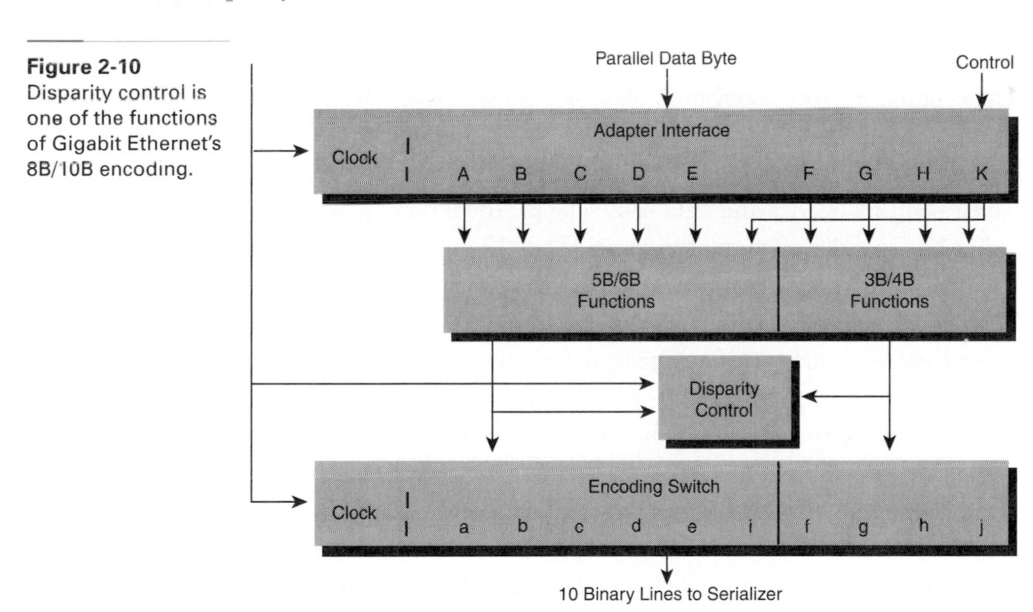

Figure 2-10
Disparity control is one of the functions of Gigabit Ethernet's 8B/10B encoding.

An unencoded information byte is composed of eight information bits A,B,C,D,E,F,G,H and the control variable Z. This information is encoded by FC-1 into the bits a,b,c,d,e,i,f,g,h,j of a 10-bit Transmission Character. The control variable has either the value D (D-type) for data characters or the value K (K-type) for special characters. The information received is recovered 10 bits at a time. Transmission Characters used for data (D-type) are decoded into one of the 256 8-bit combinations. Some of the remaining Transmission Characters (K-type; special characters) are used for protocol management functions. Codes detected at the receiver that are not D- or K-type are signaled as code violation errors. Figure 2-10 displays the information bits and a K-type value for the control variable (Z=K).

Frame Format

Gigabit Ethernet was designed to adhere to the standard Ethernet frame format. This setup maintains compatibility with the installed base of Ethernet and Fast Ethernet products, requiring no frame translation. Figure 2-11 displays the IEEE 802.3 Ethernet frame format.

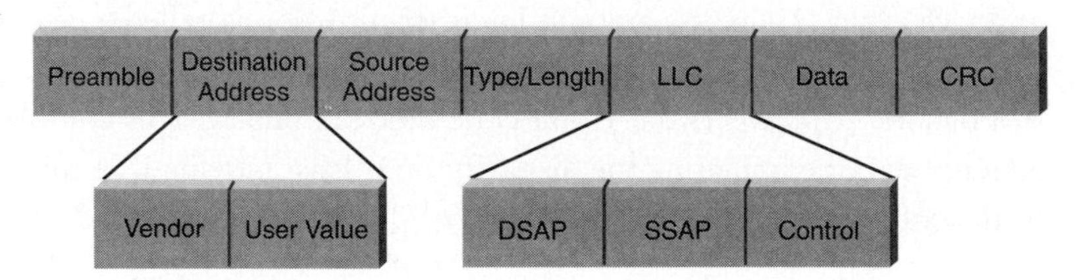

Figure 2-11 The frame format for Gigabit Ethernet matches that of IEEE 802.3 Ethernet.

The original Xerox specification identified a *type* field, which was utilized for protocol identification. The IEEE 802.3 specification eliminated the *type* field, replacing it with the *length* field. The *length* field identifies the length in bytes of the 802.2 Logical Link Control (LLC) header and data field. The protocol type in 802.3 frames is left to the 802.2 LLC header portion of the frame. The LLC layer is a sublayer of Layer 2 that is responsible for providing services to the network layer regardless of media type, such as FDDI, Ethernet, or Token Ring. The LLC layer makes use of LLC protocol data units in order to communicate between the Media Access Control (MAC) layer and the upper layers of the protocol stack. The LLC layer includes three fields used to determine access into the upper layers via the LLC PDU. The LLC PDU consists of the LLC header and the data field, which includes the Layer 3 packet and begins with the Layer 3 header. The three LLC header fields are the destination service access point (DSAP), the source service access point (SSAP), and the control variable. The DSAP address identifies the Layer 3 protocol in use by the destination device; the SSAP provides the same information for the source device.

The LLC defines service access for protocols that conform to the Open System Interconnection (OSI) model for network protocols. Unfortunately, protocols such as IP and IPX do not obey the rules for those layers. Therefore, additional information had to be added to the LLC in order to provide information regarding those protocols. The method used to provide this additional protocol information is called the Subnetwork Access Protocol (SNAP) frame. SNAP encapsulation is indicated by the SSAP and DSAP addresses being set to 0xAA. When you see that address, you know that a SNAP header follows. The SNAP header is 5 bytes long. The first 3 bytes consist of the organization code, assigned by IEEE. The following 2 bytes use the protocol *type* value from the original Ethernet specifications. Figure 2-12 compares the SNAP format to the other three Ethernet Frame formats. The second frame format displayed is sometimes referred to as "raw" Ethernet and is Novell-proprietary.

Figure 2-12
The four Ethernet frame types are still relevant for Gigabit Ethernet.

	← Frame →									
				← Packet →						
Frame Format	Layer 2 Header Fields 14-octets			← Data Field 1500-octets →						FCS
Ethernet v2 (ARPA)	MAC DA 6-octets	MAC SA 6-octets	Type 2-octets	Data						
802.3	MAC DA 6-octets	MAC SA 6-octets	Length 2-octets	Data						
802.3/802.2	MAC DA 6-octets	MAC SA 6-octets	Length 2-octets	DSAP 1-octet	SSAP 1-octet	Control	Data			
802.3/802.2 SNAP	MAC DA 6-octets	MAC SA 6-octets	Length 2-octets	0xAA 1-octet	0xAA 1-octet	0x03 1-octet	Org Code 3-octets	Type 2-octets	Data	4-octets

Quality of Service and IEEE 802.1p

Gigabit Ethernet supports QoS to reduce latency problems associated with voice, video, and data integration. QoS mechanisms provide options for prioritizing resources as a function of requirements. The convergence of mission-critical data with demanding multiservice applications can result in low-quality VoIP and video over IP. Gigabit Ethernet works hand-in-hand with QoS to optimize the convergence of data, voice, and video on your network. The QoS mechanism that Gigabit Ethernet relies on is specified by the IEEE 802.1p standard.

One question that remained unresolved until 1998 was whether a bridge should transmit buffered frames in the order in which they were received, or should a user priority be accounted for in determining which frame to transmit next? Consideration of this issue led to the development of the concept of *traffic class*. This was incorporated into the 1998 version of IEEE 802.1D and is called *IEEE 802.1p*. The goal of IEEE 802.1p is to allow Layer 2 switches and bridges to support time-critical traffic, such as voice and video, effectively. Many operating systems, such as Windows XP, now support IEEE 802.1p.

Queuing Delay and Traffic Classes

Queuing delay is how long a frame must wait until it becomes first in line for transmission on an outbound port. This delay is determined by the queuing method used by the networking device. The simplest scheme is first in, first out (FIFO). The concept of traffic classes is specifically designed to optimize queuing delay using a number of sophisticated queuing mechanisms. Some of these queuing methods include priority queuing (PQ), custom queuing (CQ), class-based weighted fair queuing (CBWFQ), low-latency queuing (LLQ), Weighted Random Early Detection (WRED), Committed Access Rate (CAR), Generic Traffic Shaping (GTS), and Resource Reservation Protocol (RSVP).

On a bridge supporting traffic classification, up to eight different traffic-class queues or buffers can be implemented on each outbound port. A traffic-class value, ranging from 0 (lowest priority) to 7 (highest priority), is associated with each queue.

On a given output port with multiple queues, the rules for transmission are as follows:

- A frame can be transmitted from a queue only if all queues corresponding to numerically higher values of traffic classes are empty. For example, if there is a frame in queue 0, it can be transmitted only if all the other queues at that port are currently empty. This is similar to the way priority queuing works.
- Within a given queue, the order of frame transmission must satisfy the following: The order of frames received by this bridge and assigned to this outbound port shall be preserved for the following:
  - Unicast frames with a given combination of destination address and source address
  - Multicast frames for a given destination (for example, CDP uses the multicast address 01-00-0C-CC-CC-CC)

In practice, the FIFO method is typically used. It follows that, during times of congestion, lower-priority frames might get stuck indefinitely at a bridge that devotes its resources to moving out the higher-priority frames.

Solution for Prioritizing Ethernet Traffic

IEEE 802.3 frames do not include a priority field, meaning that Ethernet prioritization could not be utilized prior to IEEE 802.1p. The solution agreed upon was to use something that most modern switches can understand—the priority field contained in the IEEE 802.1Q header (used with VLANs). The 802.1Q specification defines a 4-byte tag header inserted after the source and destination address fields of the Ethernet frame header. This tag header includes a 3-bit user priority field. Thus, if 802.1Q is in use by Ethernet sources, a user priority can be defined that stays with the frame from source to destination.

IEEE 802.1D provides a list of traffic types, each of which can benefit from simple segregation from the others. Do not confuse these traffic types with IP precedence values. These traffic types are mapped according to user priority by the QoS mechanism in play. For example, they can be mapped to other values to the IP precedence values ranging from 0 to 7. (Voice traffic is normally marked with an IP precedence of 5, which differs from the voice value in the following list.) In descending importance, the traffic types are as follows:

- **Network control** (7)—Both time-critical and safety-critical, consisting of traffic needed to maintain and support the network infrastructure, such as routing protocol frames.
- **Voice** (6)—Time-critical, characterized by less than a 10 ms delay, such as interactive voice.
- **Video** (5)—Time-critical, characterized by less than a 100 ms delay, such as interactive video.
- **Controlled load** (4)—Non-time-critical but loss-sensitive, such as streaming multimedia and business-critical traffic. A typical use is for business applications subject to some form of reservation or admission control, such as capacity reservation per flow.
- **Excellent effort** (3)—Also non-time-critical but loss-sensitive, but of a lower priority than controlled load. This is a best-effort type of service that an information services organization would deliver to its most important customers.
- **Best effort** (2)—Non-time-critical and loss-insensitive. This is LAN traffic handled in the traditional fashion.
- **Spare** (1)—Reserved class to be used for traffic of more importance than background but less importance than best effort.
- **Background** (0)—Non-time-critical and loss-insensitive, but of lower priority than best effort. This type includes bulk transfers and other activities that are permitted on the network but that should not affect the use of the network by other users and applications.

For example, if there are two queues, 802.1D recommends assigning network control, voice, video, and controlled load to the higher-priority queue and excellent effort, best effort, and background to the lower-priority queue. The reasoning supplied by the standard is as follows: To support a variety of services in the presence of bursty best-effort traffic, it is necessary to segregate time-critical traffic from other traffic. In addition, further traffic that is to receive superior service and that is operating under admission control also needs to be separated from the uncontrolled traffic. As you can see, the 802.1D standard allows flexibility in configuring queuing methods. In particular, it is not required to configure seven distinct queues for prioritization.

The user priority and traffic class concepts enable bridges and Layer 2 switches to implement a traffic-handling policy within a bridged collection of LANs that gives preference to certain types of traffic. These concepts are needed because bridges and Layer 2 switches cannot "see" above the MAC layer and hence cannot recognize or utilize QoS parameters specified in higher layers, such as IP. However, it is often the case that traffic from a bridged set of LANs must cross wide-area networks (WANs) that make use of QoS functionality. For example, the Internet can prioritize traffic based on IP-level QoS. A way is needed to map between traffic classes and QoS in such cases. The IP *type of service* (ToS) field provides a way to label traffic with different QoS demands. The ToS field is preserved along the entire path from source to destination through multiple routers. The ToS field includes a 3-bit IP precedence subfield. A router connecting a LAN to the Internet can be configured to read the Layer 2 traffic-class field and copy that into the ToS precedence field in one direction and copy the 3-bit precedence field into the IEEE 802.1Q user priority field in the other direction.

Gigabit Ethernet in Campus Network Design

Now that you have a handle on Gigabit Ethernet fundamentals, another look at campus network design is in order.

The availability of multigigabit campus switches, some pushing bundled link capacity to the terabit threshold, presents the opportunity to build extremely high-performance networks with high reliability. Gigabit Ethernet and Gigabit EtherChannel provide the high-capacity trunks needed to connect these gigabit switches. If you follow the right network design approach, performance and reliability are easy to achieve. Unfortunately, some alternative network design approaches can result in a network with lower performance, reliability, and manageability. With so many features available, and with so many permutations and combinations possible, it is easy to go astray. This section helps you avoid some of these pitfalls by following a commonsense design approach leading to a simple, reliable, manageable network.

EtherChannel is a technology based on the grouping of multiple full-duplex 802.3 Fast Ethernet, Gigabit Ethernet, or 10 Gigabit Ethernet links to provide fault-tolerant high-speed connections between switches, routers, and servers, as shown in Figure 2-13. EtherChannel is based on industry standards. It is an extension of the original EtherChannel technology offered by Kalpana in its switches in the early 1990s. Kalpana was acquired by Cisco in October 1994.

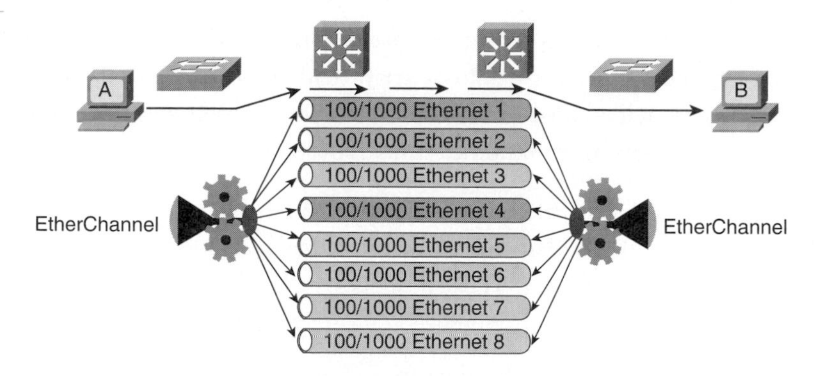

Figure 2-13 EtherChannel allows you to aggregate bandwidth by bundling multiple Ethernet, Fast Ethernet, Gigabit Ethernet, or 10 Gigabit Ethernet links into a channel.

Implementing EtherChannel has four major advantages:

- **Flexible incremental bandwidth**—EtherChannel bundles segments into groups of two to eight Fast Ethernet or Gigabit Ethernet links. The Catalyst 6500 series can bundle up to eight 10 Gigabit Ethernet links. An EtherChannel can operate as either an access or trunk link. With these options, you see that many incremental bandwidth configurations are available to the network engineer. You can estimate the bandwidth required for any Ethernet link by averaging the aggregate bandwidth of all devices utilizing the link.
- **Transparent to network applications**—EtherChannel does not require any changes to networked applications. For support of EtherChannel on enterprise-class servers and network interface cards, smart software drivers can coordinate distribution of loads across multiple network interfaces.
- **Load balancing**—When EtherChannel is used within the campus, switches and routers provide load balancing transparently across multiple links to network users. Unicast, multicast, and broadcast traffic is distributed across the links in the channel.
- **Resiliency and fast convergence**—EtherChannel provides automatic recovery for loss of a link by redistributing loads across the remaining links. If a link does fail, EtherChannel redirects traffic from the failed link to the remaining links in less than a second. This convergence is transparent to the end user. No host protocol timers expire, so no sessions are dropped.

Integrating Gigabit Ethernet in campus network design is simply a matter of extending the modular, multilayer campus design utilizing switch blocks introduced in Chapter 1. A multilayer campus intranet is highly deterministic, making it easy to troubleshoot as it scales. Intelligent Layer 3 services reduce the scope of typical problems caused by misconfigured or malfunctioning equipment. Intelligent Layer 3 routing protocols such as EIGRP, OSPF, and IS-IS support load balancing and fast convergence.

The multilayer model makes migration easier, because it preserves the existing addressing scheme of campus networks based on routers and hubs. Fault tolerance and fast convergence to the wiring closet are provided by Hot Standby Router Protocol (HSRP), which is discussed in more detail in Chapter 8. The multilayer campus design model supports all common campus protocols.

Structured Design with Multilayer Switching

The development of Layer 2 switching in hardware led to network designs that emphasized Layer 2 switching. These designs are characterized as "flat" because they avoid the logical, hierarchical structure and summarization provided by routers.

Layer 3 switching provides the same advantages as routing in campus network design, with the added performance boost from packet forwarding handled by specialized hardware. Integrating Layer 3 switching in the distribution layer and backbone of the campus segments the campus into smaller, more manageable pieces. Important services such as QoS and port-based authentication can be used in Catalyst Layer 2 switches at the access layer. The multilayer approach combines Layer 2 switching with Layer 3 switching to achieve robust, highly available campus networks.

Failure Domain

A group of connected Layer 2 switches is called a Layer 2 switched domain. A Layer 2 switched domain is also considered a failure domain because a misconfigured or malfunctioning workstation can introduce errors that affect or disable the entire domain. A jabbering network interface card might flood the entire domain with broadcasts. A workstation with the wrong IP address can become a black hole for packets. Problems of this nature are difficult to localize.

If possible, you should reduce the scope of the failure domain by restricting it to a single Layer 2 switch in a single wiring closet. To do this, you must restrict the deployment of VLANs and VLAN trunking. Ideally, one VLAN (IP subnet) is restricted to one wiring-closet switch. In this idealized topology, the gigabit uplinks from each wiring-closet switch connect directly to routed interfaces on Layer 3 switches.

Broadcast Domain

MAC-layer broadcasts flood throughout the Layer 2 switched domain. Using Layer 3 switching in a structured design reduces the scope of broadcast domains. In addition, intelligent, protocol-aware features of Layer 3 switches further contain broadcasts such as Dynamic Host Configuration Protocol (DHCP) broadcasts by converting them into directed unicasts.

Spanning-Tree Domain

Layer 2 switches execute the spanning-tree algorithm to provide a loop-free Layer 2 topology. If loops are included in the Layer 2 design, redundant links are put in blocking mode and do not forward traffic. In Gigabit Ethernet campus network design, it is preferable to avoid Layer 2 loops and have the Layer 3 protocols handle load balancing and redundancy so that all links are used for traffic.

The spanning-tree domain should be kept as simple as possible, and loops should be avoided. With loops in the Layer 2 topology, the spanning-tree protocol can take up to 50 seconds to converge. So avoiding loops is especially important in the mission-critical parts of the network, such as the campus backbone. To prevent spanning-tree protocol convergence events in the campus backbone, ensure that all links connecting backbone switches are routed links, not VLAN trunks. This also constrains the broadcast and failure domains, as explained previously.

Using Layer 3 switching in a structured design reduces the scope of spanning-tree domains. Ideally, you should let a Layer 3 routing protocol, such as EIGRP, OSPF, or IS-IS, handle load balancing, redundancy, and recovery in the backbone.

Virtual LANs

A VLAN is an extended Layer 2 switched domain that provides Layer 3 services such as broadcast containment. If several VLANs coexist across a set of Layer 2 switches, each individual VLAN has the same characteristics as a failure domain, broadcast domain, and spanning-tree domain, as just described. So, although VLANs can be used to segment the campus network logically, deploying pervasive VLANs throughout the campus adds to the complexity. Avoiding loops and restricting one VLAN to a single Layer 2 switch in one wiring closet minimizes the complexity.

One of the motivations in the development of VLAN technology was to take advantage of high-speed Layer 2 switching. With the advent of high-performance Layer 3 switching in hardware, the use of VLANs is no longer related to performance. A VLAN can be used to logically associate a workgroup with a common access policy as defined by access control lists (ACLs). Similarly, VLANs can be used within a server farm to associate a group of servers with a common access policy as defined by ACLs.

IP Subnets

An IP subnet also maps to the Layer 2 switched domain; therefore, an IP subnet is the logical Layer 3 equivalent of the Layer 2 VLAN. The IP subnet address is defined at the Layer 3 switch where the Layer 2 switch domain terminates. The advantage of subnetting is that Layer 3 switches exchange summarized reachability information rather than learning the path to every host in the switching network. Summarization is the key to the scalability benefits of routing protocols, such as EIGRP, OSPF, and IS-IS.

In an ideal, highly structured design, one IP subnet maps to a single VLAN (with routers providing inter-VLAN communication), and a single VLAN maps to a single switch in a wiring closet. This design model is somewhat restrictive, but it pays huge dividends in simplicity and ease of troubleshooting.

NOTE
Secondary IP addresses can be used to allow two or more IP subnets per VLAN. Integrated Routing and Bridging (IRB; see Appendix B) can be used to allow two or more VLANs per IP subnet.

Policy Domain

Access policy is usually defined on routers or Layer 3 switches in the campus intranet. A convenient way to define policy is with ACLs that apply to an IP subnet. Thus, a group of servers with similar access policies can be conveniently grouped in the same IP subnet and the same VLAN. Other services, such as DHCP, should be defined on an IP subnet basis.

A useful new feature of the Catalyst 6000 family of products is the VLAN access control list (VACL). A Catalyst 6000 or Catalyst 6500 can use conventional ACLs as well as VACLs. A VACL provides granular policy control applied between hosts on the same VLAN.

Modular Design

Modular design is based on building blocks. In a typical design, Gigabit Ethernet trunks connect Layer 2 switches in each wiring closet to a redundant pair of Layer 3 switches in the distribution layer, as shown in Figure 2-14. Recall that the hierarchical block concept is also applied to server farms and WAN connectivity.

Redundancy and fast failure recovery are achieved with HSRP configured on the two Layer 3 switches in the distribution layer. HSRP recovery is 10 seconds by default, but it can be tuned as required. The cost of adding redundancy is on the order of 15 to 25 percent of the overall hardware budget. The cost is limited because only the switches in the distribution layer and the backbone are fully redundant. This extra cost is a reasonable investment when the particular building block contains mission-critical servers or a large number of networked devices.

In the model shown in Figure 2-15, each IP subnet is restricted to one wiring-closet switch. This design features no spanning-tree loops and no VLAN trunking to the wiring closet. Each gigabit uplink is a native routed interface on the Layer 3 switches in the distribution layer. This model provides a hierarchical design that is becoming more common as a result of the high-speed Layer 3 switching that's now standard in multilayer switches. If one VLAN must span more than one wiring-closet switch, an alternative is laid out in the section "Alternative Building-Block Design." The solution described in that section is more general and also supports distributed workgroup servers attached to the distribution-layer switches.

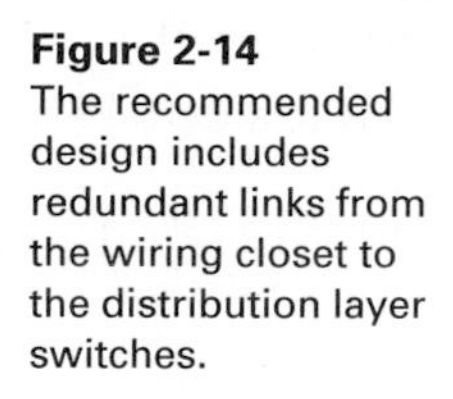

Figure 2-14
The recommended design includes redundant links from the wiring closet to the distribution layer switches.

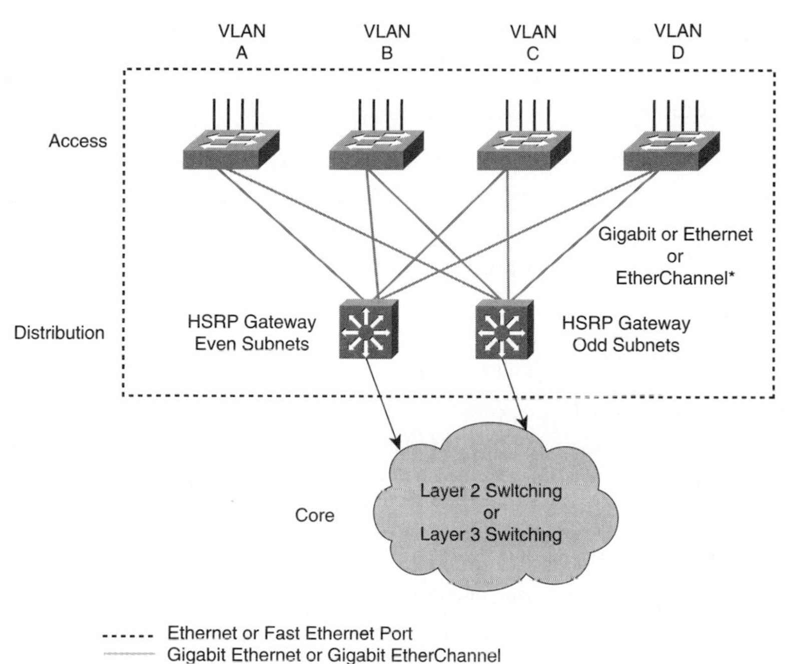

Figure 2-15
One IP subnet per wiring closet with no spanning-tree loops and no VLAN trunking is an attractive design option.

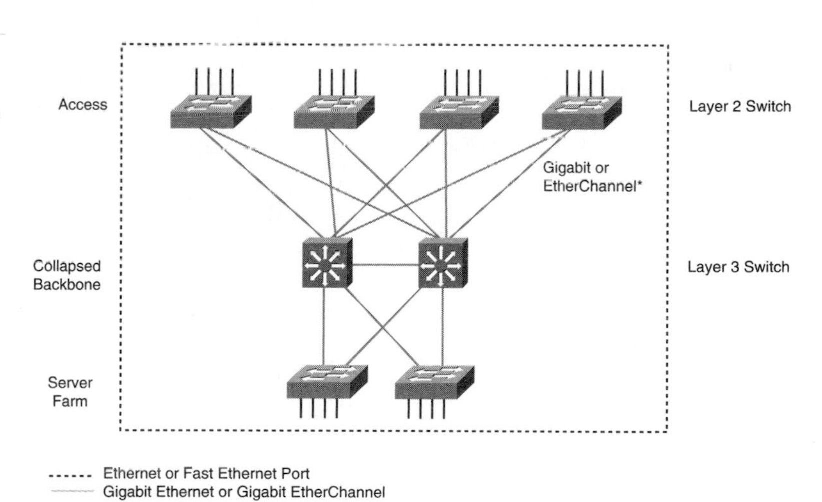

An optimal design features load balancing from the wiring-closet switch across both uplinks. Load balancing within the switch block can be achieved in several ways. For example, two IP subnets (two VLANs) can be configured on each wiring-closet switch. One distribution-layer switch is designated as the HSRP primary gateway for one subnet/VLAN, and the other distribution-layer switch is designated as the HSRP primary gateway for the other subnet/VLAN. Another option is to use HSRP with access lists to force half of the traffic within each VLAN to use one gateway on the Layer 3 switch and the other half to use the other gateway. These options are detailed in Chapter 8.

With this design, packets from a particular host always leave the building block via the active HSRP gateway. Either Layer 3 switch forwards returning packets. If symmetric routing is desired, configure a lower routing metric on the wiring closet VLAN interface of the HSRP gateway router. This metric is forwarded out to Layer 3 switches in the backbone as part of a routing update, making this the lowest-cost path for returning packets.

Building Design

The building design shown in Figure 2-16 comprises a single redundant building block. The two Layer 3 switches form a collapsed building backbone. Layer 2 switches are deployed in the wiring closets for desktop connectivity. Each Layer 2 switch has redundant gigabit uplinks to the backbone switches. Alternatively, if one VLAN must span more than one wiring closet, or if distributed servers are to be attached to the distribution layer switches, see the next section.

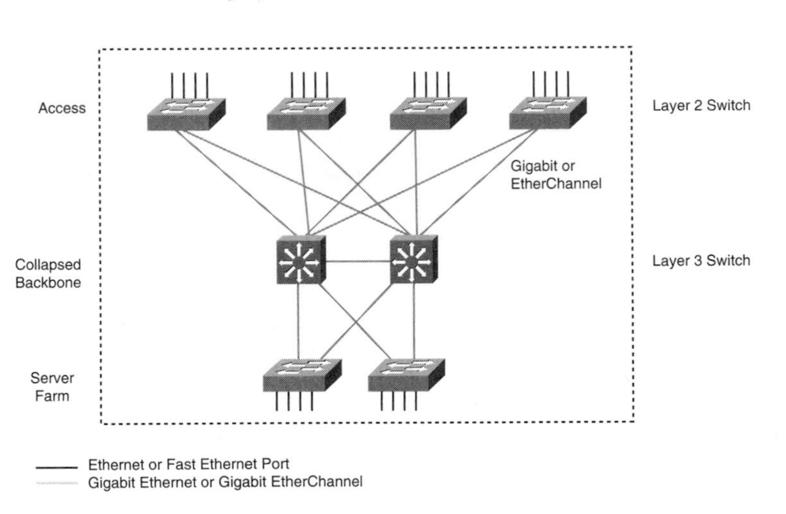

Figure 2-16
The building design illustrates the building block approach.

An optimization for the building design is to turn off routing protocol exchanges through the wiring closet subnets. To do this, use the Cisco IOS Software **passive-interface** command on the distribution-layer switches. In this configuration, the distribution-layer switches exchange routes with each other only over the direct link, not with each other across the wiring closet VLANs. Turning off routing protocol exchanges reduces CPU overhead on the distribution-layer switches. Other protocol exchanges, such as Cisco Discovery Protocol (CDP) and HSRP, are not affected.

In the building design, servers can be attached to Layer 2 switches or directly to the Layer 3 backbone switches, depending on performance and density requirements.

Alternative Building-Block Design

Figure 2-17 shows a more general building-block design with a workgroup server attached to the distribution-layer switch. This particular design assumes that more than one VLAN is serviced by the access-layer switch. This design assumes that the customer wants to have the server for workgroup A in the same subnet and VLAN as the client workstations for policy reasons. To accomplish this safely, a VLAN trunk is placed between the distribution-layer switches. The VLAN for workgroup A now forms a triangle; hence, Spanning-Tree Protocol puts one link in blocking mode. The triangle topology is required to maintain the integrity of the VLAN should one of the uplinks fail; otherwise, a discontiguous subnet would result. In this case, the VLAN trunk becomes the backup recovery path at Layer 2. The distribution-layer switch on the left is made the spanning-tree root switch for half of the VLANs (say, the even-numbered VLANs), and the distribution-layer switch on the right is made the spanning-tree root for the other half of the VLANs (say, the odd-numbered VLANs).

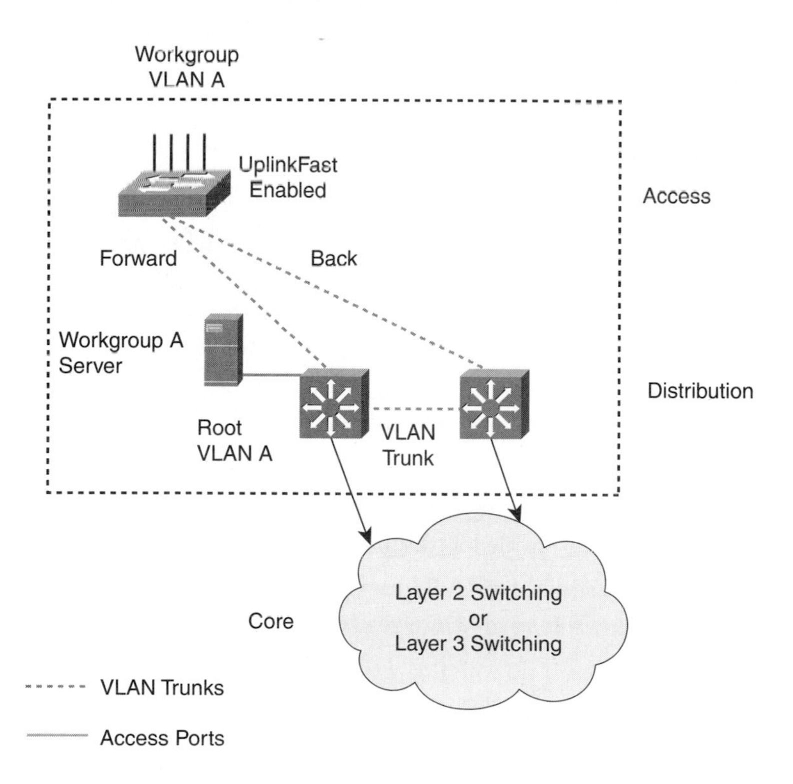

Figure 2-17 The alternative building-block design utilizes a trunk between distribution layer switches to prevent discontiguous subnets from forming when an access-distribution link fails.

It is important that the distribution-layer switch on the left is also the HSRP primary gateway for even-numbered VLANs so that symmetry between Layer 2 and Layer 3 is maintained. Fast Layer 2 spanning-tree recovery is achieved by enabling the UplinkFast feature on each wiring-closet switch. If the forwarding uplink is broken, UplinkFast puts the blocking uplink into forwarding mode in about 2 seconds.

Multilayer Campus Design

The multilayer campus design consists of a number of building blocks connected across a campus backbone. Figure 2-18 shows a generic campus design. Note the three characteristic layers: access, distribution, and core. In the most general model, Layer 2 switching is used in the access layer, Layer 3 switching in the distribution layer, and Layer 3 switching in the core.

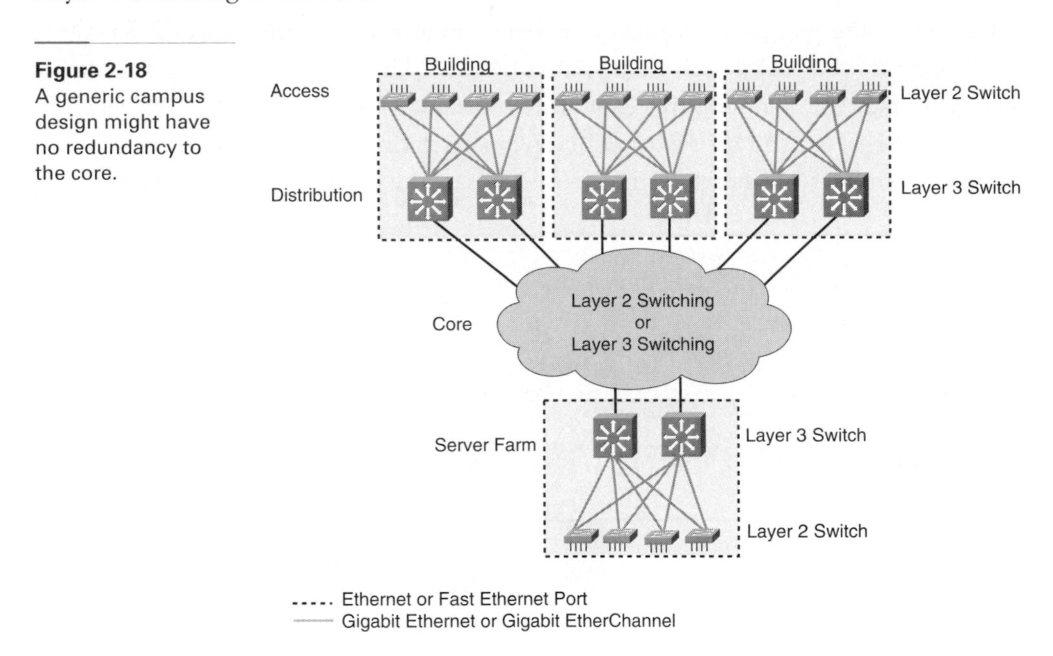

Figure 2-18
A generic campus design might have no redundancy to the core.

One advantage of the multilayer campus design is scalability. New buildings and server farms can easily be added without changing the design. The redundancy of the building block is extended with redundancy in the backbone. If a separate backbone layer is configured, it should always consist of at least two separate switches. Ideally, these switches should be located in different buildings to maximize the redundancy benefits.

The multilayer campus design takes maximum advantage of many Layer 3 services, including segmentation, load balancing, and failure recovery. IP multicast traffic is handled by Protocol Independent Multicast (PIM) routing in all the Layer 3 switches.

Access lists are applied at the distribution layer for granular policy control. Broadcasts, such as those generated by ARP, are kept off the campus backbone. Protocol-aware features such as DHCP forwarding convert broadcasts to unicasts before packets leave the building block. In the generic campus model shown in Figure 2-18, each block has two equal-cost paths to every other block. Figure 2-19 shows a more highly redundant connectivity model. In this model, each distribution-layer switch has two equal-cost paths to the backbone. This model provides fast failure recovery, because each distribution switch maintains two equal-cost paths in the routing table to every destination network. When one connection to the backbone fails, all routes immediately switch over to the remaining path in about 1 second after the link failure is detected.

Figure 2-19
A highly redundant multilayer campus network design includes redundancy to the core.

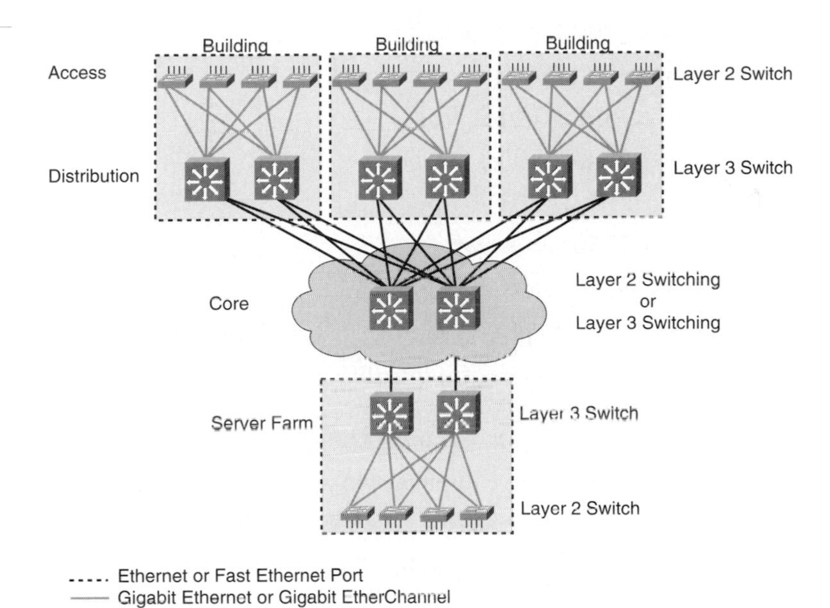

An alternative design that also achieves high availability is to use the design model shown in Figure 2-18, but to use EtherChannel links everywhere. A benefit of the Catalyst 6000 family switches is that availability can be improved further by attaching the links to different line cards in the switch. In addition, the Catalyst 6000 family and Catalyst 8500 series switches support IP-based load balancing across EtherChannel. The advantage of this approach, versus the design shown in Figure 2-19, is that the number of routing neighbors is smaller. The advantage of the design shown in Figure 2-19 is the greater physical redundancy provided by two links emanating from each distribution layer switch to the core.

Gigabit Ethernet Products

Cisco's switch and router product lines address customer needs by leveraging multiple technologies to deliver Gigabit Ethernet devices. These devices enable campus networks to scale to multigigabit rates. These product families include the following:

- Cisco 12000 series routers
- Catalyst 8500 series switching routers
- Cisco 7500 series routers
- Cisco 7200 series routers
- Catalyst 6000 family switches
- Catalyst 5000 family switches
- Catalyst 4000 family switches
- Catalyst 3500 XL switches
- Catalyst 2900 family switches
- Catalyst 2900 XL switches
- Catalyst 2950 series switches
- Catalyst 3550 series switches

In particular, Cisco offers a wide range of 1000BaseT products that allow customers to leverage existing Category 5 cabling infrastructures. The Cisco Switch Clustering technology allows network engineers to quickly expand and upgrade their networks across multiple wiring closets and various LAN media without having to add resources or replace existing switching equipment. In addition, the campus LAN can be managed with the web-based Cisco Cluster Management Suite (CMS).

The Cisco line of 1000BaseT products includes the following:

- **Cisco Catalyst 6000 family 16-port Gigabit Ethernet line card with RJ-45—** Designed to meet the growing demand of gigabit switching applications in both enterprise and service provider networks. A wide range of fabric-enabled Gigabit Ethernet modules are available for the Catalyst 6500 series 256 Gbps platform.
- **Cisco Catalyst 4000 family 48-port 10/100/1000BaseT line card**—The highest-density line card and system in the industry, with modular investment protection for desktop and server connectivity.
- **Cisco 1000BaseT Gigabit Interface Converter** (GBIC)—Provides full-duplex Gigabit Ethernet connectivity to high-end workstations and between wiring closets.

- **Cisco Catalyst 3550-12T**—The first stackable multilayer Gigabit Ethernet aggregation switch. Improves network control through Cisco IOS Software Intelligent Network Services.
- **Cisco Catalyst 2950 series**—A family of four fixed-configuration wire-speed Fast Ethernet desktop switches with 10/100/1000BaseT uplinks. Delivers enhanced QoS and multicast management.

The 2950T-24, the 3550-12T, and the 4000 family are popular solutions for delivering Gigabit Ethernet to the access layer, as shown in Figure 2-20.

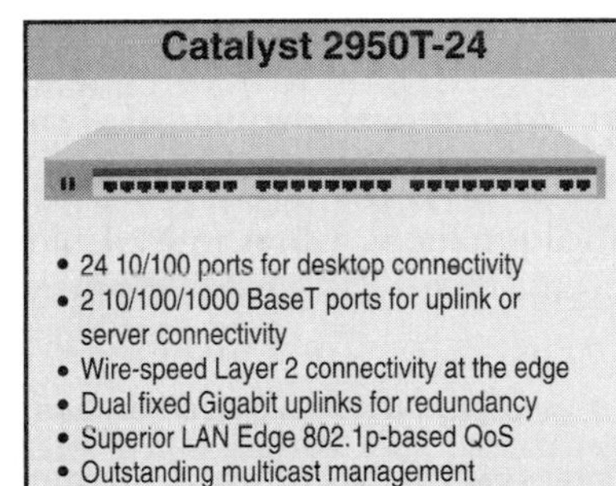

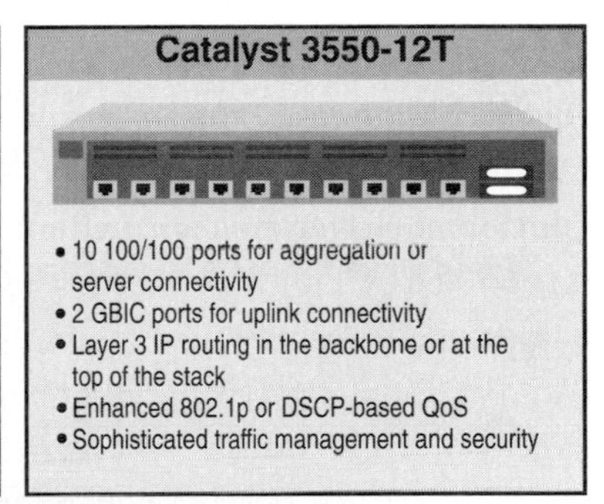

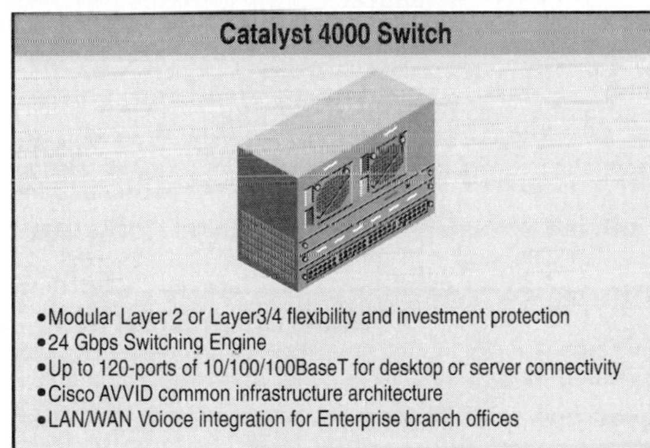

Figure 2-20 Cisco offers a number of 1000BaseT Catalyst switch solutions.

Cisco and Intel have teamed up to boost the usability, flexibility, and ease of deployment of Gigabit Ethernet solutions. Cisco 1000BaseT desktop switches offer seamless compatibility with the Intel PRO/1000 T Server Adapter, a Gigabit Ethernet adapter for Category 5 infrastructures. With the 1000BaseT standard, 1000BaseT network interface cards and switches support both 100/1000 and 10/100/1000 autonegotiation between Fast Ethernet and Gigabit Ethernet.

For the sake of completeness, the Cisco 12000 series routers cannot go without mention. The Cisco 12000 series is the industry's premier next-generation Internet routing platform, featuring the capacity, performance, and operational efficiencies that service providers require to build the most competitive IP backbone and high-speed provider edge networks. The 12416 Internet Router supports 10 Gbps (OC-192) in each of its 16 slots.

On the lighter side, two relatively new Gigabit Ethernet products include the 48-port 1000BaseLX Gigabit Ethernet line card for the Catalyst 4000 and the one-port Catalyst 6500 10GBaseEX4 Metro 10 Gigabit Ethernet module. Cisco also released the Catalyst 4000 Supervisor Engine III in early 2002. It integrates with all the existing Catalyst 4000 Gigabit Ethernet line cards and frees up a slot by integrating the Route Processor on board the Supervisor. It also is the first Supervisor Engine to use Native-mode IOS (CatIOS) as the default Catalyst image.

1000BaseT Deployment Scenarios

For branch offices of up to 100 users, migrating to 1000BaseT is easy, because 1000BaseT runs on the same existing Category 5 copper wiring, as shown in Figure 2-21. A 1000BaseT backbone provides the high-speed connectivity throughout the LAN necessary to handle high traffic and high-bandwidth applications, without the expense of rewiring.

Figure 2-21
1000BaseT is ideal for a branch office deployment.

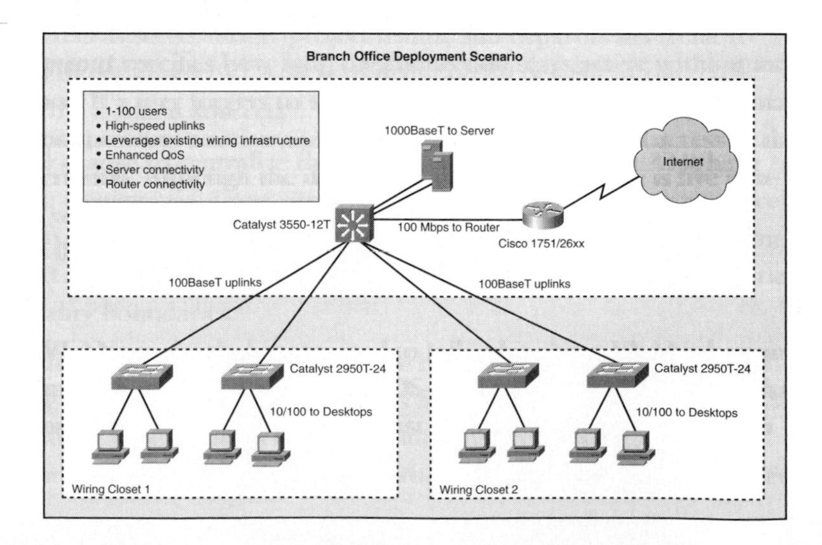

Catalyst 2950T-24 switches can be used to provide 10/100 connectivity to PCs, as well as provide 1000BaseT switch uplinks to a Catalyst 3550-12T switch at the top of the stack. The Catalyst 3550-12T also provides 1000BaseT connectivity to servers and 100 Mbps to the router. Branch offices can use such a simple 1000BaseT design to significantly relieve bottlenecks throughout the network and optimize performance and productivity.

In a campus network with more than 500 users, Gigabit Ethernet over fiber connections can be used to link different buildings throughout the campus, as shown in Figure 2-22. In the wiring closet of each building, Catalyst 3550-12T switches can be used to provide 1000BaseT connectivity to servers and to aggregate multiple Catalyst 2950T-24 switches. In addition, the GBIC ports on the Catalyst 3550-12T can be used to connect to the core using 1000BaseSX or LX GBICs.

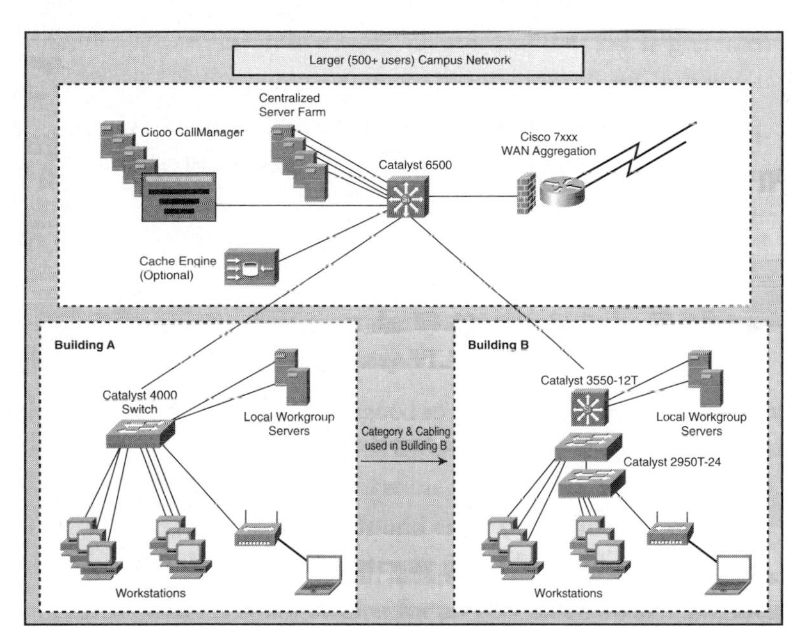

Figure 2-22
A Catalyst 6500 switch is recommended in a campus network with 500 or more users.

Catalyst 2950T-24 switches have Gigabit uplinks and deliver 10/100 speeds to desktop devices such as PCs and wireless access points. In another building, Catalyst 4000 switches with a 24-port 10/100/1000BaseT line card offer the flexibility to support 10, 100, and 1000 Mbps on the same interface. The Catalyst 4000 switch supports Gigabit speeds of up to 100 meters over Category 5 cable. Catalyst 3524-PWR XL switches can be used provide 10/100 access to desktop devices and Gigabit uplinks using Cisco GBICs.

Summary

Gigabit Ethernet and 10 Gigabit Ethernet are revolutionizing the way networks are constructed. Ethernet is now extending from the LAN to the MAN and WAN spaces. The backward compatibility of Gigabit Ethernet standards enables a smooth migration path. In addition, the cost of Gigabit Ethernet products is small relative to competing technologies such as SONET/ATM. In this sense, Gigabit Ethernet makes a network engineer's job that much easier.

This chapter detailed the Gigabit Ethernet standards, the media options for Gigabit Ethernet, and the Gigabit Ethernet protocol architecture (how it works). The QoS options available for Gigabit Ethernet were also discussed.

Considering the success of Gigabit Ethernet, we took time to revisit campus network design with a special emphasis on Gigabit Ethernet. The network design portion concluded with a look at the Cisco product lines that support Gigabit Ethernet and their role in network deployments.

In the next chapter, you begin learning how to configure Cisco Catalyst switches, beginning with basic Catalyst switch system administration.

Check Your Understanding

Test your understanding of the concepts covered in this chapter by answering these review questions. Answers are listed in Appendix A, "Check Your Understanding Answer Key."

1. Which types of quality of service are available for Gigabit Ethernet implementations?
   - **A.** EIA/TIA 1000BaseT
   - **B.** IEEE 802.3ab
   - **C.** IEEE 1000BaseT
   - **D.** IEEE 802.1p
2. Gigabit links can be used for which of the following?
   - **A.** Switch uplinks
   - **B.** Server connectivity
   - **C.** Desktop connectivity
   - **D.** All of the above
3. The serializer/deserializer is used to do what?
   - **A.** Convert serial communication to parallel communication
   - **B.** Allow presentation of encoding schemes to upper layers
   - **C.** Support serial implementations of Gigabit Ethernet
   - **D.** Encrypt Ethernet frames
4. Which body worked on the standards for Gigabit Ethernet?
   - **A.** IEEE 802 Working Group
   - **B.** IEEE 802.3 Working Group
   - **C.** IEEE 802.3z Task Force
   - **D.** IEEE 802.3ab Task Force
5. IEEE 802.3z was formally approved in what year?
   - **A.** 1996
   - **B.** 1997
   - **C.** 1998
   - **D.** 1999

6. What are the two standards for Gigabit Ethernet over fiber-optic cable?
   A. 1000BaseSX and 1000BaseLH
   B. 1000BaseX and 1000BaseSH
   C. 1000BaseSX and 1000BaseLX
   D. 1000BaseX and 1000BaseT
7. The IEEE 802.1Q header contains a 4-byte tag header inserted after the source and destination address fields in an Ethernet frame. This 4-byte header includes a 3-bit user priority field that is used for what?
   A. VLAN ID
   B. ISL compatibility
   C. Quality of service
   D. Bridge ID
8. What is the maximum cable run for 1000BaseCX?
   A. 10 m
   B. 20 m
   C. 25 m
   D. 100 m
9. List four families of Catalyst switches that support Gigabit Ethernet.
10. In modular campus network design, which protocol allows for redundant connections from an access layer switch to Layer 3 switches in the distribution layer?

Key Terms

1000BaseCX The standard for Gigabit Ethernet transmission over a special balanced 150-ohm cable. Specified in IEEE 802.3z. The distance limitation is 25 m. The connector types are DB-9 or HSSDC.

1000BaseLX The long-wavelength (1300 nm) laser standard for Gigabit Ethernet over fiber-optic cable. Specified in IEEE 802.3z. Supports multimode and single-mode fiber.

1000BaseSX The short-wavelength (850 nm) laser standard for Gigabit Ethernet over fiber-optic cable. Specified in IEEE 802.3z. Supports only multimode fiber.

1000BaseT Provides 1 Gbps bandwidth via four pairs of Category 5 UTP cable (250 Mbps per wire pair).

8B/10B encoding The encoding scheme specified in IEEE 802.3z for Gigabit Ethernet. 8B/10B combines two other codes—a 5B/6B code and a 3B/4B code. The mapping represents 8-bit raw data blocks in terms of 10-bit code blocks.

GBIC (Gigabit Interface Converter) GBICs were developed to allow network engineers to configure gigabit ports on a port-by-port basis for short-wavelength (SX), long-wavelength (LX), and long-haul (LH) interfaces. The converters fit into modular slots on a wide variety of Catalyst switches.

IEEE 802.1p An IEEE standard that was incorporated into the 1998 version of the IEEE 802.1D standard. It introduces the concept of traffic classes. Layer 2 switches and bridges supporting this standard can be configured to effectively prioritize time-critical traffic, such as voice and video.

IEEE 802.3ab The IEEE standard that specifies 1000 Mbps over Category 5 and Category 5e cable (1000BaseT).

IEEE 802.3ae The IEEE standard that specifies 10 Gigabit Ethernet implementations. 10 Gigabit Ethernet is a full-duplex technology targeting the LAN, MAN, and WAN application spaces.

IEEE 802.3z The IEEE standard that specifies 1000 Mbps over fiber-optic cable.

After completing this chapter, you will be able to perform tasks related to the following:

- Navigating the CLI on Cisco IOS-based and CatOS switches
- Managing configuration files
- Passwords and password recovery
- Configuring switches for Telnet access
- Loading images to Flash
- Setting port speed and duplex

Chapter 3

Switch Administration

Now that you have a basic handle on campus network design and the fundamentals of Ethernet technology, we turn our focus to basic system administration for Catalyst switches.

Cisco routers run the Cisco Internetwork Operating System (IOS). The command-line interface (CLI) embedded in the Cisco IOS Software is probably familiar to you by now. However, you might not be as familiar with the Catalyst switch operating systems (OS).

The most common Catalyst switch OS is referred to alternatively as Catalyst OS, *CatOS*, Catalyst set-command-based OS, or simply set-based OS. The CatOS is distinguished by its reliance on **set** commands. **set** commands are used to configure virtually everything with CatOS. For example, to configure port 2/3 as a member of VLAN 10, you type **set vlan 10 2/3**.

Some Catalyst switches, such as the 2950, use an IOS-like OS—a hybrid OS that evolved from Cisco's acquisition of Kalpana in 1994. This Catalyst switch OS is normally referred to as command-based or *IOS-based*. The term command-based comes from the fact that the routers use hundreds of commands in various modes to configure a variety of technologies. This is distinguished from the CatOS by the requirement to use the **set** command "in front of" other commands whenever technologies are configured on a Catalyst switch running CatOS. In fact, executing a **show version** command on a 2950 switch reveals "Cisco Internetwork Operating System Software" in the first line of output. The IOS-based Catalyst OS is similar to the router IOS, which you are accustomed to seeing.

Adding to the mix, the once-popular Catalyst 1900 and 2820 series switches have a menu-driven interface. Most of the 1900 and 2820 series switches will not be supported after December 2002. Cisco discontinued selling these devices in the late 1990s.

Finally, Catalyst 6000 family switches have an optional *Native IOS*. This is a single Cisco IOS image that simultaneously controls the Layer 2, 3, and 4 configuration. It is also called Catalyst 6000 IOS, CatIOS, or Supervisor IOS. In this case, the Supervisor module and the Multilayer Switch Feature Card (MSFC) both run a single, bundled Cisco IOS image. The MSFC is discussed in Chapter 6, "Inter-VLAN Routing."

NOTE
The 1900 and 2820 switches with the Enterprise feature set are menu-driven, but they support an optional CLI. We don't explore the configuration commands for 1900 and 2820 switches with the Enterprise feature set in this book, nor do we discuss the configuration of switches with menu-driven OSs.

In this chapter, you learn basic system administration for Catalyst switches: accessing the switch CLI, configuring idle timeouts, configuring passwords, recovering passwords, erasing configurations, saving configurations, naming the switch, configuring Telnet access, setting the port speed and duplex, setting port/interface descriptors, using command-history options, using help features, using the **ping** command, executing TFTP and Xmodem downloads, and using basic **show** commands. This chapter does not discuss web administration of Catalyst switches.

Preparing to Access the Switch

Before you can begin configuring a switch, you need to access its OS by using a console connection. You also need a terminal emulation program, such as HyperTerminal, to access the command line for the switch OS. After you configure the switch for basic IP connectivity, you can access and configure it through Telnet. Another option is to configure the switch for access with a modem connection. However, this is rarely used as a method to access switches, so it is not detailed in this book. The following sections discuss methods of accessing the OS of a Catalyst switch.

NOTE
Chassis-based Catalyst switches require a *Supervisor Engine*. This is the precise term, but many times this line card is called a "Supervisor module" or simply a "Supervisor." The Supervisor module can be thought of as the brains of the Catalyst switch. All other modules in the chassis can be accessed or configured from the Supervisor module. Either CatOS or CatIOS runs on the Supervisor module.

Console Access

The initial method of accessing the OS on a Catalyst switch involves the console port. When a switch is right out of the box, the first step to configuring it is to connect your workstation to it via the console port. Your switch documentation details the connectors and cables to use. The switch should arrive with the connectors and cables you need for console access. Should you need to terminate cables or adapters for console access, the pinouts for each Catalyst switch can be found on Cisco Connection Online (CCO) (www.cisco.com).

For example, to console into a Supervisor Engine on a Catalyst 4000/5000/6000 switch from a PC, use an RJ-45-to-RJ-45 rollover cable and the DB-9-to-RJ-45 serial adapter supplied with the switch, and follow these steps:

Step 1 Connect one end of the rollover cable to the console port.

Step 2 Attach the supplied DB-9-to-RJ-45 serial adapter (labeled Terminal) to a serial port on the PC.

Step 3 Attach the other end of the rollover cable to the RJ-45 port on the serial adapter.

Connecting via the console port gives you direct physical access to the switch, whether a fixed-chassis switch or a switch chassis with 13 slots. On chassis-based switches, you often have the option of consoling directly into various modules populating the chassis. Normally, though, you console into the Supervisor module on these switches and access modules indirectly via the switch's backplane using the **session** command.

Terminal Emulation and Telnet

A second method of accessing the OS on your Catalyst switch is Telnet. After you set up the basic IP configuration, you normally access the switch through Telnet, unless your security policy precludes Telnet access.

Many applications, such as Tera Term, HyperTerminal, and SecureCRT, include terminal emulation programs that can be used for console access to a Catalyst switch, as shown in Figure 3-1. These applications also provide other access options, such as Telnet (TCP port 23) and Secure Shell (TCP port 22). Secure Shell is often used instead of Telnet so that your session is encrypted (unlike Telnet, which sends your passwords as clear text). Applications such as SecureCRT and Tera Term give you a wide range of useful options, such as scrollback buffer size, function keys, scripting, window size, and line delay.

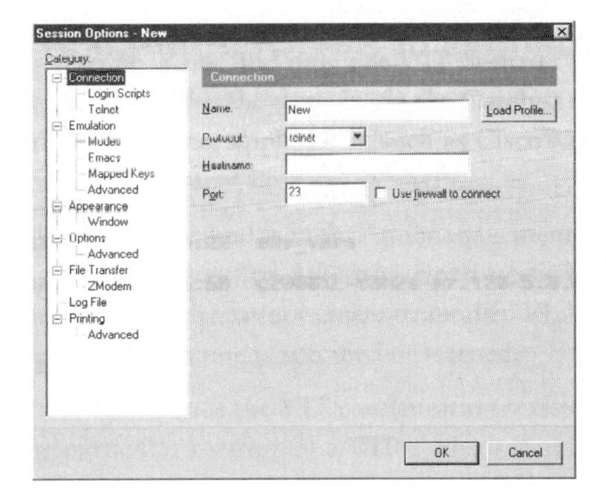

Figure 3-1 VanDyke Software's CRT program is useful for providing Telnet access to your switch. Multiple sessions can be created with a wide range of configurable options governing the window environment.

When you open the program, you create a new session, configure the IP address or host name, name the session for future reference, and click Connect (or something similar) to begin the Telnet session. For example, Figure 3-1 displays the configuration options for VanDyke Software's CRT program. The only option that is absolutely necessary for using Telnet is the IP address of the device to which you are connecting.

After you connect, you are prompted for a username and password, as shown in Figure 3-2. (The username might not be required, depending on the configuration of the device to which you are connecting.)

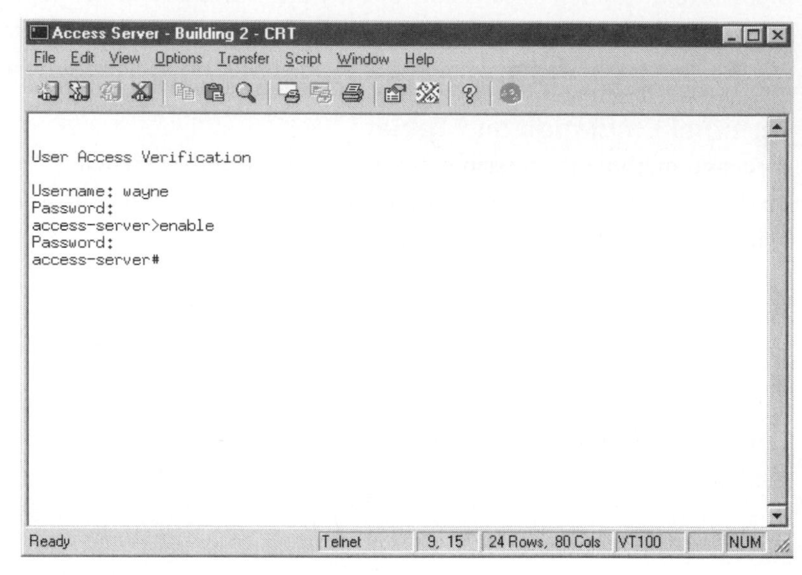

Figure 3-2 After your Telnet session initiates, you are prompted for a username and password.

Fundamentals of Switch Configuration

After you have access to the switch's OS, you need to understand how to navigate through the system and perform basic tasks, just as you would on a PC. The following sections discuss the procedures followed and commands utilized on Catalyst switches for a CatOS switch and an IOS-based switch.

> **NOTE**
> Many times, the IOS-based switch configurations match those of Cisco routers running IOS; in these cases, procedures and command sets aren't listed.

Entering the Switch and Changing Modes

After you have access to the switch, you need the ability to change between the various modes. For CatOS switches, configuration commands are entered in *privileged mode*. On IOS-based switches, there are several modes, such as interface mode, where configuration takes place. Modes for CatOS and IOS-based switches are discussed in the following sections.

CatOS Switch

On a CatOS switch, the switch CLI supports two modes of operation:

- Normal (also called login or user mode)
- Privileged (also called enable mode)

Both modes are password-protected. Normal-mode commands are used for system monitoring, similar to a Cisco router's user EXEC mode. *Normal mode* allows you to view most switch parameters, but you are unable to make configuration changes. Privileged-mode commands change the system configuration. This is unlike Cisco routers, where you need to enter global configuration mode to make system changes.

When you first connect to the switch (with the proper physical media and a terminal emulation program), you are presented with the prompt that's shown in Example 3-1.

Example 3-1 Entering Normal Mode on a CatOS Switch

```
Enter password:
```

On a new switch or a switch where the configuration has been cleared, the normal-mode password is null. If you are connecting to a new switch or one where the configuration has been cleared, press Return at the prompt. Otherwise, enter the normal-mode password for the switch.

After entering the correct password, you see the user-level command-line prompt, shown in Example 3-2.

Example 3-2 Entering Normal Mode on a CatOS Switch

```
Enter Password: <normal-mode_password>
Console>
```

To disconnect from the switch CLI, enter the **exit** command, as shown in Example 3-3.

Example 3-3 Exiting the CLI on a CatOS Switch

```
Console> exit
Session Disconnected...



Cisco Systems, Inc. Console            Fri Dec 27 2002, 02:47:02
```

Many commands (such as commands that modify the configuration) are entered only in privileged mode. From normal mode, you enter privileged mode by entering the **enable** command. On a new switch, the privileged-mode password is null. If you are connecting to a new switch, press Return at the "Enter password:" prompt. Otherwise, enter the privileged-mode password for the switch, as shown in Example 3-4.

Example 3-4 Entering Privileged Mode on a CatOS Switch

```
Console> enable

Enter password: <privileged-mode_password>
Console> (enable)
```

Note the change in the prompt for privileged mode. To exit privileged mode and return to normal mode, enter the **disable** command.

IOS-Based Switch

An IOS-based switch behaves exactly the same way as a router running the Cisco IOS with respect to entering the switch and changing modes. The switch has some modes that you don't see on Cisco routers. For example, you can enter *VLAN configuration mode*, which is accessed from privileged mode with the **vlan database** command. VLAN configuration mode allows you to make changes to the VLAN configuration, such as adding and removing VLANs. Release 12.1(9) added the global configuration command **vlan vlan-id**, permitting an alternative method for creating VLANs; entering this command puts you in config-vlan mode.

Table 3-1 describes the main command modes, how to access each one, the prompt you see in that mode, and how to exit the mode. The examples in the table use the host name Switch.

Table 3-1 Configuration Modes on IOS-Based Switches

Mode	Access Method	Prompt	Exit Method	Description
User EXEC	Begin a session with your switch.	Switch>	Enter **logout** or **quit**.	Use this mode to perform basic tests and display system information.
Privileged EXEC	While in user EXEC mode, enter the **enable** command.	Switch#	Enter **disable** or **exit**.	Use this mode to verify commands that you entered. Some configuration commands are available. Use a password to protect access to this mode.
VLAN configuration	While in privileged EXEC mode, enter the **vlan database** command.	Switch(vlan)#	To exit to privileged EXEC mode, enter **exit**.	Use this mode to configure VLAN-specific parameters.
Global configuration	While in privileged EXEC mode, enter the **configure** command.	Switch(config)#	To exit to privileged EXEC mode, enter **exit** or **end** or press Ctrl-Z.	Use this mode to configure parameters that apply to the entire switch.

Table 3-1 Configuration Modes on IOS-Based Switches (Continued)

Mode	Access Method	Prompt	Exit Method	Description
Interface configuration	While in global configuration mode, enter the **interface** command (with a specific interface).	Switch(config-if)#	To exit global configuration mode, enter **exit**. To return to privileged EXEC mode, press Ctrl-Z or enter **end**.	Use this mode to configure parameters for the Ethernet interfaces.
Line configuration	While in global configuration mode, specify a line with the **line vty** or **line console** command.	Switch(config-line)#	To exit to global configuration mode, enter **exit**. To return to privileged EXEC mode, press Ctrl-Z or enter **end**.	Use this mode to configure parameters for the terminal lines and the console line.

TECH NOTE: INTERFACE TERMINOLOGY

The term *interface* is used in reference to ports on IOS-based switches. For example, to configure parameters for port 1 on a Catalyst 2950 switch, you must first type **interface FastEthernet0/1** to enter the mode where configuration changes for port 1 can be made. The "interface" terminology is also reflected in the outputs of some of the various switch commands, such as the **show running-config** command output shown in Example 3-5.

Example 3-5 Output for the show running-config Command Displaying Interface (Instead of Ports) on an IOS-Based Switch

```
Switch#show running-config
Building configuration...

Current configuration:
!
version 12.0
no service pad
service timestamps debug uptime
```

continues

Example 3-5 Output for the show running-config Command Displaying Interface (Instead of Ports) on an IOS-Based Switch (Continued)

```
service timestamps log uptime
no service password-encryption
!
hostname Switch
!
!
!
!
!
!
!
ip subnet-zero
!
!
!
interface FastEthernet0/1
 duplex full
 speed 100
 port group 1 (replaced by channel-group command in release 12.1(6))
 switchport mode trunk
!
interface FastEthernet0/2
 duplex full
 speed 100
 port group 1 (replaced by channel-group command in release 12.1(6))
 switchport mode trunk
!
interface FastEthernet0/3
 duplex full
 speed 100
 switchport access vlan 10
!
interface FastEthernet0/4
!
interface FastEthernet0/5
!
<output omitted>
```

Command-Line Processing and Editing

A number of keystroke combinations are available to make your life easier. These shortcuts and help tools are specific to the OS in use. It is not imperative that you master all these shortcuts and help utilities. Over time, you will pick them up as you try to perform configuration tasks faster. The following sections discuss some of the command-line processing and editing shortcuts for the CatOS and IOS-based switches.

CatOS Switch

A number of specific rules come into play when you enter CatOS commands. CatOS switch commands are not case-sensitive. You can abbreviate commands and parameters as long as they contain enough letters to distinguish them from other currently available commands or parameters. You can scroll through the last 20 commands stored in the history buffer and enter or edit the command at the prompt. Table 3-2 lists command-line processing and editing keystroke shortcuts.

Table 3-2 Command-Line Processing and Editing Keystroke Shortcuts

Keystroke	Function
Ctrl-A	Jumps to the first character of the command line
Ctrl-B or left arrow key	Moves the cursor back one character
Ctrl-C	Escapes from and terminates prompts and tasks
Ctrl-D	Deletes the character at the cursor
Ctrl-E	Jumps to the end of the current command line
Ctrl-F or right arrow key	Moves the cursor forward one character
Ctrl-K	Deletes from the cursor to the end of the command line
Ctrl-L; Ctrl-R	Repeats the current command line on a new line
Ctrl-N or down arrow key	Enters the next command line in the history buffer
Ctrl-P or up arrow key	Enters the previous command line in the history buffer
Ctrl-U; Ctrl-X	Deletes from the cursor to the beginning of the command line
Ctrl-W	Deletes last word typed
Esc-B	Moves the cursor back one word
Esc-D	Deletes from the cursor to the end of the word
Esc-F	Moves the cursor forward one word
Delete key or Backspace key	Erases a mistake when you enter a command; erases one character at a time

Just as with the Cisco router IOS, if you cannot remember a complete command name, press the Tab key to allow the system to complete a partial entry. If you enter a set of characters that might indicate more than one command, the system beeps to indicate an error. Enter a question mark (?) to obtain a list of commands that begin with that set of characters. Do not put a space between the last letter and the question mark, and don't press the Return key after entering the question mark. For example, suppose that three commands in privileged mode start with "co." To see what they are, enter **co?** at the privileged prompt. The system displays the commands that start with the letters "co" and their descriptions.

When the output is longer than the terminal screen can display, a --More-- prompt appears at the bottom of the screen. To view the next line or screen, press the Return key to see the next line, or press the Spacebar to see the next screen of output. If you enter /text and press the Return key at the --More-- prompt, the display starts immediately above the line containing the text string.

IOS-Based Switch

The shortcuts and rules for command-line processing and editing with IOS-based switches are identical to those used with the Cisco router IOS. For example, if you type **show run** and then press the Tab key, the entire command **show running-config** appears on the console screen. Also, you can press Ctrl-A and Ctrl-E to move the cursor to the beginning or end of the current line, respectively.

Command History

The ability to access previously executed commands is one of the most useful features available. The command history feature permits the network engineer to retrieve previous commands, edit them, and rerun them for testing or configuration.

CatOS Switch

Several shortcuts are available for retrieving previously entered commands. After retrieving commands, you can edit them if necessary and reenter them. One of the most common examples is the **ping** command. For example, you test connectivity with **ping**, the **ping** fails, you make a configuration change, and you repeat the **ping** command to test whether the configuration change did the trick. For example, to repeat the previous command, just type **!!** followed by Enter. Table 3-3 lists the various command history combinations available.

Table 3-3 CatOS Command History Options Reference

Command	Function
To Repeat Recent Commands	
!!	Repeats the most recent command
!-*n*	Repeats the *n*th most recent command
!*n*	Repeats the command *n*
!*aaa*	Repeats the command beginning with the string *aaa*
!?*aaa*	Repeats the command containing the string *aaa*
To Modify and Repeat the Most Recent Command	
^*aaa*^*bbb*	Replaces string *aaa* with string *bbb* in the most recent command
To Add a String to the End of a Previous Command and Repeat It	
!!*aaa*	Adds string *aaa* to the end of the most recent command
!*n aaa*	Adds string *aaa* to the end of command *n*
!*aaa bbb*	Adds string *bbb* to the end of the command beginning with string *aaa*
!?*aaa bbb*	Adds string *bbb* to the end of the command containing string *aaa*

IOS-Based Switch

The shortcuts for repeating previous commands are the same as those used with the Cisco router IOS. For example, using the up arrow key or pressing Ctrl-P allows you to recall the previous command. In addition, the privileged-mode command **terminal history size** *number-of-lines* sets the number of lines that you can retrieve using Ctrl-P (10 is the default).

Help Features

Fortunately, the help features built into Cisco OSs frequently prevent you from having to look up needed commands. As with command-line processing and editing, a number of options are available that likely go beyond your day-to-day needs.

CatOS Switch

To see a list of commands and command categories, type **help** while you are in normal or privileged mode. You can view context-sensitive help (usage and syntax information) for individual commands by appending **help** to specific commands. If you enter a command using the wrong number of arguments or inappropriate arguments, usage and syntax information for that command is displayed. Additionally, appending **help** to a command category displays a list of commands in that category. These options are all shown in Figure 3-3. (Note: The output of **clear help** is truncated).

When the console displays help, the command line at the end of the display returns with strictly the switch prompt appearing (where the **help**, **copy help**, and **clear help** commands were entered in Figure 3-3). The command string you enter is not displayed for you because it is on a router.

IOS-Based Switch

The help features for IOS-based switches are the same as those used by routers running the Cisco IOS. For example, typing **show ?** lists all the keywords associated with the **show** command and the purpose of each keyword. Also, typing **show c?** lists all keywords associated with the **show** command whose first letter is c.

Erasing and Saving the Configuration

Although it is not something you do every day on a production switch, it is important to know how to clear a configuration file, which returns the switch to its default configuration. This will require subsequent reconfiguration, but you are assured that you are starting with a blank slate, and you don't have to worry about residual commands causing problems. The following sections describe how to erase and save configuration files for the CatOS switch and IOS-based switch.

CatOS Switch

On a Catalyst switch, as with a Cisco router, commands are additive. This means that adding configuration statements to an existing file does not completely overwrite the existing configuration. To ensure that a new configuration completely overwrites an existing configuration, enter the **clear config all** command in privileged mode. When you clear the configuration using the **clear config all** command, the factory default switch configuration is restored. The **all** keyword specifies that the modules and system configuration information are cleared, including the IP address (so if you are accessing the switch via Telnet, you lose your connection to the switch). With the **all** keyword, all ports are restored to VLAN 1, VTP is not configured (see Chapter 4, "Introduction to VLANs"), and all spanning-tree parameters revert to their default values.

Figure 3-3
The CatOS **help** keyword allows you to view command options associated with each command.

```
Console> (enable) show help
Show commands:
show accounting            Shows accounting information
show alias                 Shows aliases for commands
show arp                   Shows ARP table
show authentication        Shows authentication information
show authorization         Shows authorization information
show bridge                Shows bridge information
show cam                   Shows CAM table
show cdp                   Shows Cisco Discovery Protocol Information
show cgmp                  Shows CGMP info
show channel               Shows channel information
show config                Shows system configuration
show cops                  Shows COPS information
show default               Shows default status
show drip                  Shows DRIP information
show dvlan                 Shows dynamic vlan statistics
show errdisable-timeout    Shows err-disable timeout config
show errordetection        Shows error dection settings
show fddi                  Shows FDDI module entries
show fddicam               Shows FDDI module CAM table
show flash                 Shows system flash information
show garp                  Shows GARP information
Console> (enable) help
Commands:
clear                      Clear, use clear help for more info
configure                  Configures system from network
copy                       Copies code between TFTP/RCP server and module
disable                    Disables privileged mode
disconnect                 Disconnects user session
download                   Downloads code to a processor
enable                     Enables privileged mode
help                       Shows this help screen
history                    Shows contents of history substitution buffer
ping                       Sends echo packets to hosts
quit                       Exits from the Admin session
reconfirm                  Reconfirms VMPS
reload                     Forces software reload to linecard
reset                      Resets system or module
session                    Tunnels to ATM or Router module
set                        Sets commands, use set help for more info
show                       Show commands, use show help for more info
slip                       Attaches/detaches serial Line IP interface
switch                     Switches to standby <clock|supervisor>
telnet                     Telnets to a remote host
test                       Tests command, use test help for more info
traceroute                 Traces the route to a host
upload                     Uploads code from a processor
wait                       Waits for x seconds
write                      Writes system configuration to terminal/network
Console> (enable) copy help
Usage: copy <tftp|rcp> flash
       copy flash <tftp|rcp>
Console> (enable) clear help
Clear commands:
clear alias                Clears aliases of commands
clear arp                  Clears ARP table entries
clear banner               Clears Message Of The Day banner
clear cam                  Clears CAM table entries
clear cgmp                 Clears CGMP statistics
clear channel              Clears PagP statistical information
clear config               Clears configuration and resest system
clear cops                 Clears COPS information
clear counters             Clears MAC and Port counters
clear drip                 Clears DRIP statistics
clear gmrp                 Clears GMRP statistics
clear gvrp                 Clears GVRP statistics
clear igmp                 Clears IGMP statistics
clear ip                   Clears IP, use clear ip help for more info
clear kerberos             Clears Kerberos configuration information
clear key                  Clears config-key string
clear log                  Clears log information
clear logging              Clears system logging information
clear mls                  Clears multilayer switching information
clear multicast            Clears multicast router port
clear ntp                  Clears NTP servers and timezone
```

Example 3-6 shows how to clear the configuration for the entire switch.

Example 3-6 Restoring the Default Configuration on a CatOS Switch

```
Console> (enable) clear config all
This command will clear all configuration in NVRAM.
This command will cause ifIndex to be reassigned on the next system startup.
Do you want to continue (y/n) [n]? y
........
.............................


System configuration cleared.
Console> (enable)
```

To clear the configuration on an individual module, use the **clear config** *mod_num* command in privileged mode. Example 3-7 shows how to clear the configuration for module 2.

NOTE

Generally, a chassis-based Catalyst switch accommodates multiple Layer 3 modules, which are inserted to allow for load sharing, redundancy, and increased Layer 3 switching performance. Layer 3 modules, such as the Layer 3 Services module for the 4000, the RSM for the 5000, and the MSM for the 6000, are hot-swappable and can be removed or replaced without resetting other modules in the chassis.

Example 3-7 Erasing the Configuration for Module 2 in a CatOS Switch

```
Console> (enable) clear config 2
This command will clear module 2 configuration.
Do you want to continue (y/n) [n]? y
.............................
Module 2 configuration cleared.
Console> (enable)
```

The **clear config all** command affects only modules that are directly configured from the Supervisor module. To clear the configurations on router modules, you must access the modules with the **session** *module_number* command. This command performs the equivalent of an internal Telnet to the module so that you can make configuration changes. To display which slot the router module is in, use the **show module** command. The router modules on a switch use Cisco IOS commands to change, save, and clear configurations.

Configuring the switch through the console and via Telnet allows you to enter commands in real time: The switch immediately stores the entered commands in nonvolatile random-access memory (NVRAM). Commands typed into a CatOS switch are immediately stored and remembered, even through a power cycle. This presents a challenge when you attempt to reverse a series of commands. On a router, you can reverse a series of commands with a reboot, as long as you don't write the running configuration into NVRAM. You will find that, many times, the most challenging tasks with CatOS involve removing configured items.

IOS-Based Switch

An IOS-based switch behaves much like a Cisco router. You erase the configuration from NVRAM with the **erase startup-config** command in privileged mode. A subsequent reload of the switch presents you with a clean configuration. The exception to this is VLAN database settings. (For more information on VLAN database settings, see Chapter 4.) As a historical note, the 1900 and 2820 switches with the Enterprise edition use the command **delete nvram** to delete the configuration from NVRAM.

Unlike CatOS switches, the IOS-based switch does not immediately store commands in NVRAM, and it does require you to perform the **copy run start** command (or the equivalent **write memory** command). This greatly reduces the challenge of reversing a series of commands. You can reverse a series of commands with a reload as long as you don't write the running configuration into NVRAM.

Passwords

One of the first tasks you need to perform when configuring a networking device is securing it against unauthorized access. The simplest form of security in a campus network is to limit access to the switches in the switch block with passwords. By setting passwords, you can limit the level of access or completely exclude a user from logging on to an access, distribution, or core layer switch.

You can apply two types of passwords to your devices. The login password requires that you verify authorization before accessing any line, including the console. The enable password requires authentication before you can set or change the system operating parameters.

How you configure passwords depends on the particular Catalyst switch OS, as detailed in the following sections.

CatOS Switch

To set the normal (login) mode password on a CatOS switch, use the **set password** command. You are prompted to enter the old password and the new password. You are also prompted to retype the new password. Passwords are case-sensitive and may be 0 to 30 characters in length, including spaces. You do not see output as you type the passwords. Example 3-8 shows you how to change the CatOS login password.

NOTE

The default login and enable passwords are null on CatOS switches, so just press Return at the prompt (you don't have to type in a password).

Example 3-8 Changing the CatOS Login Password

```
Console> (enable) set password
Enter old password:
Enter new password:
Retype new password:
Password changed.
Console> (enable)
```

Example 3-9 demonstrates how to change the privileged (enable) mode password on the switch using the **set enablepass** command.

Example 3-9 Changing the CatOS Enable Password

```
Console> (enable) set enablepass
Enter old password:
Enter new password:
Retype new password:
Password changed.
Console> (enable)
```

The following code displays a portion of the CatOS **show config** output. This command displays the nondefault system and module configuration. You can see the login and enable passwords displayed as encrypted text:

```
Console> (enable) show config
This command shows non-default configurations only.
Use show config all to show both default and non-default configurations.
******




**

begin
!
# ***** NON-DEFAULT CONFIGURATION *****
!
!
#time: Tue Mar 1 2011, 21:20:59
!
#version 5.5(2)
!
set password $2$1KOx$rIxfbqEL1VpiB9BoZag6d.
set enablepass $2$zJba$XpzyxrESujXbDvofTP5cU/
!
```

You see that configuring passwords on a CatOS switch is straightforward. The fact that the passwords are encrypted prevents anyone from standing over your shoulder and seeing the passwords in the show config output.

IOS-Based Switch

Console and virtual terminal passwords are set on IOS-based switches just as they are on Cisco routers. To set the enable password on an IOS-based switch, you have two options: **enable secret** *password* (a secure encrypted password) and **enable password** *password* (a less secure unencrypted password).

You must enter one of these passwords to gain access to privileged mode. The enable secret password is recommended. When both are set, the enable secret password takes precedence. If you enter the **enable secret** command, the text is encrypted before it is written to the configuration file and is encrypted in the **show run** output. If you enter the **enable password** command, the text is written to the configuration file and can be read in the **show run** output exactly as it was typed.

TECH NOTE: ENABLE PASSWORD

The **enable password** command should no longer be used. (You can remove it with the command **no enable password**). All current images support the enable secret password, which is more secure. The only instance in which the **enable password** command might be used is when the device is running in a boot mode that does not support the **enable secret** command.

Enable secrets are hashed using the MD5 algorithm. The strength of the encryption used is the only significant difference between the two commands.

If you set the enable password to be the same as the enable secret, you have made the enable secret as prone to attack as the enable password. If both are set but the passwords differ, the enable secret password is required to enter privileged mode.

Enable passwords and enable secret passwords are governed by the following rules:

- The passwords must contain from 1 to 25 uppercase and lowercase alphanumeric characters. Any combination of uppercase and lowercase is acceptable.
- The passwords cannot start with a number.
- The passwords can have leading spaces, but they are ignored (not recorded as part of the password). However, intermediate and trailing spaces are recognized.

- The passwords can contain the question mark (?) character if the question mark is preceded with the key combination Ctrl-V. For example, to create the password abc?123, do the following:

  Step 1 Enter **abc**.

  Step 2 Press Ctrl-V.

  Step 3 Enter **?123**.

  When the system prompts you to enter the enable password, you do not need to precede the question mark with Ctrl-V; you can simply enter **abc?123** at the password prompt.

- Finally, to force the OS to encrypt all passwords displayed with the **show run** command, use the global configuration command **service password-encryption**.

Password Recovery

Password recovery refers to the recovery of lost passwords. Each networking device has its own procedure for recovering passwords. For Cisco devices, the procedures are conveniently organized by platform at www.cisco.com/warp/public/474/.

In the password-recovery process, you see that an attacker needs only the ability to reboot the Catalyst switch and access to the console to get into privileged mode. When in privileged mode, the attacker can make any changes he or she wants. It is important that you maintain the physical security of your equipment. It really does pay to keep wiring closets secured and console access restricted.

Next, the password recovery processes for CatOS and IOS-based Catalyst switches are detailed.

CatOS Switch

The CatOS password recovery procedure is relatively easy. Here are the steps:

Step 1 Connect to the switch with the appropriate console cable, adapter, terminal emulation program, and terminal settings.

Step 2 Turn the switch off and then back on.

Step 3 Within 30 seconds of turning the switch off and then on, perform the following sequence:

- Press Return at the password prompt to enter a null password.
- Type **enable** at the prompt to enter enable mode.
- Press Return at the password prompt to enter a null password.
- Change the password using the **set password** command or the **set enablepass** command.
- Press Return at the "Enter old password:" prompt to enter the null password you just configured.

It might take a few attempts to complete this procedure within the 30-second window. It takes practice to master this technique if you are accustomed to password recovery on Cisco routers, but you might end up finding this procedure much easier in comparison.

IOS-Based Switch

The password recovery procedure for an IOS-based switch is more complicated than the CatOS procedure. (This sequence assumes that you are using a console connection.) This procedure applies to the Catalyst 2900 XL, 3500 XL, 2950, and 3550 switches:

Step 1 Unplug the power cable.

Step 2 Hold down the mode button located on the left side of the front panel while reconnecting the power cord to the switch. You can release the mode button a second or two after the LED above port 1x is no longer illuminated. The following instructions appear:

```
The system has been interrupted prior to initializing the flash file
  system.
The following commands will initialize the flash file system, and finish
  loading the operating system software:
flash_init
load_helper
boot
```

Step 3 Type **flash_init**. This initializes the Flash file system.

Step 4 Type **load_helper**. This loads and initializes the helper image, which is a minimal IOS image stored in ROM that is typically used for disaster recovery.

Step 5 Type **dir flash:** (don't forget the colon). This displays a list of files and directories in the Flash file system. The switch file system is displayed:

```
Directory of flash:
2 -rwx 843947 Mar 01 1993 00:02:18 C2900XL-h-mz-112.8-SA
y4 drwx   3776 Mar 01 1993 01:23:24 html
66 -rwx    130 Jan 01 1970 00:01:19 env_vars
68 -rwx 1296   Mar 01 1993 06:55:51 config.text
1728000 bytes total (456704 bytes free)
```

Step 6 Type **rename flash:config.text flash:config.old** to rename the configuration file. This file contains the password definition.

Step 7 Type **boot** to boot the system.

Step 8 Enter **n** (for "no") at the prompt to circumvent the Setup program:

```
Continue with the configuration dialog? [yes/no] : n
```

Step 9 At the switch prompt, type **en** to turn on enable mode.

Step 10 Type **rename flash:config.old flash:config.text** to rename the configuration file with its original name.

Step 11 Copy the configuration file, config.text, into memory (press Return after the second and third lines):

```
switch#copy flash:config.text system:running-config
Source filename [config.text]?
Destination filename [running-config]?
```

The configuration file is now reloaded.

Step 12 Change the password:

```
switch#configure terminal
switch(config)#no enable secret
switch(config)#enable password Cisco
switch(config)#^Z
```

Step 13 Write the running configuration to the configuration file:

```
switch#write memory
```

As you might note, this is more complicated than Cisco router and CatOS switch password recovery. It is probably the most difficult type of password recovery for Cisco devices. Hopefully, the engineers will make this a little easier in future platforms!

Name, Contact, and Location

Administrative tasks include configuring the name, contact information, and location for the Catalyst switch. The name is simply the name by which you refer to your switch in your network documentation. The contact information is normally the name or e-mail address of the switch's administrator. The location is a string describing the switch's location. Next, we detail how to configure these settings for CatOS and IOS-based switches.

CatOS Switch

The system name on a CatOS switch is similar to the host name on a Cisco router. It can be configured manually or assigned through the Domain Name System (DNS). To manually set the system name, use the command **set system name** *name_string*. When you set the system name, the first 20 characters of the system name are used as the system prompt; you can override this with the **set prompt** *prompt_string* command. To clear the system name, use the command **set system name** with no arguments.

You can specify the contact name and location to help you with resource-management tasks. The system contact and location are set with the commands **set system contact** and **set system location**. Example 3-10 demonstrates the name, contact, and location system commands.

Example 3-10 Setting System Name, Contact, and Location

```
Console> (enable) set system name Cat6506
System name set.
Cat6506> (enable) set system name
System name cleared.
Console> (enable) set system contact MarcoPolo@whatta.trp
System contact set.
Console> (enable) set system location EurAsia
System location set.
```

It is recommended that you configure the system name, contact, and location settings. They help you maintain a well-organized, well-documented network.

IOS-Based Switch

On an IOS-based switch, the **hostname** command sets the system name. IOS-based switches, like routers, do not have configuration options for system contact or system location information. The interface-mode **description** command is the nearest alternative to the system location option on a CatOS switch. These **hostname** and **description** commands are illustrated in Example 3-11. The **show run** command verifies the configuration.

Example 3-11 The Switch Name and Interface Information Configured on IOS-Based Switches

```
Switch(config)#hostname Mililani
Mililani(config)#interface fa0/1
Mililani(config-if)#description Cubie 2-5
Mililani(config-if)#end
Mililani#show run
Building configuration...

Current configuration:
<output omitted>
version 12.0
no service pad
```

continues

Example 3-11 The Switch Name and Interface Information Configured on IOS-Based Switches (Continued)

```
service timestamps debug uptime
service timestamps log uptime
service password-encryption
!
hostname Mililani
!
<output omitted>
!
interface FastEthernet0/1
 description Cubie 2-5
 duplex full
 speed 100
 port group 1 (replaced by channel-group command in release 12.1(6))
 switchport mode trunk
!
<output omitted>
```

Remote Access

Before you can Telnet to, ping, or globally manage a switch, you must assign the switch an IP address and associate the switch with the *management VLAN*. The management VLAN is the VLAN on the network used for network management.

Although many LAN switches are primarily Layer 2 devices, they do maintain an IP stack for administrative purposes. Assigning an IP address to the switch associates that switch with the management VLAN, provided that the subnet mask of the switch IP address matches the subnet mask of the management VLAN. A default gateway or default route is also required to enable remote access.

Each Catalyst switch has a unique method of configuring the management VLAN and default gateway/route for IP connectivity; this connectivity is required for network management.

Of course, remote connectivity at Layer 3 is not possible unless Layer 1 connectivity is established first. Workstations and routers connect to Catalyst switch ports with straight-through cables during Layer 1 connectivity with switches. A crossover cable is required to connect a Catalyst switch to another Catalyst switch.

When the port status LED on a switch is illuminated green, it indicates that both the switch and connected device are powered up.

CatOS Switch

CatOS switches have a management interface called the *sc0 interface* that is associated with the management VLAN. The sc0 interface is a logical interface that has no physical port associated with it. This in-band management interface is connected to the switching fabric and participates in the functions of a normal switch port, such as spanning tree, Cisco Discovery Protocol (CDP), and VLAN membership.

After you configure the IP address, subnet mask, broadcast address, and VLAN membership of the sc0 interface, you can access the switch through Telnet or Simple Network Management Protocol (SNMP). SNMP is used by network management software packages to communicate information pertinent to network management, such as interface link status, environmental parameters, and failed login attempts.

IP traffic generated by the switch (for example, a Telnet session opened from the switch to a host) is forwarded according to the entries in the switch IP routing table. For network communication to occur, you must configure at least one default gateway for the sc0 interface. The default gateway is the IP address used by the switch to forward IP traffic originating from the sc0 interface, just as a workstation uses a default gateway. The switch's default gateway will likely differ from that of connected workstations (because they will be members of VLANs other than the management VLAN). The switch's IP routing table forwards traffic originating from the switch only; it does not forward traffic sent by devices connected to the switch.

The switch can be configured to obtain its IP configuration automatically via DHCP or RARP. The focus here is on manual configuration using static IP addresses.

The default IP settings for the sc0 interface are as follows:

- The IP address, subnet mask, and broadcast address are 0.0.0.0.
- The sc0 interface is assigned to VLAN 1.
- The default gateway address defaults to 0.0.0.0 with a metric of 0.

To assign an IP address to the sc0 interface, use the command **set interface sc0** [*ip_addr*[*/netmask*] [*broadcast*]]. You can specify the subnet mask (*netmask*) by using the number of subnet bits or using the subnet mask in dotted-decimal format. Example 3-12 illustrates these two options.

NOTE

The *switch fabric* is the hardware architecture in a chassis-based Catalyst switch that enables high-speed point-to-point connections to each line card. The switch fabric provides a mechanism to simultaneously forward packets on all point-to-point connections between the chassis slots. The switch fabric makes possible the simultaneous transmission and receipt of data, thus providing much higher aggregate throughput.

Example 3-12 Setting the IP for the Management Interface

```
Console> (enable) set interface sc0 128.1.2.3/24
Interface sc0 IP address and netmask set.
```

continues

Example 3-12 Setting the IP for the Management Interface (Continued)

```
Console> (enable) show interface
sl0: flags=50<DOWN,POINTOPOINT,RUNNING>
        slip 0.0.0.0 dest 0.0.0.0
sc0: flags=63<UP,BROADCAST,RUNNING>
        vlan 1 inet 128.1.2.3 netmask 255.255.255.0 broadcast 128.1.2.255
me1: flags=62<DOWN,BROADCAST,RUNNING>
        inet 0.0.0.0 netmask 0.0.0.0 broadcast 0.0.0.0

Console> (enable) set interface sc0 128.1.2.3/255.255.255.0
Interface sc0 IP address and netmask set.

Console> (enable) show interface
sl0: flags=50<DOWN,POINTOPOINT,RUNNING>
        slip 0.0.0.0 dest 0.0.0.0
sc0: flags=63<UP,BROADCAST,RUNNING>
        vlan 1 inet 128.1.2.3 netmask 255.255.255.0 broadcast 128.1.2.255
me1: flags=62<DOWN,BROADCAST,RUNNING>
        inet 0.0.0.0 netmask 0.0.0.0 broadcast 0.0.0.0
```

NOTE

Configuring the VLAN membership for the management interface (sc0) differs from the syntax used to assign physical switch ports to particular VLANs. For example, to assign ports 2/3, 2/4, 2/5, and 5/7 to VLAN 3, use the command **set vlan 3 2/3-5,5/7**. (See Chapter 4 for more details.)

To associate the in-band logical interface with a specific VLAN, enter the **set interface** sc0 *vlan* command. If you do not specify a VLAN, the system automatically defaults to VLAN 1 and the management VLAN. Example 3-13 shows how to assign interface sc0 to VLAN 1 and administratively bring it up (this is analogous to typing **no shutdown** on an Ethernet interface on a router).

Example 3-13 Assigning the Management Interface to a VLAN

```
Console> (enable) set interface sc0 1
Interface sc0 vlan set.
Console> (enable) set interface sc0 up
Interface sc0 administratively up.
```

Finally, you need to configure the default gateway, just as you would on a PC. The Supervisor engine sends IP packets destined for other IP subnets to the default gateway (a router interface in the same subnet as the switch IP address). The switch does not use the IP routing table to forward traffic from connected devices, only IP traffic generated by the switch itself (for example, Telnet, TFTP, and ping).

You can define up to three default IP gateways. Use the **primary** keyword to make a gateway the primary gateway. If you do not specify a primary default gateway, the first gateway configured is the primary gateway. If more than one gateway is designated as primary, the last one configured is the primary default gateway.

The switch sends off-network IP traffic to the primary default gateway. If connectivity to the primary gateway is lost, the switch attempts to use the backup gateways in the order in which they were configured. The switch sends periodic ping messages to determine whether each default gateway is up or down. If connectivity to the primary gateway is restored, the switch resumes sending traffic to the primary gateway.

To specify one or more default gateways, use the command **set ip route default** *gateway* [*metric*] [**primary**]. To remove default gateway entries, use the command **clear ip route default** *gateway*, or use **clear ip route all** to remove all default gateways and static routes. Example 3-14 provides an example of configuring and removing default gateways.

Example 3-14 Setting and Removing a Default Gateway

```
Console> (enable) set ip route default 128.1.2.1 primary
Route added.
Console> (enable) set ip route default 128.1.2.2
Route added.
Console> (enable) show ip route
Fragmentation   Redirect   Unreachable
-------------   --------   -----------
enabled         enabled    enabled

The primary gateway: 128.1.2.1
Destination      Gateway          RouteMask    Flags   Use       Interface
---------------  ---------------  ----------   -----   --------  ---------
default          128.1.2.2        0x0          G       0           sc0
default          128.1.2.1        0x0          UG      0           sc0
128.1.2.0        128.1.2.3        0xffffff00   U       14          sc0

Console> (enable) clear ip route default 128.1.2.1
Route deleted.
Console> (enable) sh ip route
Fragmentation   Redirect   Unreachable
-------------   --------   -----------
```

continues

Example 3-14 Setting and Removing a Default Gateway (Continued)

```
enabled           enabled    enabled


The primary gateway: 128.1.2.2
Destination        Gateway            RouteMask     Flags   Use         Interface
---------------    ---------------    ----------    -----   --------    ---------
default            128.1.2.2          0x0           UG      0             sc0
128.1.2.0          128.1.2.3          0xffffff00    U       332           sc0
```

After the switch is configured for IP, it can communicate with other nodes on the network (beyond simply switching Layer 2 traffic for connected hosts). To test connectivity to remote hosts, use the command **ping** *destination ip address*. The **ping** command returns one of the following responses:

- **Success rate is 100 percent or** ***ip_address*** **is alive or exclamation marks**—This response occurs in 1 to 10 seconds, depending on network traffic and the number of Internet Control Message Protocol (ICMP) packets sent. The type of output depends on the particular CatOS release.
- **Destination does not respond**—No answer message is returned if the host does not respond.
- **Unknown host**—This response occurs if the targeted host does not exist.
- **Destination unreachable**—This response occurs if the default gateway cannot reach the specified network.
- **Network or host unreachable**—This response occurs if there is no entry in the route table for the host or network.

Finally, because we are talking about remote access, it's useful to be able to adjust idle timeouts. The *idle timeout* specifies how long the connection stays active without any keystrokes taking place. If a user forgets to log out and leaves his or her terminal unattended, the idle timeout prevents another user from gaining unauthorized access to the switch through the terminal. Although the default setting for this feature is five minutes, you can alter it with the **set logout** *number of minutes* command. With the 0 argument (**set logout** 0), sessions are not automatically logged out.

IOS-Based Switch

Chapter 4 discusses VLANs in depth, but we need to talk a bit about VLANs here in the context of assigning a switch IP address to an IOS-based switch. Keep in mind that VLANs are logical constructs and to some sense exist independently of the ports on the switch.

Communication with the switch management interfaces is done through the switch IP address. The IP address is associated with the management VLAN, which, by default, is VLAN 1. A switch can have only one IP address. The switch's IP address can be accessed only by nodes connected to ports that belong to the management VLAN, unless inter-VLAN routing is configured. (You will find that, in some cases, configuring inter-VLAN routing is still insufficient to guarantee communication with the switch from devices on other VLANs).

Again, by default, the management VLAN is VLAN 1; however, you can configure a different VLAN as the management VLAN on switches running Cisco IOS Software Release 12.0(5) or later. When managing VLANs, keep the following in mind:

- With the exception of VLAN 1, the management VLAN can be deleted.
- When a management VLAN other than VLAN 1 is created, the management VLAN is administratively down. If VLAN 1 is the management VLAN, it is automatically up.
- Only one management VLAN can be active at a time.

Thus, using the management VLAN, you can manually assign an IP address, mask, and default gateway to the switch. The CPU sends traffic destined for an unknown IP network to the default gateway.

Beginning in privileged EXEC mode, follow these steps to enter the IP information:

Step 1 In global configuration mode, enter the VLAN to which the IP information is assigned. You can configure any VLAN from 1 to 1001. Use the command **interface vlan** *number*.

Step 2 Enter the IP address and subnet mask just as you would on a router interface.

Step 3 Return to global configuration mode and enter the IP address of the default router with the **ip default-gateway** *ip_address* command.

NOTE

To remove the IP address and disable the IP stack, use **no ip address** in management VLAN interface configuration mode.

Example 3-15 displays the IP address configuration for an IOS-based switch and the corresponding **show run** and **show ip interface** output.

Example 3-15 Configuring IP for Network Management

```
Switch(config)#interface vlan 1
Switch(config-if)#ip address 128.1.2.3 255.255.255.0
Switch(config-if)#exit
Switch(config)#ip default-gateway 128.1.2.1
```

continues

Example 3-15 Configuring IP for Network Management (Continued)

```
Switch#show run
<output omitted>
!
interface VLAN1
 ip address 128.1.2.3 255.255.255.0
 no ip directed-broadcast
 no ip route-cache
!
ip default-gateway 128.1.2.1
<output omitted>
Switch#show ip interface
VLAN1 is up, line protocol is up
  Internet address is 128.1.2.3/24
  Broadcast address is 255.255.255.255
  Address determined by setup command
  MTU is 1500 bytes
<output omitted>
```

Use the **ping** command in user mode or privileged mode to test your IP configuration.

Finally, to set idle timeouts on an IOS-based switch (optional), use the **exec-timeout** command in virtual terminal line mode or line console mode (just as you would on a Cisco router). The default idle timeout is 10 minutes for IOS-based switches.

Loading Images to Flash

One of the more common administrative tasks is loading images to Flash. The purpose of doing this is to upgrade the OS or to restore a stable version of the OS. The procedure is similar for all Cisco network devices. The idea is to set up a Trivial File Transfer Protocol (TFTP) server and download a new image to the device via Ethernet using the IP address configuration of the appropriate management interface. In a worst-case scenario, you need to download an image over a console cable via Xmodem.

The following sections describe the methods of loading Flash images to Catalyst switches.

CatOS Switch

The Supervisor of these switches implements a file system and can hold several images. Supervisors have at least a Flash device named *bootflash:*. The *bootflash* is a Flash memory device that stores OS images. It can also store a boot helper image or system configuration information.

Depending on the number of PCMCIA slots (also called Flash PC card slots) on the Supervisor, you can have a slot0: and a slot1: Flash device available. Most basic operations are available, such as listing, copying, and deleting files on these devices. The commands are similar in name and syntax to DOS commands. Here is a list of common commands:

- **Formatting Flash—format** *device*:
- **Listing files on Flash—dir** [*device*:] [**all**]
- **Changing the default Flash device—cd** *device*
- **Copying files—copy** [*device*:] *filename* [*device*:]*filename*
- **Marking files as deleted—delete** [*device*:]*filename*
- **Squeezing Flash—squeeze** *device*:

NOTE

device: refers to the device referenced by the respective command, such as bootflash: or slot0: or slot1:.

The Catalyst 5000 family, 6000 family, and 2926G series switches support two IP management interfaces—the in-band management interface (sc0) and the SLIP interface (sl0). The slip interface, sl0, is an out-of-band management port because it is not attached to the switching fabric, and no traffic is switched over it.

The Catalyst 4000 family, 2948G, and 2980G switches support three IP management interfaces: sc0, sl0, and an out-of-band management Ethernet interface (me1). The 10/100 *me1 interface* is not attached to the switching fabric. If both the sc0 and me1 interfaces are configured, the Supervisor engine software determines which interface to use when performing standard transmission of IP packets based on the local routing table. Operations that utilize this functionality include TFTP, ping, Telnet, and SNMP.

When you configure the IP address, subnet mask, and broadcast address (and, on the sc0 interface, VLAN membership) of the sc0 or me1 interface, you can access the switch through Telnet or SNMP.

When you configure the SLIP (sl0) interface, you can open a point-to-point connection to the switch from a workstation through the console port. A network engineer can use a remote PC to dial up switches anywhere in the world and manage them using SNMP or Telnet over SLIP (sl0). It is also possible to upload system software over the *sl0 interface* using TFTP. This option requires connecting a modem to the Catalyst switch console port. This is called out-of-band management, which means that the device is being accessed by a management terminal via a path that does not include the network to which the switch is connected.

The sc0 interface does not have an external port for a direct connection. It exists as a logical interface inside the switch and can be accessed via any of the physical ports on the switch.

NOTE

Use a straight-through patch cable to connect the me1 port to the TFTP server's NIC. In this sense, the me1 port acts like a regular switch port. This requirement is often confusing, considering that TFTP downloads to routers are normally done with crossover cables.

The me1 interface is actually a physical Ethernet port on 2948G switches, on 2980G switches, and on the Supervisor module of the Catalyst 4000 family of switches. This interface is used for network management only and does not support network switching.

The me1 or sc0 management interface must be assigned an IP address in the same subnet as the TFTP server.

The following example demonstrates a TFTP download via the me1 interface. The procedure is the same for the sc0 interface. The commands to configure IP for the me1 interface from the enable prompt are shown in Example 3-16.

Example 3-16 Configuring IP for the me1 Interface on a Catalyst 4006

```
Console> (enable) set interface me1 172.16.0.5 255.255.255.0
Interface me1 IP address and netmask set.
Console> (enable)
```

You need to confirm IP connectivity with the TFTP server by pinging the server, as shown in Example 3-17.

NOTE

On some versions of the CatOS, you get a "172.16.0.2 is alive" message instead of the typical successful Cisco ping output. These both indicate that IP connectivity has been achieved.

Example 3-17 Confirming Connectivity to the TFTP Server

```
Console> (enable) ping 172.16.0.2
!!!!!
----172.16.0.2 PING Statistics----
5 packets transmitted, 5 packets received, 0% packet loss
round-trip (ms)  min/avg/max = 14/15/17
Console> (enable)
```

Now, use the **show flash** command to check the contents of Flash to confirm that space is available for the new image. In this case, the new image is 4,089,736 bytes. Example 3-18 shows you how to check for available memory.

Example 3-18 Checking the Contents of Flash

```
Console> (enable) show flash
-#- ED --type-- --crc--- -seek-- nlen -length- -----date/time------ name
  1 .. ffffffff 548c8f9c  39cf70   17  3526384 --- -- ---- --:--:--
  cat4000.5-4-2.bin
12071928 bytes available (3526384 bytes used)
Console> (enable)
```

You are now ready to copy the image from the TFTP server. Example 3-19 illustrates the process.

Example 3-19 Copying the Image to Flash

```
Console> (enable) copy tftp flash
IP address or name of remote host [172.16.0.2]?
Name of file to copy from []? cat4000.6-2-1.bin
Flash device [bootflash]?
Name of file to copy to []? cat4000.6-2-1.bin
7981064 bytes available on device bootflash, proceed (y/n) [n]? y
CCCCCCCCCCCCCCCCCCCCCCCCCCCCCCCCCCCCCCCCCCCCCCCCCCCCCCCCCCCCCCCCCCCCCCCCCCCCCC
CCCCCCCCCCCCCCCCCCCCCCCCCCCCCCCCCCCCCCCCCCCCCCCC
File has been copied successfully.

Console> (enable)
```

To confirm that the file was downloaded, use the **show flash** or **dir** command. You can see that both images are now present, as shown in Example 3-20.

Example 3-20 Verifying the New Image in Flash

```
Console> (enable) show flash
-#- ED --type-- --crc--- -seek-- nlen -length- -----date/time------ name
  1 .. ffffffff 548c8f9c  39cf70   17  3526384 --- -- ---- --:--:--
  cat4000.5-4-2.bin
  2 .. ffffffff d39d5c46  783778   17  4089736 Apr 17 2001 14:40:15
  cat4000.6-2-1.bin

7981192 bytes available (7616376 bytes used)

Console> dir
-#- -length- -----date/time------ name
  1  3526384 --- -- ---- --:--:-- cat4000.5-4-2.bin
  2  4089736 Apr 17 2001 14:40:15 cat4000.6-2-1.bin

7981192 bytes available (7616376 bytes used)
```

Use the **set boot system flash bootflash:** *image_name* **prepend** command to tell the switch which image to use. It is critical that you use the **prepend** option to force the switch to load the new image by default; otherwise, the switch will not use the new image you loaded. Example 3-21 illustrates the impact of the **prepend** option.

Example 3-21 Reordering the Images in Flash

```
Console> (enable) set boot system flash bootflash:cat4000.6-2-1.bin prepend


Console> (enable) show config
<output omitted>
#set boot command
set boot config-register 0x2
set boot system flash bootflash:cat4000.6-2-1.bin
set boot system flash bootflash:cat4000.5-4-2.bin
<output omitted>
```

To delete the old image, use the **delete** command. Using the **dir** command shows that the file is gone, but actually it's just marked as deleted (note that the "bytes available" and "bytes used" have not changed). The file still appears in the output of the **dir deleted** command. To get rid of the image, you need to execute the command **squeeze bootflash:**. Each of these commands is illustrated in Example 3-22.

Example 3-22 Deleting an Old Image

```
Console> (enable) delete cat4000.5-4-2.bin
Console> (enable) dir
-#- -length- -----date/time------ name
  2  4089736 Apr 17 2001 14:40:15 cat4000.6-2-1.bin


7981192 bytes available (7616376 bytes used)
Console> (enable) dir deleted
-#- ED --type-- --crc--- -seek-- nlen -length- ----date/time---- name
  1 .. ffffffff 548c8f9c  39cf70   17  3526384 -- -- ---- --:-:- cat4000.5-4-2.bin


7981192 bytes available (7616376 bytes used)


Console> (enable) squeeze bootflash:


All deleted files will be removed, proceed (y/n) [n]? y
```

Example 3-22 Deleting an Old Image (Continued)

```
Squeeze operation may take a while, proceed (y/n) y

Console> (enable) dir
-#- -length- -----date/time------ name
  1  4089736 Apr 17 2001 14:40:15 cat4000.6-2-1.bin
12070928 bytes available (4089736 bytes used)
```

The same method can be used to download configuration files from a TFTP server or to upload configuration files to a TFTP server. If you want to restore a previous configuration to a CatOS switch, enter **clear config all** and then load the previous configuration.

> **NOTE**
> The command **write network** has the same effect as the **copy config tftp** command on a CatOS switch.

IOS-Based Switch

Images are downloaded via TFTP onto IOS-based switches just as they are on Cisco routers: by using the **copy tftp flash** command. This method also works for downloading configuration files to IOS-based switches.

Because we are discussing loading Flash images, now is an opportune time to illustrate a useful disaster-recovery option: recovering from a corrupted or missing Flash image using Xmodem. The following scenario describes the procedure for Catalyst 2900 XL, 3500 XL, and 2950 switches. When the Flash image is corrupted or missing, you'll see the error message "Error Loading Flash" when booting up after a power failure or after an incorrect software upgrade.

In this case, you need to load a valid Flash image to the Catalyst switch. In describing the procedure for recovering from a corrupted Flash image, we assume that you have a console connection to the switch, that you are running HyperTerminal in a Windows environment, that you are in ROM Monitor mode, and that you have a TFTP server set up with the appropriate image ready for transfer.

If there's not enough memory capacity (as verified by the **dir flash:** command), first delete the existing file(s) with the command **delete flash:***filename,* where *filename* is the name of the file to be deleted.

Now, perform the commands shown in Example 3-23, using the appropriate filenames.

Example 3-23 Preparing for an Xmodem Transfer

```
switch: load_helper
switch: flash_init
switch: copy xmodem:c2900XL-c3h2s-mz.120-5.3.WC.1.bin flash:
  c2900XL-c3h2s-mz.120-5.3.WC.1.bin
Begin the Xmodem or Xmodem-1K transfer now...
```

At this point, you need to use the HyperTerminal program to download the image (see Figure 3-4). Select Transfer, Send File, and click the Browse button. Highlight the desired file.

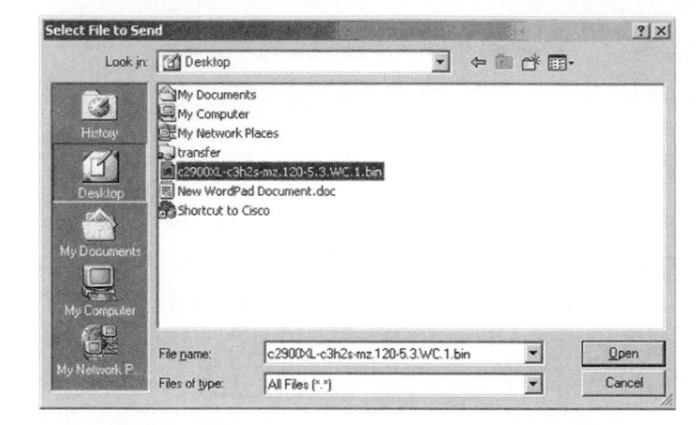

Figure 3-4
The first step in using the Xmodem option in HyperTerminal to recover from a corrupted Flash image is to select the image to send.

Next, select Open (see Figure 3-5) and then select Xmodem from the drop-down menu.

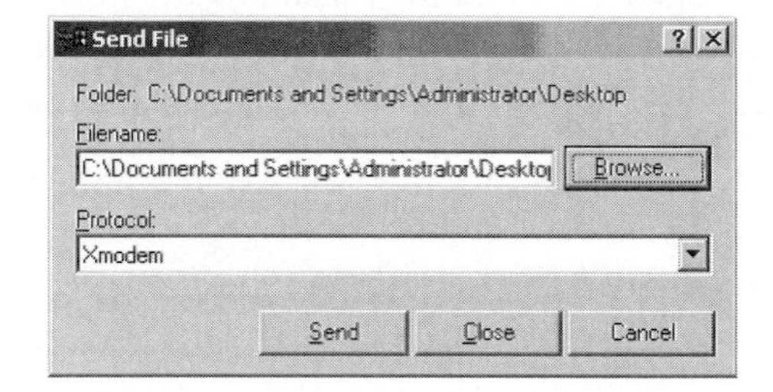

Figure 3-5
The second step with HyperTerminal is to open the image and choose the Xmodem protocol option.

Finally, click Send. You see a dialog box similar to the one shown in Figure 3-6. When you see this dialog box, you'll know that the image is in the process of downloading.

Figure 3-6
After clicking Send in HyperTerminal, you have about 30 minutes to wait for the download to complete.

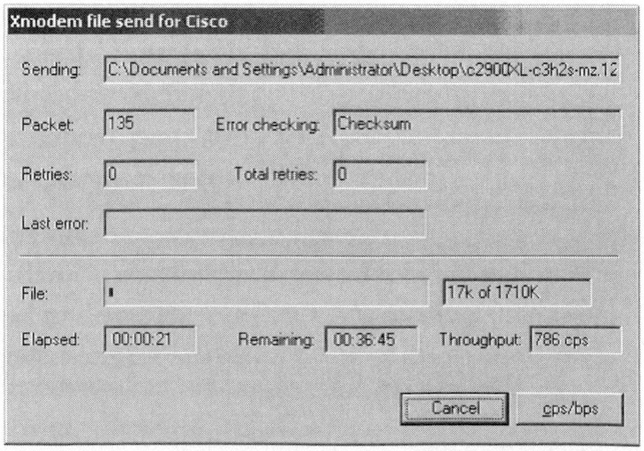

The download takes approximately 30 minutes and concludes with a message such as this:

```
File "xmodem:c2900XL-c3h2s-mz.120-5.3.WC.1.bin" successfully copied to "flash:c".
```

At the conclusion of the download, reload the switch with the **reset** command and, after booting into normal mode, check the results with the **dir flash:** command. Example 3-24 shows how to reset the switch after the download is complete.

Example 3-24 Resetting the Switch After the Download Is Complete

```
switch: reset
Are you sure you want to reset the system (y/n)?y
System resetting...
         --- System Configuration Dialog ---
At any point you may enter a question mark '?' for help.
Use ctrl-c to abort configuration dialog at any prompt.
Default settings are in square brackets '[]'.

Continue with configuration dialog? [yes/no]:no

Switch#dir flash:
Directory of flash:/

  2  -rwx          3   Jan 01 1970 00:00:22  env_vars
  3  -rwx    1750400   Jan 01 1970 00:45:58  c2900XL-c3h2s-mz.120-5.3.WC.1.bin

3612672 bytes total (1860608 bytes free)
Switch#
```

You have now seen the details of how to load Flash images to Catalyst switches. This is just part of the job for a network engineer. You need to be familiar with the procedures for loading Flash images via TFTP and via Xmodem for Flash recovery.

Port Descriptors

You can add a description to an interface or port to help facilitate switch administration, such as documenting what access- or distribution-layer device the interface services. This description is useful in an environment where a switch has numerous connections and the network engineer needs to check a link to a specific location. This description is meant solely as a comment to help identify how the interface is being used or where it is connected (such as which floor, which office, and so on). This description appears in the output when you display the configuration information.

Configuring port descriptors for Catalyst switches is described in the following sections.

CatOS Switch

To assign a name to a port, use the command **set port name** *mod_num/port_num* [*name_string*], and verify the configuration with the **show port** [*mod_num*[/*port_num*]] command. The **show port** command can have a module number as an argument; in this case, output is displayed for all ports on the referenced module.

Example 3-25 demonstrates how to set the name for ports 1/1 and 1/2. It also shows how to verify that the port names are configured correctly.

Example 3-25 Configuring and Verifying Descriptive Identifiers for Ports

```
Console> (enable) set port name 1/1 Router Connection
Port 1/1 name set.
Console> (enable) set port name 1/2 Server Link
Port 1/2 name set.
Console> (enable) show port 1
Port  Name               Status     Vlan       Level  Duplex Speed Type
----- ------------------ ---------- ---------- ------ ------ ----- ------------
 1/1  Router Connection  connected  1          normal   full  1000 1000BaseSX
 1/2  Server Link        connected  1          normal   full  1000 1000BaseSX

<output omitted>

Last-Time-Cleared
--------------------------
Tue Dec 28 2002, 16:25:57
Console> (enable)
```

You can easily configure port descriptors on CatOS switches. They provide a useful and ready reference for a network engineer.

IOS-Based Switch

To add a unique comment to an interface on an IOS-based switch, use the interface configuration command **description** *string*. The actual description appears in the **show interface** output. To clear a description, enter the **no description** command in interface configuration mode. Example 3-26 shows how to use the **description** command.

Example 3-26 Using the description Command

```
Switch(config)#interface vlan 1
Switch(config-if)#description Managing the Farm
Switch(config-if)#end
Switch#show interface vlan 1
VLAN1 is up, line protocol is up
  Hardware is CPU Interface, address is 0006.d6a9.5ec0 (bia 0006.d6a9.5ec0)
  Description: Managing the Farm
  Internet address is 128.1.2.3/24
  MTU 1500 bytes, BW 10000 Kbit, DLY 1000 usec,
<output omitted>
Switch#show run
<output omitted>
interface VLAN1
 description Managing the Farm
 ip address 128.1.2.3 255.255.255.0
 no ip directed-broadcast
 no ip route-cache
<output omitted>
```

Port descriptors allow a network engineer to quickly identify the function of a link connected to a given port without having to trace the cable. The configuration for IOS-based switches mirrors that of Cisco routers.

Port Speed and Duplex

Setting the port speed and duplex is one of the most common configuration requirements for a network engineer. It can be frustrating to see messages constantly appear on your console screen, complaining about speed or duplex mismatches. As a rule, you want to immediately address these error messages by configuring the port speed and duplex appropriately. If you do not, a "spillover effect" causes other technologies configured for the ports in question to fail.

Although autonegotiation is generally supposed to take care of speed and duplex configuration, you will frequently find that you have to hard-code the settings. You should get in the habit of not depending on autonegotiation. You can avoid much frustration by proactively configuring port speed and duplex.

The order of configuration for speed and duplex is important. When configuring speed and duplex, you need to first configure the speed for a port and follow that with the duplex configuration.

Next, we detail the commands used on CatOS and IOS-based switches to configure port speed and duplex.

> **NOTE**
> If the port speed is set to **auto** on a 10/100-Mbps Fast Ethernet port, both speed and duplex are autonegotiated. Make sure that the device on the other end of the link is configured for autonegotiation or half-duplex; otherwise, a port speed mismatch or duplex mismatch will likely result. If there is any question about the recommended settings for port speed and duplex for a specific Catalyst switch platform, refer to the platform's Release Notes at www.cisco.com.

CatOS Switch

You can configure the port speed on 10/100/1000-Mbps Fast Ethernet modules. To set the port speed for a 10/100/1000-Mbps port, perform this task in privileged mode. Use the command **set port speed** *mod_num/port_num* {**10** | **100** | **1000** | **auto** | **nonegotiate**}, and verify it with the command **show port** [*mod_num*[*/port_num*]].

Example 3-27 shows how to set the port speed to 100 Mbps on port 2/3.

Example 3-27 Setting and Verifying Port Speed

```
Console> (enable) set port speed 2/3 100
Port(s)  2/3 speed set to 100Mbps.
Console> (enable) show port 2/3
Port  Name                 Status     Vlan       Level  Duplex Speed Type
----- ------------------ ---------- ---------- ------ ------ ----- ------------
 2/3                       connected  1          normal   half   100 10/100BaseTX
<output omitted>

Console> (enable) set port speed 2/3 auto
Port(s)  2/3 speed set to auto detect.
Console> (enable) show port 2/3
Port  Name                 Status     Vlan       Level  Duplex Speed Type
----- ------------------ ---------- ---------- ------ ------ ----- ------------
 2/3                       connected  1          normal a-half a-100 10/100BaseTX
<output omitted>
```

You can set the port duplex mode to full or half duplex for Ethernet and Fast Ethernet ports (the default is half duplex). You cannot change the duplex mode of ports configured for autonegotiation. To set a port's duplex mode, use the command **set port duplex** *mod num/port num* {**full** | **half**} and verify with the **show port** command, as shown in Example 3-28.

Example 3-28 Setting and Verifying Port Duplex

```
Console> (enable) set port duplex 2/3 half
Port(s)  2/3 set to half-duplex.
Console> (enable) show port 2/3
```

Example 3-28 Setting and Verifying Port Duplex (Continued)

```
Port  Name               Status     Vlan       Level  Duplex Speed Type
----- ------------------ ---------- ---------- ------ ------ ----- ------------
 2/3                     connected  1          normal   half   100 10/100BaseTX
<output omitted>
```

GIGABIT ETHERNET PORTS

On 1000BaseT Gigabit Ethernet ports, you cannot configure speed or duplex mode. 1000BaseT ports only operate in the default configuration where the speed is 1000 and the duplex mode is full. You cannot disable autonegotiation.

In general, for Catalyst switches, Gigabit Ethernet and 10-Gigabit Ethernet are full duplex only. You cannot change the duplex mode on Gigabit Ethernet and 10-Gigabit Ethernet ports. To determine which features an Ethernet port supports, enter the **show port capabilities** *mod_num* command, as shown in Example 3-29.

Example 3-29 Viewing and Verifying the Capabilities of Switch Ports

```
Console> (enable) show port capabilities
Model                     WS-X5509
Port                      1/1
Type                      100BaseTX
Speed                     100
Duplex                    half,full
Trunk encap type          ISL
Trunk mode                on,off,desirable,auto,nonegotiate
Channel                   1/1-2
Broadcast suppression     percentage(0-100)
Flow control              no
Security                  yes
Membership                static,dynamic
Fast start                yes
QOS scheduling            rx-(none),tx-(none)
CoS rewrite               no
ToS rewrite               no
Rewrite                   no
UDLD                      yes
AuxiliaryVlan             no
SPAN                      source,destination
<output omitted>
```

IOS-Based Switch

Setting port speed and duplex on switches is not rocket science, but you must understand some not-so-obvious rules. It pays to have a solid understanding of these rules to prevent the frustration of trial-and-error configuration. The following guidelines can serve as a reference when you configure the duplex and speed settings on IOS-based switches:

- Gigabit Ethernet ports should always be set to 1000 Mbps, but they can negotiate full duplex with the attached device. To connect to a remote Gigabit Ethernet device that does not autonegotiate, disable autonegotiation on the local device, and set the duplex parameters to be compatible with the other device. Gigabit Ethernet ports that do not match the settings of an attached device lose connectivity and do not generate statistics.
- To connect to a remote 100BaseT device that does not autonegotiate, set the duplex setting to **full** or **half**, and set the speed setting to **auto**. Autonegotiation for the speed setting selects the correct speed even if the attached device does not autonegotiate, but the duplex setting must be explicitly set.
- 10/100/1000 ports can operate at 10 or 100 Mbps when they are set to half- or full-duplex mode, but they operate in full-duplex mode only when set to 1000 Mbps.
- Gigabit Interface Converter (GBIC) module ports operate only at 1000 Mbps.
- 100BaseFX ports operate at only 100 Mbps in full duplex.

To set the speed and duplex settings, first set the speed in interface configuration mode with the command **speed** {**10** | **100** | **1000** | **auto**}. Second, use the command **duplex** {**full** | **half** | **auto**} to set the duplex mode. Example 3-30 demonstrates how to set the speed and duplex; the **show interface** and **show run** commands verify the settings.

Example 3-30 Setting Port Speed and Duplex

```
Switch(config-if)#speed 100
Switch(config-if)#
3d19h: %LINK-3-UPDOWN: Interface FastEthernet0/1, changed state to down
3d19h: %LINEPROTO-5-UPDOWN: Line protocol on Interface FastEthernet0/1, changed
  state to down
3d19h: %LINK-3-UPDOWN: Interface FastEthernet0/1, changed state to up
3d19h: %LINEPROTO-5-UPDOWN: Line protocol on Interface FastEthernet0/1, changed
  state to up
Switch(config-if)#duplex full
Switch(config-if)#
3d19h: %LINK-3-UPDOWN: Interface FastEthernet0/1, changed state to down
```

Example 3-30 Setting Port Speed and Duplex (Continued)

```
3d19h: %LINEPROTO-5-UPDOWN: Line protocol on Interface FastEthernet0/1, changed
  state to down
3d19h: %LINK-3-UPDOWN: Interface FastEthernet0/1, changed state to up
3d19h: %LINEPROTO-5-UPDOWN: Line protocol on Interface FastEthernet0/1, changed
  state to up
Switch(config-if)#end

Switch#show interface fa0/1
FastEthernet0/1 is up, line protocol is up
  Hardware is Fast Ethernet, address is 0006.d6a9.5ec1 (bia 0006.d6a9.5ec1)
  Description: testing blank spaces
  MTU 1500 bytes, BW 100000 Kbit, DLY 100 usec,
     reliability 255/255, txload 1/255, rxload 1/255
  Encapsulation ARPA, loopback not set
  Keepalive not set
  Full-duplex, 100Mb/s, 100BaseTX/FX
  ARP type: ARPA, ARP Timeout 04:00:00
<output omitted>

Switch#show run
<output omitted>
interface FastEthernet0/1
 description testing blank spaces
 duplex full
 speed 100
<output omitted>
```

Setting the speed and duplex for Catalyst switches is one of the most basic yet most important configuration tasks in switch administration. Learning early on the commands and rules for configuring speed and duplex settings will go a long way toward preventing frustration.

Summary

This chapter provided an exhaustive introduction to Catalyst switch administration. You have seen how to configure the bread-and-butter tasks required for switch administration, such as preparing for remote access, downloading Flash images, and setting port speed and duplex.

Developing expertise with Catalyst switch configuration depends on your having a solid background in the fundamentals of switch administration. The concepts explored in the remainder of this book rely on the concepts and procedures discussed in this chapter. In particular, you will consistently utilize the IP configuration options for management interfaces and the port speed and duplex options for Catalyst switches. Remember to refer to this chapter when you are configuring port speed and port duplex if you have any problems with these settings.

The next two chapters get to the heart of modern campus switch deployment: implementing VLANs and STP.

Check Your Understanding

Test your understanding of the concepts covered in this chapter by answering these review questions. Answers to these questions are listed in Appendix A, "Check Your Understanding Answer Key."

1. Which of the following is true of accessing a switch by means of an Ethernet port?
   - **A.** The console port must be configured.
   - **B.** A crossover cable connects the workstation to the switch port.
   - **C.** The switch can be managed only through the console port.
   - **D.** The switch must be configured with an IP address.
2. Select the true statement regarding CatOS switches.
   - **A.** Configuration changes are made from global configuration mode.
   - **B.** The **copy run start** command is used to save the configuration to NVRAM.
   - **C.** The **clear config all** command restores the switch to the factory default configuration.
   - **D.** The privileged-mode prompt is Switch#.
3. What are the two major classifications of Cisco Catalyst switches?
   - **A.** IOS-based and CatOS
   - **B.** CLI-based and IOS-based
   - **C.** Menu-driven and IOS-based
   - **D.** Menu-driven and CLI-based
4. What is the default management interface on an IOS-based switch?
   - **A.** sc0
   - **B.** me1
   - **C.** sl0
   - **D.** interface vlan 1
5. Select the best description of the command **set port speed 10/1 auto.**
   - **A.** Both the speed and duplex of the specified port are automatically negotiated.
   - **B.** The speed of the first port on module 10 is automatically negotiated.
   - **C.** Ports operating at 10 Mbps automatically negotiate their duplex mode.
   - **D.** Ports can negotiate a speed up to ten times their base rate.

6. In which of the following ways does the help feature on a CatOS switch differ from that on a Cisco router?

   A. Typing **help** in privileged mode does not return an error.

   B. The switch displays help on a parameter-by-parameter basis.

   C. The switch does not automatically display the command in question at the end of the help output.

   D. Because it does not use the IOS, the switch has no help feature.

7. Which command is used to view the active configuration of a CatOS switch?

   A. **show config**

   B. **show running-config**

   C. **show active-config**

   D. **show current-config**

8. On a Catalyst 5505 switch, how much time do you have to perform password recovery after the initial command prompt is displayed?

   A. 15 seconds

   B. 30 seconds

   C. 45 seconds

   D. 60 seconds

9. Which commands are used to set the idle timeout on a Catalyst switch?

   A. **set timeout** or **idle-timeout**

   B. **set idle-timeout** or **idle-logout**

   C. **set exec-timeout** or **idle-logout**

   D. **set logout** or **exec-timeout**

10. What is indicated when the port status LED on a switch is illuminated green?

    A. There is a problem with the adapter on the attached device.

    B. Both the switch and the connected device are powered up.

    C. The wrong type of cable has been used to connect the switch to another device.

    D. There is a problem with the connection on the switch end of the cable.

Key Terms

bootflash: A Flash memory device that stores OS images. It can also store a boot helper image or system configuration information.

CatOS The OS used by the Supervisor Engine on chassis-based Catalyst switches (with two exceptions), such as the 4000, 5000, and 6000 series. Two exceptions to this are the Native IOS loaded on a Catalyst 6000 (optional) and the Native OS loaded on the Catalyst 4000 with Supervisor Engine III. CatOS is distinguished by its reliance on set commands.

device: Hardware that supports a Flash image. This includes bootflash: and PCMCIA cards (Flash PC cards) for CatOS switches.

idle timeout A timer that specifies how long a connection stays active without any keystrokes taking place.

IOS-based switch A switch running the Cisco IOS. The OS on these switches mirrors that of Cisco routers, except for the addition of the VLAN database configuration mode and the config-vlan mode.

management VLAN The VLAN on the network used for network management.

me1 interface A physical, out-of-band, Ethernet management interface/port on the Catalyst 2948G, 2980G, and 4000 families of switches. This interface is used for network management only and does not support network switching. It acts like a switch port in that it is accessed from a PC's NIC with straight-through cable.

Native IOS A single (optional) IOS image that simultaneously controls the Layer 2, 3, and 4 configuration of a Catalyst family 6000 switch. It is also called Catalyst 6000 IOS, Cat IOS, or Supervisor IOS. In this case, the Supervisor module and the Multilayer Switch Feature Card (MSFC) both run a single bundled Cisco IOS image.

normal mode The mode on a CatOS switch that allows you to view most switch parameters. Configuration changes are not allowed in this mode.

privileged mode The CatOS switch mode used to change the system configuration. This differs from Cisco routers, in which you need to enter global configuration mode to make system changes.

sc0 interface The logical, in-band, management interface on a CatOS switch that is associated with the management VLAN. The sc0 interface participates in the functions of a normal switch port, such as spanning tree, CDP, and VLAN membership.

sl0 interface The out-of-band management port that relies on SLIP. This interface is found on the Catalyst 2926G, 2948G, 2980G, 4000, 5000, and 6000 families of switches. By connecting a modem to the console port, a network engineer can remotely dial up the switch using SLIP.

Supervisor Engine Also called Supervisor module or Supervisor. Can be thought of as the brains of a chassis-based Catalyst switch. All other modules in the chassis can be accessed or configured from the Supervisor module. The CatOS or Cisco IOS runs on the Supervisor module, depending on the particular switch.

switch fabric The hardware architecture in a chassis-based Catalyst switch that enables high-speed point-to-point connections between each line card. The switch fabric provides a mechanism to simultaneously forward packets on all point-to-point connections between the chassis slots.

VLAN configuration mode The mode on an IOS-based switch that is accessed from privileged mode with the **vlan database** command. VLAN configuration mode is used to make changes to the VLAN configuration, such as adding and removing VLANs.

After completing this chapter, you will be able to perform tasks related to the following:

- Advantages of VLANs
- End-to-end and local VLANs and their role in modern switched network design
- Configuring port-based VLANs
- Dynamic VLANs and VMPS operation
- Operation of the four VLAN trunking technologies
- Configuring ISL and IEEE 802.1Q trunking
- VTP operation and configuration
- VTP pruning and configuration

Chapter 4

Introduction to VLANs

Virtual LAN (VLAN) design has undergone a dramatic shift over the last decade as a result of advances in technology. There was a time in the mid-1990s when the deployment of campus-wide VLANs was promoted as the best design methodology. However, it turned out that the lack of hierarchy inherent in such an approach was a major limiting factor. A driving force behind campus-wide VLAN design was the speed of Layer 2 switching relative to that of packet switching through traditional routers. This argument eventually went away as a result of the widespread adoption of multilayer switches and their capability to perform Layer 3 switching at rates approximating that of Layer 2 switching. With hardware-based Layer 3 switching speeds of up to 210 million pps now available, one of the primary arguments for campus-wide VLANs became irrelevant. Although campus-wide VLANs play a role in special circumstances, designing a stable, scalable campus network generally precludes this as an option.

Chapter 5, "Spanning-Tree Protocol," discusses the role of Spanning-Tree Protocol (STP) in VLAN design and why STP can be problematic in campus-wide VLAN deployments. This chapter clarifies the role of VLANs and VLAN protocols in modern campus networks. A host of protocols were developed to facilitate VLAN deployment, including both proprietary and industry-standard protocols. Many times, a vendor develops a protocol, such as Cisco's Inter-Switch Link (ISL), only for it to be replaced over time by an industry-standard protocol, such as IEEE 802.1Q. This same theme repeats itself continuously in the world of networking.

We will take a close look at the IEEE 802.1Q standard and its role in a VLAN deployment. In addition to ISL, other Cisco-proprietary LAN-switching protocols explored in this chapter include VLAN Query Protocol (VQP), Dynamic Trunking Protocol (DTP), and VLAN Trunking Protocol (VTP).

Several industry-standard VLAN protocols are now available for network engineers to utilize. Generic Attribute Registration Protocol (GARP) and GARP VLAN Registration Protocol (GVRP) are examples of industry-standard protocols supported by modern switches to enable VLAN implementation in a multivendor environment. GVRP is a GARP application that serves as an open-standard analog to VTP. These protocols are not discussed in this book. Instead, refer to the IEEE 802.1D standard. IEEE 802.1D now encompasses the projects associated with IEEE 802.1p, where GARP and GVRP were originally described.

This chapter covers a great deal of ground in its treatment of VLANs. First we define VLANs, and then we expand on the VLAN design concepts touched on in Chapters 1 and 2. We follow that with a discussion of VLAN trunking, VLAN Management Policy Server (VMPS), and VTP. Finally, Catalyst switch configuration is explained in context.

VLANs Defined

A *virtual LAN (VLAN)* is a group of end stations with a common set of requirements, independent of their physical location, that communicate as if they were attached to the same wire. A VLAN has the same attributes as a physical LAN but allows you to group end stations even if they are not located on the same LAN segment (see Figure 4-1).

Figure 4-1 VLANs permit flexibility in designing networks, allowing broadcast domains to be shaped according to the needs of the users.

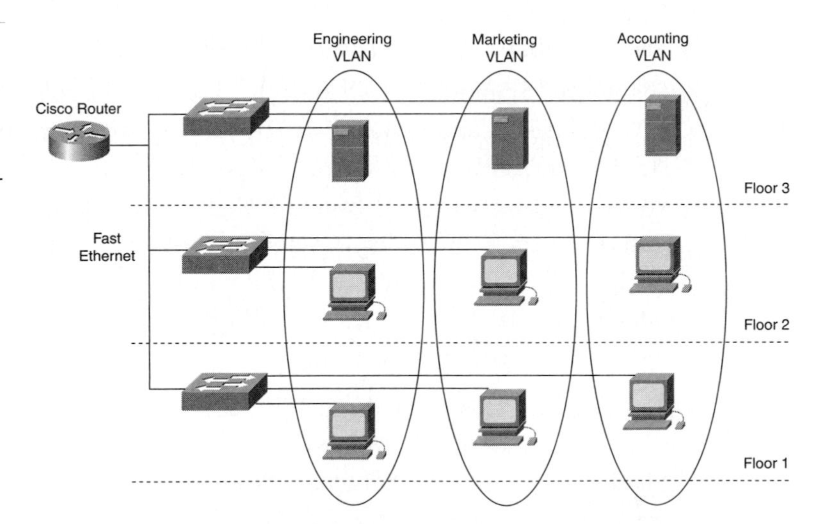

It is ironic that, as network design has evolved with advances in technology, VLANs are now likely to be geographically isolated. This appears to contradict the definition of VLANs. It is now common to implement only one or two VLANs unique to each access layer switch. The "independent of their physical location" part of the VLAN definition is still functionally accurate, but it is decreasingly realized in modern switch deployments.

In any case, VLANs allow you to group ports on a switch to limit unicast, multicast, and broadcast traffic flooding. Flooded traffic originating from a particular VLAN is only flooded out ports belonging to that VLAN, including trunk ports.

Specifically, any switch port can belong to a VLAN. Unicast, broadcast, and multicast packets are forwarded and flooded only to stations in the same VLAN. Each VLAN is a logical network, and packets destined for stations that do not belong to the VLAN must be forwarded through a router or bridge. (Bridging between VLANs is detailed in Appendix B, "Transparent Bridging with Routers.") Because a VLAN is considered a separate logical network, a switch contains its own Management Information Base (MIB) for each VLAN. Also, Cisco switches support an independent implementation of the STP for each VLAN.

VLANs are essentially Layer 2 constructs, and IP subnets are Layer 3 constructs. In a campus LAN employing VLANs, often, a one-to-one relationship exists between VLANs and IP subnets. (The relationship is not one-to-one if bridging between VLANs allows more than VLAN per subnet or if using secondary addresses within a single VLAN allows more than one subnet per VLAN.) So VLANs and IP subnets provide an interesting example of independent Layer 2 and Layer 3 constructs that often map one onto the other. As you can imagine, this correspondence is practical when it comes to network design.

NOTE

The current campus network design philosophy is that, in an ideal, highly structured network, one IP subnet maps to a single VLAN, which in turn maps to a single switch in a wiring closet. This design model is restrictive, but it pays huge dividends in simplicity and ease of troubleshooting. A corollary of this design philosophy is that a set of servers with similar access policies is grouped in the same IP subnet and VLAN.

Motivation for VLANs

VLANs were invented primarily to allow for flexibility in LAN design. VLANs were conceived to allow network engineers to assign users to workgroups independent of location within a campus network. Having said that, the preceding discussion might have you wondering what, if any, motivation there is to implement VLANs. We will focus on these additional motivations for VLANs:

- Security
- Broadcast control
- Bandwidth utilization
- Latency

These motivations are discussed in the following sections.

Security

Security is, and will continue to be, one of the most pervasive aspects of networking. Almost everything you do in networking has a connection to security. When it comes to VLAN technology, security is essentially a side effect (security was not a motivating factor for the creation of VLANs). One exception to this is *private VLANs*, which are available on Catalyst 4000s and 6000s.

TECH NOTE: PRIVATE VLANS

A private VLAN is a VLAN that you configure to effect Layer 2 isolation from other ports within the same private VLAN. Ports belonging to a private VLAN are associated with a common set of supporting VLANs that create the private VLAN structure. You can configure private VLANs and normal VLANs on the same Catalyst 4000 or 6000 switch.

There are three types of private VLAN ports:

- A promiscuous port communicates with all other private VLAN ports. It is the port you use to communicate with routers, backup servers, and administrative workstations.
- An isolated port has complete Layer 2 separation, including broadcasts, from other ports within the same private VLAN, with the exception of the promiscuous port.
- Community ports communicate among themselves and with their promiscuous ports. These ports are isolated at Layer 2 from all other ports in other communities or from isolated ports within their private VLAN. Broadcasts propagate only between associated community ports and the promiscuous port.

Privacy is granted at the Layer 2 level by blocking outgoing traffic to all isolated ports. All isolated ports are assigned to an isolated VLAN where this hardware function occurs. Traffic received from an isolated port is forwarded to all promiscuous ports only.

In an Ethernet-switched environment, you can assign an individual VLAN and associated IP subnet to groups of stations. The servers in a VLAN only require the ability to communicate with a default gateway to gain access to end points outside the VLAN itself. By incorporating these stations into one private VLAN, you can do the following:

- Designate the server ports as isolated to prevent any interserver communication at Layer 2. (This is a common security precaution because you don't want a hacker breaching one server to automatically have access to other servers.)
- Designate as promiscuous the ports to which the default gateway(s) or backup servers are attached, to allow all stations to have access to these gateways or servers.
- Reduce VLAN consumption. You need to allocate only one IP subnet to the entire group of stations because all stations reside in one common private VLAN.
- Conserve public address space. Servers are now isolated from one another using private VLANs. This eliminates the necessity of creating multiple IP subnets, which wastes public IP addresses on multiple subnet and broadcast addresses. As a result, all servers can be members of the same IP subnet but remain isolated from one another.

When you create VLANs, you are creating independent networks with built-in firewalls separating them. Perhaps *firewall* is too strong a term, but the point is that hosts on different VLANs cannot communicate with each other unless you explicitly configure your network to allow it. If you have a one-to-one mapping between IP subnets and VLANs, hosts on different VLANs must talk to each other through IP routing. After you configure routing, you have inter-VLAN communication (see Chapter 6, "Inter-VLAN Routing" for more information). In this case, access control lists (ACLs) filter traffic between VLANs. Now we are definitely talking about security.

Theoretically, you could configure VLANs without inter-VLAN routing as a security solution. Only users within a particular VLAN would be able to communicate with each other. In reality, this is rarely done. You might isolate some VLANs this way for security, but normally you configure full inter-VLAN routing and then secure particular VLANs or portions thereof using ACLs.

Another consideration arises with shared-media technologies. Whenever a station transmits data in a hub-based LAN, all stations attached on the segment receive a copy of the frame, even if they are not the intended recipients. Any host running protocol analyzer software can monitor the traffic. This enables the capture of passwords, the reading of sensitive e-mail, and so on. If the users on a segment belong to the same department, this might not be a problem, but when users from mixed departments or divisions share a segment, sensitive information can be compromised. If someone from Human Resources or Accounting sends data such as salary information or health records on the shared network, anyone with a network-monitoring tool on that LAN can intercept the information.

In fact, this problem is not restricted to a single LAN segment. If hosts reside on any shared-media LAN segment connecting the source and destination, the data is compromised from a security perspective.

Or, if a user has physical or logical access to a switch connecting the source and destination, again the data can be intercepted. This particular security breach can be solved only by physically securing the switch and using passwords to prevent remote access.

Realistically, assuming physical security of a switch and password-protected remote access, the issue of intercepting data due to shared-media technology is primarily a thing of the past. Virtually all hosts today reside on switched ports, so unicast data is not visible to any hosts except the sender and receiver. VLANs add one additional layer of isolation between hosts and therefore make it much more difficult for data to be intercepted.

Another case where VLANs play a definitive role in network security is the ability to configure *VLAN ACLs (VACLs)* on Catalyst 6000s.

TECH NOTE: VACLS

A useful feature of the Catalyst 6000 family is the VLAN access control list (VACL). A VACL is a generalized access control list applied on a Catalyst 6000 that permits filtering of both intra-VLAN and inter-VLAN packets.

The Catalyst 6000 family of switches, along with the Multilayer Switch Feature Card (MSFC), can accelerate packet routing between VLANs by using Layer 3 switching (see Chapter 7 for more information). The switch first bridges the packet, the packet is then routed internally without going to the router, and the packet is bridged again to send it to its destination. During this process, the switch can access-control all packets it switches, including packets bridged within a VLAN.

Cisco IOS Software ACLs provide access control for routed traffic between VLANs, and VACLs provide access control for all packets.

Standard and extended Cisco IOS Software ACLs classify packets. Classified packets can be subject to a number of features, such as security, encryption, policy-based routing, and so on. Standard and extended Cisco IOS Software ACLs are configured only on router interfaces and are applied to routed packets.

VACLs can provide access control based on Layer 3 addresses for IP and IPX protocols. Unsupported protocols are access-controlled through MAC addresses. A VACL is applied to all packets (bridged and routed) and can be configured on any VLAN interfiace. As soon as a VACL is configured on a VLAN, all packets (routed or bridged) entering the VLAN are checked against the VACL. Packets can enter the VLAN either through a switch port or through a router port after being routed. Figures 4-2 and 4-3 illustrate the relative order of operations for VACLs and ACLs. For routed or Layer 3-switched packets, the ACLs are applied in the following order:

1. VACL for input VLAN
2. Input Cisco IOS Software ACL
3. Output Cisco IOS Software ACL
4. VACL for output VLAN

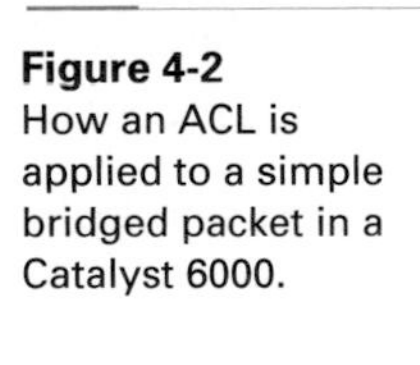

Figure 4-2 How an ACL is applied to a simple bridged packet in a Catalyst 6000.

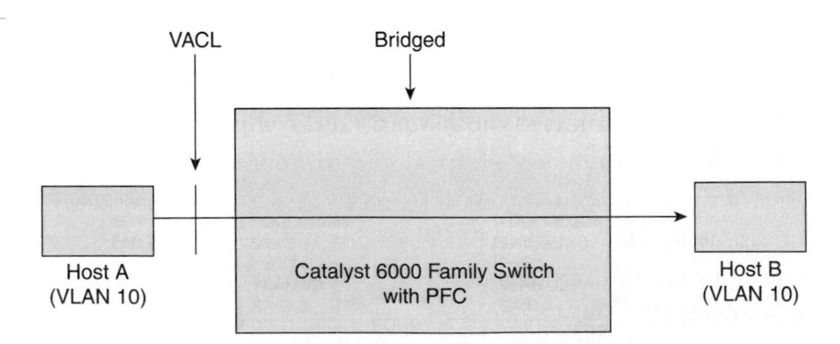

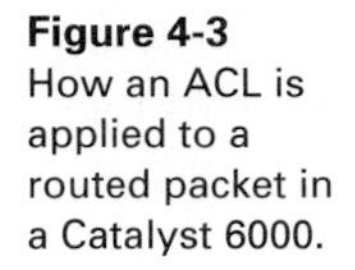

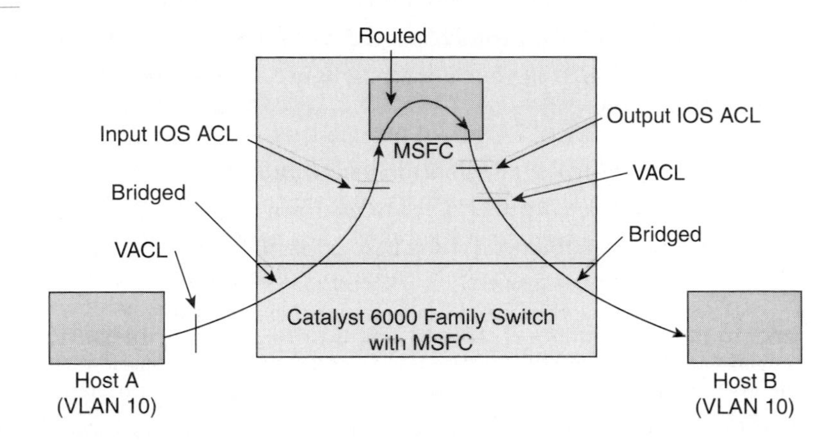

Figure 4-3 How an ACL is applied to a routed packet in a Catalyst 6000.

Broadcast Control

Practically every network protocol creates broadcast traffic. For example, AppleTalk routers generate routing updates in the form of broadcast frames every 10 seconds. Also, AppleTalk hosts, Microsoft hosts, and Novell end stations create broadcasts to announce or request services. Further, multimedia applications create broadcast and multicast frames that are distributed across the broadcast domain.

Broadcasts are received by all devices in a broadcast domain and must be processed by these devices. Broadcasts can have a profound effect on workstation performance. Any broadcast received by a workstation interrupts the CPU and, as a result, temporarily prevents it from processing user applications. As the number of broadcasts per second increases at the interface, effective CPU utilization diminishes. The actual level of degradation depends on the applications running on the workstation, the type of network interface card and drivers, the operating system, and the workstation platform.

In addition to degrading workstation performance, broadcasts and multicasts consume real bandwidth and can seriously degrade network performance if not controlled. The idea, of course, is to minimize any unnecessary broadcast and multicast traffic on your network. In fact, many protocols, such as Internet Group Management Protocol (IGMP) and Cisco Group Multicast Protocol (CGMP), were created primarily to minimize extraneous multicast and broadcast traffic on LANs (see Chapter 9, "Multicasting" for more information on constraining multicast traffic). Sometimes, routers, hosts, and switches are not intelligent enough to decide which traffic is garbage and which traffic is useful or necessary. As time goes on, new protocols are developed, and network devices are manufactured to support these new protocols, with the end goal of automating traffic optimization on your network. To the extent that these protocols do not exist and current protocols and devices are limited in their ability to automate traffic prioritization, it is your job to configure network devices to minimize extraneous broadcast and multicast traffic.

What does this have to do with VLANs? The answer is simple: VLANs form broadcast domains (see Figure 4-4). Broadcasts and multicasts are not propagated between VLANs unless the intermediary network devices are specifically configured to do this. If broadcasts and multicasts are creating problems in the network, creating smaller VLANs can mitigate the negative effects of these transmissions. This means dividing existing VLANs into smaller VLANs to reduce the number of devices on each VLAN. The effectiveness of this action depends on the source of the broadcast/multicast traffic. If the source is a local server, you might simply need to isolate the server in another domain. If the broadcasts/multicasts originate from end stations, creating additional VLANs helps reduce the number of broadcasts/multicasts in each domain.

Figure 4-4
VLANs comprise broadcast domains in a campus network.

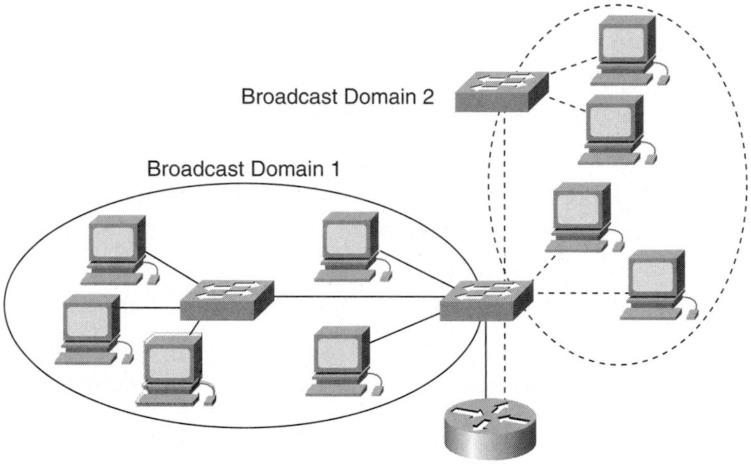

Bandwidth Utilization

If a workstation is connected to a switch port, the only unicast traffic it typically receives is that destined specifically for it. The switch, just like a bridge, filters the remaining unicast traffic. The only exception is this: When a unicast frame with an unknown destination is received on an ingress switch port for a VLAN, the frame is flooded out all the other ports in the VLAN. This continues until the switch discovers the mapping between the destination MAC address and the associated switch port. What is new in this explanation is that the unicast traffic with an unknown destination is only flooded out ports in the same VLAN. Configuring more VLANs on a switch translates to less flooding of traffic on switch ports. It follows that implementing VLANs improves bandwidth utilization.

Latency

Latency measures the amount of time necessary to transport a frame from the ingress port to the egress port of a networking device. Latency can be reduced by implementing VLANs. Where traffic previously had to travel through routers (Layer 3) to get from one workgroup to another or to move between geographically separated members of a common workgroup, VLANs provide a streamlined Layer 2 solution. Routers impose increased network latency relative to switches, so using VLANs can speed the flow of traffic between users on a campus LAN. This argument assumes that the routers cannot perform pervasive hardware-based routing (as with Cisco Express Forwarding on an 8500 series router). With hardware-based routing, the difference in latency would be nominal. So, if your network has routers in place that cannot perform pervasive hardware-based routing, implementing VLANs reduces latency for cross-campus intra-VLAN communication by eliminating intermediary routers.

Using VLANs in a traditional network (one that is incapable of Layer 3 switching) reduces the number of segments traversed between hosts on a common VLAN. Intra-VLAN traffic travels only between switches, eliminating segments involving routers. This reduces end-to-end network latency, especially if reliable protocols requiring acknowledgments are in play (such as TCP).

Again, in a modern campus deployment, these latency arguments become somewhat moot as a result of Layer 3 switching (hardware-based routing). A more likely scenario in this day and age is to have one or two VLANs dedicated to each access layer switch cluster (see the sidebar, "Tech Note: Clustering"), with Layer 3 switching between wiring closets performed at the distribution layer or Main Distribution Facility (MDF).

TECH NOTE: CLUSTERING

Cisco switch clustering technology is a set of software features available to all Catalyst 3500XL, 2900XL, 2950, 3550, and Catalyst 1900/2820 Standard and Enterprise Edition switches. Clustering technology lets up to 16 maximum interconnected switches form a managed, single-IP address network. *Clustering* is a method of managing a group of switches without having to assign an IP address to every switch.

A *command switch* provides the primary management interface for the entire cluster. It is typically the only switch within the switch cluster configured with an IP address. Using Cisco Discovery Protocol Version 2 (CDPV2), all switches, including the command switch, discover their CDP neighbors and store this information in their respective CDP neighbor cache. Switches running cluster-capable software pass information about themselves and their respective neighbors to the command switch using the Intracluster Communication (ICC) mechanism, which runs on top of User Datagram Protocol (UDP).

Clusters are managed through a web interface integrated with the Cisco Visual Switch Manager (VSM) or the Cisco Cluster Management Suite (CMS).

Classifying VLANs

Just as routing protocols are classified in various ways, VLANs can be classified in a variety of ways. Routing protocols are either distance-vector, hybrid, or link-state; alternatively, routing protocols are interior gateway protocols or exterior gateway protocols. VLANs are classified as local (geographic) or end-to-end (campus-wide); alternatively, VLANs are classified as either port-based (static) or dynamic. Each of these is discussed in the following sections.

End-to-End VLANs Versus Local VLANs

End-to-end VLANs span a campus network, whereas *local VLANs* operate within a limited geographic area.

An end-to-end VLAN network has the following characteristics:

- Users are grouped into VLANs independent of physical location and dependent on group or job function.
- The (old) 80/20 traffic flow pattern is in effect.
- As a user moves around the campus, VLAN membership for that user should not change.
- Each VLAN has a common set of security requirements for all members.

In this model, servers tend to be dedicated to workgroups, which in turn are associated with VLANs. This ensures that the majority of client/server traffic (approximately 80 percent) is Layer 2 switched traffic. Campus enterprise servers require inter-VLAN routing; traffic destined for these servers comprises a significant portion of the remaining 20 percent of LAN traffic.

End-to-end VLANs allow devices to be grouped based on resource usage. This includes such parameters as server usage, project teams, and departments (refer to Figure 4-1). The goal of end-to-end VLANs is to maintain 80 percent of the traffic on the VLAN where the traffic is sourced.

As corporate networks move to centralize their resources, end-to-end VLANs have become more difficult to maintain. Users are required to use many different resources, many of which are no longer in "their" VLAN. Because of this shift in placement and usage of resources, VLANs now are frequently created around geographic boundaries rather than commonality boundaries.

This geographic location can be as large as an entire building or as small as a single switch inside a wiring closet. A floor of a building is also a common manifestation of a local VLAN. A local, or geographic, VLAN is defined by a restricted geographic location, such as a wiring closet. In a local VLAN structure, it is typical to find the new

20/80 rule in effect, with 80 percent of the traffic remote to the user and 20 percent local to the user. Although this topology means that the user must cross a Layer 3 device in order to reach 80 percent of the resources, this design allows the network to provide for a deterministic, consistent method of accessing resources.

Local VLANs are also considerably easier to manage and conceptualize than end-to-end VLANs. Local VLANs permit a geographic definition of VLANs, which are easier to document and monitor with network management tools.

End-to-end VLANs were originally Cisco's recommended approach to configuring VLANs in the switch block. This helped facilitate the old 80/20 rule. That is, 80 percent of your traffic should be local, and 20 percent should be remote. As the corporate community began to move to server farms, application servers, and enterprise servers, this became increasingly difficult to manage. Cisco no longer recommends using end-to-end VLANs in general due to the associated management and STP overhead. (STP's role in VLAN design is discussed in Chapter 5.)

Port-Based VLANs Versus Dynamic VLANs

VLANs are also classified according to the mechanism by which ports become members of particular VLANs. By far, the most common implementation of VLANs is port-based (also called static). In a *port-based VLAN*, the port-to-VLAN mappings are manually configured, one-by-one, specifying which ports are associated with that VLAN. This hard-codes the mapping between ports and VLANs directly on each switch. The mapping between ports and VLANs is only of local significance—the switches do not share this information with each other.

In port-based VLAN membership, the port is assigned to a specific VLAN independent of the user or system attached to it. This means that all users attached to the port must be members of the same VLAN. The port configuration is static and cannot be automatically changed to another VLAN without manual reconfiguration.

The device that is attached to the port likely has no understanding that a VLAN exists. The device simply knows that it is a member of a subnet and that the device should be able to talk to all other members of the subnet by simply sending data on the cable segment. The switch is responsible for identifying that the information came from a specific VLAN and for ensuring that the information is sent to all other members of the VLAN. The switch is further responsible for ensuring that ports in a different VLAN do not receive the information.

Port-based VLANs have the advantage of reducing the complexity of VLAN membership administration. The number of variables involved in implementing VLAN memberships is minimized with this approach.

On the other hand, if users frequently move their workstations or laptops around within a large campus network, port-based VLANs might prove untenable. In this case, it might prove worthwhile to implement *dynamic VLANs*. With dynamic VLANs, end stations are automatically assigned to the appropriate VLAN based on their MAC address. This is made possible via a MAC address-to-VLAN mapping table contained in a *VLAN Management Policy Server (VMPS)* database. VMPS is a Cisco-proprietary solution for enabling dynamic VLAN assignments to switch ports within a VTP domain. A VMPS server is responsible for delivering the information contained in the VMPS database to the Catalyst switches.

Dynamic VLANs also provide a form of security. The usual configuration for a VMPS database is to deny users access to the network or assign them to a restricted VLAN if their MAC address is not in the database. One downside of dynamic VLANs is the administrative overhead of continually having to update a VMPS database. However, the alternative is to rely on port-based VLANs and a significant amount of work manually reconfiguring port-to-VLAN mappings on all your switches. VMPS configuration is discussed in detail in the section, "Configuring VMPS and Dynamic Ports." With the many options now available to a network administrator, such as mobile IP and Layer 3 switching, the argument for deploying VMPS in a campus network is not as convincing as it once was.

Another alternative for easing VLAN management is to use network-management software. For Catalyst switches, the CiscoWorks2000 LAN Management Solution (LMS) serves this purpose. With LMS's Web-based interface, you can remotely assign switch ports to particular VLANs. LMS includes six packages designed to organize and streamline your LAN-management tasks. For more details, check out www.cisco.com/univercd/cc/td/doc/pcat/wlwnmn.htm. Of course, a significant learning curve is always associated with network-management software.

The following section explains VMPS operation in a campus network.

VMPS Operation

With VMPS, you can assign switch ports to VLANs dynamically, based on the source Media Access Control (MAC) address of the device connected to the port. When you move a host from a port on one switch in the network to a port on another switch in the network, the switch dynamically assigns the new port to the proper VLAN for that host.

When you enable VMPS, a MAC address-to-VLAN mapping database downloads from a Trivial File Transfer Protocol (TFTP) server to the VMPS server (a Catalyst switch). (Note the redundancy in the phrase "VMPS server"—similar to "NIC card.")

The VMPS server then begins accepting client requests. If you reset or power-cycle the VMPS server, the VMPS database downloads from the TFTP server automatically, and VMPS is reenabled.

VMPS opens a UDP socket to communicate and listen to client requests. A VMPS client communicates with a VMPS server through the VLAN Query Protocol (VQP). When the VMPS server receives a valid VQP request from a client switch, it searches its database for a MAC address-to-VLAN mapping.

The server response is based on this mapping and whether the server is in secure mode. Secure mode determines whether the server shuts down the port when a VLAN is not allowed on it or just denies the port access to the VLAN.

In response to a request, the VMPS server takes one of these actions:

- If the assigned VLAN is restricted to a group of ports, the VMPS server verifies the requesting port against this group and responds as follows:
  - If the VLAN is allowed on the port, the VMPS server sends the VLAN name to the client in response.
  - If the VLAN is not allowed on the port, and the VMPS server is not in secure mode, the VMPS server sends an access-denied response.
  - If the VLAN is not allowed on the port, and the VMPS server is in secure mode, the VMPS sends a port-shutdown response.
- If the VLAN in the database does not match the current VLAN on the port and active hosts exist on the port, the VMPS server sends an access-denied or port-shutdown response, depending on the secure mode of the VMPS database.

If the VMPS client receives an access-denied response from the VMPS server, it continues to block traffic based on that source MAC address to or from the port. The switch continues to monitor the packets directed to the port and sends a query to the VMPS server when it identifies a new address. If the switch receives a port-shutdown response from the VMPS server, it disables the port. The port must then be reenabled manually.

You can also configure a fallback VLAN name with VMPS. If you connect a device with a MAC address that is not in the database, VMPS sends the fallback VLAN name to the client. If you do not configure a fallback VLAN and the MAC address does not exist in the database, VMPS sends an access-denied response or a port-shutdown response, depending on the secure mode configured in the VMPS database.

You can also make an explicit entry in the configuration table to deny access to specific MAC addresses for security reasons by specifying a **--NONE--** keyword for the VLAN name. In this case, VMPS sends an access-denied or port-shutdown response.

Next, we describe dynamic port operation.

Dynamic Port VLAN Membership

A dynamic (nontrunking) port on the switch can belong to only one VLAN. When the link comes up, the switch does not forward traffic to or from this port until the VMPS server provides the VLAN assignment. The VMPS client receives the source MAC address from the first packet of a new host connected to the dynamic port and attempts to match the MAC address to a VLAN in the VMPS database via a VQP request to the VMPS server.

If a match occurs, the VMPS server sends the VLAN number for that port. If there is no match, the VMPS either denies the request or shuts down the port (depending on the VMPS secure-mode setting).

Multiple hosts (MAC addresses) can be active on a dynamic port if they are all in the same VLAN; however, the VMPS shuts down a dynamic port after a threshold is met (particular to the platform).

If the link goes down on a dynamic port, the port returns to an isolated state and does not belong to a VLAN. Any hosts that come online through the port are checked again with the VMPS server before the port is assigned to a VLAN.

There are several restrictions for configuring VMPS client switch ports as dynamic. The following are some guidelines that apply to dynamic port VLAN membership:

- You must configure VMPS before you configure ports as dynamic.
- The VTP management domain of VMPS clients and the VMPS server must be the same.
- The management VLAN of VMPS clients and the VMPS server must be the same.
- When you configure a port as dynamic, spanning-tree PortFast is enabled automatically for that port. PortFast is an STP feature that allows a port to start forwarding data frames as soon as the physical link is active. (PortFast is discussed in Chapter 5.) You can disable spanning-tree PortFast mode on a dynamic port.
- If you reconfigure a port from static (the default setting) to dynamic on the same VLAN, the port connects immediately to that VLAN. However, VMPS checks its database for the legality of the specific host on the dynamic port after a certain period.
- Static secure ports cannot become dynamic ports. You must turn off security on the static secure port before it can become dynamic. (Port security is discussed in Chapter 5.)

- Static ports that are trunking cannot become dynamic ports. You must turn off trunking on the trunk port before changing it from static to dynamic.
- Physical ports in a port channel cannot be configured as dynamic ports. A port channel is a set of Ethernet links bundled to form a single link with a greater aggregate bandwidth. (Port channels are discussed in Chapter 5.)
- VMPS shuts down a dynamic port when a certain number of active hosts is reached for a given port. For example, 20 is the maximum allowed on a Catalyst 2950.

TECH NOTE: DYNAMIC PORTS AND VOICE OVER IP

On Catalyst 4000 and 6000 switches, prior to Cisco IOS Software Release 6.2(1), dynamic ports could belong to only one VLAN. In short, native VLANs are traditional VLANs used for data traffic, and auxiliary VLANs are VLANs dedicated for use with IP phones. Prior to 6.2(1), you could not enable the dynamic-port VLAN feature on ports that carried a native VLAN and an auxiliary VLAN.

With Cisco IOS Software Releases 6.2(1) and later, dynamic ports can belong to two VLANs. The switch port configured for connecting an IP phone can have separate VLANs configured to carry the following:

- Voice traffic to and from the IP phone (auxiliary VLAN)
- Data traffic to and from the PC connected to the switch through the IP phone's access port (native VLAN)

Configuring VLANs

Chapter 3, "Switch Administration" detailed the configuration of switch administration in terms of the Catalyst switch OS type: CatOS or IOS-based. Here, this methodology continues. We begin with port-based VLAN configuration, followed by VMPS configuration.

Later in this chapter, you explore *VLAN Trunking Protocol (VTP)*. VTP is a Cisco-proprietary protocol that communicates information about VLANs between Catalyst switches. VTP permits three modes for a given switch: server, client, and transparent. When configuring VLANs, keep the following rule in mind: VLANs can be created or deleted only in VTP server mode or VTP transparent mode.

Configuring Port-Based VLANs

The configuration of port-based, or static, VLANs differs considerably between the CatOS- and IOS-based platforms. The following section details CatOS switch and IOS-based switch configuration. You must configure a VTP domain before assigning ports to a VLAN. VTP is discussed in the section, "VLAN Trunking Protocol."

CatOS Switch

To create a new Ethernet VLAN, use the privileged-mode command **set vlan**. Verify the creation of the VLAN with the **show vlan** command. The default VLAN type is Ethernet, so you do not have to specify the VLAN type. Example 4-1 illustrates creating an Ethernet VLAN and verifying the configuration.

Example 4-1 Creating an Ethernet VLAN and Verifying the Configuration

```
Console> (enable) set vlan 500 name Engineering

Vlan 500 configuration successful
Console> (enable) show vlan 500

VLAN Name                             Status    IfIndex Mod/Ports, Vlans
---- -------------------------------- --------- ------- ------------------------
500  Engineering                      active    344
VLAN Type  SAID       MTU   Parent RingNo BrdgNo Stp  BrdgMode Trans1 Trans2
---- ----- ---------- ----- ------ ------ ------ ---- -------- ------ ------
500  enet  100500     1500  -      -      -      -    -        0      0
VLAN AREHops STEHops Backup CRF
---- ------- ------- ----------
Console> (enable)
```

It is also possible to create a range of VLANs with a single instance of the **set vlan** command (for example, **set vlan 200-210**). If you create a range of VLANs, you cannot specify a name. VLAN names must be unique. The **set vlan** command also modifies the VLAN parameters on an existing Ethernet VLAN.

A VLAN created in a management domain remains unused until you assign one or more switch ports to it. You can create a new VLAN and then specify the module and ports later, or you can create the VLAN and specify the module and ports in a single step.

Use the command **set vlan** to assign one or more switch ports to a VLAN. You verify the port VLAN membership with the commands **show vlan** and **show port**. Example 4-2 illustrates this.

Example 4-2 Configuring Ports to Join a VLAN and Verifying the Configuration

```
Console> (enable) set vlan 560 4/10

VLAN 560 modified.
```

Example 4-2 Configuring Ports to Join a VLAN and Verifying the Configuration (Continued)

```
VLAN 1 modified.
VLAN  Mod/Ports
---- -----------------------

560   4/10

Console> (enable) show vlan 560

VLAN Name                             Status    IfIndex Mod/Ports, Vlans
---- -------------------------------- --------- ------- ------------------------
560  Engineering                      active    348     4/10
VLAN Type  SAID       MTU   Parent RingNo BrdgNo Stp  BrdgMode Trans1 Trans2
---- ----- ---------- ----- ------ ------ ------ ---- -------- ------ ------
560  enet  100560     1500  -      -      -      -    -        0      0
VLAN AREHops STEHops Backup CRF
---- ------- ------- ----------
Console> (enable) show port 4/10

Port  Name               Status     Vlan       Duplex Speed Type
----- ------------------ ---------- ---------- ------ ----- ------------
4/10                     connected  560        a-half a-100 10/100BaseTX

Port  AuxiliaryVlan AuxVlan-Status
----- ------------- --------------
 4/10  none          none<output omitted>
```

As you might guess, the **set vlan** and **show vlan** commands are some of the most ubiquitous CatOS commands.

It is not uncommon to delete and create VLANs. Use the **clear vlan** *vlan_number* command to remove a VLAN. If the switch is in VTP transparent mode, the VLAN is removed only on that particular switch. If the switch is in VTP server mode, the VLAN is removed throughout the VTP domain.

When you attempt to delete a VLAN, the switch warns you that all ports in the management domain belonging to that VLAN will be deactivated. See Example 4-3.

Example 4-3 Clearing a VLAN on a CatOS Switch

```
cat4> (enable) clear vlan 980
This command will deactivate all ports on vlan(s) 980
Do you want to continue(y/n) [n]? y
VTP advertisements transmitting temporarily stopped,
and will resume after the command finishes.
Vlan 980 deleted
```

If you recreate the VLAN, the ports previously associated with that VLAN automatically become active again because the switch remembers the previous VLAN membership for the ports.

IOS-Based Switch

VLAN configuration is different on IOS-based switches than it is on CatOS switches.

Use the **vlan database** privileged-mode command to enter VLAN configuration mode. Prior to Cisco IOS Software Release 12.1(6), VLAN configuration mode was called VLAN database mode.

> **NOTE**
> VLAN configuration mode differs from other modes because it is session-oriented. When you add, delete, or modify VLAN parameters, the changes aren't applied until you exit the session by entering the **apply** or **exit** command. When changes are applied, the VTP configuration version is incremented.

VLAN additions, modifications, and deletions are written to the vlan.dat file. You can display them by entering the privileged-mode command **show vlan**. The vlan.dat file is stored in nonvolatile RAM. You can delete this file, but beware of the consequences: Deleting vlan.dat will likely create inconsistencies in the VLAN database. If you want to modify the VLAN configuration, the preferred action is to use VLAN configuration-mode commands.

In VLAN configuration mode, you can add or modify VLANs using the command **vlan** *vlan-id* [**name** *vlan-name*]. Use the **no** form of this command to delete a VLAN.

The commands **show vlan name** *vlan-name*, **show vlan id** *vlan-id*, and **show vlan brief** are all options to verify the configuration.

The next step is to associate specific ports with the respective VLANs. You use interface configuration command mode to define the port membership mode (access or trunk) and to add ports to or remove ports from VLANs. You enter interface configuration mode for a switch port exactly as you would on a Cisco router.

Here, we focus on assigning a port to or removing a port from a particular VLAN; this means that we are dealing with an access link as opposed to a trunk link. The interface configuration mode command to define the port's VLAN membership mode (as an access port) is **switchport mode access**. Use the **no** form of this command to remove a port from a VLAN.

Next, you assign a port to a particular VLAN using the interface command **switchport access vlan** *vlan-id*. Use the **no** form of this command to remove a port from a VLAN.

Finally, you exit interface mode with the **exit** command.

The results of these commands are written to the running-configuration file, which you can verify with the command **show running-config**. In addition, you can verify the VLAN configuration with the command **show interface** *interface-id* **switchport**. This command is extremely useful.

Pulling all this together, Example 4-4 demonstrates the process.

Example 4-4 Creating a VLAN, Adding Ports to a VLAN, and Checking VLAN Configuration

```
Switch#vlan database
Switch(vlan)#vlan 20 name Accounting
VLAN 20 modified:
    Name: Accounting
Switch(vlan)#exit
APPLY completed.
Exiting....

Switch#show vlan ?
  brief   VTP all VLAN status in brief
  id      VTP VLAN status by VLAN id
  name    VTP VLAN status by VLAN name
  |       Output modifiers
  <cr>

Switch#show vlan id ?
  <1-1005>  A VTP VLAN number

Switch#show vlan id 20
VLAN Name                             Status    Ports
---- -------------------------------- --------- -------------------------------
20   Accounting                       active

VLAN Type  SAID       MTU   Parent RingNo BridgeNo Stp  BrdgMode Trans1 Trans2
---- ----- ---------- ----- ------ ------ -------- ---- -------- ------ ------
20   enet  100020     1500  -      -      -        -    -        0      0
```

continues

Example 4-4 Creating a VLAN, Adding Ports to a VLAN, and Checking VLAN Configuration (Continued)

```
Switch#show vlan name Accounting
VLAN Name                             Status    Ports
---- -------------------------------- --------- -------------------------------
20   Accounting                       active

VLAN Type  SAID       MTU   Parent RingNo BridgeNo Stp  BrdgMode Trans1 Trans2
---- ----- ---------- ----- ------ ------ -------- ---- -------- ------ ------
20   enet  100020     1500  -      -      -        -    -        0      0

Switch#show vlan brief
VLAN Name                             Status    Ports
---- -------------------------------- --------- -------------------------------
1    default                          active    Fa0/1, Fa0/2, Fa0/3, Fa0/4,
                                                Fa0/5, Fa0/6, Fa0/7, Fa0/8,
                                                Fa0/9, Fa0/10, Fa0/11, Fa0/12,
                                                Fa0/13, Fa0/14, Fa0/15, Fa0/16,
                                                Fa0/17, Fa0/18, Fa0/19, Fa0/20,
                                                Fa0/21, Fa0/22, Fa0/23, Fa0/24
20   Accounting                       active
1002 fddi-default                     active
1003 token-ring-default               active
1004 fddinet-default                  active
1005 trnet-default                    active

Switch#config t
Enter configuration commands, one per line.  End with CNTL/Z.
Switch(config)#interface fa0/1
Switch(config-if)#switchport mode access
Switch(config-if)#switchport access vlan 20
Switch(config-if)#end
Switch#show vlan brief
VLAN Name                             Status    Ports
---- -------------------------------- --------- -------------------------------
1    default                          active    Fa0/2, Fa0/3, Fa0/4, Fa0/5,
```

Example 4-4 Creating a VLAN, Adding Ports to a VLAN, and Checking VLAN Configuration (Continued)

```
                                                Fa0/6, Fa0/7, Fa0/8, Fa0/9,
                                                Fa0/10, Fa0/11, Fa0/12, Fa0/13,
                                                Fa0/14, Fa0/15, Fa0/16, Fa0/17,
                                                Fa0/18, Fa0/19, Fa0/20, Fa0/21,
                                                Fa0/22, Fa0/23, Fa0/24
20   Accounting                       active    Fa0/1
1002 fddi-default                     active
1003 token-ring-default               active
1004 fddinet-default                  active
1005 trnet-default                    active
Switch#vlan database
Switch(vlan)#no vlan 20
Deleting VLAN 20...
Switch(vlan)#exit
APPLY completed.
Exiting....
Switch#show vlan brief
VLAN Name                             Status    Ports
---- -------------------------------- --------- -------------------------------
1    default                          active    Fa0/2, Fa0/3, Fa0/4, Fa0/5,
                                                Fa0/6, Fa0/7, Fa0/8, Fa0/9,
                                                Fa0/10, Fa0/11, Fa0/12, Fa0/13,
                                                Fa0/14, Fa0/15, Fa0/16, Fa0/17,
                                                Fa0/18, Fa0/19, Fa0/20, Fa0/21,
                                                Fa0/22, Fa0/23, Fa0/24
1002 fddi-default                     active
1003 token-ring-default               active
1004 fddinet-default                  active
1005 trnet-default                    active
Switch#vlan database
Switch(vlan)#vlan 20
VLAN 20 added:
    Name: VLAN0020
Switch(vlan)#exit
APPLY completed.
```

continues

Example 4-4 Creating a VLAN, Adding Ports to a VLAN, and Checking VLAN Configuration (Continued)

```
Exiting....
Switch#show vlan brief
VLAN Name                             Status    Ports
---- -------------------------------- --------- -------------------------------
1    default                          active    Fa0/2, Fa0/3, Fa0/4, Fa0/5,
                                                Fa0/6, Fa0/7, Fa0/8, Fa0/9,
                                                Fa0/10, Fa0/11, Fa0/12, Fa0/13,
                                                Fa0/14, Fa0/15, Fa0/16, Fa0/17,
                                                Fa0/18, Fa0/19, Fa0/20, Fa0/21,
                                                Fa0/22, Fa0/23, Fa0/24
20   VLAN0020                         active    Fa0/1
1002 fddi-default                     active
1003 token-ring-default               active
1004 fddinet-default                  active
1005 trnet-default                    active
```

Notice in Example 4-4 that, when VLAN 20 is deleted, interface Fa0/1 does not show up in the **show vlan brief** command output. (It showed up before VLAN 20 was deleted.) Also notice that when VLAN 20 is recreated, Fa0/1 shows up again as being a member of VLAN 20 without having to be reassigned. The switch retains the association between VLAN 20 and interface Fa0/1. However, the name Accounting assigned to VLAN 20 is not retained in the process of deleting and re-creating VLAN 20. It now appears with the default name of VLAN0020.

Configuring VMPS and Dynamic Ports

To use VMPS, you must first create a VMPS database and store it on a TFTP server. The next section shows you how to configure the VMPS database.

Creating a VMPS Database

A VMPS database configuration file is an ASCII text file that is stored on a TFTP server accessible to the switch configured as the VMPS server.

The VMPS parser is line-based. You start each entry in the text file on a new line. Ranges are not allowed for the switch port numbers. It is probably easiest to take a sample VMPS database file and edit it to meet your needs.

If you are hard-core, you need to follow these rules to create a VMPS database file from scratch:

- **Begin the configuration file with the word "VMPS"**—To prevent other types of configuration files from being read incorrectly by the VMPS server, you must begin the configuration file with the word "VMPS."
- **Define the VMPS domain**—The VMPS domain should correspond to the VTP domain name configured on the switch!
- **Define the security mode**—VMPS can operate in open or secure mode. The default mode is open.
- **Define a fallback VLAN** (**optional**)—The fallback VLAN is assigned if the MAC address of the connected host is not defined in the database.
- **Define the MAC address-to-VLAN name mappings**—Enter the MAC address of each host and the VLAN to which each should belong. Use the **--NONE--** keyword as the VLAN name to explicitly deny the specified host network connectivity. A port is identified by the switch's IP address and the port's module/port number in the form *mod/port*.
- **Define port groups**—A port group is a logical group of ports. You can apply VMPS policies to individual ports or to port groups. The keyword **all-ports** specifies all the ports in the specified switch.
- **Define VLAN groups**—A VLAN group defines a logical group of VLANs. These logical groups define the VLAN port policies.
- **Define VLAN port policies**—VLAN port policies define the ports associated with a restricted VLAN. You can configure a restricted VLAN by defining the set of dynamic ports on which it can exist.

To review the general mechanism by which VMPS makes policy decisions affecting switch ports and VLAN assignments, review the earlier section "VMPS Operation." Example 4-5 illustrates how the rules just listed work with a sample VMPS database configuration file. The following is a summary of the configuration:

- The security mode is open.
- The default is used for the fallback VLAN.
- MAC address-to-VLAN name mappings provide the MAC address of each host and the VLAN to which each host belongs.
- Port groups are defined.
- VLAN groups are defined.
- VLAN port policies are defined for the ports associated with restricted VLANs.

Example 4-5 Sample VMPS Database File Residing on a TFTP Server

```
!VMPS File Format, version 1.1
! Always begin the configuration file with
! the word "VMPS"
!
!vmps domain <domain-name>
! The VMPS domain must be defined.
!vmps mode {open ¦ secure}
! The default mode is open.
!vmps fallback <vlan-name>
!vmps no-domain-req { allow ¦ deny }
!
! The default value is allow. Deny means requests from clients with no domain
  name are
! rejected.
vmps domain WBU
vmps mode open
vmps fallback default
vmps no-domain-req deny
!
!
!MAC Addresses
!
vmps-mac-addrs
!
! address <addr> vlan-name <vlan_name>
!
address 0012.2233.4455 vlan-name hardware
address 0000.6509.a080 vlan-name hardware
address aabb.ccdd.eeff vlan-name Green
address 1223.5678.9abc vlan-name ExecStaff
address fedc.ba98.7654 vlan-name --NONE--
address fedc.ba23.1245 vlan-name Purple
!
!Port Groups
!
```

Example 4-5 Sample VMPS Database File Residing on a TFTP Server (Continued)

```
!vmps-port-group <group-name>
! device <device-id> { port <port-name> ¦ all-ports }
!
vmps-port-group WiringCloset1
 device 198.92.30.32 port 3/2
 device 172.20.26.141 port 2/8
vmps-port-group "Executive Row"
 device 198.4.254.222 port 1/2
 device 198.4.254.222 port 1/3
 device 198.4.254.223 all-ports
!
!
!VLAN groups
!
!vmps-vlan-group <group-name>
! vlan-name <vlan-name>
!
vmps-vlan-group Engineering
vlan-name hardware
vlan-name software
!
!
!VLAN port Policies
!
!vmps-port-policies {vlan-name <vlan_name> ¦ vlan-group <group-name> }
! { port-group <group-name> ¦ device <device-id> port <port-name> }
!
vmps-port-policies vlan-group Engineering
 port-group WiringCloset1
vmps-port-policies vlan-name Green
 device 198.92.30.32 port 4/8
vmps-port-policies vlan-name Purple
 device 198.4.254.22 port 1/2
 port-group "Executive Row"
```

NOTE

The VMPS database configuration file on the server must use the naming convention for the respective VMPS client switch. For example, on a Catalyst 2950, FastEthernet0/5 is fixed-port number 5 and would be referenced in the VMPS database file as port Fa0/5.

When you are ready to create a VMPS database, first determine the MAC addresses of the hosts you want to be assigned to VLANs dynamically, determine the VMPS mode, and then map out your port groups, VLAN groups, port policies, and fallback VLAN. Second, create an ASCII text file on your workstation that contains the MAC address-to-VLAN mappings and policies. Finally, move the ASCII text file to a TFTP server so that it can be downloaded to the switch (VMPS server).

Configuring the VMPS Server and the VMPS Client

As soon as the VMPS database is in place on the TFTP server, you need to configure the VMPS server and clients.

Recall that, when you enable VMPS, the switch downloads the VMPS database from the TFTP server and begins accepting VMPS requests from VMPS clients.

In general, the role of VMPS server is assumed by a CatOS switch, so we make that assumption here. On a CatOS switch, you enter the following commands to configure a VMPS server:

- **set vmps downloadserver** *ip_addr* [*filename*]—*ip_addr* is the IP address of the TFTP server, and *filename* is the name of the VMPS database file.
- **set vmps state enable**—This enables VMPS on the switch.

Use the **set vmps state disable** command to disable VMPS on the switch.

The last task for VMPS is to configure the VMPS clients. For a CatOS switch, enter the following commands, where *ip_addr* is the IP address of the VMPS server switch:

- **set vmps server** *ip_addr* [**primary**]—**primary** is used only if you plan to have backup VMPS servers.
- set port membership *mod/port* **dynamic**—For the appropriate ports.

On an IOS-based switch, configure VMPS clients with the following commands, starting in global configuration mode:

- **vmps server** *ipaddress* **primary**
- **vmps server** *ipaddress* for up to three secondary VMPS servers (optional)
- **interface** *interface*
- **switchport mode access**
- **switchport access vlan dynamic**

Repeat the third, fourth, and fifth commands for each dynamic switch port. To reenable a shutdown dynamic port, enter the interface configuration command **no shutdown**. Next, you need to verify the VMPS installation.

Verifying VMPS Installation

Now that you have all the VMPS components configured and in place, you want to verify that all is working as desired. The following sections explain how to do this based on the type of switch you are using.

CatOS Switch To view VQP information, to verify VMPS configuration, or to check that VMPS is disabled, use the **show vmps** command. Example 4-6 shows the output when VMPS is disabled.

Example 4-6 Verifying the VMPS Server Configuration

```
Cat5> (enable) show vmps
VMPS Server Status:
-------------------
Management Domain:     (null)
State:                 disabled
Operational Status:    inactive
TFTP Server:           default
TFTP File:             vmps-config-database.1
Fallback VLAN:         (null)
Secure Mode:           open
VMPS No Domain Req:    allow

VMPS Client Status:
---------------------
VMPS VQP Version:      1
Reconfirm Interval:    60 min
Server Retry Count:    3
VMPS domain server:

No dynamic ports configured.
```

Use the **show vmps server** command to verify the VMPS server identity, as shown in Example 4-7.

Example 4-7 Verifying the VMPS Server Specification

```
Console> (enable) show vmps server


VMPS domain server            VMPS Status
```

continues

Example 4-7 Verifying the VMPS Server Specification (Continued)

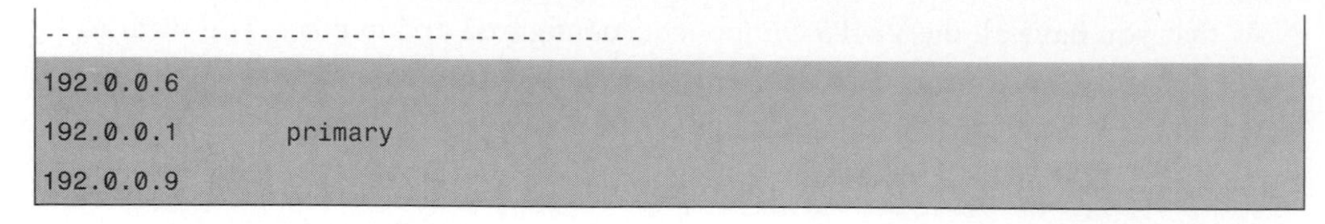

```
-------------------------------------
192.0.0.6
192.0.0.1          primary
192.0.0.9
```

Use the **show port** command to verify dynamic port assignment. This command displays **dyn-** under the **Vlan** column when the port has not yet been assigned a VLAN. This is illustrated in Example 4-8 (see port 3/1).

Example 4-8 Verifying Dynamic Port Assignment

```
Console> show port

Port  Name   Status   Vlan   Level   Duplex   Speed   Type
1/1          connect  dyn-3  normal  full     100     100 BASE-TX
1/2          connect  trunk  normal  half     100     100 BASE-TX
3/1          connect  dyn-   normal  half     10      10 BASE-T
3/2          connect  dyn-5  normal  half     10      10 BASE-T
3/3          connect  dyn-5  normal  half     10      10 BASE-T
```

Next, we consider IOS-based switch commands to verify the VMPS configuration.

IOS-Based Switch Use the show vmps command to verify the VMPS server entry. This command did not become available on some IOS-based switch platforms until Cisco IOS Software Release 12.1(6). This command displays the VQP version and the VMPS server IP addresses (including which one is primary). Use the statistics keyword to display client-side statistics.

Use the command **show interface** *interface* **switchport** to verify dynamic assignment of VLAN membership to a given port.

Sample VMPS Network Installation

Figure 4-5 shows a network with a VMPS server switch and VMPS client switches with dynamic ports. The following assumptions apply to this scenario:

- The VMPS servers and the VMPS clients are separate switches.
- Switch 1 is the primary VMPS server.
- Switch 3 and Switch 10 are secondary VMPS servers.

- End stations are connected to these clients:
  — Switch 2
  — Switch 9
- The database configuration file is called Bldg-G.db. It is stored on a TFTP server with IP address 172.20.22.7.

Figure 4-5
VMPS components include a TFTP server, a VMPS server, and VMPS clients.

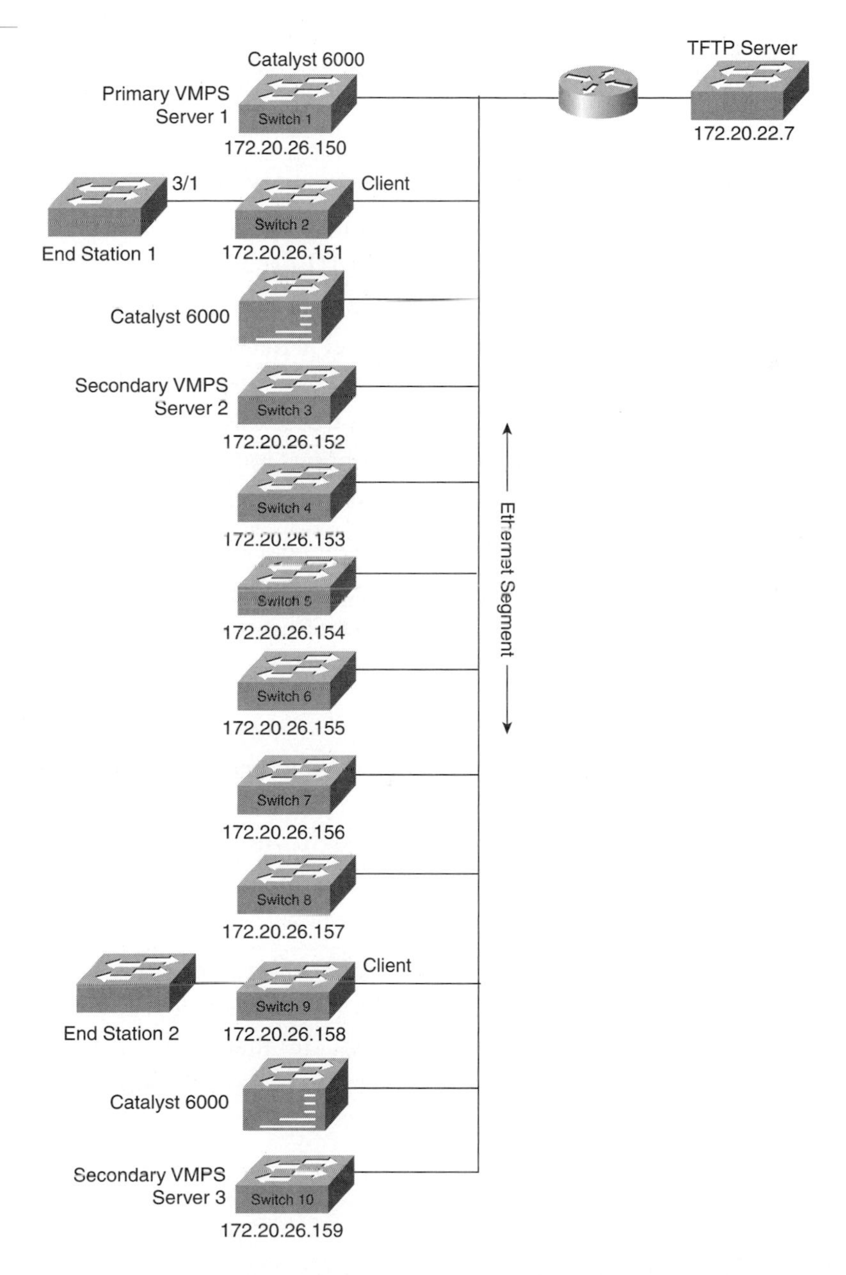

To configure VMPS as pictured, proceed as follows:

Step 1 On Switch 1, enter the commands **set vmps downloadserver 172.20.22.7 Bldg-G.db** and **set vmps state enable**. After you enter these commands, the file Bldg-G.db is downloaded to Switch 1, and Switch 1 becomes the VMPS server.

Step 2 Configure the VMPS server addresses on each VMPS client. For example, on Switch 2, enter the command **set vmps server 172.20.26.150 primary**, followed by **set vmps server 172.20.26.152** and **set vmps server 172.20.26.159**.

Step 3 Verify the VMPS server addresses with the command **show vmps server**.

Step 4 Configure the dynamic ports. For example, on Switch 2, enter the command **set port membership 3/1 dynamic**.

Step 5 Physically connect End Station 1 on Port 3/1. When End Station 1 sends a packet, Switch 2 sends a VQP query to the primary VMPS server, Switch 1. Switch 1 responds with the appropriate VLAN to assign to Port 3/1. Because spanning-tree PortFast mode is enabled by default on dynamic ports, Port 3/1 connects immediately and enters forwarding mode. See Chapter 5 for more information on PortFast.

Repeat these steps for each VMPS client and appropriate switch ports.

This completes the discussion of VMPS configuration. Now consider the concept of VLAN trunking, which plays a fundamental role in campus networks.

Trunking

Switch ports run in either access mode or trunk mode. The links connected to these switch ports are accordingly referred to as access or trunk links. In access mode, the interface belongs to one and only one VLAN. An *access port* is a switch port that connects to an end-user device or a server; the frames transmitted on an access link look like any other Ethernet frames.

Trunks, on the other hand, multiplex traffic for multiple VLANs over the same point-to-point link. A trunk is a point-to-point link connecting a switch to another switch, a router, or a server (with a special adapter card) that carries traffic for multiple VLANs over the same link. The VLANs are multiplexed over the link with a trunking protocol.

You must understand that a trunk link does not belong to a specific VLAN. A trunk link acts as a conduit for VLANs between switches and routers. Trunks can extend VLANs across an entire network. A trunk link can be configured to transport all VLANs or a restricted set of VLANs.

Trunk links are typically configured on ports that support the greatest bandwidth for a given switch. Catalyst switches support trunking on Fast Ethernet, Gigabit Ethernet, and 10 Gigabit Ethernet ports.

To multiplex VLAN traffic, special protocols exist that encapsulate or tag the frames so that the receiving device knows to which VLAN the frame belongs. Ethernet trunking protocols are either proprietary or based on IEEE 802.1Q. *ISL* is a proprietary trunking protocol that lets Cisco devices multiplex VLANs between Cisco devices. IEEE 802.1Q, on the other hand, is an industry-standard protocol that permits multiplexing of VLANs over a trunk link connecting different vendors' switches.

Without trunk links, multiple access links would have to be installed to support multiple VLANs between switches (one link per VLAN). This is clearly neither a cost-effective nor scalable solution. Trunking is absolutely essential for interconnecting switches in a campus network.

NOTE

A *native VLAN* is the VLAN that a trunk port reverts to if trunking is disabled on the port.

In Figure 4-6, Port A and Port B are defined as access links on the same VLAN. By definition, they can belong to only this one VLAN and cannot receive frames with a VLAN identifier. As Switch Y receives traffic from Port A destined for Port B, Switch Y does not encapsulate the frame with an ISL header and CRC. (The ISL encapsulation process is detailed later in this chapter.)

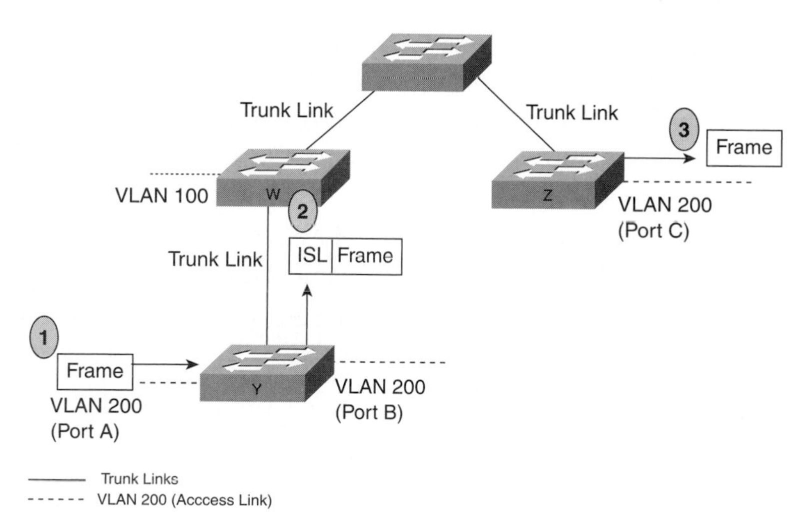

Figure 4-6
Access and trunk links interconnect end devices, permitting intra-VLAN communication. Here, ISL trunks use frame tagging to multiplex Ethernet frames for multiple VLANs.

Port C is also an access link. As you can see, Port C has been defined as a member of VLAN 200. If Port A sends a frame destined for Port C, the following things happen:

1. Switch Y receives the frame and identifies it as traffic destined for VLAN 200 by the VLAN and port number association.
2. Switch Y encapsulates the frame with an ISL header identifying VLAN 200 and sends the frame through the intermediate switch on a trunk link.
3. This process is repeated for every switch that the frame must transit as it moves to its final destination of Port C.
4. Switch Z receives the frame, removes the ISL header, and forwards the frame to Port C.

The process outlined in these four steps is, in a nutshell, how trunking works. You might want to keep this in mind when you find yourself engrossed in a myriad of convoluted trunking configuration options.

Trunking Technologies

VLAN identification logically identifies which packets belong to which VLAN group. Multiple trunking methodologies exist:

- **Inter-Switch Link (ISL)**—This is a Cisco-proprietary encapsulation protocol for trunking between switches. ISL prepends a 26-byte header and appends a 4-byte CRC to each data frame. ISL is supported in switches and routers.
- *IEEE 802.1Q*—This IEEE standard for identifying VLANs works by inserting a VLAN identifier into the frame header. This process is called frame tagging or internal tagging.
- *IEEE 802.10*—This standard provides a method for transporting VLAN information inside the standard 802.10 Fiber Distributed Data Interface (FDDI) frame. The VLAN information is written to the security association identifier (SAID) portion of the 802.10 frame. This method transports VLANs across FDDI backbones.
- **ATM LAN emulation (LANE)**—LANE is an ATM Forum standard that can transport VLANs over Asynchronous Transfer Mode (ATM) networks.

These technologies are summarized in Figure 4-7. The next sections describe each of these trunking technologies in more detail.

Figure 4-7
Four major technologies handle all the trunking requirements for switches and routers.

Identification Method	Encapsulation	Tagging (Insertion into Frame)	Media
802.1Q	No	Yes	Ethernet
ISL	Yes	No	Ethernet
802.10	No	No	FDDI
LANE	No	No	ATM

Inter-Switch Link

Inter-Switch Link (ISL) is a Cisco-proprietary protocol for interconnecting multiple switches and maintaining VLAN information as data moves between switches. This technology provides one method for multiplexing bridge groups (VLANs) over a high-speed backbone. It is defined for Fast Ethernet and Gigabit Ethernet, as is IEEE 802.1Q. ISL has been available on Cisco routers since Cisco IOS Software Release 11.1.

With ISL, an Ethernet frame is encapsulated with a header that maintains VLAN IDs between switches. A 26-byte header is prepended to the Ethernet frame; the header includes a 10-bit VLAN ID. In addition, a 4-byte cyclic redundancy check (CRC) is appended to the end of each frame. This CRC is in addition to any frame checking that the Ethernet frame requires. Figure 4-8 illustrates ISL encapsulation. The encapsulation is what makes ISL proprietary.

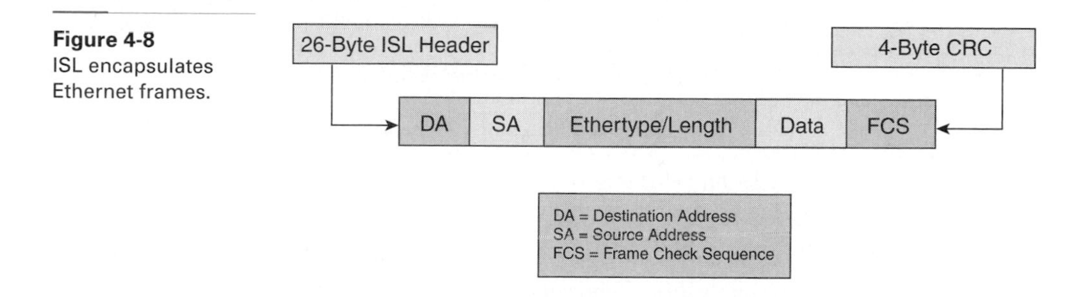

Figure 4-8
ISL encapsulates Ethernet frames.

Figure 4-9 shows the details of the fields in the ISL header. Table 4-1 describes each field's function.

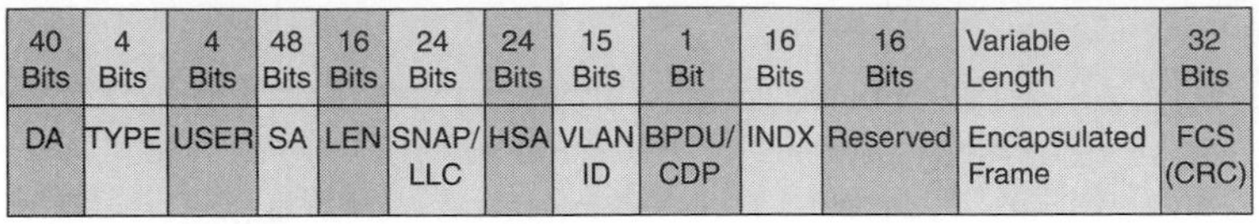

Figure 4-9 The ISL header includes a 15-bit VLAN field, 10 bits of which indicate the numerical value of the source VLAN for the encapsulated frame.

40 Bits	4 Bits	4 Bits	48 Bits	16 Bits	24 Bits	24 Bits	15 Bits	1 Bit	16 Bits	16 Bits	Variable Length	32 Bits
DA	TYPE	USER	SA	LEN	SNAP/LLC	HSA	VLAN ID	BPDU/CDP	INDX	Reserved	Encapsulated Frame	FCS (CRC)

Table 4-1 ISL Field Descriptors

Label	Description
DA	A 40-bit multicast address with a value of 0x01-00-0C-00-00. Indicates to the receiving Catalyst that the frame is an ISL encapsulated frame.
Type	A 4-bit value indicating the source frame type. Values include 0000 (Ethernet), 0001 (Token Ring), 0010 (FDDI), and 0011 (ATM).
User	A 4-bit value usually set to 0, but can be used for special situations when transporting Token Ring.
SA	The 802.3 MAC address of the transmitting Catalyst switch. This is a 48-bit value.
Length	The LEN field is a 16-bit value indicating the length of the user data and ISL header. Excludes the DA, Type, User, SA, Length, and ISL CRC bytes.
SNAP	A 24-bit value with a fixed value of 0xAA-AA-03.
HSA	This 24-bit value duplicates the high-order bytes of the ISL SA field.
VLAN	A 15-bit value that reflects the numerical value of the source VLAN that the user frame belongs to. Only 10 bits are used.
BPDU	A single-bit value that, when set to 1, indicates that the receiving Catalyst should immediately examine the frame because the data contains a spanning tree, ISL, VTP, or CDP message.
Index	Indicates what port the frame exited from the source Catalyst.
Reserved	Token Ring and FDDI frames have special values that need to be transported over the ISL link. These values, such as Access Control (AC) and Frame Copied (FC), are carried in this field. The value of this field is 0 for Ethernet frames.

Table 4-1 ISL Field Descriptors (Continued)

Label	Description
Encapsulated Frame	The original user data frame is inserted here, including the frame's FCS.
CRC	ISL calculates a 32-bit CRC for the header and user frame. This double-checks the integrity of the message as it crosses an ISL trunk. It does not replace the User Frame CRC.

ROUTERS AND ISL TRUNKING

The following Cisco routers support ISL trunking: 2600, 3600, 4500, 5300, 7200, and 7500.

The following Cisco routers support IEEE 802.1Q trunking: 1751, 2600, 3600, 4500, 7200, and 7500.

Catalyst 2900XL and 3500XL switches support both ISL and IEEE 802.1Q trunking when configured with Cisco IOS Software Release 12.0(5)WC(1) or later.

Catalyst 2926G, 5000, and 6000 currently support both ISL and IEEE 802.1Q. CatOS 4.2 or higher is required on 2926G and 5000.

Catalyst 2950, 2948G, and 4000 support only IEEE 802.1Q.

Catalyst 4000 with Supervisor Engine II supports ISL on external Gigabit Ethernet ports on the Layer 3 Services Module.

Catalyst 4006 with Supervisor Engine III supports ISL trunking.

To determine whether a specific piece of hardware supports trunking, and to determine which trunking encapsulations are supported, see your hardware documentation or use the **show port capabilities** command, as shown in Example 4-9.

Example 4-9 Determining Which Trunking Technologies Are Supported on a Switch Port

```
cat4> (enable) show port capabilities
Model                   WS X4013
Port                    1/1
Type                    1000BaseSX
Speed                   1000
Duplex                  full
Trunk encap type        802.1Q
Trunk mode              on,off,desirable,auto,nonegotiate
Channel                 1/1-2
Flow control            receive-(off,on,desired),send-(off,on,desired)
```

continues

Example 4-9 Determining Which Trunking Technologies Are Supported on a Switch Port (Continued)

```
Security                yes
Dot1x                   yes
Membership              static,dynamic
Fast start              yes
QOS scheduling          rx-(none),tx-(2q1t)
CoS rewrite             no
ToS rewrite             no
Rewrite                 no
UDLD                    yes
Inline power            no
AuxiliaryVlan           no
SPAN                    source,destination

<output omitted>
```

IEEE 802.1Q

The official name of the IEEE 802.1Q standard is Standard for Virtual Bridged Local-Area Networks. It relates to the capability to carry the traffic of more than one VLAN on a single segment. The 802.1Q committee defined this method of multiplexing VLANs in an effort to provide multivendor VLAN support.

802.1Q uses an internal tagging mechanism. Internal means that a tag is inserted within the frame; the tagging mechanism implies a modification of the frame. The trunking device inserts a 4-byte tag and recomputes the frame check sequence (FCS). The 802.1Q embedded tag is illustrated in Figure 4-10. The fields are described in Table 4-2.

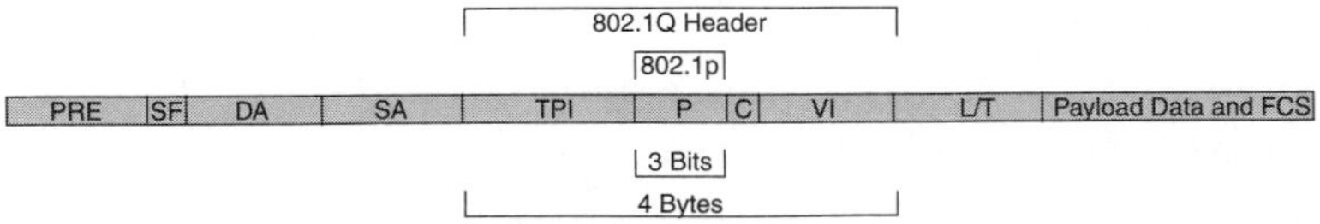

Figure 4-10 IEEE 802.1Q trunking is enabled by embedding a 4-byte tag in the Ethernet header.

802.1Q Header
802.1p
PRE | SF | DA | SA | TPI | P | C | VI | L/T | Payload Data and FCS
3 Bits
4 Bytes

Table 4-2 IEEE 802.1Q Field Descriptors

Label	Description
PRE	Preamble. Used to synchronize traffic between nodes.
SF	Start of Frame delimiter. Marks the beginning of the header.
DA	The destination's 48-bit 802.3 MAC address.
SA	The source's 48-bit 802.3 MAC address.
TPI	Two-byte Tag Protocol Identifier. Set to 0x8100 for Ethernet. This is the IEEE 802.3ac tag format.
P	Three-bit IEEE 802.1p priority level. Decimal values range from 0 to 7.
C	One-bit canonical identifier. Indicates whether the MAC address is in canonical format. Ethernet uses 0 (canonical).
VI	Twelve-bit VLAN ID. Indicates which VLAN this frame belongs to. Values are between 0 and 4095.
L/T	Two-byte Length/Type Field. Either IEEE 802.3 length information or Ethernet II type information.
Payload	User data. Less than or equal to 1500 bytes.
FCS	Four-byte Frame Check Sequence. Used for error checking of frame contents. Also known as cyclic redundancy check (CRC).

Table 4-2 references the IEEE 802.1p standard for traffic prioritization. This standard was created to allow for Layer 2 prioritization of traffic (see Chapter 2, "Gigabit Ethernet"). It defines how network frames are tagged, with user priority levels ranging from 0 (lowest) to 7 (highest). This allows time-critical data to receive preferential treatment over non-time-critical data. Some user devices can now tag frames with IEEE 802.1p QoS information, so access ports as well as trunk ports need to be 802.1p-aware.

You might see a reference to the 802.3ac standard. It was ratified in 1998. It extends the normal 802.3 maximum frame size (1518 bytes) to 1522 bytes for the specific purpose of accommodating the additional octets of the VLAN tag header. These *extended* frames are also called "baby giant" frames. Older networking devices might recognize such frames as oversized and drop them, or they might pass the frames without a priority benefit and report them as anomalies. IEEE 802.1p relies on the IEEE 802.1Q tagging mechanism for Layer 2 prioritization. Most network device manufacturers now support these extended frames.

NOTE

The IEEE 802.1Q standard specifies much more than VLAN identification, including spanning-tree enhancements, GARP, and 802.1p QoS tagging.

The 802.1Q frame-tagging scheme has significantly less overhead than ISL encapsulation. Compared to the 30 bytes added by ISL, 802.1Q adds only 4 bytes to the Ethernet frame, as shown in Figure 4-10.

IEEE 802.10

VLANs can be multiplexed across a FDDI backbone supporting the IEEE 802.10 protocol. (FDDI is sometimes pronounced "fiddy.") Catalyst 5000 and 6000 family switches support 802.10 trunking.

FDDI interfaces that support 802.10 make selective forwarding decisions within a network domain based on a VLAN identifier. This VLAN identifier is the user-configurable 4-byte IEEE 802.10 Security Association Identifier (SAID). The SAID identifies traffic as belonging to a particular VLAN.

VTP enables FDDI module configuration for 802.10-based VLANs. VTP requires a protocol type (Ethernet, FDDI, or Token Ring) to be configured for each VLAN. A VLAN can have only one type associated with it. Each VLAN type must have its own unique identifier, and translations between different identifiers must be mapped. VTP advertises VLAN translation mappings to all switches in a management domain.

On a Catalyst 5000 or 6000, if a FDDI module receives a packet containing a VLAN SAID that maps to a locally supported Ethernet VLAN on the switch, the FDDI module translates the packet into Ethernet format and forwards it across the switch backplane to the Ethernet module. FDDI modules filter the packets they receive from reaching the backplane if the VLAN SAIDs in the packets do not map to a locally supported VLAN.

Figure 4-11 illustrates the configuration for forwarding a packet from the Ethernet Module, Port 1, Slot 2, to the FDDI Module, Port 1, Slot 5. For this example, you specify the translation of Ethernet VLAN 2 to FDDI VLAN 22. The VLAN SAID must be identical on both FDDI modules. Because 802.10 FDDI interface links can also operate as ISL trunks, you can configure multiple VLAN translations over a single link.

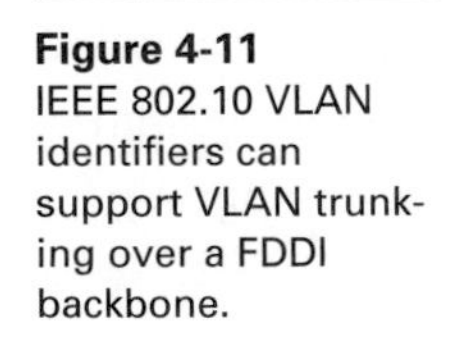

Figure 4-11 IEEE 802.10 VLAN identifiers can support VLAN trunking over a FDDI backbone.

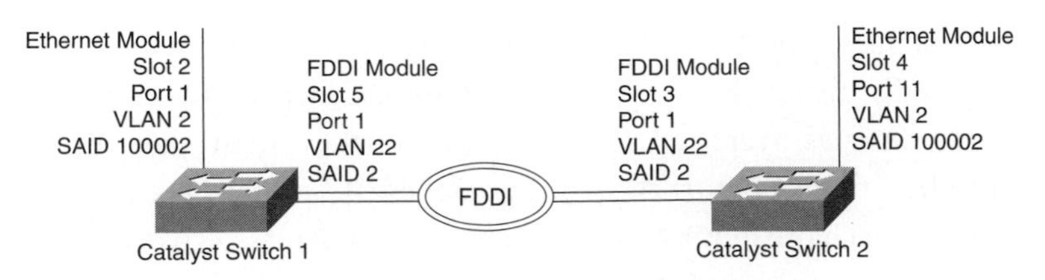

FDDI modules also support one native (nontrunk) VLAN, which handles all non-802.10 encapsulated FDDI traffic.

ATM LAN Emulation

ATM LAN emulation (LANE) is a standard defined by the ATM Forum that gives ATM-attached devices the same capabilities they normally have with Ethernet and Token Ring. As the name suggests, the function of the LANE protocol is to emulate a LAN on top of an ATM network. Specifically, the LANE protocol defines mechanisms for emulating either an IEEE 802.3 Ethernet or an 802.5 Token Ring LAN.

LANE defines an interface for network-layer protocols that is identical to that of existing LANs. Data sent across the ATM network is encapsulated in the appropriate LAN MAC format. In other words, LANE makes an ATM network look and behave like an Ethernet or Token Ring LAN.

An *emulated LAN (ELAN)* is a logical construct, implemented with switches, that provides Layer 2 communication between a set of hosts in a LANE network. One or more ELANs can run on the same ATM network. However, each ELAN is independent of the others, and users on separate ELANs cannot communicate directly. Just like a VLAN, communication between ELANs is possible only through a Layer 3 device. ELANs also emulate broadcast domains.

Because an ELAN provides Layer 2 communication analogous to VLANs, it can be equated to a broadcast domain. This makes it possible to map an ELAN to a VLAN on Layer 2 switches with different VLAN multiplexing technologies, such as ISL or 802.10. Layer 3-capable devices can also map IP subnets to ELANs and route between ELANs.

LANE does not attempt to emulate the access method of the specific LAN concerned (CSMA/CD for Ethernet and token passing for IEEE 802.5). LANE requires no modifications to higher-layer protocols in order for them to function over an ATM network. Because the LANE service presents the same service interface of existing MAC protocols to network-layer drivers (such as a Network Driver Interface Specification [NDIS] or Open Data-Link Interface [ODI] driver interface), no changes are required for these drivers.

The virtual LAN provided by LANE is transparent to applications. Applications can use normal LAN functions without the underlying complexities of the ATM implementation. For example, a station can send broadcasts and multicasts even though ATM does not directly support any-to-any services.

To accomplish this, special low-level software is implemented on an ATM switch or on a router or on an ATM client workstation, called the LAN Emulation Client (LEC). The client software communicates with a central control point called a LAN Emulation Server (LES). A Broadcast and Unknown Server (BUS) acts as a central point to distribute broadcasts and multicasts. The LAN Emulation Configuration Server (LECS) holds a database of LECs and the ELANs they belong to. A network administrator maintains this database. The LECS and BUS must be configured on the same device.

An ATM LANE network incorporating all the necessary components is illustrated in Figure 4-12.

Figure 4-12 ATM LANE enables the multiplexing of Ethernet VLANs over an ATM infrastructure.

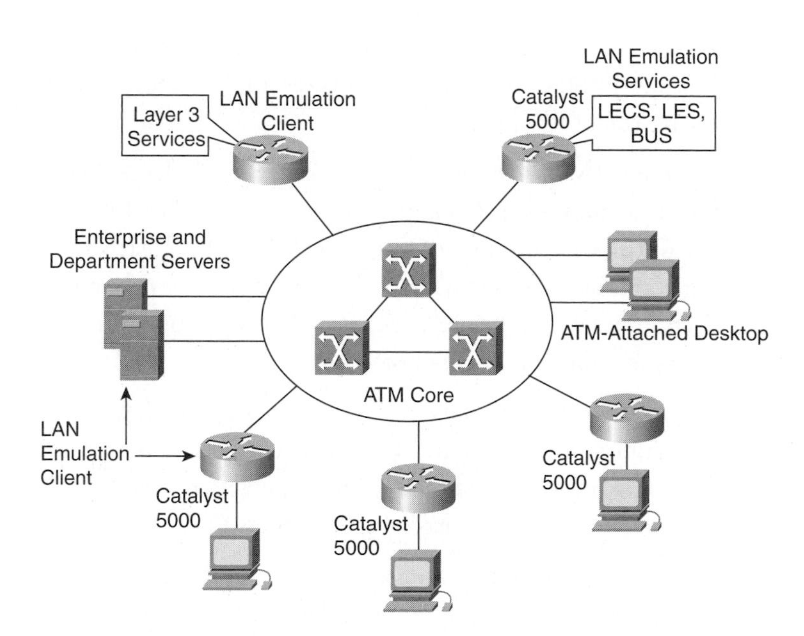

Configuring Ethernet Trunks

Configuring ISL and IEEE 802.1Q trunks varies, depending on the switch OS. You can specify whether the trunk will use ISL encapsulation or 802.1Q encapsulation or whether the encapsulation type will be autonegotiated (if both encapsulation types are supported).

For trunking to be autonegotiated, the ports must be in the same VTP domain. (VTP operation is detailed in the next section.) However, you can use on or nonegotiate mode to force a port to become a trunk, even if it is in a different domain. Trunk negotiation is managed by ***Dynamic Trunking Protocol (DTP)***. DTP is a Cisco-proprietary protocol that autonegotiates trunk formation for either ISL or 802.1Q trunks.

DTP is a strategic replacement for its precursor, Dynamic ISL (DISL). DISL's function is to negotiate between two devices whether a link should be trunking with ISL or not. DTP can negotiate the method of trunking encapsulation that will be used when any combination of ISL and 802.1Q technologies are in play. See the earlier section, "Inter-Switch Link" for a summary of Catalyst device support for ISL and 802.1Q trunking.

The following are five different states for which DTP can be configured:

- **Auto**—Makes the port willing to convert the link to a trunk link. The port becomes a trunk port if the neighboring port is set to on or desirable mode. This is the default mode for all Ethernet ports.
- **Desirable**—Makes the port actively attempt to convert the link to a trunk link. The port becomes a trunk port if the neighboring port is set to on, desirable, or auto mode.
- **On**—Puts the port into permanent trunking mode and negotiates to convert the link into a trunk link. The port becomes a trunk port even if the neighboring port does not agree to the change.
- **Nonegotiate**—Puts the port into permanent trunking mode but prevents the port from generating DTP frames. You must configure the neighboring port manually as a trunk port to establish a trunk link.
- **Off**—Puts the port into permanent nontrunking mode and negotiates to convert the link into a nontrunk link. The port becomes a nontrunk port even if the neighboring port does not agree to the change.

In addition to the configuration options for DTP, three trunk encapsulation options can be specified when you're configuring a port for a trunk link: **isl**, **dot1q**, and **negotiate**. The negotiate option is available on Catalyst switches supporting both ISL and 802.1Q. It specifies that the port should negotiate with the neighboring port to become an ISL trunk (preferred) or an 802.1Q trunk, depending on the configuration and capabilities of the neighboring port. For some switches, the DTP configuration options are available, but only one encapsulation choice is available.

When both ISL and 802.1Q encapsulation are supported on a piece of Cisco hardware, the default setting for trunk encapsulation is **negotiate.**

The trunking mode, the trunk encapsulation type, and the hardware capabilities of the two connected ports determine whether a trunk link comes up and the type of trunk the link becomes. Table 4-3 shows the result of the possible trunking configurations.

Table 4-3 Trunk Mode and Trunk Encapsulation Negotiation

Neighbor Port Trunk Mode and Trunk Encapsulation	Local Port Trunk Mode and Trunk Encapsulation								
	Off isl or dot1q	**On isl**	**Desirable isl**	**Auto isl**	**On dot1q**	**Desirable dot1q**	**Auto dot1q**	**Desirable negotiate**	**Auto negotiate**
Off isl or dot1q	Local: nontrunk Neighbor: nontrunk	Local: ISL trunk Neighbor: nontrunk	Local: nontrunk Neighbor: nontrunk	Local: nontrunk Neighbor: nontrunk	Local: 1Q trunk Neighbor: nontrunk	Local: nontrunk Neighbor: nontrunk	Local: nontrunk Neighbor: nontrunk	Local: nontrunk Neighbor: nontrunk	Local: nontrunk Neighbor: nontrunk
On isl	Local: nontrunk Neighbor: ISL trunk	Local: ISL trunk Neighbor: ISL trunk	Local: ISL trunk Neighbor: ISL trunk	Local: ISL trunk Neighbor: ISL trunk	Local: 1Q trunk Neighbor: ISL trunk	Local: nontrunk Neighbor: ISL trunk	Local: nontrunk Neighbor: ISL trunk	Local: ISL trunk Neighbor: ISL trunk	Local: ISL trunk Neighbor: ISL trunk
Desir-able isl	Local: nontrunk Neighbor: nontrunk	Local: ISL trunk Neighbor: ISL trunk	Local: ISL trunk Neighbor: ISL trunk	Local: ISL trunk Neighbor: ISL trunk	Local: 1Q trunk Neighbor: nontrunk	Local: nontrunk Neighbor: nontrunk	Local: nontrunk Neighbor: nontrunk	Local: ISL trunk Neighbor: ISL trunk	Local: ISL trunk Neighbor: ISL trunk
Auto isl	Local: nontrunk Neighbor: nontrunk	Local: ISL trunk Neighbor: ISL trunk	Local: ISL trunk Neighbor: ISL trunk	Local: nontrunk Neighbor: nontrunk	Local: 1Q trunk Neighbor: nontrunk	Local: nontrunk Neighbor: nontrunk	Local: nontrunk Neighbor: nontrunk	Local: ISL trunk Neighbor: ISL trunk	Local: nontrunk Neighbor: nontrunk
On dot1q	Local: nontrunk Neighbor: 1Q trunk	Local: ISL trunk Neighbor: 1Q trunk	Local: nontrunk Neighbor: 1Q trunk	Local: nontrunk Neighbor: 1Q trunk	Local: 1Q trunk Neighbor: 1Q trunk	Local: 1Q trunk Neighbor: 1Q trunk	Local: 1Q trunk Neighbor: 1Q trunk	Local: 1Q trunk Neighbor: 1Q trunk	Local: 1Q trunk Neighbor: 1Q trunk
Desir-able dot1q	Local: nontrunk Neighbor: nontrunk	Local: ISL trunk Neighbor: nontrunk	Local: nontrunk Neighbor: nontrunk	Local: nontrunk Neighbor: nontrunk	Local: 1Q trunk Neighbor: 1Q trunk	Local: 1Q trunk Neighbor: 1Q trunk	Local: 1Q trunk Neighbor: 1Q trunk	Local: 1Q trunk Neighbor: 1Q trunk	Local: 1Q trunk Neighbor: 1Q trunk
Auto dot1q	Local: nontrunk Neighbor: nontrunk	Local: ISL trunk Neighbor: nontrunk	Local: nontrunk Neighbor: nontrunk	Local: nontrunk Neighbor: nontrunk	Local: 1Q trunk Neighbor: 1Q trunk	Local: 1Q trunk Neighbor: 1Q trunk	Local: nontrunk Neighbor: nontrunk	Local: 1Q trunk Neighbor: 1Q trunk	Local: nontrunk Neighbor: nontrunk
Desir-able negotiate	Local: nontrunk Neighbor: nontrunk	Local: ISL trunk Neighbor: ISL trunk	Local: ISL trunk Neighbor: ISL trunk	Local: ISL trunk Neighbor: ISL trunk	Local: 1Q trunk Neighbor: 1Q trunk	Local: 1Q trunk Neighbor: 1Q trunk	Local: 1Q trunk Neighbor: 1Q trunk	Local: ISL trunk Neighbor: ISL trunk	Local: ISL trunk Neighbor: ISL trunk
Auto negotiate	Local: nontrunk Neighbor: nontrunk	Local: ISL trunk Neighbor: ISL trunk	Local: ISL trunk Neighbor: ISL trunk	Local: nontrunk Neighbor: nontrunk	Local: 1Q trunk Neighbor: 1Q trunk	Local: 1Q trunk Neighbor: 1Q trunk	Local: nontrunk Neighbor: nontrunk	Local: ISL trunk Neighbor: ISL trunk	Local: nontrunk Neighbor: nontrunk

With all the DTP and trunk encapsulation machinery behind you, you are ready to configure trunks on Catalyst switches. The default configuration for a trunk permits all VLANs to be propagated over a trunk (VLANs 1 to 1005 on non-6000s and VLANs 1025 to 4094 on 6000s).

CatOS Switch

To configure an ISL trunk port, use the **isl** option with the command **set trunk** *mod/port* {**on** | **off** | **desirable** | **auto** | **nonegotiate**} [*vlans*] [**isl** | **dot1q** | **negotiate**]. For an 802.1Q trunk, use **dot1q** in this command. For a negotiated trunk encapsulation type, use the **negotiate** option (again assuming that the hardware supports both ISL and 802.1Q).

Verify settings with the **show trunk** command.

Example 4-10 shows how to configure a port to negotiate the encapsulation type and verify the trunk configuration. This example assumes that the neighboring port is in DTP auto mode with encapsulation set to **isl** or **negotiate** (recall that ISL is preferred when trunk encapsulation is negotiated). Cross-reference Table 4-3 to verify negotiated trunking values.

Example 4-10 Configuring a Trunk on a CatOS Switch

```
Console> (enable) set trunk 4/11 desirable negotiate
Port(s) 4/11 trunk mode set to desirable.
Port(s) 4/11 trunk type set to negotiate.
Console> (enable) show trunk 4/11

Port      Mode         Encapsulation  Status       Native vlan
--------  -----------  -------------  -----------  -----------
 4/11     desirable    n-isl          trunking     1
Port      Vlans allowed on trunk
--------  ----------------------------------------------------------------------
 4/11     1-1005,1025-4094

Port      Vlans allowed and active in management domain
--------  ----------------------------------------------------------------------
 4/11     1,5,10-32,55,101-120,998-1000

Port      Vlans in spanning tree forwarding state and not pruned
--------  ----------------------------------------------------------------------
 4/11     1,5,10-32,55,101-120,998-1000
Console> (enable)
```

NOTE

When manually enabling trunking on a link to a Cisco router, use the **nonegotiate** DTP keyword to cause the port to become a trunk but not generate DTP frames.

Also, DTP is a point-to-point protocol, but some internetworking devices might forward DTP frames improperly. To avoid this problem, ensure that trunking is turned *off* on ports connected to nonswitch devices if you do not intend to trunk across those links. See Chapter 5 for restrictions on configuring trunk links pertaining to STP.

> **NOTE**
> On a Catalyst 6000 family switch, VLANs 1001 to 1024 have a special purpose. You cannot create or use VLAN 1001. It might be available in the future. VLANs 1002 to 1005 are Cisco defaults for FDDI and Token Ring; you cannot delete them. VLANs 1006 to 1009 are Cisco defaults. They are not currently used, but might be used for defaults in the future. You can map unreserved VLANs to these reserved VLANs when necessary. VLANs 1010 to 1024 are not visible and you cannot use them, but you can map nonreserved VLANs to these reserved VLANs when necessary.

Note the **n** for the negotiated ISL trunk encapsulation type in the output. On a 4000 switch, the keyword **negotiate** in Example 4-10 is not an option for many module/port combinations. In this case, the only keyword option is **dot1q**. The Supervisor III module for the 4006 switch supports ISL encapsulation and trunk encapsulation negotiation, but this module uses Cat IOS (not CatOS).

You also need to know how to restrict the set of VLANs allowed on a trunk port.

Use the command **clear trunk** *mod/port vlans* to remove VLANs from the allowed list of VLANs for a trunk. Example 4-11 shows how to define the VLAN list for trunk port 1/1 to allow VLANs 1 to 100, VLAN 250, and VLANs 500 to 1005. It also shows how to verify the allowed VLAN list for the trunk.

Example 4-11 Configuring a Restricted Set of Allowed VLANs on a Trunk

```
Console> (enable) clear trunk 1/1 101-499
Removing Vlan(s) 101-499 from allowed list.
Port 1/1 allowed vlans modified to 1-100,500-1005.
Console> (enable) set trunk 1/1 250
Adding vlans 250 to allowed list.
Port(s) 1/1 allowed vlans modified to 1-100,250,500-1005.
Console> (enable) show trunk 1/1

Port      Mode         Encapsulation  Status        Native vlan
--------  -----------  -------------  ------------  -----------
 1/1      desirable    isl            trunking      1
Port      Vlans allowed on trunk
--------  ---------------------------------------------------------------------
 1/1      1-100,250,500-1005
Port      Vlans allowed and active in management domain
--------  ---------------------------------------------------------------------
 1/1      1,521-524
Port      Vlans in spanning tree forwarding state and not pruned
--------  ---------------------------------------------------------------------
 1/1      1,521-524
Console> (enable)
```

Finally, to turn off trunking for a port, use the command **set trunk** *mod/port* **off**.

Use the command **clear trunk** *mod/port* to return the port to the default trunking type and mode for that particular port. (This command does not disable trunking for the port.)

We will look at some more trunking examples after we spend some time with Spanning Tree and port channeling. These topics are also covered in Chapter 5.

NOTE

When you first configure a port as a trunk, entering the **set trunk** command always adds all VLANs to the allowed VLAN list for the trunk, even if you specify a VLAN range. Any specified VLAN range is ignored. To modify the allowed VLANs list, use a combination of the **clear trunk** and **set trunk** commands to specify the allowed VLANs.

IOS-Based Switch

A layout involving trunking with IOS-based switches is shown in Figure 4-13.

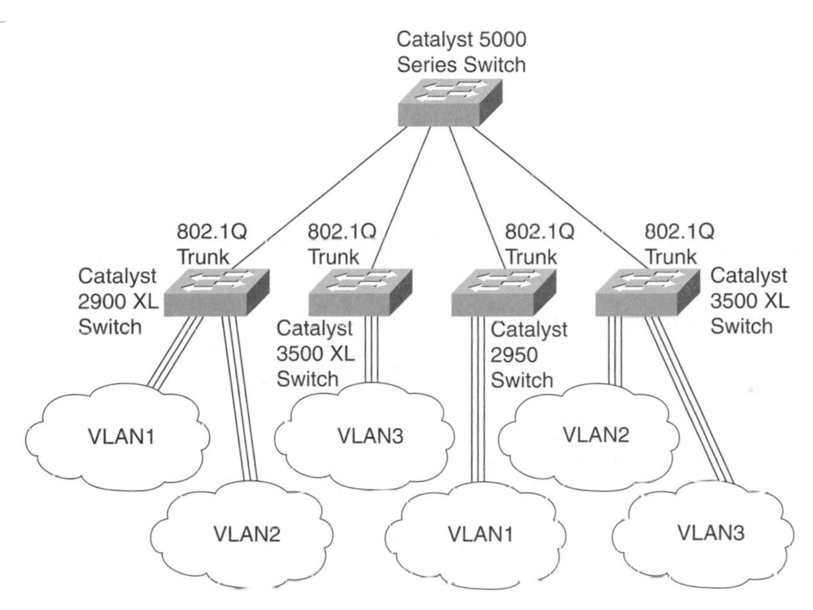

Figure 4-13
IEEE 802.1Q trunking is enabled here between a variety of IOS-based switches and a Catalyst 5000.

Some IOS-based switches, such as the 2950, support trunk negotiation via DTP. Table 4-4 summarizes the possible interface modes for a Catalyst 2950, Cisco IOS Software Release 12.1(6); these switches do not support ISL.

Table 4-4 Interface Modes on a Catalyst 2950

Mode	Description
switchport mode access	Puts the interface (access port) into permanent nontrunking mode and negotiates to convert the link to a nontrunk link. The interface becomes a nontrunk interface even if the neighboring interface is not a trunk interface.
switchport mode dynamic desirable	Makes the interface actively attempt to convert the link to a trunk link. The interface becomes a trunk interface if the neighboring interface is set to trunk, desirable, or auto mode. The default mode for all Ethernet interfaces is desirable.

continues

NOTE

Interfaces connected to devices that do not support DTP should be configured with the **access** keyword if you do not intend to trunk across those links. To enable trunking to a device that does not support DTP, use the **nonegotiate** keyword to cause the interface to become a trunk port but not generate DTP frames.

Trunking issues affected by spanning tree and port channeling are discussed in Chapter 5.

Table 4-4 Interface Modes on a Catalyst 2950 (Continued)

Mode	Description
switchport mode dynamic auto	Allows the interface to convert the link to a trunk link. The interface becomes a trunk interface if the neighboring interface is set to trunk or desirable mode.
switchport mode trunk	Puts the interface into permanent trunking mode and negotiates to convert the link into a trunk link. The interface becomes a trunk interface even if the neighboring interface is not a trunk interface.
switchport nonegotiate	Puts the interface into permanent trunking mode but prevents it from generating DTP frames. You must manually configure the neighboring interface as a trunk interface to establish a trunk link.

To configure a trunk on an IOS-based switch and verify the configuration, proceed as follows:

Step 1 Enter global configuration mode by typing **configure terminal** in privileged mode.

Step 2 Enter interface configuration mode for the interface that you want to configure as a trunk using the command **interface** *interface_id*.

Step 3 Configure the port as a VLAN trunk with the command **switchport mode trunk.**

Step 4 Configure the port to support ISL or 802.1Q encapsulation (if the switch supports both, such as the 2900XL or 3500XL). You must configure each end of the link with the same encapsulation type. Use the command **switchport trunk encapsulation** {**isl** | **dot1q**}.

Step 5 Or, on an 802.1Q-only, DTP-enabled switch, use the appropriate command from Table 4-4.

Step 6 Return to privileged mode with the **end** command or by pressing Ctrl-Z.

Step 7 Verify your configuration with the command **show interface** *interface-id* **switchport.** Be sure to include the **switchport** keyword, or your output will look like the output of a **show interface** command on a router (which isn't useful here).

Step 8 Save the configuration with the command **copy running-config startup-config** (or **wr** if you prefer shortcuts).

Putting all this together, Example 4-12 demonstrates the configuration sequence on a Catalyst 2950, with no dynamic trunk negotiation.

Example 4-12 Configuring and Verifying Trunking on an IOS-Based Switch

```
Switch#config t
Enter configuration commands, one per line.  End with CNTL-Z.
Switch(config)#interface fa0/24
Switch(config-if)#switchport mode trunk
Switch(config-if)#end
Switch#show interface fa0/24 switchport
Name: Fa0/24
Switchport: Enabled
Administrative mode: trunk
Operational Mode: trunk
Administrative Trunking Encapsulation: dot1q
Operational Trunking Encapsulation: dot1q
Negotiation of Trunking: Disabled
Access Mode VLAN: 0 ((Inactive))
Trunking Native Mode VLAN: 1 (default)
Trunking VLANs Enabled: ALL
Trunking VLANs Active: 1,20
Pruning VLANs Enabled: 2-1001

Priority for untagged frames: 0
Override vlan tag priority: FALSE
Voice VLAN: none
Appliance trust: none
Switch#
```

As with CatOS switches, by default, a trunk port carries traffic for VLANs 1 to 1005. To restrict the traffic that a trunk carries, use the **switchport trunk allowed vlan remove** *vlan-list* interface configuration command to remove specific VLANs from the allowed list.

When VTP detects a newly enabled VLAN and the VLAN is in the allowed list for a trunk port, the trunk port automatically begins to forward traffic sourced from or destined for this VLAN. When VTP detects a new VLAN and the VLAN is not in the allowed list for a trunk port, the trunk port does not become a member of the new VLAN.

To verify the allowed list of VLANs for a trunk port, use the command **show interface** *interface-id* **switchport allowed-vlan**.

Example 4-13 illustrates how to remove VLAN 2 from the allowed list of VLANs on a Catalyst 3550 trunk port and verify the configuration.

Example 4-13 Restricting VLANs on an IOS-Based Switch Trunk Port

```
Switch(config)#interface gigabitethernet0/1
Switch(config-if)#switchport trunk allowed vlan remove 2
Switch(config-if)#end
Switch#show interface gigabitethernet0/1 switchport
Name: Gi0/1
Switchport: Enabled
Administrative Mode: dynamic desirable
Operational Mode: static access
Administrative Trunking Encapsulation: negotiate
Operational Trunking Encapsulation: native
Negotiation of Trunking: On
Access Mode VLAN: 1 (default)
Trunking Native Mode VLAN: 1 (default)
Trunking VLANs Enabled: 1,3-1005
Pruning VLANs Enabled: 2-1001

Protected: false
Unknown unicast blocked: false
Unknown multicast blocked: false

Broadcast Suppression Level: 100
Multicast Suppression Level: 100
Unicast Suppression Level: 100
```

Finally, to disable trunking on an IOS-based switch port, simply return it to its default static-access mode by using either the **no switchport mode** or **switchport mode access** command. Example 4-14 illustrates changing a trunk port to an access port.

Example 4-14 Changing a Trunk Port to an Access Port

```
Switch#show interface fa0/24 switchport
Name: Fa0/24
Switchport: Enabled
Administrative mode: trunk
```

Example 4-14 Changing a Trunk Port to an Access Port (Continued)

```
Operational Mode: trunk
<output omitted>
Switch#configure terminal
Enter configuration commands, one per line.  End with CNTL/Z.
Switch(config)#interface fa0/24
Switch(config-if)#switchport mode access
Switch(config-if)#end
Switch#show interface fa0/24 switchport
Name: Fa0/24
Switchport: Enabled
Administrative mode: static access
Operational Mode: static access
<output omitted>
```

TECH NOTE: NATIVE VLANS ON 802.1Q TRUNKS

A trunk port configured with 802.1Q tagging can receive both tagged and untagged traffic. By default, the switch forwards untagged traffic with the native VLAN configured for the port. The native VLAN is VLAN 1 by default.

The native VLAN can be assigned any VLAN ID. It is not dependent on the management VLAN.

To configure the native VLAN for an 802.1Q trunk on an IOS-based switch, enter the interface configuration command **switchport trunk native vlan** *vlan-id*, where *vlan-id* is between 1 and 1001. As before, verify your settings with the command **show interface** *interface-id* **switchport**.

If a frame has a VLAN ID that is the same as the outgoing port native VLAN ID, the packet is transmitted untagged; otherwise, the switch transmits the packet with a tag.

On a CatOS switch, the command **set vlan** assigns a port to a specific VLAN. In the case of a trunking port, it can be used to change the native VLAN. For example, **set vlan 2 5/24** changes the native VLAN of trunk port **5/24** to VLAN 2.

The native VLAN configured on each end of an 802.1Q trunk must be the same. Remember that a switch receiving a nontagged frame assigns it to the trunk's native VLAN. A Cisco device warns you if there is a native VLAN mismatch. Here is such a warning on a Catalyst 2950:

```
1w4d: %CDP-4-NATIVE_VLAN_MISMATCH: Native VLAN mismatch discovered on FastEthernet0/
  24 (4), with JAB04210C5T 2/27 (1).
```

Here is the corresponding warning on a Catalyst 4006:

```
2002 Jan 19 16:09:03 %CDP-4-NVLANMISMATCH:Native vlan mismatch detected on port
  2/27.
```

VLAN Trunking Protocol

VLAN Trunking Protocol (VTP) is a Cisco-proprietary protocol available on most Catalyst switches that reduces VLAN-related administration. When you configure a new VLAN on a VTP server, the VLAN is distributed to all switches in the VTP domain. In general, VTP is responsible for synchronizing VLAN information within a VTP domain. This reduces the need to configure the same VLAN information on each switch.

VTP also provides a mapping scheme to facilitate traffic flow across mixed-media backbones, allowing for the mapping of Ethernet VLANs to ATM LANE ELANs or FDDI 802.10 VLANs. This VTP mapping scheme enables seamless trunking within a network employing mixed-media technologies.

Before you get too excited about the wonders of VTP, it does have some disadvantages. These disadvantages normally relate to Spanning-Tree Protocol (STP)—the greatest risk being an STP loop's propagating throughout the entire campus network. The STP implications of using VTP are considered in Chapter 5. For now, suffice it to say that you have to balance the ease of administration made possible by VTP with the risk of a large, potentially unstable STP domain.

Basic Mechanics of VTP

VTP is a Layer 2 messaging protocol that maintains VLAN configuration consistency by managing the addition, deletion, and renaming of VLANs on a network-wide basis. VTP minimizes misconfigurations that can result in a number of problems, such as duplicate VLAN names, incorrect VLAN-type specifications, and security violations.

You can use VTP to manage VLANs 1 to 1005 in your network (VTP does not support VLANs 1025 to 4094 on a Catalyst 6000). With VTP, you can make configuration changes centrally on one switch and have those changes automatically communicated to all other switches in the network (assuming that they are in the same VTP domain).

Keep these important facts in mind regarding VTP:

- VTP operates through multicast VTP messages sent to a particular MAC address: 01-00-0C-CC-CC-CC.
- VTP advertisements travel only through trunk ports.
- VTP messages are carried only through VLAN 1. This is why VLAN 1 cannot be removed from a trunk. However, there is a way to limit the extent of VLAN 1. This feature is called "VLAN 1 disable on trunk." It has been available on Catalyst 4000, 5000, and 6000 family switches since Cisco IOS Software Release 5.4. It allows you to prune VLAN 1 from a trunk as you would do for any other VLAN, but this pruning does not include all the control protocol traffic that will still be allowed on the trunk (DTP, PAgP, CDP, VTP, and so on).

However, you will block all user traffic on that trunk. Using this feature, you can completely avoid the VLAN spanning the entire campus. As such, STP loops are limited in extent for VLAN 1. For example, if port 2/1 is a trunk, you perform "VLAN 1 disable on trunk" with the command **clear trunk 2/1 1**.

- VTP information flows through an 802.1Q trunk only when the trunk is up, after DTP negotiation. Recall that DTP automatically negotiates trunk status for a port (not to be confused with trunk encapsulation negotiation).

Each switch in the VTP domain sends periodic advertisements out each trunk port to the reserved VTP multicast address. VTP advertisements are received by neighboring switches, which update their VTP and VLAN configurations as necessary. VTP messages are encapsulated in either ISL frames or IEEE 802.1Q frames.

Next, we discuss the role of VTP domains.

VTP Domains

A *VTP domain*, also called a VLAN management domain, is made up of one or more interconnected switches that share the same VTP domain name. A VTP domain for a network is the set of all contiguously trunked switches with the same VTP domain name. A switch can be configured to be in one and only one VTP domain. You can make global VLAN configuration changes for the domain via the Catalyst switch CLI or with the appropriate network management application, such as CiscoWorks2000 LMS, employing Simple Network Management Protocol (SNMP).

By default, a Catalyst is in VTP server mode and is in the "no management domain" state until the switch receives an advertisement for a domain over a trunk link or you configure a VLAN management domain. You cannot create or modify VLANs on a VTP server until the management domain name is specified or learned.

A critical parameter governing VTP function is the VTP configuration revision number. This 32-bit number indicates the particular revision of a VTP configuration. A configuration revision number starts at 0 and increments by 1 with each modification until it reaches the value 4294947295, at which point it cycles back to 0 and starts incrementing again. Each VTP device tracks its own VTP configuration revision number; VTP packets contain the sender's VTP configuration revision number. This information determines whether the received information is more recent than the current version. In order to reset a switch's configuration revision to 0, simply disable trunking, change the VTP domain name, change it back to the original name, and reenable trunking.

If the switch receives a VTP advertisement over a trunk link, it inherits the management domain name and the VTP configuration revision number. The switch ignores advertisements that have a different management domain name or an earlier configuration revision number.

If you configure the switch as VTP transparent, you can create and modify VLANs, but the changes affect only the individual switch (changes are not propagated to other switches).

When you make a change to the VLAN configuration on a VTP server, the change is propagated to all switches in the VTP domain. VTP advertisements are transmitted out all trunk connections, including ISL, IEEE 802.1Q, IEEE 802.10, and ATM LANE.

VTP maps VLANs dynamically across multiple LAN types with unique names and internal index associations. Mapping eliminates excessive device administration required from network administrators.

In order to use VTP, you must assign a VTP domain name to each switch. VTP information remains in the VLAN management domain. The following are conditions for a VTP domain:

- Each Catalyst switch in a domain must have the same VTP domain name, whether learned or configured.
- The Catalyst switches must be adjacent, meaning that all switches in the VTP domain form a contiguous tree in which every switch is connected to every other switch in the domain via the tree.
- Trunking must be enabled between all Catalyst switches.

If any one of the previous conditions is not met, the VTP domain becomes disconnected, and information will not travel between the separate parts.

Next, the four VTP modes are discussed: server, client, transparent, and off.

VTP Modes

You can configure a switch to operate in any one of these VTP modes:

- **Server**—The VTP server maintains a full list of all VLANs within the VTP domain. In VTP server mode, you can create, modify, and delete VLANs and specify other configuration parameters (such as VTP version and VTP pruning) for the entire VTP domain. VTP servers advertise their VLAN configuration to other switches in the same VTP domain and synchronize their VLAN configuration with other switches based on advertisements received over trunk links. VTP server is the default mode for all Catalyst switches. VTP information is stored in nonvolatile random-access memory (NVRAM).
- **Client**—The VTP client also maintains a full list of all VLANs within the VTP domain. However, it does not store the information in NVRAM. VTP clients behave the same way as VTP servers, but you cannot create, change, or delete VLANs on a VTP client. Any changes made must be received from a VTP server advertisement.

- **Transparent**—Switches in VTP transparent mode do not participate in VTP. A VTP transparent switch does not advertise its VLAN configuration and does not synchronize its VLAN configuration based on received advertisements. However, VTP transparent switches do forward VTP advertisements that they receive out their trunk ports. VLANs can be configured on a switch in VTP transparent mode, but the information is local to the switch and is stored in NVRAM.
- **Off**—In VTP off mode, switches behave the same as in VTP transparent mode, except that VTP advertisements are not forwarded.

Table 4-5 summarizes the differences between the VTP modes.

Table 4-5 VTP Modes Determine How VTP Information Is Handled

Feature	Server?	Client?	Transparent?
Source VTP messages	Yes	Yes	No
Listens to VTP messages	Yes	Yes	No
Creates VLANs	Yes	No	Yes (locally significant only)
Remembers VLANs	Yes	No	Yes (locally significant only)

In Table 4-5, "Source VTP messages" refers to the sending of VTP messages to all trunks. "Listens to VTP messages" refers to listening for the multicast MAC address 01-00-0C-CC-CC-CC and processing the VTP update.

There is no difference in how the server and client source listen to VTP messages. The only differences between the server and the client are that VLANs cannot be configured directly on a client, and the client does not remember VLAN information after rebooting (no VLAN information is written to NVRAM).

Transparent mode indicates that a switch does not participate in VTP. Therefore, the switch ignores all received VTP messages. However, the switch does forward messages to all other outgoing trunks.

VTP advertisements communicate VTP information between members of a VTP domain. This is the next topic that's explored.

VTP Advertisements

Periodic VTP advertisements are sent out each trunk port with the multicast destination MAC address 01-00-0C-CC-CC-CC. VTP advertisements contain the following configuration information:

- VLAN IDs (ISL and 802.1Q)
- Emulated LAN names (ATM LANE)

- 802.10 SAID values (FDDI)
- VTP domain name
- VTP configuration revision number
- VLAN configuration, including the maximum transmission unit (MTU) size for each VLAN
- Frame format

VTP messages are encapsulated in either ISL frames or IEEE 802.1Q frames.

VTP messages are sent with the following Ethernet frame field values:

- Multicast destination MAC address 01-00-0C-CC-CC-CC
- Destination Service Access Point (DSAP) 0xAA in the Logical Link Control (LLC) header
- Source Service Access Point (SSAP) 0xAA in the LLC header
- Organizational Unique Identifier (OUI) of 00-00-0C (for Cisco) in the Subnetwork Access Protocol (SNAP) header
- Ethertype of 2003 in the SNAP header

Figure 4-14 shows a VTP packet encapsulated in an ISL frame.

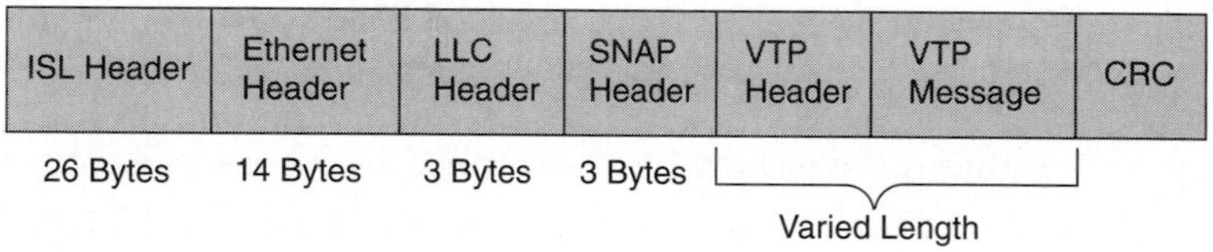

Figure 4-14 ISL and IEEE 802.1Q protocols encapsulate VTP messages. ISL encapsulation is pictured.

ISL Header	Ethernet Header	LLC Header	SNAP Header	VTP Header	VTP Message	CRC
26 Bytes	14 Bytes	3 Bytes	3 Bytes	Varied Length		

Now you can look inside a VTP packet. The VTP header's format varies, depending on the type of VTP message. However, all packets contain the following fields in the header:

- VTP protocol version: 1 or 2
- VTP message type
- Management domain length
- Management domain name

The following VTP message types are used by VTP:

- Summary advertisements
- Subset advertisements

- Advertisement requests
- VTP join messages

Next, we detail the function of each of the first three VTP message types: summary advertisements, subset advertisements, and advertisement requests. VTP join messages, which rely on the VTP Pruning Protocol, are explored in the section, "VTP Pruning."

Summary Advertisements

Summary advertisements inform adjacent Catalyst switches of the current VTP domain name and the configuration revision number. By default, Catalyst switches issue summary advertisements in five-minute intervals.

When the switch receives a summary advertisement packet, it compares the VTP domain name. If the name is different, the switch simply ignores the packet. If the name is the same, the switch then compares the configuration revision. If its own configuration revision is higher or equal, the packet is ignored. If it is lower, an advertisement request is sent.

TECH NOTE: INADVERTENT VLAN DELETION

A problem that occurs all too often is the introduction of a Catalyst switch into a VTP domain that has a higher VTP configuration revision number than the other Catalysts in the VTP domain. What happens is exactly what is supposed to happen with VTP: The other VTP clients and servers update their VLAN databases to reflect that of the introduced switch. The problem is that the introduced switch likely will not have the same VLANs configured as the other switches in the network. The result is that you bring down the network for all hosts residing in VLANs not configured on the introduced Catalyst. All the switch access ports associated with VLANs not configured on the introduced Catalyst are immediately transitioned to inactive status. This happens whether the introduced switch is a VTP client or a VTP server. A VTP client can force a VTP server to erase VLAN information.

You know that this problem has occurred on your network when many of the ports in your network suddenly go into inactive state. The solution to this problem is to quickly reconfigure all the VLANs on one of the VTP servers.

The lesson to be learned here is always make sure that the configuration revision number of a switch inserted into the VTP domain is lower than the configuration revision number of the switches already in the VTP domain.

To reset the configuration revision number to 0 on the switch you are introducing into the VTP domain, change the switch's VTP domain name to some arbitrary name, and then change it back to the desired name. You might have to delete the vlan.dat file on an IOS-based switch, followed by a reboot, to reset the revision number to 0.

Figure 4-15 illustrates a VTP summary advertisement.

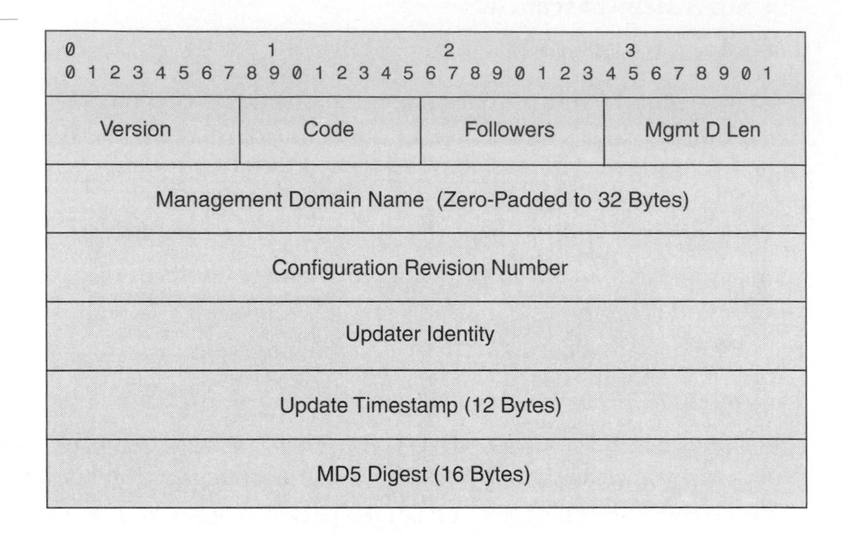

Figure 4-15 VTP summary advertisement message format.

The following list describes each of the fields in the summary advertisement packet:

- The VTP version is either 1 or 2.
- Code indicates which of the four VTP message types is included—here, 0x01 or type 1.
- Followers indicates how many VTP subset advertisement messages (type 2) follow the summary advertisement frame. The value can range from 0 to 255; 0 indicates that no subset advertisements follow. A Catalyst transmits the subset advertisement only if there is a change in the system or as a response to an advertisement request.
- MgmtD Len specifies the length of the VTP domain name.
- Management Domain Name specifies the VTP domain name.
- The Configuration Revision Number field is 32 bytes.
- The Updater Identity is the IP address of the last switch that incremented the configuration revision.
- Update Timestamps are the date and time of the last increment of the configuration revision.
- MD5 Digest consists of a message-digest hash—a function of the VTP password and the VTP header contents (excluding the MD5 Digest field). If the receiving Catalyst hash computation does not match, the packet is discarded. See the Tech Note in the section "VTP Passwords."

Subset Advertisements

When you add, delete, or change a VLAN on a VTP server, the configuration revision number is incremented, and a summary advertisement is issued, followed by one or several subset advertisements. Subset advertisements are also triggered by suspending or activating a VLAN, changing its name, or changing its MTU.

A subset advertisement contains a list of VLANs and corresponding VLAN information. If there are several VLANs, more than one subset advertisement might be required in order to advertise all the information.

Figure 4-16 illustrates a VTP subset advertisement and the contents of each VLAN information field (ordered with lower-valued VLAN IDs occurring first).

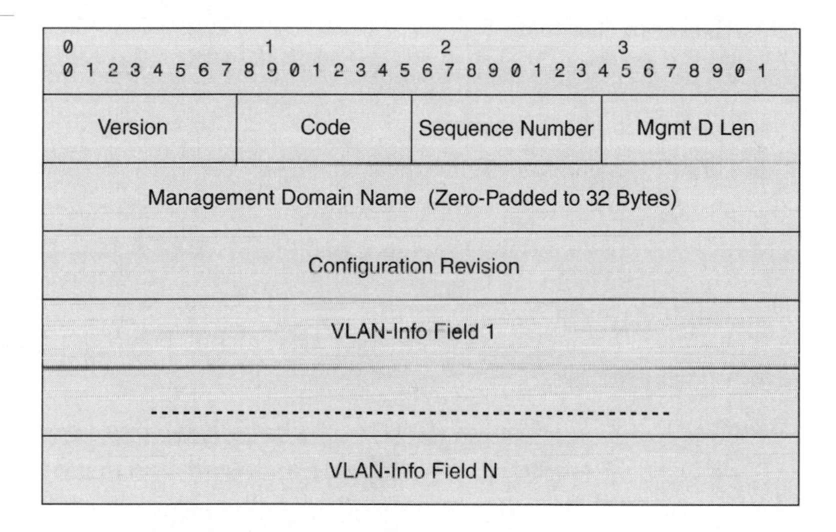

Figure 4-16 VTP subset advertisement message format.

Most of the fields in this packet were described previously (refer to the summary advertisement field descriptions). Here are clarifications for the remaining descriptors:

- Code is 0x02 for subset advertisement (type 2).
- Seq-Number represents the sequence number of the packet in the stream of subset advertisements following a summary advertisement. The sequence starts with 1. The receiving Catalyst uses this value to ensure that it receives all subset advertisements. If it doesn't, it requests a resend, starting with a specific subset advertisement.

- VLAN-Info fields each contain the following information:
  — The VLAN's status (active or suspended)
  — VLAN-Type (Ethernet, Token Ring, FDDI, or otherwise)
  — VLAN-Name Len—Length of the VLAN name
  — ISL VLAN-ID—VLAN number of this named VLAN
  — MTU Size—Maximum frame size supported for this VLAN
  — 802.10 Index—SAID value used if the frame passed over a FDDI trunk
  — VLAN-name

The VTP subset advertisement lists this information for each individual VLAN, including default VLANs.

Advertisement Requests

A switch issues a VTP advertisement request in the following situations:

- The switch has been reset.
- The VTP domain name has been changed.
- The switch has received a VTP summary advertisement with a higher configuration revision number than its own.

Upon receipt of an advertisement request, a VTP device sends a summary advertisement, followed by one or more subset advertisements. Figure 4-17 illustrates a VTP advertisement request.

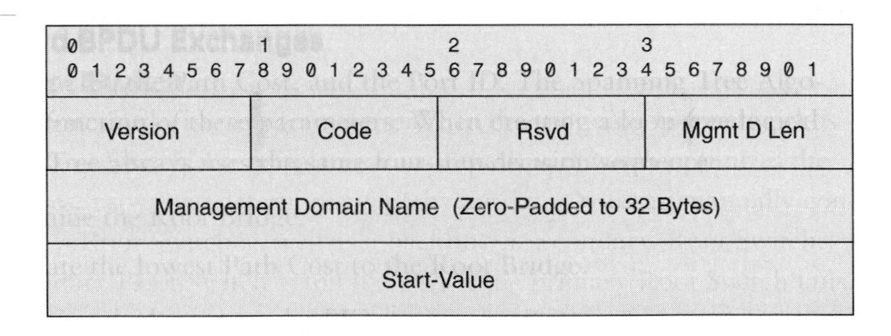

Figure 4-17 VTP advertisement request format.

The fields in the advertisement request are described next. Fields that were already described for summary and subset advertisements are not included:

- **Code**—0x03 for advertisement request (type 3)
- **Rsvd**—Reserved. Always set to 0.

- **Start-Value**—Used in cases in which there are several subset advertisements. If subset advertisement *n* is the first advertisement that has not been received in a sequence of advertisements, an advertisement request is sent for advertisements starting with *n*. If the start value is 0, all subset advertisements are sent for the particular management domain.

Configuring a VTP Server

When a switch is in VTP server mode, you can change the VLAN configuration, and the VLAN information propagates throughout the management domain. VTP server is the default VTP mode for Catalyst switches.

Make sure you configure at least one switch as the VTP server in the VTP domain. If you configure all the switches in a VTP domain as clients, it is impossible to make changes to that domain's VLAN configuration.

CatOS Switch

To configure the switch as a VTP server, enter the commands **set vtp domain** *name* and **set vtp mode server**. To verify the configuration, use the command **show vtp domain**.

Example 4-15 shows how to configure a switch as a VTP server and verify the configuration.

Example 4-15 Configuring a CatOS Switch as a VTP Server

```
Console> (enable) set vtp domain kauai
VTP domain kauai modified
Console> (enable) set vtp mode server
VTP domain kauai modified
Console> (enable) show vtp domain

Domain Name                      Domain Index VTP Version Local Mode  Password
-------------------------------- ------------ ----------- ----------- ----------
kauai                            1            2           server      -
Vlan-count Max-vlan-storage Config Revision Notifications
---------- ---------------- --------------- -------------
10         1023             40              enabled
Last Updater    V2 Mode  Pruning  PruneEligible on Vlans
--------------- -------- -------- -------------------------
172.20.52.70    disabled disabled 2-1000
Console> (enable)
```

IOS-Based Switch

To configure an IOS-based switch as a VTP server, go to VLAN configuration mode and enter the commands **vtp domain** *domain-name* and **vtp server**. The VTP domain name can be from 1 to 32 characters long. To verify the configuration, use the command **show vtp status**. Example 4-16 illustrates these steps.

Example 4-16 Configuring an IOS-Based Switch as a VTP Server

```
Switch(vlan)#vtp domain oahu
Changing VTP domain name from maui to oahu
Switch(vlan)#vtp server
Setting device to VTP SERVER mode.
Switch(vlan)#exit
APPLY completed.
Exiting....
Switch#show vtp status
VTP Version                     : 2
Configuration Revision          : 0
Maximum VLANs supported locally : 68
Number of existing VLANs        : 8
VTP Operating Mode              : Server
VTP Domain Name                 : oahu
VTP Pruning Mode                : Disabled
VTP V2 Mode                     : Enabled
VTP Traps Generation            : Disabled
MD5 digest                      : 0x06 0xAE 0x15 0x28 0xB7 0x68 0xA8 0xB5
Configuration last modified by 156.20.10.4 at 1-20-02 14:53:34
```

Configuring a VTP Client

When a switch is in VTP client mode, you cannot change its VLAN configuration. The client switch receives VTP updates from a VTP server in the management domain and modifies its configuration accordingly.

We now detail the configuration of VTP clients on CatOS and IOS-based switches.

CatOS Switch

To configure a CatOS switch as a VTP client, simply enter the commands **set vtp domain** *name* and **set vtp mode client**. Verify the configuration with the **show vtp domain** command, as shown in Example 4-17.

Example 4-17 Configuring a CatOS Switch as a VTP Client

```
Console> (enable) set vtp domain kauai
VTP domain kauai modified
Console> (enable) set vtp mode client
VTP domain kauai modified
Console> (enable) show vtp domain

Domain Name                      Domain Index VTP Version Local Mode  Password
-------------------------------- ------------ ----------- ----------- ----------
kauai                            1            2           client      -
Vlan-count Max-vlan-storage Config Revision Notifications
---------- ---------------- --------------- -------------
10         1023             40              enabled
Last Updater    V2 Mode  Pruning  PruneEligible on Vlans
--------------- -------- -------- -------------------------
172.20.52.70    disabled disabled 2-1000
Console> (enable)
```

IOS-Based Switch

To configure an IOS-based switch as a VTP client, go to VLAN configuration mode and enter the commands **vtp domain** *domain-name* and **vtp client**. Verify the configuration with the command **show vtp status**. Example 4-18 illustrates these steps.

Example 4-18 Configuring an IOS-Based Switch as a VTP Client

```
Switch(vlan)#vtp domain maui
Changing VTP domain name from oahu to maui
Switch(vlan)#vtp client
Setting device to VTP CLIENT mode.
Switch(vlan)#exit
In CLIENT state, no apply attempted.
Exiting....
Switch#show vtp status
VTP Version                     : 2
Configuration Revision          : 0
Maximum VLANs supported locally : 68
Number of existing VLANs        : 8
```

continues

Example 4-18 Configuring an IOS-Based Switch as a VTP Client (Continued)

```
VTP Operating Mode              : Client
VTP Domain Name                 : maui
VTP Pruning Mode                : Disabled
VTP V2 Mode                     : Enabled
VTP Traps Generation            : Disabled
MD5 digest                      : 0xE7 0x3F 0xB9 0x38 0xB8 0x67 0x37 0xAF
Configuration last modified by 156.20.10.4 at 1-20-02 14:53:34
Switch#
```

Configuring VTP Transparent Mode

When you configure the switch in VTP transparent mode, you disable VTP on the switch. A VTP transparent switch does not send VTP updates and does not act on VTP updates received from other switches. However, a VTP transparent switch does forward received VTP advertisements out all its trunk links.

Next, we describe the configuration of VTP transparent mode on Catalyst switches.

CatOS Switch

Use the command **set vtp mode transparent** to configure VTP transparent mode on a CatOS switch, as shown in Example 4-19.

Example 4-19 Configuring a CatOS Switch for VTP Transparent Mode

```
Console> (enable) set vtp mode transparent
VTP domain kauai modified
Console> (enable) show vtp domain

Domain Name                      Domain Index VTP Version Local Mode  Password
-------------------------------- ------------ ----------- ----------- ----------
kauai                            1            2           Transparent -
Vlan-count Max-vlan-storage Config Revision Notifications
---------- ---------------- --------------- -------------
10         1023             0               enabled
Last Updater    V2 Mode  Pruning  PruneEligible on Vlans
--------------- -------- -------- -------------------------
172.20.52.70    disabled disabled 2-1000
Console> (enable)
```

You can also disable VTP using the command **set vtp mode off**. When you disable VTP using off mode, the switch behaves the same as in VTP transparent mode, except that VTP advertisements are not forwarded. See Example 4-20.

Example 4-20 Configuring a CatOS Switch in VTP Off Mode

```
Console> (enable) set vtp mode off
VTP domain kauai modified
Console> (enable) show vtp domain

Domain Name                      Domain Index VTP Version Local Mode  Password
-------------------------------- ------------ ----------- ----------- ----------
kauai                            1            2           off         -
Vlan-count Max-vlan-storage Config Revision Notifications
---------- ---------------- --------------- -------------
10         1023             0               enabled
Last Updater    V2 Mode  Pruning  PruneEligible on Vlans
--------------- -------- -------- -------------------------
172.20.52.70    disabled disabled 2-1000
Console> (enable)
```

IOS-Based Switch

To configure an IOS-based switch in VTP transparent mode, enter the command **vtp transparent** in VLAN configuration mode, as shown in Example 4-21. Unlike CatOS switches, when you configure an IOS-based switch for VTP transparent mode, you disable VTP on the switch. The switch then does not send VTP updates and does not act on VTP updates received from other switches. However, a VTP transparent switch does forward received VTP advertisements on all its trunk links.

Example 4-21 Configuring an IOS-Based Switch in VTP Transparent Mode

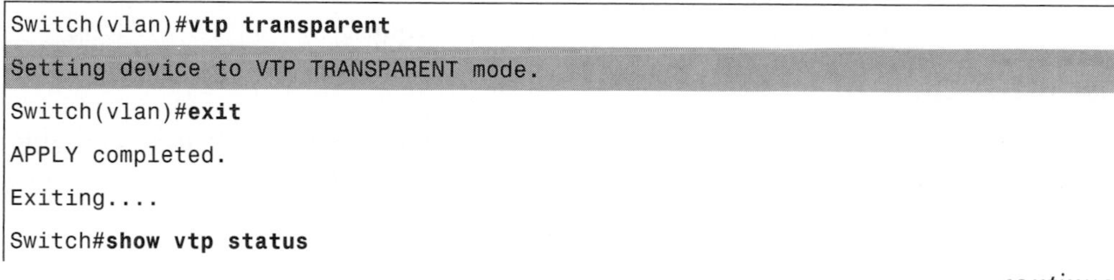

```
Switch(vlan)#vtp transparent
Setting device to VTP TRANSPARENT mode.
Switch(vlan)#exit
APPLY completed.
Exiting....
Switch#show vtp status
```

continues

Example 4-21 Configuring an IOS-Based Switch in VTP Transparent Mode (Continued)

```
VTP Version                     : 2
Configuration Revision          : 0
Maximum VLANs supported locally : 68
Number of existing VLANs        : 8
VTP Operating Mode              : Transparent
VTP Domain Name                 : maui
VTP Pruning Mode                : Disabled
VTP V2 Mode                     : Enabled
VTP Traps Generation            : Disabled
MD5 digest                      : 0xE7 0x3F 0xB9 0x38 0xB8 0x67 0x37 0xAF
Configuration last modified by 156.20.10.4 at 1-20-02 14:53:34
```

This completes our look at VTP mode configuration. Next we consider VTP version configuration.

VTP Version 2

If you use VTP in your network, you must decide whether to use VTP version 1 or version 2. If you are using VTP in a Token Ring environment, you must use version 2.

VTP version 2 is not much different from VTP version 1. The major difference is that version 2 introduces support for Token Ring VLANs. Again, if you are using Token Ring VLANs, you need to enable version 2; otherwise, there is no compelling reason to use it.

If all switches in a VTP domain can run VTP version 2, you need to enable VTP version 2 on only one switch. The version number is propagated to the other VTP version 2-capable switches in the VTP domain. VTP versions 1 and 2 are not interoperable on switches in the same VTP domain. Every switch in the VTP domain must use the same VTP version. You should not enable VTP version 2 unless every switch in the VTP domain supports version 2.

The following sections describe VTP version configuration on Catalyst switches.

CatOS Switch

To enable version 2 on a CatOS switch, enter the command **set vtp v2 enable**. Disable version 2 with the command **set vtp v2 disable**. Verify the VTP version with the command **show vtp domain**. Example 4-22 illustrates this.

Example 4-22 Enabling and Disabling VTP Version 2

```
Console> (enable) show vtp domain
Domain Name                      Domain Index VTP Version Local Mode  Password
-------------------------------- ------------ ----------- ----------- ----------
cisco                            1            2           server      -

Vlan-count Max-vlan-storage Config Revision Notifications
---------- ---------------- --------------- -------------
8          1023             3               disabled
Last Updater    V2 Mode  Pruning  PruneEligible on Vlans
--------------- -------- -------- -------------------------
0.0.0.0         disabled disabled 2-1000
Console> (enable) set vtp v2 enable
This command will enable the version 2 function in the entire management domain.
All devices in the management domain should be version2-capable before enabling.
Do you want to continue (y/n) [n]? y
VTP domain cisco modified
Console> (enable) show vtp domain
Domain Name                      Domain Index VTP Version Local Mode  Password
-------------------------------- ------------ ----------- ----------- ----------
cisco                            1            2           server      -

Vlan-count Max-vlan-storage Config Revision Notifications
---------- ---------------- --------------- -------------
8          1023             4               disabled
Last Updater    V2 Mode  Pruning  PruneEligible on Vlans
--------------- -------- -------- -------------------------
156.20.10.4     enabled  disabled 2-1000
Console> (enable) set vtp v2 disable
This command will disable the version 2 function in the entire management domain.
Warning: trbrf & trcrf vlans will not work properly in this mode.
Do you want to continue (y/n) [n]? y
VTP domain cisco modified
Console> (enable) show vtp domain
```

continues

Example 4-22 Enabling and Disabling VTP Version 2 (Continued)

```
Domain Name                      Domain Index VTP Version Local Mode  Password
-------------------------------- ------------ ----------- ----------- ----------
cisco                            1            2           server      -

Vlan-count Max-vlan-storage Config Revision Notifications
---------- ---------------- --------------- -------------
8          1023             5               disabled
Last Updater   V2 Mode  Pruning  PruneEligible on Vlans
-------------- -------- -------- -------------------------
156.20.10.4    disabled disabled 2-1000
```

The number under **VTP Version** reads 2 both before and after you change the VTP version. This output just means that the switch is version 2-capable. The only way to tell whether version 2 is in effect is to check for **enabled** or **disabled** under **V2 Mode** in the output.

IOS-Based Switch

On an IOS-based switch in VTP server mode, enable VTP version 2 using the VLAN configuration mode command **vtp v2-mode** (or disable it with the **no** form of this command). Verify the configuration with the command **show vtp status**.

Example 4-23 demonstrates this process. Again, the **VTP Version** output remains **2** throughout (meaning that the switch is version 2-capable). You need to look for **Enabled** or **Disabled** next to **VTP V2 Mode**.

Example 4-23 Changing the VTP Version on an IOS-Based Switch

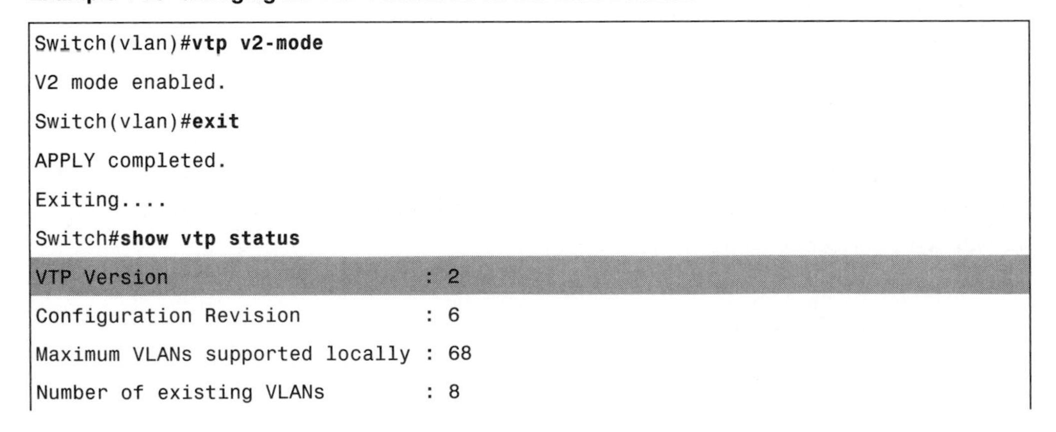

```
Switch(vlan)#vtp v2-mode
V2 mode enabled.
Switch(vlan)#exit
APPLY completed.
Exiting....
Switch#show vtp status
VTP Version                     : 2
Configuration Revision          : 6
Maximum VLANs supported locally : 68
Number of existing VLANs        : 8
```

Example 4-23 Changing the VTP Version on an IOS-Based Switch (Continued)

```
VTP Operating Mode              : Server
VTP Domain Name                 : cisco
VTP Pruning Mode                : Disabled
VTP V2 Mode                     : Enabled
VTP Traps Generation            : Disabled
MD5 digest                      : 0x9B 0x20 0x3A 0xE6 0xD9 0x26 0x64 0x29
Configuration last modified by 156.20.10.4 at 1-20-02 14:19:38
Switch#vlan database
Switch(vlan)#no vtp v2-mode
V2 mode disabled.
Switch(vlan)#exit
APPLY completed.
Exiting....
Switch#show vtp status
VTP Version                     : 2
Configuration Revision          : 7
Maximum VLANs supported locally : 68
Number of existing VLANs        : 8
VTP Operating Mode              : Server
VTP Domain Name                 : cisco
VTP Pruning Mode                : Disabled
VTP V2 Mode                     : Disabled
VTP Traps Generation            : Disabled
MD5 digest                      : 0xC8 0x9B 0x34 0x92 0x23 0x7C 0xD7 0x60
Configuration last modified by 0.0.0.0 at 3-13-93 20:54:56
```

Note the two lines of output beginning with **Configuration last modified.** The first instance reflects a consistent VTP version 2 setting between two switches trunking via 802.1Q; this is apparent due to the presence of the neighbor's management IP address and the time/date stamp. The second instance reflects an inconsistent setting relative to the connected switch: No IP address appears to be set for the neighbor's management VLAN, and the date is set back to the default of 1993!

NOTE

VTP server mode is required to change the VTP version!

Next, we show how to configure VTP passwords.

VTP Passwords

If you configure a password for VTP, it needs to be configured on all switches in the VTP domain, and it needs to be the same password. The VTP password you configure is translated using an algorithm resulting in a 16-byte MD5 value carried in all summary advertisements.

TECH NOTE: MD5

MD5 is a one-way hash function created in 1991 by Professor Ronald Rivest. RFC 1321 details the MD5 algorithm.

A hash is a number generated from a string of text. The hash is significantly smaller than the text itself. It is generated by a formula in such a way that it is extremely unlikely that some other text will produce the same hash value.

MD5's one-way hash function takes a chunk of data and converts it into a 128-bit string of digits called a hash, or message digest. The message digest is a representation of text in the form of a single string of digits.

The sender uses the one-way hash function to generate a 128-bit message digest from a string of data and the sender's password. This message digest is sent along with the string of data to the receiver. The receiver recomputes the hash code from the same string of data and the receiver's password. If the sender's message digest does not match the receiver's message digest, it is safe to assume that the passwords don't match and that the data is compromised.

On a CatOS switch, you set the VTP password with the command **set vtp passwd** *passwd*. If the VTP password has already been defined, entering **set vtp passwd 0** clears the VTP password.

On an IOS-based switch, set the VTP password in VLAN configuration mode with the command **vtp password** *password-value*. The password can be from 8 to 64 characters. Clear the VTP password with the command **no vtp password**.

Monitoring VTP

On a CatOS switch, you can monitor VTP activity with the command **show vtp statistics**. The comparable command on an IOS-based switch is **show vtp counters**. These commands are particularly useful for analyzing VTP join messages associated with VTP pruning (discussed in the next section).

Example 4-24 shows the output of these commands on a Catalyst 4006 and Catalyst 2950, respectively.

Example 4-24 Monitoring VTP Activity

```
Console> (enable) show vtp statistics
VTP statistics:
summary advts received          40
subset  advts received          24
request advts received          2
summary advts transmitted       181
subset  advts transmitted       39
request advts transmitted       15
No of config revision errors    18
No of config digest errors      0


VTP pruning statistics:

Trunk    Join Transmitted Join Received Summary advts received from GVRP PDU
                                        non-pruning-capable device  Received
-------- ---------------- ------------- --------------------------- ---------
 2/4     0                0             0                           0
 2/27    0                0             0                           0
 2/28    0                0             0                           0
Console> (enable)

Switch#show vtp counters
VTP statistics:
Summary advertisements received    : 484
Subset advertisements received     : 27
Request advertisements received    : 6
Summary advertisements transmitted : 208
Subset advertisements transmitted  : 42
Request advertisements transmitted : 62
Number of config revision errors   : 14
Number of config digest errors     : 0
Number of V1 summary errors        : 0
```

continues

Example 4-24 Monitoring VTP Activity (Continued)

```
VTP pruning statistics:

Trunk            Join Transmitted Join Received    Summary advts received from
                                                   non-pruning-capable device
---------------- ---------------- ---------------- ---------------------------
Fa0/23             0                0                0
Fa0/24             0                0                0
```

Revision errors (highlighted in the Example 4-24 output) increment whenever the switch receives an advertisement and the revision number matches the switch's revision number, but the MD5 digest values do not match. This error means that the VTP passwords on the two switches are different or that the switches have inconsistent VTP configurations. In this case, the switch filters incoming advertisements. The result is a network that has an unsynchronized VTP database.

VTP Pruning

VTP ensures that switches in the VTP domain are aware of all VLANs. There are occasions, however, when VTP can create unnecessary traffic. All multicasts, unknown unicasts, and broadcasts in a VLAN are flooded to all ports in the broadcast domain, including trunk ports. Trunks transport traffic for all VLANs unless they are explicitly restricted (refer to Examples 4-11 and 4-13, which illustrate what you might call static pruning). When neighbor switches receive these multicasts, unknown unicasts, and broadcasts, they in turn flood the traffic out all their trunk ports. In this fashion, traffic other than known unicast traffic propagates across the switched network, even to Catalyst switches with no ports configured in the VLAN that sourced the traffic. Chapters 9 and 10 explore switch and router features designed to limit the propagation of multicast traffic. *VTP pruning* helps reduce the flooding of unknown unicast and broadcast traffic. It is a feature used to dynamically eliminate, or prune, this unnecessary VLAN traffic. In other words, VTP performs dynamic pruning of unnecessary VLAN traffic.

If you look at the effect of VTP pruning in terms of network bandwidth, you see that VTP pruning increases available bandwidth by reducing unnecessary flooded traffic. VTP pruning increases available bandwidth by restricting flooded traffic to trunk links that must be traversed for the traffic to reach the destination device.

The mechanics for VTP pruning depend on the fourth type of VTP message. VTP pruning protocol is an extension to VTP that utilizes this fourth VTP message type: the VTP join message (referenced in the section "VTP Advertisements"). A join message announces VLAN membership whenever you associate Catalyst ports to a VLAN. The message informs the Catalyst neighbors that this Catalyst is interested in receiving traffic for the VLAN associated with the configured ports. The neighbor Catalyst uses this information to decide whether flooded traffic should transit the trunk. VLAN traffic is not sent down the trunk unless an appropriate join message has been received on the trunk link.

Specifically, in order to support pruning, a new state variable was defined per VLAN per forwarding trunk port. This state variable indicates either the pruned state or the joined state of that VLAN on that port. This state affects only the sending of messages on that port; it has no effect on the receipt of messages. In the joined state, the port sends frames exactly as it currently does. In the pruned state, no frames are sent on that VLAN on that port, except possibly for STP, Cisco Discovery Protocol (CDP), or VTP packets.

A nontrunk port is in the joined state for each VLAN for which traffic is allowed to be sent on that port and is pruned for other VLANs. For a trunk port, a subset of VLANs is always in the joined state. This subset includes the factory-default VLANs. Other VLANs can be included in this subset through configuration of the port via the CLI or network-management software. Each VLAN not in this subset is termed "pruning-eligible" on that port, and its state is set to joined or pruned, according to the contents of the join messages received on the port.

Each of the VTP join messages contains information about the VLANs in question and a bit indicating whether this VLAN should be pruned for this trunk (a 1 indicates that it should not be pruned). With pruning enabled, VLAN traffic is not normally sent across the trunk link unless an appropriate join message with the corresponding VLAN's bit enabled is received on the trunk link. This is important because it tells you that when using VTP pruning, you must make sure that the correct information and configuration exist. For example, when you're using VTP pruning, it is critical that all switches in the VTP domain are configured for VTP pruning. By default, VTP pruning is disabled for Catalyst switches. When pruning negotiation is complete, the VLAN state is set as either pruned or joined for that trunk.

To illustrate the function of VTP pruning, let's look at a sample switched network. Figure 4-18 shows a switched network without VTP pruning enabled. All switches are connected with trunk links, which act as conduits for propagating VLAN traffic. Port

1 on Switch 1 and Port 2 on Switch 4 are assigned to the Red VLAN. A broadcast is sent from the host connected to Switch 1. Switch 1 floods the broadcast, and every switch in the network receives it via the trunk links, even though Switches 3, 5, and 6 have no ports in the Red VLAN.

Figure 4-18
VLAN traffic without VTP pruning enabled.

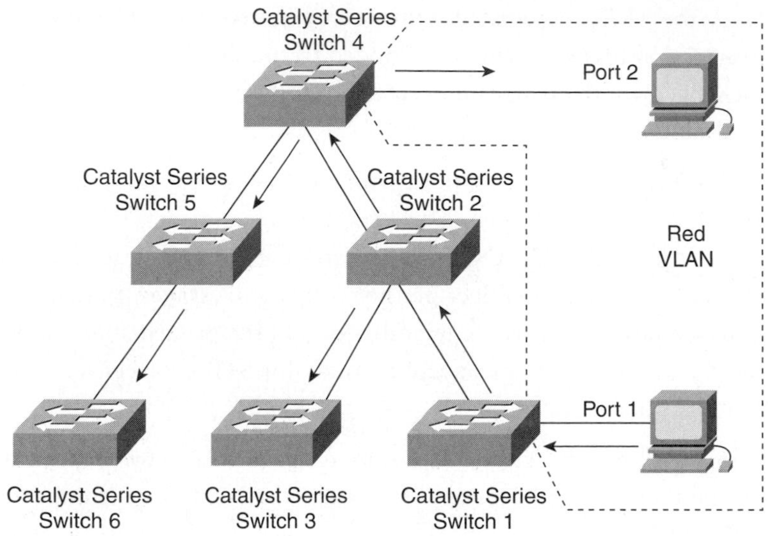

Figure 4-19 shows the same switched network with VTP pruning enabled. The broadcast traffic from Switch 1 is not forwarded to Switches 3, 5, and 6, because traffic for the Red VLAN has been pruned on Port 5 of Switch 2 and Port 4 of Switch 4.

Figure 4-19
VLAN traffic with VTP pruning enabled.

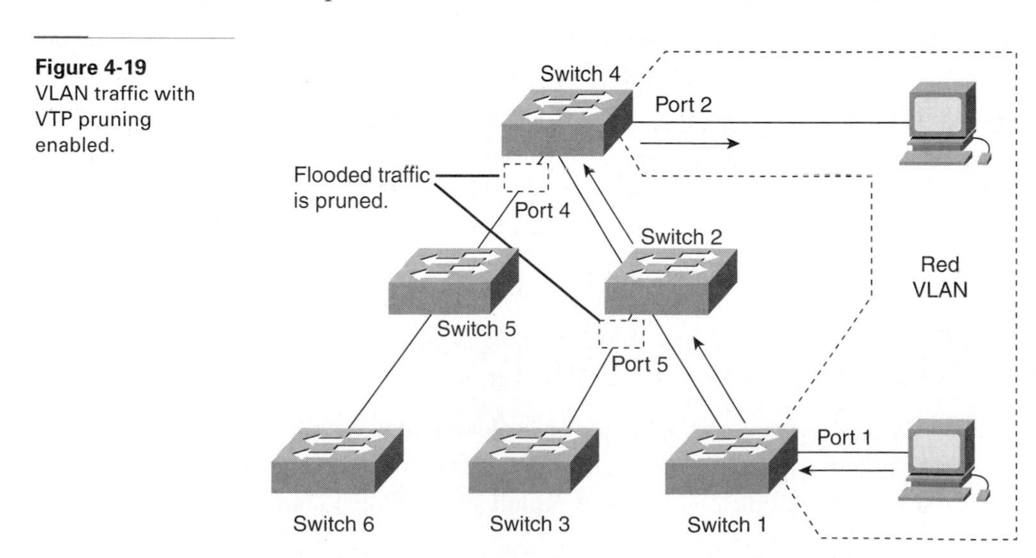

Configuring VTP Pruning

Enabling VTP pruning on a VTP server enables pruning for the entire management domain. VTP pruning takes effect several seconds after you enable it. By default, VLANs 2 through 1000 or 2 through 1001 are pruning-eligible, depending on the platform. VTP pruning does not prune traffic from VLANs that are pruning-ineligible. VLAN 1 is always pruning-ineligible. VLAN 1 cannot be removed from a trunk. However, the "VLAN 1 disable on trunk" feature, available on Catalyst 4000, 5000, and 6000 family switches, allows you to prune user traffic, but not control protocol traffic (DTP, PAgP, CDP, VTP, and so on) for VLAN 1 from a trunk.

CatOS Switch

By default, VLANs 2 through 1000 are pruning-eligible on CatOS switch trunk ports. To enable VTP pruning in the management domain, enter the command **set vtp pruning enable**. To make a VLAN pruning-ineligible, enter the **clear vtp pruneeligible** command. To make a VLAN pruning eligible again, enter the **set vtp pruneeligible** command. You can set VLAN pruning eligibility regardless of whether VTP pruning is enabled or disabled for the domain. Pruning eligibility always applies to the local device only, not to the entire VTP domain.

To verify the VTP pruning information, use the **show vtp domain** and **show trunk** commands.

Example 4-25 demonstrates how to enable VTP pruning in the management domain and how to make VLANs 2 to 99, 250 to 255, and 501 to 1000 pruning-eligible on the particular device.

Example 4-25 Enabling Pruning and Pruning Eligibility on a CatOS Switch

```
Console> (enable) set vtp pruning enable
This command will enable the pruning function in the entire management domain.
All devices in the management domain should be pruning-capable before enabling.
Do you want to continue (y/n) [n]? y

VTP domain kauai modified
Console> (enable) clear vtp pruneeligible 100-500
Vlans 1,100-500,1001-1005 will not be pruned on this device.
VTP domain kauai modified.
Console> (enable) set vtp pruneeligible 250-255
Vlans 2-99,250-255,501-1000 eligible for pruning on this device.
VTP domain kauai modified.
```

continues

Example 4-25 Enabling Pruning and Pruning Eligibility on a CatOS Switch (Continued)

```
Console> (enable) show vtp domain

Domain Name                        Domain Index VTP Version Local Mode  Password
---------------------------------- ------------ ----------- ----------- ----------
kauai                              1            2           server      -

Vlan-count Max-vlan-storage Config Revision Notifications
---------- ---------------- --------------- -------------
8          1023             16              disabled
Last Updater    V2 Mode  Pruning  PruneEligible on Vlans
--------------- -------- -------- -------------------------
172.20.52.2     disabled enabled  2-99,250-255,501-1000
172.20.52.3
Console> (enable) show trunk

Port      Mode         Encapsulation  Status        Native vlan
--------  -----------  -------------  ------------  -----------
 1/1      auto         isl            trunking      523
Port      Vlans allowed on trunk
--------  ---------------------------------------------------------------------
 1/1      1-1005
Port      Vlans allowed and active in management domain
--------  ---------------------------------------------------------------------
 1/1      1,522-524
Port      Vlans in spanning tree forwarding state and not pruned
--------  ---------------------------------------------------------------------
 1/1      1,522-524
Console> (enable)
```

IOS-Based Switch

By default, VLANs 2 through 1001 are pruning-eligible on IOS-based switch trunk ports. To enable VTP pruning in the management domain, enter the VLAN configuration mode command **vtp pruning**. After you type **exit**, the VLAN database is updated and is propagated throughout the VTP domain.

To remove VLANs from the pruning-eligible list on a trunk port, use the interface configuration command **switchport trunk pruning vlan remove** *vlan-id*. Separate nonconsecutive VLAN IDs with a comma and no spaces, and use a hyphen to designate a range of IDs.

To verify the VTP pruning configuration, use the commands **show vtp status** and **show interface** *interface-id* **switchport**.

Example 4-26 illustrates the VTP pruning configuration on an IOS-based switch. Notice the change in the **Pruning VLANs Enabled** output.

Example 4-26 Enabling Pruning and Pruning Eligibility on an IOS-Based Switch

```
Switch(vlan)#vtp pruning
Pruning switched ON
Switch(vlan)#exit
APPLY completed.
Exiting....
Switch#show vtp status
02:35:29: %SYS-5-CONFIG_I: Configured from console by consoleatus
VTP Version                     : 2
Configuration Revision          : 8
Maximum VLANs supported locally : 68
Number of existing VLANs        : 11
VTP Operating Mode              : Client
VTP Domain Name                 : cisco
VTP Pruning Mode                : Enabled
VTP V2 Mode                     : Disabled
VTP Traps Generation            : Disabled
MD5 digest                      : 0x78 0xE6 0x01 0x64 0xF8 0xE8 0x56 0x3A
Configuration last modified by 156.20.10.4 at 1-22-02 05:53:30
Switch#show interface fa0/3 switchport
Name: Fa0/3
Switchport: Enabled
Administrative mode: trunk
Operational Mode: trunk
Administrative Trunking Encapsulation: dot1q
```

continues

Example 4-26 Enabling Pruning and Pruning Eligibility on an IOS-Based Switch (Continued)

```
Operational Trunking Encapsulation: dot1q
Negotiation of Trunking: Disabled
Access Mode VLAN: 0 ((Inactive))
Trunking Native Mode VLAN: 1 (default)
Trunking VLANs Enabled: ALL
Trunking VLANs Active: 1-4,6,7,200
Pruning VLANs Enabled: 2-1001

Priority for untagged frames: 0
Override vlan tag priority: FALSE
Voice VLAN: none
Appliance trust: none
Switch#configure terminal
Enter configuration commands, one per line.  End with CNTL/Z.
Switch(config)#interface fa0/3
Switch(config-if)#switchport trunk pruning vlan remove 2-3,7
Switch(config-if)#end
Switch#show interface fa0/3 switchport
Name: Fa0/3
Switchport: Enabled
Administrative mode: trunk
Operational Mode: trunk
Administrative Trunking Encapsulation: dot1q
Operational Trunking Encapsulation: dot1q
Negotiation of Trunking: Disabled
Access Mode VLAN: 0 ((Inactive))
Trunking Native Mode VLAN: 1 (default)
Trunking VLANs Enabled: ALL
Trunking VLANs Active: 1-4,6,7,200
Pruning VLANs Enabled: 4-6,8-1001

Priority for untagged frames: 0
Override vlan tag priority: FALSE
Voice VLAN: none
Appliance trust: none
```

Summary

This chapter began by describing the advantages of deploying VLANs. VLANs can be classified as end-to-end versus local or static versus dynamic. VLAN Management Policy Server (VMPS) technology relies on VLAN Query Protocol (VQP) to implement dynamic VLANs. We detailed the configuration of static VLANs, dynamic VLANs, and VMPS.

Four major trunking technologies exist: ISL, IEEE 802.1Q, IEEE 802.10, and ATM LANE. ISL encapsulates Ethernet frames, and IEEE 802.1Q embeds a tag in the Ethernet header. ISL and IEEE 802.1Q allow for trunk negotiation on certain Catalyst platforms, utilizing Dynamic ISL (DISL) or Dynamic Trunking Protocol (DTP). On Catalysts supporting both ISL and 802.1Q, the trunking encapsulation itself can be negotiated as well.

VTP is a Cisco-proprietary protocol used to partially automate VLAN configuration. VTP relies on four message types to perform this function. VTP also reduces the chance of inconsistent VLAN configurations. VTP servers and VTP clients work together to distribute VLAN information throughout a VTP domain. VTP pruning permits the dynamic pruning of unnecessary VLAN traffic within the management domain. We discussed in detail the configuration of Ethernet trunks, all the VTP components used to build a VTP domain, and VTP pruning.

The next chapter provides the final piece of the Layer 2 switching puzzle. Spanning-Tree Protocol enables Layer 2 redundancy, load balancing, and loop avoidance.

Check Your Understanding

Test your understanding of the concepts covered in this chapter by answering these review questions. Answers are listed in Appendix A, "Check Your Understanding Answer Key."

1. Which of the following are VLAN trunking technologies? Choose all that apply.

   A. IEEE 802.1Q

   B. ISL

   C. DTP

   D. MD5

2. Which protocol does VMPS use for client requests?

   A. GVRP

   B. IEEE 802.10

   C. VQP

   D. VMS

3. Can the native VLAN for a trunk port be VLAN 2?

   A. Yes

   B. No

4. What is another name for a VLAN management domain?

   A. Native VLAN

   B. VLAN 1

   C. VMS

   D. VTP domain

5. If a functioning VTP domain has four Catalyst switches, what is the minimum number of trunks that must be configured? What is the minimum number of VTP clients?

   A. A. 3; 1

   B. B. 2; 2

   C. C. 3; 0

   D. D. 4; 4

6. Which of the following options cannot be configured in a VMPS database?
   A. The security mode
   B. The fallback VLAN
   C. VLAN port policies
   D. VMPS servers
7. VTP pruning prevents the unnecessary propagation of what type of VLAN traffic? Choose all that apply.
   A. Broadcast
   B. Anycast
   C. Forecast
   D. Unknown unicast
8. What is the VLAN configuration mode command to configure an IOS-based switch as a VTP client?
   A. **set vtp mode client**
   B. **vtp client**
   C. **vtp mode client**
   D. **set vtp client**
9. What CatOS command displays whether VTP version 2 mode is enabled?
   A. **show vlan**
   B. **show trunk**
   C. **show vtp domain**
   D. **show vtp status**
10. What CatOS command creates VLAN 500 with a name of Engineering?
    A. **set vlan Engineering 500**
    B. **set vlan 500 Engineering**
    C. **set vlan 500 name Engineering**
    D. This is not possible in a single command.

Key Terms

access port A switch port that connects to an end user device or a server.

ATM LANE (ATM LAN emulation LANE) A standard defined by the ATM Forum that gives two stations attached via ATM the same capabilities they normally have with Ethernet and Token Ring.

clustering A method of managing a group of switches without having to assign an IP address to every switch.

DTP (Dynamic Trunking Protocol) A Cisco-proprietary protocol that autonegotiates trunk formation for either ISL or 802.1Q trunks.

dynamic VLAN A VLAN in which end stations are automatically assigned to the appropriate VLAN based on their MAC address. This is made possible via a MAC address-to-VLAN mapping table contained in a VLAN Management Policy Server (VMPS) database.

ELAN (emulated LAN) A logical construct, implemented with switches, that provides Layer 2 communication between a set of hosts in a LANE network. See ATM LANE.

end-to-end VLAN Also known as a campus-wide VLAN. An end-to-end VLAN spans a campus network. It is characterized by the mapping to a group of users carrying out a similar job function (independent of physical location).

IEEE 802.10 The IEEE standard that provides a method for transporting VLAN information inside the IEEE 802.10 frame (FDDI). The VLAN information is written to the security association identifier (SAID) portion of the 802.10 frame. This allows for transporting VLANs across FDDI backbones.

IEEE 802.1Q The IEEE standard for identifying VLANs associated with Ethernet frames. IEEE 802.1Q trunking works by inserting a VLAN identifier into the Ethernet frame header.

ISL (Inter-Switch Link) A Cisco-proprietary encapsulation protocol for creating trunks. ISL prepends a 26-byte header and appends a 4-byte CRC to each data frame.

LANE (LAN emulation) An ATM Forum standard used to transport VLANs over ATM networks.

local VLAN Also known as a geographic VLAN. A local VLAN is defined by a restricted geographic location, such as a wiring closet.

native VLAN The VLAN that a trunk port reverts to if trunking is disabled on the port.

port-based VLAN Also known as a static VLAN. A port-based VLAN is configured manually on a switch, where ports are mapped, one-by-one, to the configured VLAN. This hard-codes the mapping between ports and VLANs directly on each switch.

private VLAN A VLAN you configure to have some Layer 2 isolation from other ports within the same private VLAN. Ports belonging to a private VLAN are associated with a common set of supporting VLANs that create the private VLAN structure.

trunk A point-to-point link connecting a switch to another switch, a router, or a server. A trunk carries traffic for multiple VLANs over the same link. The VLANs are multiplexed over the link with a trunking protocol.

VACL (VLAN access control list) A generalized access control list applied on a Catalyst 6000 switch that permits filtering of both intra-VLAN and inter-VLAN packets.

VLAN (virtual LAN) A group of end stations with a common set of requirements, independent of their physical location, that communicate as if they were attached to the same wire. A VLAN has the same attributes as a physical LAN but allows you to group end stations even if they are not located physically on the same LAN segment.

VMPS (VLAN Management Policy Server) A Cisco-proprietary solution for enabling dynamic VLAN assignments to switch ports within a VTP domain.

VTP (VLAN Trunking Protocol) A Cisco-proprietary protocol used to communicate information about VLANs between Catalyst switches.

VTP domain Also called a VLAN management domain. A network's VTP domain is the set of all contiguously trunked switches with the same VTP domain name.

VTP pruning A switch feature used to dynamically eliminate, or prune, unnecessary VLAN traffic.

After completing this chapter, you will be able to perform tasks related to the following:

- Purpose of the Spanning-Tree Protocol
- Variables used by spanning tree to make decisions
- Spanning-tree decision sequence
- Concepts of root bridge, designated bridges, root ports, and designated ports
- Spanning-tree states
- Spanning-tree timers
- Spanning-tree configuration and monitoring
- Technologies to enhance spanning-tree convergence
- EtherChannel operation
- EtherChannel configuration

Chapter 5

Spanning-Tree Protocol

Of all the protocols that network engineers spend time learning about and planning for, the one that is probably overlooked the most is Spanning-Tree Protocol (STP). As routers became popular in the early 1990s, STP faded into the background as a "less-important protocol that just worked." However, with switching technology now in the forefront, STP has once again earned its place as an impo;rtant consideration in network design.

A poorly planned initial implementation of STP can mean that an inordinate proportion of the configuration, troubleshooting, and maintenance effort on a campus network becomes devoted to STP. This chapter explains the mechanics of Spanning-Tree Protocol, detailing its loop-prevention function in switched networks. Understanding these mechanics enables a proactive rather than reactive approach to configuring spanning tree on your switched network.

STP is definitely one of the most technical subjects in LAN switching. The challenge of understanding STP is similar to that of understanding the underlying operation of OSPF or EIGRP in a campus network (timers, packet types, algorithms, and so on). But keep in mind that STP serves as a fundamental logical building block in every campus network, so the time and effort you put into it are well worth the trouble. It is essential that a network engineer have a basic mastery of STP and the role it plays in network design and implementation.

This chapter discusses Layer 2 techniques and technologies designed to optimize network reliability, resiliency, and redundancy within a campus network. You become familiar with Cisco-specific solutions designed to ensure network availability. In particular, this chapter explores scaling issues with STP as they pertain to virtual LANs (VLANs) and VLAN Trunking Protocol (VTP).

Basic STP Operation

Spanning-Tree Protocol (STP) is a Layer 2 protocol that utilizes a special-purpose algorithm to discover physical loops in a network and effect a logical loop-free topology.

STP creates a loop-free tree structure consisting of leaves and branches that span the entire Layer 2 network. The actual mechanics of how bridges communicate and how the STP algorithm works are the subject of this chapter. Note that the terms bridge and switch are used interchangeably when discussing STP. Also, unless otherwise indicated, connections between switches are assumed to be trunks. The greater discussion of STP concerns its behavior on these trunk connections, which form the spanning tree.

Loops can occur in a network for a variety of reasons. Usually, loops in a network are the result of a deliberate attempt to provide redundancy. However, loops can also result from configuration errors. Figure 5-1 shows a typical switched network and how loops can be intentionally used to provide redundancy.

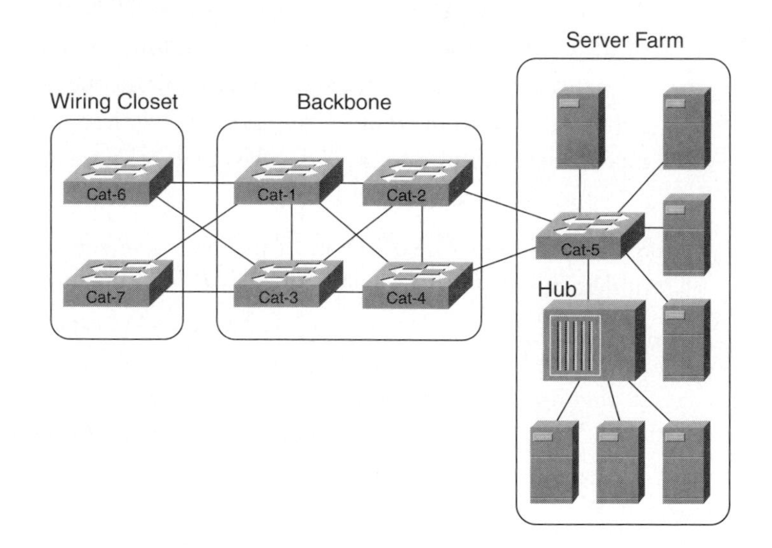

Figure 5-1 Several physical loops appear in this network. Spanning Tree Protocol blocks some of the ports to solve this problem.

Physical loops without proper STP design can be absolutely disastrous. Two problems that can result are broadcast loops and bridge-table corruption. We discuss these phenomena in the next two sections.

Broadcast Loops

Broadcasts and physical loops are a dangerous combination. Figure 5-2 shows how broadcast loops are generated. Two switches, Cat-1 and Cat-2, have STP disabled. Host A and Host B are connected to Cat-1 and Cat-2 by a hub. The hub, a physical-layer equivalent to a LAN segment, is indicated as a segment in Figure 5-2. Host A begins by sending a frame to the broadcast MAC address FF-FF-FF-FF-FF-FF, which travels to both Cat-1 and Cat-2.

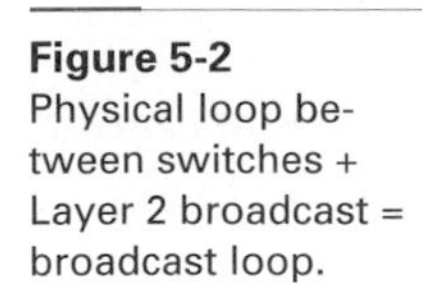

Figure 5-2
Physical loop between switches + Layer 2 broadcast = broadcast loop.

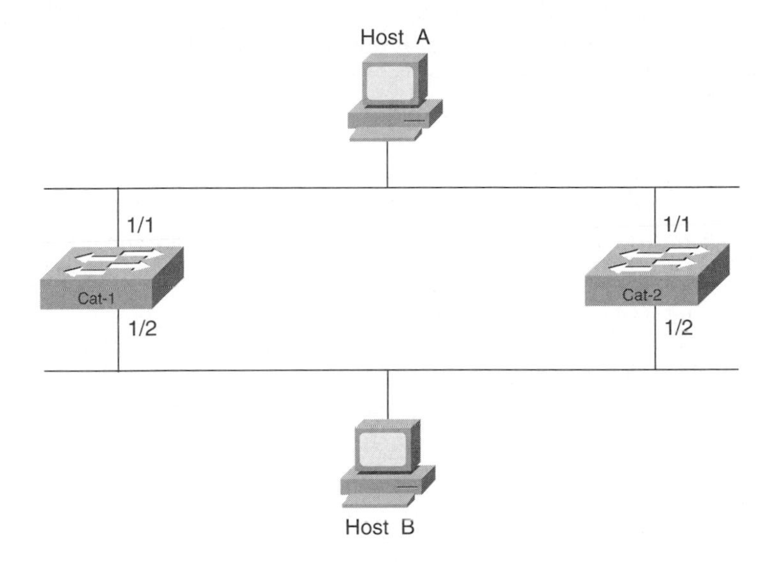

When the frame arrives at Cat-1:Port-1/1, Cat-1 follows the standard transparent bridging algorithm and floods the frame out all the other ports, including Port 1/2. The frame exiting Port 1/2 travels to all nodes on the lower Ethernet segment, including Cat-2:Port1/2. Cat-2 floods the broadcast frame out all the other ports, including Port 1/1, and, once again, the frame shows up at Cat-1:Port-1/1. Oblivious to the loop, Cat-1 sends the frame out Port 1/2 for the second time. By now you see the pattern—a broadcast loop is now in effect.

Additionally, notice that the discussion so far has not mentioned that not only would Cat-1 have received Host A's initial broadcast, but Cat-2 as well. The same problem would propagate in the reverse direction as a result. The "feedback" loop would occur in both directions.

An important conclusion to be drawn is that bridging loops are much more dangerous than routing loops. Why is this? Suppose the initial broadcast frame were an Ethernet Version II frame, as shown in Figure 5-3.

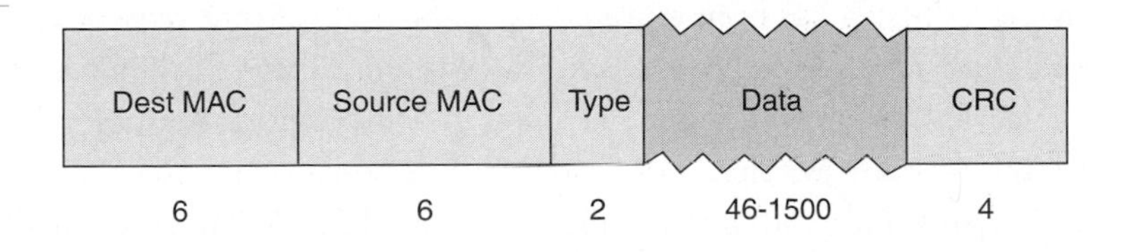

Figure 5-3
Ethernet Version II frames include a Type field that references the protocol type carried in the data field.

Recall that the Ethernet Version II frame contains only two MAC addresses, a Type field, a cyclic redundancy check (CRC), and the network-layer packet as data. In contrast, an IP header contains a Time To Live (TTL) field that is set by the source and is decremented at each router. By discarding packets that reach TTL = 0, routers prevent "runaway" datagrams. Unlike IP, Ethernet does not have a TTL field. Therefore, after a frame starts to loop in the network as just described, it continues until a switch is turned off or a link is broken.

The network shown in Figure 5-2 is extremely simple. Feedback loops of this type grow exponentially as additional switches and links are incorporated into the network. A single broadcast can render a network unusable, all because STP is disabled or misconfigured.

Bridge-Table Corruption

A less-understood problem than broadcast storms is unicast frames causing network bottlenecks. See Figure 5-4 as we describe how a unicast packet can cause such a problem.

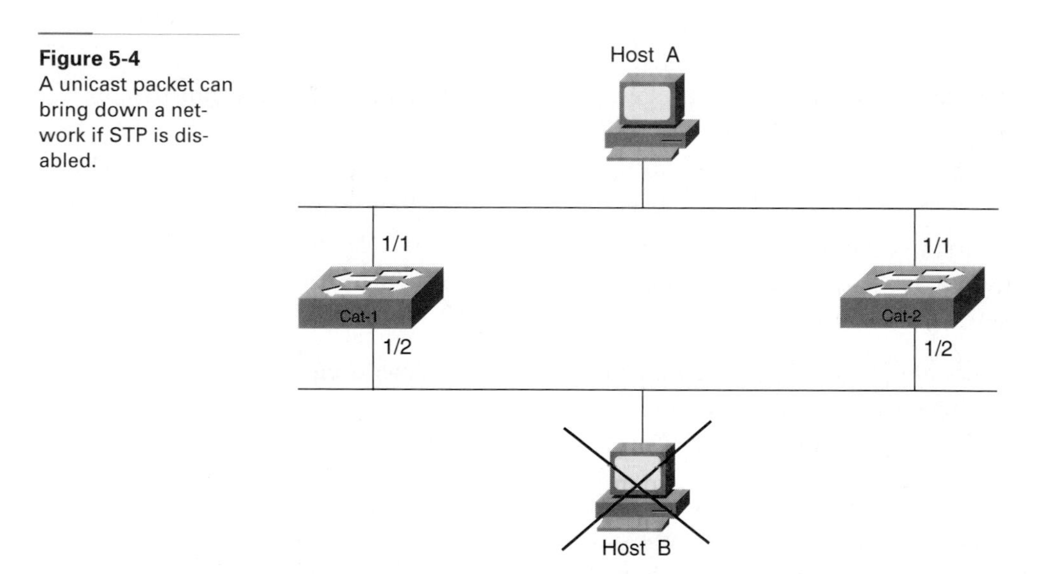

Figure 5-4 A unicast packet can bring down a network if STP is disabled.

Suppose that Host A, possessing a prior ARP entry for Host B, wants to ping Host B. However, Host B has been temporarily removed from the network, and the corresponding bridge-table entries in the switches have been flushed for Host B. Assume that neither switch is running STP. As with the previous example, the frame travels to Port 1/1 on both switches. For simplicity, we consider this scenario only relative to Cat-1. Cat-1 does not have an entry for Host B's MAC address, BB-BB-BB-BB-BB-BB, in its bridging table; as a result, it floods the frame to all other ports. Cat-2 then receives the frame on Port 1/2. Two things happen at this point.

First, Cat-2 floods the frame because MAC address BB-BB-BB-BB-BB-BB is not in its bridging table. Second, Cat-2 notices that it received a frame on Port 1/2 with a source MAC address of AA-AA-AA-AA-AA-AA (from Host A). As a result, it changes its bridging table entry for Host A's MAC address to the wrong port!

As frames loop in the reverse direction (the feedback loop goes both ways), Host A's MAC address flip-flops between Port 1/1 and Port 1/2 on the switches. In short, not only does this permanently saturate the network with the unicast ping packet, but it also corrupts the bridging tables and drives the switches' CPU utilization to 100%. So you see that broadcasts are not the only types of frames that can bring down a network.

Spanning-Tree Algorithm

An *algorithm* is a formula or set of steps for solving a particular problem. Algorithms rely on a set of rules. They have a clear beginning and end. The Spanning Tree Algorithm is no exception.

The Spanning-Tree Algorithm is defined in the IEEE 802.1D standard: a957.g.akamai.net/7/957/3680/v0001/standards.ieee.org/reading/ieee/std/lanman/802.1D-1998.pdf. The parameters used by the algorithm include Bridge ID, Path Cost, and Port ID. We explore these parameters next. After that, we briefly describe the mechanism of the spanning-tree process: exchanging bridge protocol data units (BPDUs).

Bridge ID

STP is characterized by the Spanning-Tree Algorithm. The Spanning-Tree Algorithm relies on a set of parameters to make decisions. The first parameter we discuss is the Bridge ID. A ***Bridge ID (BID)*** is an 8-byte field consisting of an ordered pair of numbers, as shown in Figure 5-5. The first is a 2-byte decimal number called the Bridge Priority, and the second is a 6-byte (hexadecimal) MAC address.

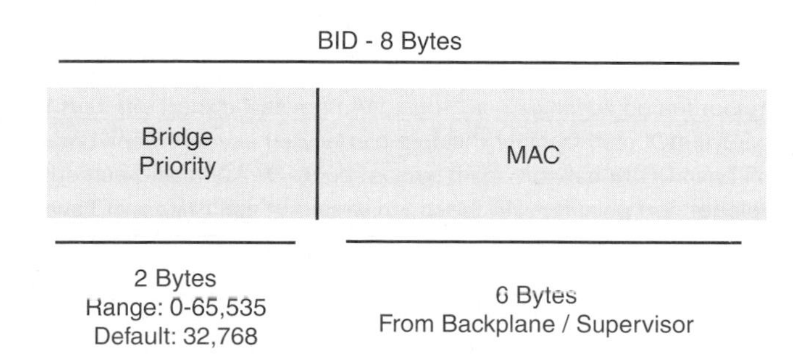

Figure 5-5
The Bridge ID is an STP parameter.

The *Bridge Priority* is a decimal number used to measure the preference of a bridge in the Spanning-Tree Algorithm. The possible values range between 0 and 65,535. The default setting is 32,768.

The MAC address in the BID is one of the switch's MAC addresses. Each switch has a pool of MAC addresses, one for each instance of STP, used as BIDs for the VLAN spanning-tree instances (one per VLAN). For example, Catalyst 4000, 5000, and 6000 switches each have a pool of 1024 MAC addresses assigned to the supervisor module or backplane for this purpose.

If (s,t) and (u,v) represent two BIDs (the first coordinate being the Bridge Priority and the second coordinate being the MAC address), (s,t)<(u,v) if and only if (i) s<u or (ii) s=u and t<v. Two BIDs cannot be equal, because Catalyst switches are assigned unique MAC addresses.

In terms of the Spanning-Tree Algorithm, when a comparison is made between two values of a given STP parameter, the lower value is always preferred.

Path Cost

The IEEE 802.1D standard originally defined the cost of a link as 1000 Mbps divided by the bandwidth of the link in Mbps. For example, a 10BaseT link has a cost of 100 (1000/10), and Fast Ethernet and FDDI each have a cost of 10 (1000/100). With the advent of Gigabit Ethernet and other high-speed technologies, a problem with this definition evolved: The cost is stored as an integer value, not as a floating-point value. For example, 10 Gbps results in 1000 Mbps ÷ 10000 Mbps = .1, an invalid cost value. To solve this problem, the IEEE changed the "inverse proportion" definition. Table 5-1 shows the updated definition of a link's *cost*.

Table 5-1 Updated STP Cost Definition to Reflect Link Speeds Exceeding 1 Gbps

Bandwidth	STP Cost
4 Mbps	250
10 Mbps	100
16 Mbps	62
45 Mbps	39
100 Mbps	19
155 Mbps	14
622 Mbps	6
1 Gbps	4
10 Gbps	2

BID was the first Spanning Tree parameter we discussed. The second STP parameter we consider is Path Cost. *Path Cost* is a measure of how close bridges are to each other. Path Cost is the sum of the costs of the links in a path between two bridges. It is not a measure of hop count. The hop count for path A might be greater than the hop count for path B, while the cost of path A is less than the cost of path B. Closeness is not necessarily reflected by hop count.

Table 5-2 STP Cost Definition Adjusted for Switches with Active Ports Running at 10 Gbps or Greater

Bandwidth	STP Cost
100 Kbps	200,000,000
1 Mbps	20,000,000
10 Mbps	2,000,000
100 Mbps	200,000
1 Gbps	20,000
10 Gbps	2000
100 Gbps	200
1 Tbps (terabits per second)	20
10 Tbps	2

NOTE

Catalyst 4000s and 6000s allow for differentiating between link costs when bandwidths greater than or equal to 10 Gbps are in play. For this option, use the command set spantree defaultcost-mode long. Use this option if any of your links are 10 Gbps or greater. With this setting, the cost definition is as shown in Table 5-2.

Port ID

The third STP parameter we consider is the Port ID, shown in Figure 5-6. The *Port ID* is a 2-byte STP parameter consisting of an ordered pair of numbers. The first is called the Port Priority, and the second is called the Port Number. On a CatOS switch, the first number is 6 bits, and the second is 10 bits. On an IOS-based switch, the first and second numbers are both 8 bits.

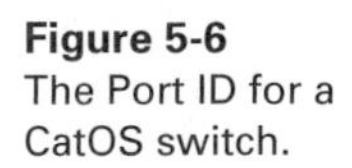
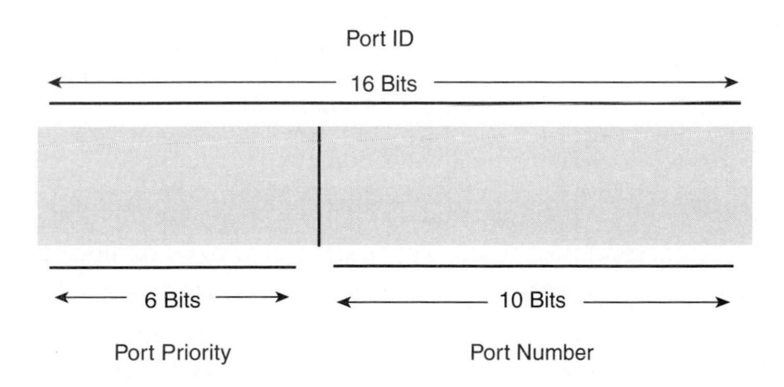

Figure 5-6
The Port ID for a CatOS switch.

Do not confuse the Port ID with the Port Number. The Port Number is just part of the Port ID.

Lower Port IDs are preferred over higher Port IDs in the STP decision process. The order relation for Port IDs is as follows:

If (s,t) and (u,v) represent two Port IDs (the first coordinate being the Port Priority and the second coordinate being the Port Number), $(s,t)<(u,v)$ if and only if (i) $s<u$ or (ii) $s=u$ and $t<v$. Two Port IDs cannot be equal, because Port Numbers uniquely identify the switch ports on a Catalyst switch.

The *Port Priority* is a configurable STP parameter (unlike the Port Number). The decimal values range from 0 to 63 on a CatOS switch, with a default value of 32. The values range from 0 to 255 on an IOS-based switch, with a default value of 128.

Port Numbers are numerical identifiers used by Catalyst switches to enumerate the ports. The Port Number leaves room for 2^{10} - 1024 ports on a CatOS switch and 2^{8} - 256 ports on an IOS-based switch (Cisco routers also use the 8-bit/8-bit Port ID definition). For example, the Port Numbers for the 10/100 ports on a 4006 L3 module in slot 2—namely, 2/3 to 2/34—are 0x43 to 0x62, respectively. Port Numbers are consecutive on a given module, but not necessarily between modules; there is no overlap with Port Numbers on a given Catalyst switch. The Port Numbers do not follow a predictable pattern other than the fact that they are consecutive numbers for a given module. Some devices start with a first Port Number of 2, some with 7, some with 13, and so on.

Next, we consider how the Bridge ID, Path Cost, and Port ID are used in STP operation.

STP Decisions and BPDU Exchanges

We defined the Bridge ID, the Path Cost, and the Port ID. The Spanning-Tree Algorithm operates as a function of these parameters. When creating a loop-free logical topology, Spanning Tree always uses the same four-step decision sequence:

Step 1 Determine the Root Bridge.

Step 2 Calculate the lowest Path Cost to the Root Bridge.

Step 3 Determine the lowest sender BID.

Step 4 Determine the lowest Port ID.

In order to make good decisions, STP needs to ensure that the participating bridges have the correct information. The bridges need to communicate the STP information between them. Bridges pass spanning-tree information between them using Layer 2 frames called *Bridge Protocol Data Units (BPDUs)*. A bridge uses the four-step decision sequence to determine the "best" BPDU seen on each port. The bridge determines the best BPDU based on the four parameters just listed. The bridge stores the best BPDU it receives for each port. When making this evaluation, it considers all the

BPDUs received on the port, as well as the BPDU it would send on that same port. As each BPDU arrives, it is checked against this four-step sequence to see if it is more attractive than the existing BPDU saved for that port. If the new BPDU (or the locally generated BPDU) is more attractive, the old value is replaced.

Additionally, this "saving-the-best-BPDU" process also controls the sending of BPDUs. When a bridge first becomes active, all its ports send BPDUs every 2 seconds (the default Hello Time). However, if a port hears about a BPDU from another bridge that is more attractive than the BPDU it has been sending, the local port eventually stops sending BPDUs. If the more-attractive BPDU stops arriving from a neighbor for 20 seconds (the default Max Age), the local port resumes sending BPDUs. Max Age is the time it takes for the best BPDU to time out.

Three Steps of STP Convergence

The Spanning-Tree Algorithm is somewhat complex, but the initial process used to converge on a loop-free topology consists of three steps:

Step 1 Elect a Root Bridge.

Step 2 Elect Root Ports.

Step 3 Elect Designated Ports.

When the network first "starts," all the bridges flood the network with a mixture of BPDU information. The bridges begin applying the four-step decision sequence discussed in the preceding section. This allows the bridges to hone in on a set of BPDUs that enable the formation of a single spanning tree for the entire network or VLAN. (The default configuration for Catalyst switches is to run one instance of the Spanning-Tree Algorithm per VLAN.) A single Root Bridge is elected to act as the central point of this network (Step 1). All the remaining bridges calculate a set of Root Ports (Step 2) and Designated Ports (Step 3) to build a loop-free topology. The resulting topology is a tree, with the Root Bridge as the trunk and loop-free active paths radiating outward as tree branches. In a steady-state network, BPDUs flow from the Root Bridge outward along these loop-free branches to every segment in the network.

After the network has converged on a loop-free active topology utilizing this three-step process, changes are handled using the spanning-tree topology change process.

The following sections consider the three steps of STP convergence in detail.

Electing the Root Bridge

As a first step in the STP process, the switches need to elect a single Root Bridge by looking for the bridge with the lowest BID. This process of selecting the bridge with the lowest BID is sometimes called the "root war."

As discussed in the earlier "Bridge ID" section, a BID is an 8-byte identifier that is composed of two subfields: the Bridge Priority and a MAC address. Referring to Figure 5-7, you see that Cat-A has a default BID of 32,768.AA-AA-AA-AA-AA-AA, Cat-B assumes a default BID of 32,768.BB-BB-BB-BB-BB-BB, and Cat-C uses 32,768.CC-CC-CC-CC-CC-CC. Because all three bridges use the default Bridge Priority of 32,768, the lowest MAC address, AA-AA-AA-AA-AA-AA, serves as the tiebreaker, and Cat-A becomes the Root Bridge. Later we show how to configure a switch to become the Root Bridge. Normally, you do not let the default settings determine the location of the Root Bridge.

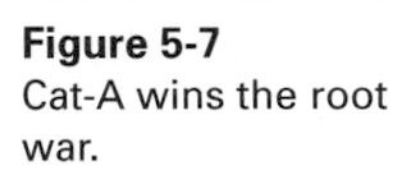

Figure 5-7
Cat-A wins the root war.

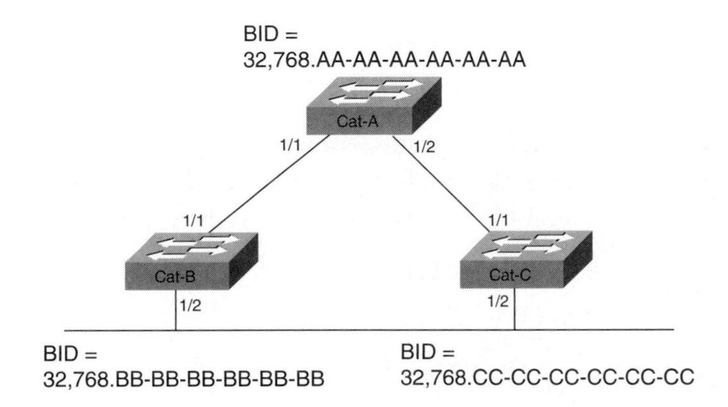

How did the bridges learn that Cat-A has the lowest BID? This is accomplished through the exchange of BPDUs. As discussed earlier, BPDUs are special frames that bridges use to exchange spanning-tree information with each other. By default, BPDUs are sent out every two seconds. BPDUs propagate between bridges, which includes switches and all routers configured for bridging. BPDUs do not carry end-user traffic. Figure 5-8 illustrates the layout of a BPDU.

Figure 5-8
BPDUs propagate spanning-tree information. Four of the 12 fields in a BPDU frame appear here.

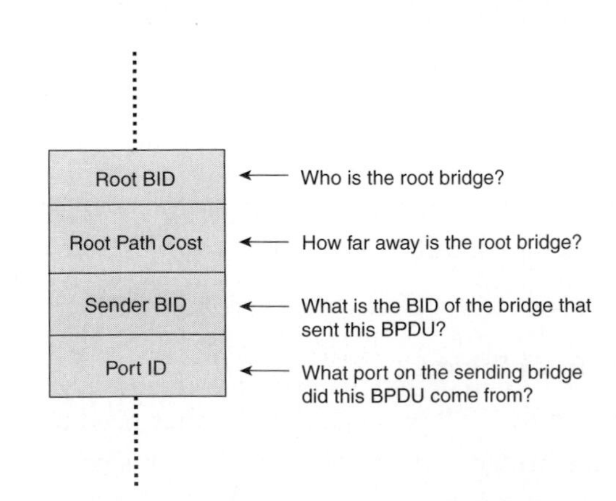

For the purposes of the root war, this discussion involves only the Root BID and Sender BID fields. When a bridge generates a BPDU every two seconds, it places what it thinks is the Root Bridge at that instant in the Root BID field. The bridge always places its own BID in the Sender BID field.

Initially, before the switch knows any better, it populates the Root BID field with its own BID. Suppose that Cat-B boots first and starts sending out BPDUs announcing itself as the Root Bridge every two seconds. A few minutes later, Cat-C boots and announces itself as the Root Bridge. When the Cat-C BPDU arrives at Cat-B, Cat-B discards the BPDU because it has a lower BID saved on its ports (its own BID). As soon as Cat-B transmits a BPDU, Cat-C learns that it is not as important as it initially thought. At this point, Cat-C starts sending BPDUs that list Cat-B as the Root BID and Cat-C as the sender BID. The network now agrees that Cat-B is the Root Bridge.

Five minutes later, Cat-A boots. Cat-A initially thinks it is the Root Bridge and starts advertising this fact via BPDUs. As soon as these BPDUs arrive at Cat-B and Cat-C, these switches hand over the Root Bridge position to Cat-A. All three switches now send out BPDUs that announce Cat-A as the Root Bridge and themselves as the sender BID.

Electing Root Ports

At the conclusion of the root war, the switches move on to selecting Root Ports. A bridge's ***Root Port*** is the port that is closest to the Root Bridge in terms of Path Cost. Every non-Root Bridge must select one Root Port.

Again, bridges use the concept of cost to measure closeness. As with some routing metrics, STP's measure of closeness is not necessarily reflected by hop count. Specifically, bridges track what is referred to as ***Root Path Cost***—the cumulative cost of all links to the Root Bridge. Figure 5-9 illustrates how this value is calculated across multiple bridges and the resulting Root Port election process.

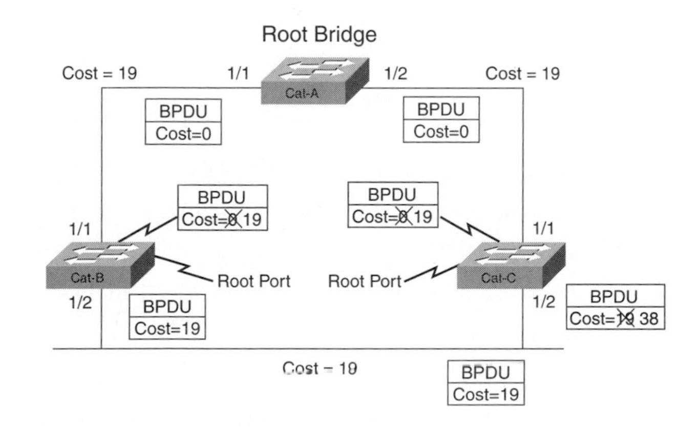

Figure 5-9
The Root Port is the port with the lowest Path Cost to the Root Bridge.

When Cat-A (the Root Bridge) sends out BPDUs, they contain a Root Path Cost of 0. When Cat-B receives these BPDUs, it adds the Path Cost of Port 1/1 to the Root Path Cost contained in the received BPDU. Assume that the network is running Fast Ethernet. Cat-B receives a Root Path Cost of 0 and adds in the Port 1/1 cost of 19. Cat-B then uses the value of 19 internally and sends BPDUs with a Root Path Cost of 19 out Port 1/2.

When Cat-C receives these BPDUs from Cat-B, it increases the Root Path Cost to 38 (19 + 19). However, Cat-C is also receiving BPDUs from the Root Bridge on Port 1/1. These enter Cat-C:Port-1/1 with a cost of 0, and Cat-C increases the cost to 19 internally. Cat-C has a decision to make: It must select a single Root Port—the port that is closest to the Root Bridge. Cat-C sees a Root Path Cost of 19 on Port 1/1 and 38 on Port 1/2. Cat-C:Port-1/1 becomes the Root Port. Cat-C then begins advertising this Root Path Cost of 19 to downstream switches.

Although it isn't detailed in Figure 5-9, Cat-B goes through a similar set of calculations. Cat-B:Port-1/1 can reach the Root Bridge at a cost of 19, whereas Cat-B:Port-1/2 calculates a cost of 38; Port-1/1 becomes the Root Port for Cat-B. Notice that costs incremented as BPDUs are received on a port, not as they are sent out of a port. For example, BPDUs arrive on Cat-B:Port-1/1 with a cost of 0 and get increased to 19 "inside" Cat-B.

Electing Designated Ports

At this point, the Spanning-Tree Algorithm still has not eliminated any loops. This is taken care of after the Designated Ports and non-Designated Ports are determined.

Each segment in a bridged network has one Designated Port. A ***Designated Port*** for a segment is the bridge port connected to that segment that both sends traffic toward the Root Bridge and receives traffic from the Root Bridge over that segment. The idea behind this is that if only one port handles traffic for each link, all the loops have been broken. A segment's ***Designated Bridge*** is the bridge containing the Designated Port for that segment.

As with the Root Port selection, the Designated Ports are chosen based on Root Path Cost to the Root Bridge, as shown in Figure 5-10. In the figure, Segment 1 forms a link between Cat-A and Cat-B. This segment has two bridge ports: Cat-A:Port-1/1 and Cat-B:Port-1/1. Cat-A:Port-1/1 has a Root Path Cost of 0 (because Cat-A is the Root Bridge), whereas Cat-B:Port-1/1 has a Root Path Cost of 19 (the value 0 received in BPDUs from Cat-A plus the path cost of 19 assigned to Cat-B:Port1/1). Because Cat-A:Port-1/1 has the lower Root Path Cost, it becomes the designated port for this link.

For Segment 2 (the Cat-A to Cat-C link), a similar election takes place. Cat-A:Port-1/2 has a Root Path Cost of 0, whereas Cat-C:Port-1/1 has a Root Path Cost of 19. Cat-A:Port-1/2 has the lower cost and becomes the designated port. Notice that every active port on the Root Bridge becomes a Designated Port.

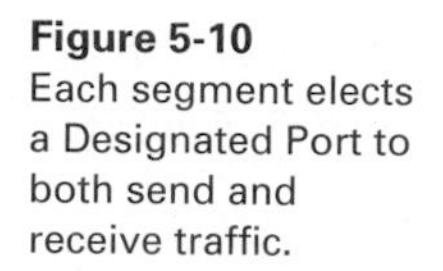

Figure 5-10 Each segment elects a Designated Port to both send and receive traffic.

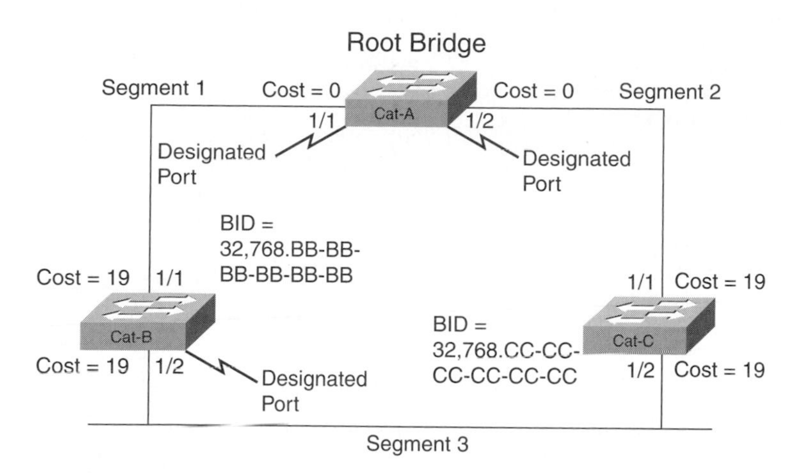

Now look at Segment 3 (Cat-B to Cat-C): Both Cat-B:Port-1/2 and Cat-C:Port-1/2 have a Root Path Cost of 19, resulting in a tie. When faced with a tie, STP always uses the four-step decision sequence discussed in the "STP Decisions and BPDU Exchanges" section. Recall that the four steps are as follows:

Step 1 Determine the Root Bridge.

Step 2 Calculate the lowest Root Path Cost.

Step 3 Determine the lowest sender BID.

Step 4 Determine the lowest Port ID.

In Figure 5-10, all three bridges agree that Cat-A is the Root Bridge, so the algorithm checks Root Path Cost next. But both Cat-B and Cat-C have a cost of 19, so the algorithm resorts to the lowest sender BID. Because the Cat-B BID (32,768.BB-BB-BB-BB-BB-BB) is lower than the Cat-C BID (32,768.CC-CC-CC-CC-CC-CC), Cat-B:Port-1/2 becomes the Designated Port for Segment 3. Cat-C:Port-1/2 becomes a non-Designated Port.

Note that access ports do not play a role in the election of Designated Ports. The process we described involves communication of STP parameters over trunk links.

STP States

After the bridges have determined which ports are Root Ports, Designated Ports, and non-Designated Ports, STP is ready to create a loop-free topology. To do this, STP configures Root Ports and Designated Ports to forward traffic; STP sets non-Designated Ports to block traffic. Although Forwarding and Blocking are the only two states commonly seen in a stable network, Table 5-3 shows that there are actually five STP states.

Table 5-3 Five STP States That Determine Switch Port Status

State	Purpose
Forwarding	Sends/receives user data
Learning	Builds bridging table
Listening	Builds "active" topology
Blocking	Receives BPDUs only
Disabled	Administratively down

This list can be viewed hierarchically in that bridge ports start at the Blocking state and work their way up to the Forwarding state. The ***Disabled state*** is the administratively shut-down STP state. It is not part of the normal STP port processing. After the switch is initialized, ports start in the Blocking state. The ***Blocking state*** is the STP state in which a bridge listens for BPDUs.

A port in the Blocking state does the following:

- Discards data frames received from the attached segment.
- Discards data frames switched from another port for forwarding.
- Does not incorporate station location into its address database. (There is no learning on a blocking port, so there is no address database update.)
- Receives BPDUs and directs them to the system module.
- Does not transmit BPDUs received from the system module.
- Receives and responds to network management messages.

If a bridge thinks it is the Root Bridge immediately after booting, or if there is an absence of BPDUs for a certain period of time, the port transitions into the Listening state. The ***Listening state*** is the STP state in which no user data is being passed, but the port is sending and receiving BPDUs in an effort to determine the active topology.

A port in the Listening state does the following:

- Discards frames received from the attached segment.
- Discards frames switched from another port for forwarding.
- Does not incorporate station location into its address database. (There is no learning at this point, so there is no address database update.)
- Receives BPDUs and directs them to the system module.

- Processes BPDUs received from the system module. Processing BPDUs is a separate action from receiving or transmitting BPDUs.
- Receives and responds to network management messages.

It is during the Listening state that the three initial convergence steps take place—elect a Root Bridge, elect Root Ports, and elect Designated Ports. Ports that lose the Designated Port election become non-Designated Ports and drop back to the Blocking state. Ports that remain Designated Ports or Root Ports after 15 seconds—the default Forward Delay STP timer value—progress into the Learning state. The Learning state's lifetime is also governed by the Forward Delay timer of 15 seconds (the default setting). The *Learning state* is the STP state in which the bridge is not passing user data frames but is building the bridging table and gathering information, such as the source VLANs of data frames. As the bridge receives a frame, it places the source MAC address and port into the bridging table. The Learning state reduces the amount of flooding required when data forwarding begins.

A port in the Learning state does the following:

- Discards frames received from the attached segment.
- Discards frames switched from another port for forwarding.
- Incorporates station location into its address database.
- Receives BPDUs and directs them to the system module.
- Receives, processes, and transmits BPDUs received from the system module.
- Receives and responds to network management messages.

If a port is still a Designated Port or Root Port after the Forward Delay timer expires for the Learning state, the port transitions into the Forwarding state. The *Forwarding state* is the STP state in which data traffic is both sent and received on a port. It is the "last" STP state. At this stage, it finally starts forwarding user data frames.

A port in the Forwarding state does the following:

- Forwards frames received from the attached segment.
- Forwards frames switched from another port for forwarding.
- Incorporates station location information into its address database.
- Receives BPDUs and directs them to the system module.
- Processes BPDUs received from the system module.
- Receives and responds to network management messages.

Figure 5-11 illustrates the STP states and possible transitions.

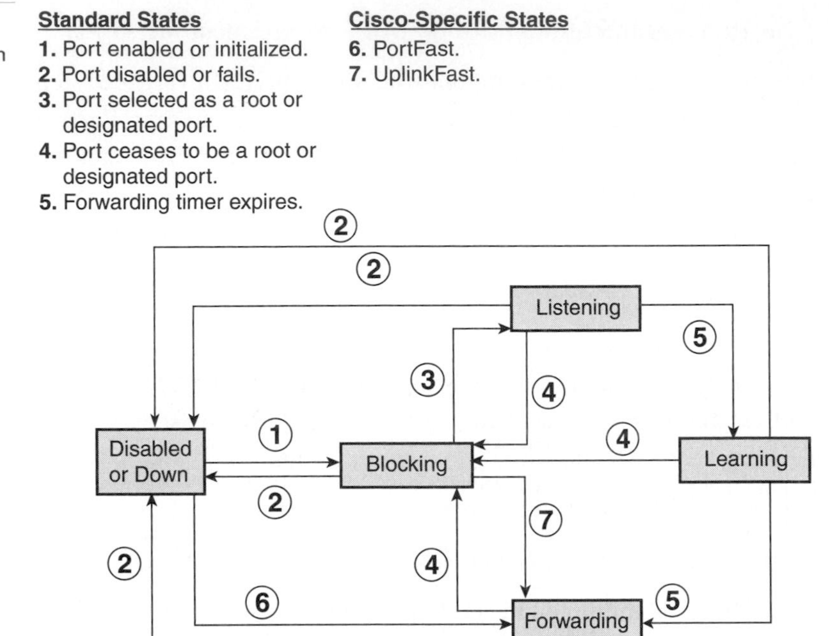

Figure 5-11 Transitions between STP states result from a number of distinct events.

Figure 5-12 illustrates a sample network with port classifications and STP states listed. Notice that all ports are forwarding except Cat-C:Port-1/2.

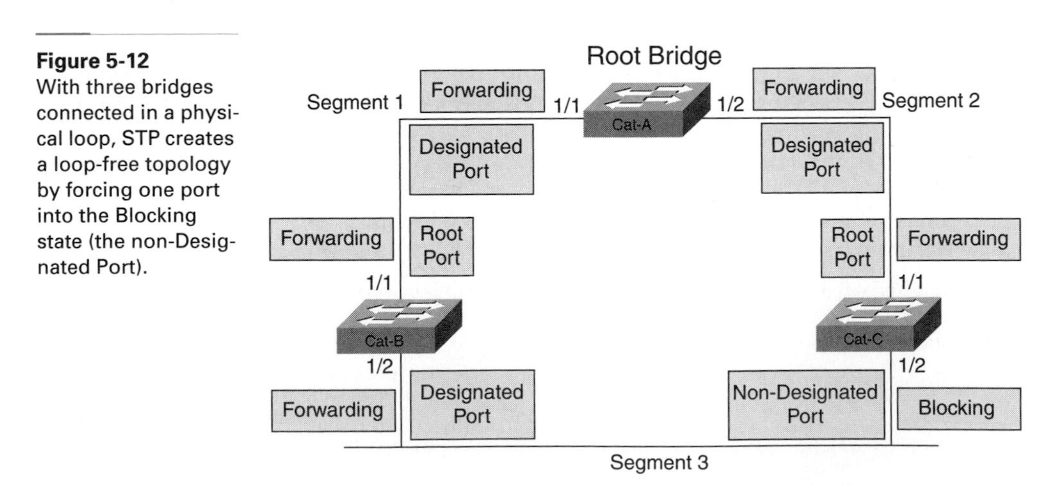

Figure 5-12 With three bridges connected in a physical loop, STP creates a loop-free topology by forcing one port into the Blocking state (the non-Designated Port).

Next, we investigate the role of timers in the Spanning-Tree Algorithm.

STP Timers

STP operation is controlled by three timers, as shown in Table 5-4.

Table 5-4 STP Timers

Timer	Primary Purpose	Default
Hello Time	The amount of time between the sending of Configuration BPDUs by the Root Bridge	2 seconds
Forward Delay	The duration of the Listening and Learning states	15 seconds
Max Age	How long a BPDU is stored	20 seconds

The *Hello Time* is the amount of time between the sending of Configuration BPDUs.

The 802.1D standard specifies a default value of 2 seconds. This value controls Configuration BPDUs as they are generated by the Root Bridge. Other bridges propagate BPDUs from the Root Bridge as they are received. In other words, if BPDUs stop arriving for the time interval ranging from 2 to 20 seconds because of a network disturbance, non-Root Bridges stop sending periodic BPDUs during this time. 2 to 20 seconds is the range between the expected receipt of a BPDU and the expiration of the Max Age time. If the outage lasts for more than 20 seconds, the default Max Age time, the bridge invalidates the saved BPDUs and begins looking for a new Root Port.

NOTE

There are two types of BPDUs: Configuration BPDUs and Topology Change Notification (TCN) BPDUs (see the next section). Configuration BPDUs govern the initial STP convergence process. TCN BPDUs are responsible for STP topology changes.

Forward Delay is the amount of time the bridge spends in the Listening and Learning states. This is a single value that controls both states. The default value of 15 seconds was originally derived assuming a maximum network size of seven bridge hops, a maximum of three lost BPDUs, and a Hello Time of 2 seconds. The Forward Delay timer also controls the bridge table age-out period after a change in the active topology.

Max Age is the STP timer that controls how long a bridge stores a BPDU before discarding it. Recall that each port saves a copy of the best BPDU it has seen. As long as the bridge receives a continuous stream of BPDUs every 2 seconds, the receiving bridge maintains a continuous copy of the BPDU values. However, if the device sending this best BPDU fails, a mechanism must exist to allow other bridges to take over.

For example, assume that the Segment 3 link in Figure 5-12 uses a hub and that the Cat-B:Port-1/2 transceiver fails. Cat-C has no immediate notification of the failure, because it still has an active Ethernet link to the hub. The only thing Cat-C knows is that BPDUs stop arriving. 20 seconds (Max Age) after the failure, Cat-C:Port-1/2 ages out the BPDU information that lists Cat-B as having the best Designated Port for Segment 3. This forces Cat-C:Port-1/2 to transition to the Listening state in an effort to become the Designated Port. Because Cat-C:Port-1/2 now offers the most attractive access from the Root Bridge to this link, it eventually transitions all the way into Forwarding mode. In practice, it takes approximately 50 seconds (20 Max Age + 15 Listening + 15 Learning) for Cat-C to take over after the failure of Port 1/2 on Cat-B.

In some situations, switches can detect topology changes on directly connected links and immediately transition into the Listening state without waiting Max Age seconds. For example, consider Figure 5-13.

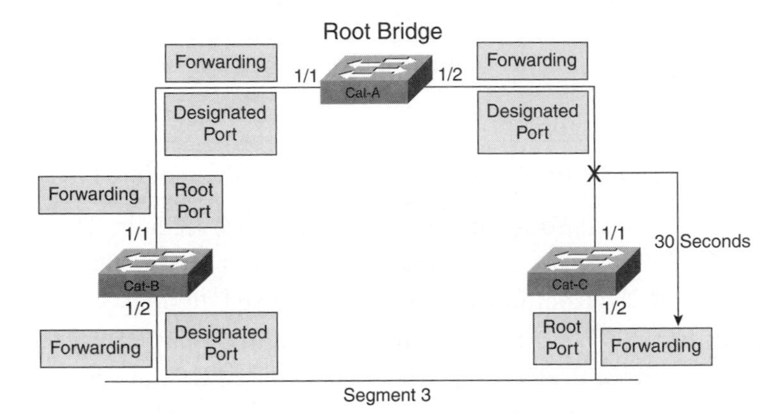

Figure 5-13
In this case, Cat-C forgoes the Max Age of 20 seconds when the link is lost on Cat-C:Port-1/1.

In this example, Cat-C:Port-1/1 fails. Because the failure results in a loss of link on the Root Port, there is no need to wait 20 seconds for the old information to age out. Note the difference between this scenario and the preceding one, in which link integrity was maintained due to the presence of a hub. When Cat-C:Port-1/1 fails, Cat-C:Port-1/2 immediately goes into Listening mode in an attempt to become the new Root Port. This has the effect of reducing the STP convergence time from 50 seconds to 30 seconds (15 Listening + 15 Learning).

There are two key points to remember about using the STP timers. First, do not change the default timer values without careful consideration. Second, assuming that you are confident enough to attempt timer tuning, you should modify the STP timers only from the Root Bridge, because the BPDUs contain three fields with timer values that can be passed from the Root Bridge to all other bridges in the network. The alternative is not good: If every bridge were locally configured, some bridges could work their way up to the Forwarding state before other bridges ever left the Listening state. This chaotic approach could quickly lead to an unstable network. By providing timer fields in the BPDUs, the single bridge acting as the Root Bridge can dictate the timing parameters for the entire bridged network.

Topology Changes and STP

As you have seen, STP implements a series of timers to prevent bridging loops from occurring within the network. You saw that it can take from 30 to 50 seconds for a network to converge to a new topology when there is a change in an otherwise stable STP process.

While the network is converging, physical addresses that can no longer be reached are still listed in the switch table. Because these addresses are in the table, the switch attempts to forward frames to devices it cannot reach. Fortunately, the STP change process requires the switch to clear the table faster in order to get rid of unreachable physical addresses.

Figure 5-14 illustrates a link failure between Switches D and E, which in turn triggers a topology change condition. This topology change condition triggers a Topology Change Notification BPDU to be generated and sent *toward* the Root Bridge. In order for this BPDU to reach the Root Bridge, each switch forwards the update out the Root Port (RP) and to the Designated Port (DP) of each Designated Bridge along the path to the Root Bridge.

Figure 5-14 The topology change process is facilitated by STP TCN BPDUs.

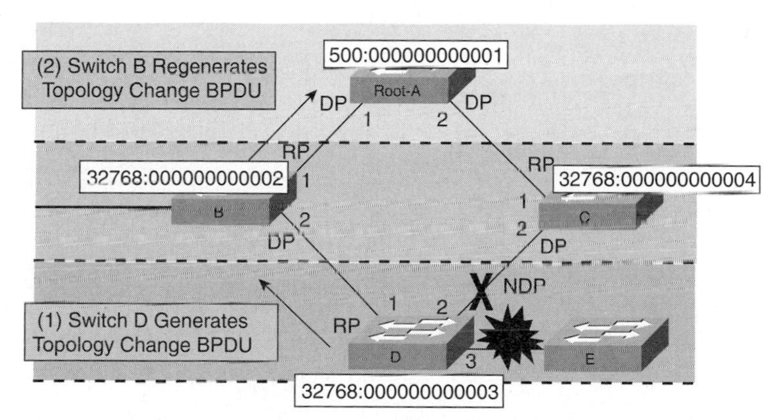

The topology change in this Spanning Tree network triggers the following steps:

Step 1 Switch D notices that a change has occurred to a link.

Step 2 Switch D sends a TCN BPDU out the Root Port destined ultimately for the Root Bridge. The TCN BPDU is indicated by the value 0x80 in the BPDU's 1-byte Type field. The bridge sends out the TCN BPDU until the Designated Bridge for that segment responds with a Topology Change Acknowledgment (TCA) Configuration BPDU, indicated by the high-order bit in the 1-byte Flag field.

Step 3 The Designated Bridge (Switch B) for that segment sends out a TCA Configuration BPDU to the originating bridge (Switch D). Switch B also sends a TCN BPDU out its Root Port destined for the Root Bridge.

Step 4 When the Root Bridge receives the (upstream) topology-change message, it changes its Configuration BPDU to indicate that a topology change is occurring (using the low-order bit in the Flag field). The Root Bridge sets the topology change in the configuration for a period of time equal to the sum of the Forward Delay and Max Age parameters.

Step 5 A bridge receiving the (downstream) topology-change configuration message from the Root Bridge uses the Forward Delay timer (15 seconds) to age out entries in the address table. This allows the device to age out entries faster than the normal 5-minute default so that stations that are no longer available are aged out faster. The bridge continues this process until it no longer receives topology-change configuration messages from the Root Bridge.

Now that the process of building a spanning tree is established, we are ready to discuss STP configuration.

> **NOTE**
>
> Most spanning-tree commands support an optional VLAN parameter. Get in the habit of always using that parameter. If you leave it out, VLAN 1 is implied, which often is not your intention.
>
> For example, on a CatOS switch, the command **set spantree priority 8192** changes the Bridge Priority to 8192 for the switch on VLAN 1. If you want to change the Bridge Priority to 8192 for VLAN 1, you should use the command **set spantree priority 8192 1**. This way, if you decide to change the Bridge Priority for VLAN 2 to 8192, it is a reflex action to type in **set spantree priority 8192 2**.

Basic STP Configuration

By default, STP is enabled for all VLANs on a Catalyst switch. It is recommended that STP be enabled on all switches, even in an environment with no physical loops. This is a precaution in the event that new links are added later.

Per-VLAN Spanning Tree (PVST or PVST+) is discussed in the section "PVST (Per-VLAN Spanning Tree), PVST+, and Mono Spanning-Tree Modes." PVST and PVST+ are used according to the switch's support for IEEE 802.1Q trunking. More-advanced STP configuration is explored later in this chapter.

The following sections describe the basic STP configuration for CatOS and IOS-based switches.

CatOS Switch

If, for some reason, STP has been disabled, you can easily reenable it. You reenable it on a CatOS switch with the **set spantree enable** *vlans* or **set spantree all** command. Disable STP with the command **set spantree disable** *vlan* or **set spantree disable all**. Example 5-1 illustrates these commands.

Example 5-1 Disabling and Enabling Spanning Tree

```
Console> (enable) set spantree disable all
Spantrees 1-1005 disabled.
Console> (enable) set spantree enable all
Spantrees 1-1005 enabled.
```

Example 5-1 Disabling and Enabling Spanning Tree (Continued)

```
Console> (enable) set spantree disable 2
Spantree 2 disabled.
Console> (enable) set spantree enable 2
Spantree 2 enabled.
Console> (enable)
```

IOS-Based Switch

To reenable STP on a per-VLAN basis, use the **spanning-tree vlan** *vlan-id* global configuration command. Use the **no** form of this command to disable STP. Example 5-2 illustrates disabling and reenabling STP for VLANs 2 and 3.

Example 5-2 Disabling and Enabling Spanning Tree on an IOS-Based Switch

```
Switch(config)#no spanning-tree vlan 2 3
Switch(config)#spanning-tree vlan 2 3
```

show spantree **and** show spanning-tree

This section looks at the commands used to verify STP configuration and monitor STP operation. For this purpose, you use the **show spantree** command on CatOS switches and the **show spanning-tree** command on IOS-based switches. As you can see, it is easy to confuse these commands. On a CatOS switch, the **show span** command analyzes network traffic; the fact that it uses the same first four letters as **spantree** can add to the possible confusion.

CatOS Switch

The most important switch command for working with STP is the **show spantree** command. Depending on your switch, there are a number of parameters available for prescribing the desired output, as shown in Example 5-3. Many of these parameters are discussed later in this chapter.

Example 5-3 show spantree Command Has a Number of Useful Variants

```
Console> (enable)  show spantree ?
  active                      Show active ports in spanning tree
  backbonefast                Show spanning tree backbone fast
  blockedports                Show ports that are blocked
  bpdu-skewing                Show spanning tree bpdu skewing statistics
  conflicts                   Show spanning tree conflicting info for vlan
```

continues

Example 5-3 show spantree Command Has a Number of Useful Variants (Continued)

```
defaultcostmode           Show spanning tree port cost mode
guard                     Show spanning tree guard info
mapping                   Show spanning tree vlan and instance mapping
mistp-instance            Show spantree info for MIST instance
portfast                  Show spanning tree port fast info
portinstancecost          Show spanning tree port instance cost
portvlancost              Show spanning tree port vlan cost
statistics                Show spanning tree statistic info
summary                   Show spanning tree summary
uplinkfast                Show spanning tree uplink fast
<mod/port>                Module number and Port number(s)
<vlan>                    VLAN number for PVST+
<cr>
```

Example 5-4 displays sample output for the **show spantree** command.

Example 5-4 show spantree Output Is Extremely Useful for Monitoring and Troubleshooting STP

```
Console> (enable)  show spantree 2
VLAN 2
Spanning tree mode          PVST+
Spanning tree type          ieee
Spanning tree enabled

Designated Root             00-02-fd-43-62-01
Designated Root Priority    8192
Designated Root Cost        0
Designated Root Port        1/0
Root Max Age   20 sec   Hello Time 2  sec   Forward Delay 15 sec

Bridge ID MAC ADDR          00-02-fd-43-62-01
Bridge ID Priority          8192
Bridge Max Age 20 sec   Hello Time 2  sec   Forward Delay 15 sec

Port                     Vlan Port-State    Cost      Prio Portfast Channel_id
------------------------ ---- ------------- --------- ---- -------- ---------
```

Example 5-4 show spantree Output Is Extremely Useful for Monitoring and Troubleshooting STP

```
2/1                        2     forwarding        20000    32 disabled 0
2/4                        2     forwarding       200000    32 disabled 0
2/27                       2     forwarding       200000    32 disabled 0
2/28                       2     forwarding       200000    32 disabled 0
2/29                       2     forwarding       200000    32 disabled 0
2/30                       2     forwarding       200000    32 disabled 0
```

This **show spantree** output is separated by blank lines and can be broken into four sections:

- Global statistics for the current switch/bridge (lines 2 through 4)
- Root Bridge statistics (lines 5 through 9)
- Local bridge statistics (lines 10 through 12)
- Port statistics (lines 13 through 20)

The global statistics appear at the top of the output. The first line of this section (VLAN 2) indicates that the output contains information only for VLAN 2. The second line indicates the STP mode (see the "Spanning-Tree Modes" section). The third line displays the STP type (IEEE). Cisco routers support additional options for STP type, such as DEC and VLAN-bridge. The fourth line indicates that STP is enabled on this switch for this VLAN.

The first line of the Root Bridge statistics displays the MAC address of the current Root Bridge. The second line displays the Bridge Priority of the Root Bridge. The Path Cost to the Root Bridge, 0, is displayed in the Designated Root Cost field. The fourth line of this section shows the Root Port of the local device. The last line of the Root Bridge statistics section shows the timer values currently set on the Root Bridge. These values are used throughout the entire network to provide consistency.

The local bridge statistics section displays the BID of the current bridge in the first two lines. The locally configured timer values are shown in the third line of this section. These timer values are not utilized unless the current bridge becomes the Root Bridge at some point (this bridge happens to be a Root Bridge).

The port statistics section is displayed at the bottom of the output. Depending on the number of ports present on the switch, this display can continue for many screens. The output includes the cost value associated with each port. This value is the cost that is added to the Root Path Cost field contained in BPDUs received on this port. With respect to STP, this value is relevant only for trunk ports.

To view only spanning-tree output pertinent to active ports, use the **show spantree active** command, as illustrated in Example 5-5.

Example 5-5 active Option of the show spantree Command Focuses on Active Ports

```
Console> (enable) show spantree 1 active

VLAN 1
Spanning tree mode          MISTP
Spanning tree type          ieee
Spanning tree enabled
VLAN mapped to MISTP Instance: 1

Port                     Vlan Port-State    Cost      Prio Portfast Channel_id
------------------------ ---- ------------- --------- ---- -------- ----------
 2/3                     1    forwarding       200000   32 disabled 0
 2/12                    1    forwarding       200000   32 disabled 0
Console> (enable)
```

For a snapshot of the **show spantree** output, use the **summary** keyword, as demonstrated in Example 5-6.

Example 5-6 summary Option of the show spantree Command Gives You a Snapshot of Spanning-Tree Information

```
Console> (enable) show spantree summary
MAC address reduction: disabled
Root switch for vlans: 1-8,200.
BPDU skewing detection disabled for the bridge
BPDU skewed for vlans:  none.
Portfast bpdu-guard disabled for bridge.
Portfast bpdu-filter disabled for bridge.
Uplinkfast disabled for bridge.
Backbonefast disabled for bridge.

Summary of connected spanning tree ports by vlan

VLAN  Blocking Listening Learning Forwarding STP Active
----- -------- --------- -------- ---------- ----------
```

Example 5-6 summary Option of the show spantree Command Gives You a Snapshot of Spanning-Tree Information (Continued)

```
   1        0        0       0         18         18
   2        0        0       0          6          6
   3        0        0       0          6          6
   4        0        0       0          6          6
   5        0        0       1          6          7
   6        0        0       0          6          6
   7        0        0       0          6          6
   8        0        0       0          6          6
 200        0        0       0          6          6

     Blocking Listening Learning Forwarding STP Active
----- -------- --------- -------- ---------- ----------
Total       0        0       1         66         67

Console>
```

If the switch is not the root for any VLANs, **none** is displayed in the **Root switch for vlans** field. You also see a quick summary of how many ports are currently in a given STP state. For example, VLAN 5 currently has one port in the Forwarding state.

If you have a real need for detailed, nitty-gritty spanning-tree information, the **show spantree statistics** *mod/port* [*vlan*] command is your best bet. This is a great tool for STP troubleshooting. Example 5-7 illustrates this.

Example 5-7 statistics Option of the show spantree Command Gives Very Detailed Information on Spanning Tree Statistics

```
Console> (enable) show spantree statistics 2/27 3
Port  2/27  VLAN 3

SpanningTree enabled for vlan = 3

                BPDU-related parameters
port spanning tree                   enabled
state                                forwarding
port_id                              0x805b
port number                          0x5b
```

continues

Example 5-7 statistics Option of the show spantree Command Gives Very Detailed Information on Spanning Tree Statistics (Continued)

```
path cost                               200000
message age (port/VLAN)                 0(20)
designated_root                         00-02-fd-43-62-02
designated_cost                         0
designated_bridge                       00-02-fd-43-62-02
designated_port                         0x805b
top_change_ack                          FALSE
config_pending                          FALSE
port_inconsistency                      none

                PORT based information & statistics
config bpdu's xmitted (port/VLAN)       141511(388495)
config bpdu's received (port/VLAN)      4(21)
tcn bpdu's xmitted (port/VLAN)          0(0)
tcn bpdu's received (port/VLAN)         5(6)
forward trans count                     1
scp failure count                       0
root inc trans count (port/VLAN)        0(0)
inhibit loopguard                       FALSE
loop inc trans count (port/VLAN)        0(0)

                Status of Port Timers
forward delay timer                     INACTIVE
forward delay timer value               15
message age timer                       INACTIVE
message age timer value                 0
topology change timer                   INACTIVE
topology change timer value             35
hold timer                              INACTIVE
hold timer value                        1
delay root port timer                   INACTIVE
delay root port timer value             0
delay root port timer restarted is      FALSE

                Vlan based information & statistics
spanningtree type                       ieee
```

Example 5-7 statistics Option of the show spantree Command Gives Very Detailed Information on Spanning Tree Statistics (Continued)

```
spanningtree multicast address       01-80-c2-00-00-00
bridge priority                      8192
bridge mac address                   00-02-fd-43-62-02
bridge hello time                    2 sec
bridge forward delay                 15(15) sec
topology change initiator:           2/27
last topology change occured:        Tue Jan 29 2002, 00:32:19
topology change                      FALSE
topology change time                 35
topology change detected             FALSE
topology change count                14
topology change last recvd. from     00-07-84-f9-f6-98

                Other port-specific info
dynamic max age transitions          0
port bpdu ok count                   0
msg age expiry count                 0
link loading                         1
bpdu in processing                   FALSE
num of similar bpdus to process      0
received_inferior_bpdu               FALSE
next state                           3
src mac count:                       0
total src mac count                  0
curr_src_mac                         00-00-00-00-00-00
next_src_mac                         00-00-00-00-00-00
channel_src_mac                      00-00-00-00-00-00
channel src count                    0
channel ok count                     0
Console> (enable)
```

In the output subsection titled **PORT based information & statistics**, the output has several entries with a number followed by a number in parentheses. In this case, the first number represents the number of packets of the specified type for the port specified in the **show spantree** command. The number in parentheses represents the number of packets of that type for the entire switch.

This command can also be used to identify port configuration mismatches (such as ISL versus IEEE 802.1Q). If you see output other than **none** for the **port inconsistency** output under the subsection titled **BPDU-related parameters**, you have some reconfiguration to do.

IOS-Based Switch

Example 5-8 illustrates the **show spanning-tree** options available on an IOS-based switch.

Example 5-8 show spanning-tree Command Has Fewer Options Than Its CatOS Equivalent

```
Switch#show spanning-tree ?
  brief       VLAN Switch Spanning Brief
  interface   Interface spanning tree information
  summary     VLAN Switch Spanning Summary
  vlan        VLAN Switch Spanning Trees
  |           Output modifiers
  <cr>
```

Example 5-9 shows a portion of the output of the **show spanning-tree** command. As you can see, this output is formatted differently but contains most of the same information as the CatOS **show spantree** command. If you leave off the **vlan** modifier, output is restricted to VLAN 1. Output is displayed for each interface, so the output extends for many screens.

Example 5-9 show spanning-tree Command Displays Spanning-Tree Information for Each Interface

```
Switch#show spanning-tree


Spanning tree 1 is executing the IEEE compatible Spanning Tree protocol
  Bridge Identifier has priority 8000, address 0007.84f9.f680
  Configured hello time 2, max age 20, forward delay 15
  Current root has priority 7000, address 0002.fd43.6200
  Root port is 2, cost of root path is 12
  Topology change flag set, detected flag not set, changes 55
  Times:  hold 1, topology change 35, notification 2
          hello 2, max age 20, forward delay 15
  Timers: hello 0, topology change 0, notification 0

```

Example 5-9 show spanning-tree Command Displays Spanning-Tree Information for Each Interface (Continued)

```
Interface Fa0/24 (port 2) in Spanning tree 1 is FORWARDING
   Port path cost 12, Port priority 80
   Designated root has priority 7000, address 0002.fd43.6200
   Designated bridge has priority 7000, address 0002.fd43.6200
   Designated port is 46, path cost 0
   Timers: message age 1, forward delay 0, hold 0
   BPDU: sent 1, received 346

Interface Fa0/2 (port 8) in Spanning tree 1 is FORWARDING
   Port path cost 19, Port priority 128
   Designated root has priority 7000, address 0002.fd43.6200
   Designated bridge has priority 8000, address 0007.84f9.f680
   Designated port is 8, path cost 12
   Timers: message age 0, forward delay 0, hold 0
   BPDU: sent 9573, received 150647

Interface Fa0/4 (port 10) in Spanning tree 1 is down
   Port path cost 100, Port priority 128
   Designated root has priority 7000, address 0002.fd43.6200
   Designated bridge has priority 8000, address 0007.84f9.f680
   Designated port is 10, path cost 12
   Timers: message age 0, forward delay 0, hold 0
   BPDU: sent 1, received 0
<output omitted>
```

Again, this output is for VLAN 1 only, contrary to some of the Command Reference documentation at www.cisco.com. Note that output for EtherChannels appears first. Here, although you do not see the output for Fa0/23 in Example 5-9, Fa0/23 and Fa0/24 participate in a Fast EtherChannel, referred to on IOS-based switches as a *port group* or a *channel group*, depending on the version of Cisco IOS Software. Channel groups replaced port groups in Cisco IOS Software Release 12.1(6). Example 5-10 shows that Fa0/23 and Fa0/24 form an EtherChannel. EtherChannel technology is explored later in this chapter.

Example 5-10 Interfaces Fa0/23 and Fa0/24 Participate in a Fast EtherChannel

```
Switch#show port group
Group  Interface              Transmit Distribution
-----  ---------------------  ---------------------
    2  FastEthernet0/23       source address
    2  FastEthernet0/24       source address
```

No output appears for interface Fa0/1 in Example 5-9. In this case, Fa0/1 is configured as an access port for VLAN 20, which is not part of the VTP domain, as shown in Example 5-11.

Example 5-11 VLAN 20 Is Not in the VTP Domain

```
Switch#show vlan brief
VLAN Name                             Status    Ports
---- -------------------------------- --------- -------------------------------
1    default                          active    Fa0/4, Fa0/5, Fa0/7, Fa0/8,
                                                Fa0/9, Fa0/11, Fa0/12, Fa0/13,
                                                Fa0/14, Fa0/15, Fa0/16, Fa0/17,
                                                Fa0/18, Fa0/19, Fa0/20, Fa0/21,
                                                Fa0/22
2    VLAN0002                         active
3    VLAN0003                         active
4    VLAN0004                         active    Fa0/3, Fa0/6
5    VLAN0005                         active
6    VLAN0006                         active
7    VLAN0007                         active    Fa0/10
8    VLAN0008                         active
200  VLAN0200                         active
1002 fddi-default                     active
1003 token-ring-default               active
1004 fddinet-default                  active
1005 trnet-default                    active
```

Also note that no output for Fa0/3 appears in Example 5-9 because Fa0/3 is an access port for VLAN 4 (see Example 5-12), and the output is restricted to VLAN 1.

Example 5-12 Interface Fa0/3 Is an Access Port in VLAN 4

```
Switch#show running-config
<output omitted>
interface FastEthernet0/1
 switchport access vlan 20
!
interface FastEthernet0/2
switchport mode trunk
!
interface FastEthernet0/3
 switchport access vlan 4
!
interface FastEthernet0/4
<output omitted>
```

The **brief** keyword appended to the **show spanning-tree** command provides a snapshot of spanning-tree information, with one "chunk" of output displayed per VLAN, as shown in Example 5-13.

Example 5-13 Interface Fa0/3 Is an Access Port in VLAN 4

```
Switch#show spanning-tree brief

VLAN1
  Spanning tree enabled protocol IEEE
  ROOT ID     Priority 7000
              Address 0002.fd43.6200
              Hello Time   2 sec  Max Age 20 sec  Forward Delay 15 sec

  Bridge ID  Priority    8000
              Address     0007.84f9.f680
              Hello Time   2 sec  Max Age 20 sec  Forward Delay 15 sec


Port                          Designated
Name    Port ID Prio Cost Sts  Cost  Bridge ID      Port ID
------- ------- ---- ---- ---  ----  -------------- -------
```

continues

Example 5-13 Interface Fa0/3 Is an Access Port in VLAN 4 (Continued)

```
Fa0/2    128.8   128  19   FWD  19    0007.84f9.f680 128.8
Fa0/4    128.10  128  100  BLK  19    0007.84f9.f680 128.10
Fa0/5    128.11  128  100  BLK  19    0007.84f9.f680 128.11
Fa0/6    128.12  128  100  BLK  19    0007.84f9.f680 128.12
Fa0/7    128.13  128  100  BLK  19    0007.84f9.f680 128.13
Fa0/8    128.14  128  100  BLK  19    0007.84f9.f680 128.14
Fa0/9    128.15  128  100  BLK  19    0007.84f9.f680 128.15
Fa0/11   128.17  128  100  BLK  19    0007.84f9.f680 128.17
Fa0/12   128.18  128  100  BLK  19    0007.84f9.f680 128.18
Fa0/13   128.19  128  100  BLK  19    0007.84f9.f680 128.19


Port                                  Designated
Name     Port ID Prio Cost Sts  Cost  Bridge ID      Port ID
-------  ------- ---- ---- ---  ----  -------------- -------
Fa0/14   128.20  128  100  BLK  19    0007.84f9.f680 128.20
Fa0/15   128.21  128  100  BLK  19    0007.84f9.f680 128.21
Fa0/16   128.22  128  100  BLK  19    0007.84f9.f680 128.22
Fa0/17   128.23  128  100  BLK  19    0007.84f9.f680 128.23
Fa0/18   128.24  128  100  BLK  19    0007.84f9.f680 128.24
Fa0/19   128.25  128  100  BLK  19    0007.84f9.f680 128.25
Fa0/20   128.26  128  100  BLK  19    0007.84f9.f680 128.26
Fa0/21   128.27  128  100  BLK  19    0007.84f9.f680 128.27
Fa0/22   128.28  128  100  BLK  19    0007.84f9.f680 128.28
Fa0/23   128.29  128  19   BLK  0     0002.fd43.6200 128.92
Fa0/24   80 .30  80   19   FWD  0     0002.fd43.6200 128.91

VLAN2
  Spanning tree enabled protocol IEEE
  ROOT ID     Priority 8192
              Address 0002.fd43.6201
              Hello Time   2 sec  Max Age 20 sec  Forward Delay 15 sec

  Bridge ID   Priority    32768
              Address     0007.84f9.f681
              Hello Time   2 sec  Max Age 20 sec  Forward Delay 15 sec
```

Example 5-13 Interface Fa0/3 Is an Access Port in VLAN 4 (Continued)

```
Port                           Designated
Name    Port ID Prio Cost Sts  Cost  Bridge ID      Port ID
------- ------- ---- ---- ---  ----  -------------- -------
Fa0/2   128.8   128  19   FWD  19    0007.84f9.f681 128.8
Fa0/23  128.29  128  19   BLK  0     0002.fd43.6201 128.92
Fa0/24  80 .30  80   19   FWD  0     0002.fd43.6201 128.91
<output omitted>
```

Finally, the **summary** option for the **show spanning-tree** command provides even more compact output than the **brief** option. Sample output is shown in Example 5-14. The output indicates that one port is in the Learning state and one port is in the Forwarding state. Note that an EtherChannel counts as one port—it is a single logical port constructed from two (or more) physical ports. In this case, Fa0/23 and Fa0/24 form a Fast EtherChannel (called a port group or channel group in IOS switch-speak), as shown in Example 5-10.

Example 5-14 show spanning-tree summary Gives Compact STP Output with Port Counts for Each VLAN for Each STP State

```
Switch#show spanning-tree summary

UplinkFast is disabled


Name                 Blocking Listening Learning Forwarding STP Active
-------------------- -------- --------- -------- ---------- ----------
VLAN1                17       0         1        1          19
VLAN2                0        0         1        1          2
VLAN3                0        0         1        1          2
VLAN4                2        0         1        1          4
VLAN5                0        0         1        1          2
VLAN6                0        0         1        1          2
VLAN7                1        0         1        1          3
VLAN8                0        0         1        1          2
VLAN200              0        0         1        1          2
-------------------- -------- --------- -------- ---------- ----------
             9 VLANs 20       0         9        9          38
```

The next section explores the various options for spanning-tree modes.

Spanning-Tree Modes

An important decision to make when implementing VLANs is which spanning-tree mode to use on your Catalyst switches. STP modes determine how STP interacts with your VLAN infrastructure. The default Cisco STP mode defines one instance of STP per VLAN. The IEEE 802.1Q standard references a single STP instance for a switched VLAN infrastructure, but it does not preclude the option of implementing one instance of STP per VLAN. The standard does not specify whether or how this should be done. In a mixed-vendor environment, you have some design issues to address in reconciling the distinct STP implementations. This section describes the various STP modes available on Catalyst switches and how Cisco solutions resolve some of the differences in STP implementations.

PVST (Per-VLAN Spanning Tree), PVST+, and Mono Spanning Tree Modes

Common Spanning Tree (CST) is specified in the IEEE 802.1Q standard. CST defines a single instance of Spanning Tree for all VLANs. BPDUs are transmitted over VLAN 1.

Per-VLAN Spanning Tree (PVST) is a Cisco-proprietary implementation requiring ISL trunk encapsulation. PVST runs a separate instance of STP for each VLAN. Among other things, this gives you the ability to configure distinct Root Switches for each VLAN and configure Layer 2 load balancing. Figure 5-15 shows a typical implementation of PVST in a campus network.

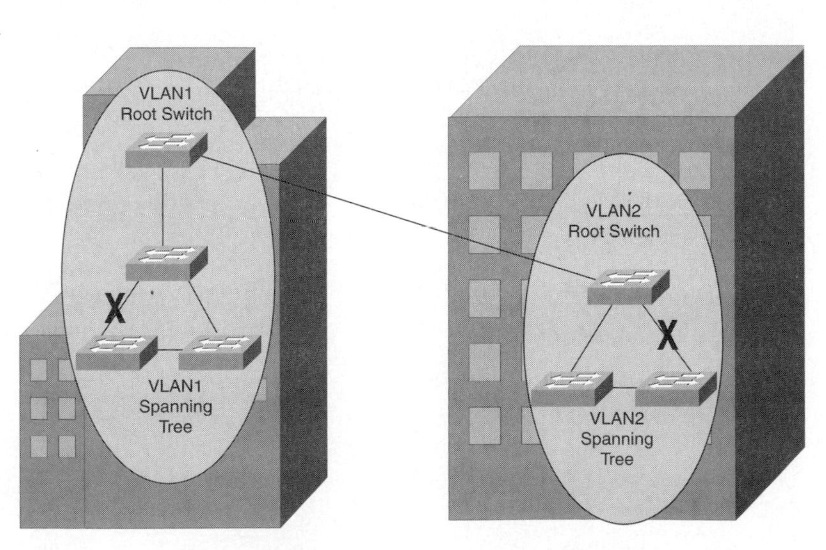

Figure 5-15 PVST enhances STP scalability and allows for Layer 2 load balancing.

Having a separate instance of spanning tree for each VLAN reduces the recovery time for STP recalculation and hence increases your network's reliability in the following ways:

- Reduces the overall size of the spanning-tree topology
- Improves scalability and decreases convergence time
- Provides faster recovery and better reliability

Disadvantages of a spanning tree for each VLAN include the following:

- Utilization of switches (such as CPU load) to support spanning tree maintenance for multiple VLANs
- Utilization of bandwidth on trunk links to support BPDUs for each VLAN

PVST+ is a Cisco-proprietary STP mode that allows CST and PVST to exist on the same network. PVST+ provides support for 802.1Q trunks and the mapping of multiple spanning trees to the single spanning tree of non-Cisco 802.1Q switches. PVST+ is automatically enabled on Catalyst 802.1Q trunks. It runs one instance of STP per VLAN when Catalyst switches are connected by 802.1Q trunks.

NOTE

The IEEE 802.1s standard is currently in draft form. This standard provides "a facility for VLAN bridges to use multiple spanning trees, providing for traffic belonging to different VLANs to flow over potentially different paths within the virtual bridged LAN."

PVST+ is the default Spanning-Tree Protocol used on all Ethernet, Fast Ethernet, and Gigabit Ethernet port-based VLANs on Catalyst 4000 and 6000 family switches and Catalyst 5000 switches starting with CatOS 4.1. PVST+ runs on each VLAN on the switch, ensuring that each has a loop-free path through the network.

PVST+ provides Layer 2 load balancing for the VLAN on which it runs. You can create different logical topologies using the VLANs on your network to ensure that all your links will be used but no one link will be oversubscribed.

Each instance of spanning tree has a single Root Switch. This Root Switch propagates the spanning-tree information associated with that VLAN to all other switches in the network. Because each switch has the same knowledge about the network, this process ensures that the network topology is maintained.

The PVST+ architecture distinguishes three types of zones or regions:

- A PVST zone or region
- A PVST+ zone or region
- A Mono Spanning Tree zone or region

Mono Spanning Tree (MST) is the spanning-tree implementation used by non-Cisco 802.1Q switches. One instance of STP is responsible for all VLAN traffic.

Each zone or region consists of a homogenous type of switch. A PVST region can be connected to a PVST+ region by connecting two ISL ports. Similarly, a PVST+ region can be connected to an MST region by connecting two 802.1Q ports. Notice that an MST and PVST region cannot be connected via a trunk link. Although it is possible to provide a nontrunk connection between the two regions by using an access link, this is of limited use in real-world networks. Figure 5-16 illustrates the three types of STP regions and how they are linked.

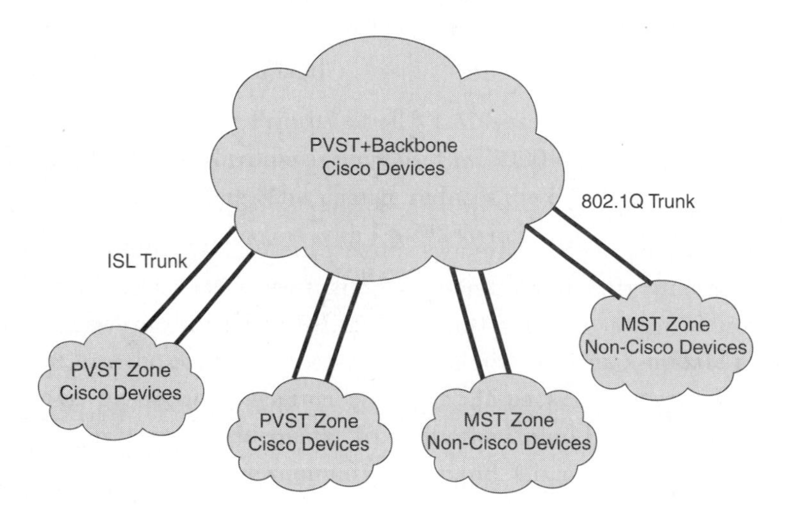

Figure 5-16
STP regions are defined by STP modes.

At the boundary between a PVST region and a PVST+ region, the mapping of spanning trees is one-to-one. At the boundary between an MST region and a PVST+ region, the spanning tree in the MST region maps to one PVST in the PVST+ region. The one it maps to is the CST. The default CST is VLAN 1.

All PVSTs, except the CST, are *tunneled* through the MST region. Tunneling means that BPDUs are flooded through the MST region along the single spanning tree present in the MST region. Do not confuse this use of the term tunnel with the point-to-point tunnels configured on Cisco routers and PIX firewalls. In general, tunnel refers to a mechanism that enables the propagation of packets or frames through a series of devices in such a way that the contents of the packets or frames are not changed in transit.

TECH NOTE: IEEE 802.1Q, NATIVE VLANS, VLAN IDS, UNTAGGED FRAMES, AND TUNNELING

The 12-bit VLAN ID (VID) field in the IEEE 802.1Q tag indicates the VLAN associated with the Ethernet frame, with possible values between 0 and 4095. Refer to Table 4-2.

Each physical port has a parameter called the port VLAN ID (PVID). Every 802.1Q port is assigned a PVID value equal to its Native VLAN ID (the default is VLAN 1). All untagged frames are assigned to the VLAN specified in the PVID parameter. When a port receives a tagged frame, the tag is respected. If the frame is untagged, the value contained in the PVID is used as the tag value. This allows the coexistence, on the same cable, of VLAN-aware bridges/stations and of VLAN-unaware bridges/stations (see Figure 5-17). Consider, for example, the two stations connected to the central trunk link in the lower part of Figure 5-17. They are VLAN-unaware. They are associated with VLAN C because the PVIDs of the VLAN-aware bridges equal VLAN C. Because the VLAN-unaware stations send only untagged frames, when the VLAN-aware bridge devices receive these untagged frames, they assign them to VLAN C.

When an 802.1Q VLAN is configured on an interface, a default of VLAN 1 is automatically created to process the CST. The default VLAN 1 that is created is used only to process spanning-tree BPDU frames. Untagged non-BPDU 802.1Q data frames are not processed by VLAN 1. All untagged data frames are processed by the explicitly defined Native VLAN. However, if no Native VLAN is defined, VLAN 1 becomes the default Native VLAN to handle all untagged frames, including both CST BPDUs and data frames.

Note the following considerations when configuring 802.1Q trunks in networks with mixed PVST+ and MST STP regions:

- When connecting Cisco switches through an 802.1Q trunk, make sure the native VLAN for an 802.1Q trunk is the same on both ends of the trunk link. If the Native VLAN on one end of the trunk is different from the Native VLAN on the other end, spanning-tree loops might result.
- Disabling spanning tree on the Native VLAN of an 802.1Q trunk without disabling spanning tree on every VLAN in the network can cause spanning-tree loops. It is recommended that you leave spanning tree enabled on the native VLAN of an 802.1Q trunk. If this is not possible, disable spanning tree on every VLAN in the network. Make sure your network is free of physical loops before disabling spanning tree.
- When you connect two Cisco switches through 802.1Q trunks, the switches exchange spanning-tree BPDUs on each VLAN allowed on the trunks. The BPDUs on the trunk's Native VLAN are sent untagged to the reserved IEEE 802.1D spanning-tree multicast MAC address, 01-80-C2-00-00-00. The BPDUs on all other VLANs on the trunk are sent tagged with a VID to the reserved Cisco Shared Spanning Tree multicast MAC address, 01-00-0C-CC-CC-CD. Note that the Cisco multicast MAC address, 01-00-0C-CC-CC-CC, is used for CDP, VTP, DISL, DTP, and PAgP (see the "EtherChannel Operation: PAgP and LACP" section later in this chapter).
- Non-Cisco 802.1Q switches maintain only a single instance of spanning tree (MST) that defines the spanning-tree topology for all VLANs. When you connect a Cisco switch to a non-Cisco switch through an 802.1Q trunk, the MST of the non-Cisco switch and the Native VLAN spanning tree of the Cisco switch combine to form the CST. CST BPDUs are sent untagged to the reserved IEEE 802.1D spanning-tree multicast MAC address 01-80-C2-00-00-00.
- Because Cisco switches transmit BPDUs to the Cisco Shared Spanning Tree multicast MAC address on VLANs other than the trunk's Native VLAN, non-Cisco switches do not recognize these frames as BPDUs and flood them as they would regular multicast data. Other Cisco switches connected to the non-Cisco 802.1Q cloud receive these flooded BPDUs and flood the frames as well. This allows Cisco switches to maintain a per-VLAN spanning-tree topology across a cloud of non-Cisco 802.1Q switches. The non-Cisco 802.1Q cloud separating the Cisco switches is treated as a single broadcast segment between all switches connected to the non-Cisco 802.1Q cloud through 802.1Q trunks. This process is called *tunneling*.
- Make certain that the Native VLAN is the same on all the 802.1Q trunks connecting the Cisco switches to the non-Cisco 802.1Q cloud.

- If you are connecting multiple Cisco switches to a non-Cisco 802.1Q cloud, all the connections must be through 802.1Q trunks. You cannot connect Cisco switches to a non-Cisco 802.1Q cloud through ISL trunks or access ports. Doing so would cause the switch to place the ISL trunk port or access port into the spanning tree "port inconsistent" state, and no traffic passes through the port.

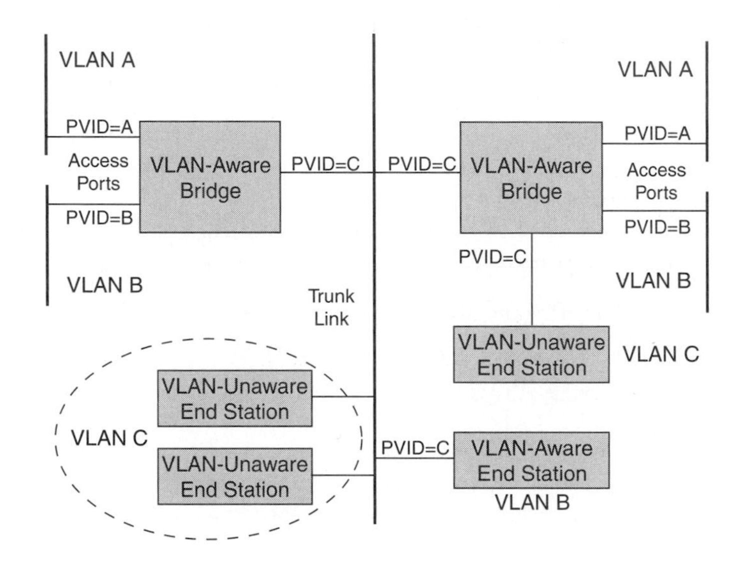

Figure 5-17
The Native VLAN is the VLAN assigned to untagged 802.1Q frames.

MISTP Mode

After the development of PVST+, someone got the idea that it would be useful to have an STP mode that allows for a compromise between PVST+ and MST. *Multiple Instance of Spanning Tree Protocol (MISTP)* allows you to group multiple VLANs under a single instance of spanning tree. MISTP combines the Layer 2 load-balancing benefits of PVST+ with the lower CPU load of IEEE 802.1Q.

MISTP is an optional STP mode that runs on Catalyst 4000 and 6000 family switches. A MISTP instance is a virtual logical topology defined by a set of bridge and port parameters. A MISTP instance becomes a real topology when VLANs are mapped to it. Each MISTP instance has its own Root Switch and a different set of forwarding links (that is, different bridge and port parameters). This Root Switch propagates the information associated with that instance of MISTP to all other switches in the network. This process ensures that the network topology is maintained, because each switch has the same knowledge about the network.

MISTP builds MISTP instances by exchanging MISTP BPDUs with peer entities in the network. There is only one BPDU for each MISTP instance, rather than for each VLAN as in PVST+. There are fewer BPDUs in a MISTP network, so there is less overhead in the network. MISTP discards any PVST+ BPDUs it sees.

A MISTP instance can have any number of VLANs mapped to it, but a VLAN can be mapped to only a single MISTP instance. You can easily move a VLAN (or VLANs) in a MISTP topology to another MISTP instance if it has converged. However, if ports are added at the same time the VLAN is moved, convergence time is required.

MISTP-PVST+ Mode

MISTP-PVST+ is a transition STP mode that allows you to use the MISTP functionality of Catalyst 4000 and 6000 series switches while continuing to communicate with the older Catalyst switches in your network that use PVST+. A switch using PVST+ mode and a switch using MISTP mode that are connected cannot see the BPDUs of the other switch, a condition that can cause loops in the network. MISTP-PVST+ allows interoperability between PVST+ and MISTP, because it detects the BPDUs of both modes. If you want to convert your network to MISTP, you can use MISTP-PVST+ to transition the network from PVST+ to MISTP in order to avoid problems.

MISTP-PVST+ conforms to the limits of PVST+. For example, you can configure only the number of VLAN ports on your MISTP-PVST+ switches that you configure on your PVST+ switches.

Multiple Spanning Tree Mode

The acronym MST stands for two things: Mono Spanning Tree and Multiple Spanning Tree. Be careful to differentiate between the two in context. The ***Multiple Spanning Tree*** feature was originally released in CatOS 7.1 for Catalyst 4000 and 6000 series switches. MST is specified in IEEE 802.1s, an amendment to IEEE 802.1Q. MST extends the IEEE 802.1w Rapid Spanning Tree (RST) algorithm to multiple spanning trees, as opposed to the single CST of the original IEEE 802.1Q specification. This extension provides for both rapid convergence and load balancing in a VLAN environment. Cisco's implementation of MST is backward-compatible with 802.1D STP, the 802.1w Rapid Spanning Tree Protocol (RSTP), and the Cisco PVST+ architecture. You can think of MST as a standards-based MISTP.

The IEEE 802.1w specification, Rapid Spanning Tree Protocol, provides for subsecond reconvergence of STP after failure of one of the uplinks in a bridged environment. 802.1w provides the structure on which the 802.1s feature operates.

The IEEE 802.1s Multiple Spanning Tree specification allows a user to build multiple spanning trees over VLAN trunks. Fast convergence of the MST topology is achieved using a modified version of the RSTP protocol. You can group and associate VLANs to spanning-tree instances. Each instance can have a topology independent of other spanning-tree instances. This new architecture provides multiple forwarding paths for data traffic and enables load balancing. Network fault tolerance is improved, because a failure in one instance does not affect other instances.

In large networks, having different VLAN spanning-tree instance assignments located in different parts of the network makes it easier to administer and utilize redundant paths. However, a spanning-tree instance can exist only on bridges that have compatible VLAN instance assignments. Therefore, MST requires that you configure a set of bridges with the same MST configuration information, allowing them to participate in a given set of spanning-tree instances. Interconnected bridges that have the same MST configuration are called MST regions.

MST, like MISTP, provides interoperability with PVST+ regions.

Advanced STP Configuration

At this point, we have covered the essential details of STP operation. Now you have the foundation necessary to make informed decisions about how best to deploy STP in a campus network.

Earlier in this chapter, you saw how to enable and disable STP on a per-VLAN and wholesale basis. Here we detail the STP configuration of bridge priority, port priority, per-VLAN port priority, port cost, per-VLAN port cost, and spanning-tree timers. These options let you control Layer 2 redundancy, STP convergence, and Layer 2 load balancing. Although these options can be configured for PVST+ VLANs, MISTP instances, and Multiple Spanning Tree instances, we will cover just the PVST+ configuration. For the MISTP and MST configurations, refer to the respective Catalyst Family Configuration Guide at www.cisco.com/univercd/cc/td/doc/product/lan.

For the remainder of this chapter, all configurations are performed in PVST+ STP mode.

Configuring a Root Switch

One of the most important decisions you must make in the spanning-tree network is the location of the Root Switch. Proper placement of the Root Switch optimizes the path that STP chooses, providing deterministic paths for data. You can manually configure a bridge to be a Root Switch as well as a backup, or secondary, Root Switch. The job of the secondary Root Switch is to take over if the primary Root Switch fails.

With PVST+, you can configure a distinct Root Switch for each VLAN. This allows you to optimize Layer 2 traffic flow on your network. The configurations vary with the Catalyst platforms, but the idea is the same: Adjust the Bridge Priority and (optionally) the STP timers.

It is strongly recommended that you do not configure an access-layer switch to be a Root Switch. An STP Root Switch should always be a distribution-layer or core-layer switch.

Next, we detail the configuration of Root Switches on CatOS and IOS-based switches.

CatOS Switch

There are two ways to configure a CatOS switch to be a Root Switch: Use the **set spantree root** command, or manually set the Bridge Priority.

The **set spantree root** command lowers the Bridge Priority to below the default of 32768. When you specify a switch as the primary Root Switch, the default Bridge Priority is modified so that it becomes the root for the specified VLANs. The command sets the bridge priority to 8192. If this setting does not result in the switch's becoming a root, it modifies the Bridge Priority to be 1 less or the same as the Bridge Priority of the current Root Switch. If reducing the Bridge Priority as low as 1 still does not make the switch the Root Switch, the system displays a message. This command is really a macro, as you can see by analyzing the **show config** output, which shows only changes reflected by **set spantree priority** commands.

The syntax for this command is **set spantree root** [*vlans*] [**dia** *network_diameter*] [**hello** *hello_time*]. **dia** specifies the maximum number of bridges (diameter) between any two points of attachment of end stations; valid values range from 2 to 7. Example 5-15 shows how to configure the primary Root Switch for VLAN 200.

Example 5-15 Setting the Root Switch with CatOS

```
Console> (enable) set spantree root 100-500 dia 5
VLAN 200 bridge priority set to 8192.
VLAN 200 bridge max aging time set to 16.
VLAN 200 bridge hello time set to 2.
VLAN 200 bridge forward delay set to 12.
Switch is now the root switch for active VLAN 200.
Console> (enable)
```

You can configure a secondary Root Switch for a VLAN with the **set spantree root secondary** command. This command reduces the Bridge Priority to 16384, making it the probable candidate to become the Root Switch if the primary Root Switch fails. You can run this command on more than one switch to create multiple backup switches in case the primary Root Switch fails.

The syntax for this command is **set spantree root** [**secondary**] *vlans* [**dia** *network_diameter*] [**hello** *hello_time*]. Example 5-16 illustrates how to configure the secondary Root Switch for VLANs 22 and 24, also cutting the default Hello Time in half. Note that, in general, it is best to leave the default settings for the STP timers.

NOTE

Catalyst 6000 family switches have a pool of 1024 MAC addresses that can be used as bridge identifiers for VLANs running under PVST+ or for MISTP instances. You can use the **show module** command to view the MAC address range.

MAC addresses are allocated sequentially, with the first MAC address in the range assigned to VLAN 1, the second MAC address in the range assigned to VLAN 2, and so on. The last MAC address in the range is assigned to the supervisor engine in-band management interface, sc0.

For example, if the MAC address range is 00-e0-1e-9b-2e-00 to 00-e0-1e-9b-31-ff, the VLAN 1 bridge ID is 00-e0-1e-9b-2e-00, the VLAN 2 bridge ID is 00-e0-1e-9b-2e-01, the VLAN 3 bridge ID is 00-e0-1e-9b-2e-02, and so forth. The in-band (sc0) interface MAC address is 00-e0-1e-9b-31-ff.

Example 5-16 Configuring the Secondary Root Switch with CatOS

```
Console> (enable) set spantree root secondary 22,24 dia 5 hello 1
VLANs 22,24 bridge priority set to 16384.
VLANs 22,24 bridge max aging time set to 10 seconds.
VLANs 22,24 bridge hello time set to 1 second.
VLANs 22,24 bridge forward delay set to 7 seconds.
Console> (enable)
```

Note that, in Example 5-16, the **set spantree root** macro dynamically adjusts the STP timers. This is an advantage over manually configuring the Bridge Priority, where you have to tweak the STP timers yourself.

The second way to configure a Root Switch is to use the **set spantree priority** command. Before giving the details, we need to explain the concepts of MAC address reduction and Bridge ID Priority. (Bridge ID Priority is not the same as Bridge Priority.)

For Catalyst 6000 family switches that support 4096 VLANs, MAC address reduction lets up to 4096 VLANs running under PVST+ or 16 MISTP instances have unique identifiers without increasing the number of MAC addresses required on the switch.

MAC address reduction reduces the number of MAC addresses required by STP from one per VLAN or MISTP instance to one per switch. However, VLANs running under PVST+ and MISTP instances running under MISTP-PVST+ or MISTP are considered "logical bridges," so each bridge must have its own unique identifier in the network.

When you enable MAC address reduction, the bridge identifier stored in the spanning-tree BPDU contains an additional field called the System ID Extension. Combined with the Bridge Priority, the System ID Extension functions as the unique identifier for a VLAN or a MISTP instance. The *System ID Extension* is always the 12-bit number of the VLAN or the MISTP instance. For example, the System ID Extension for VLAN 100 is 100, and the System ID Extension for MISTP instance 2 is 2. Note that 12 bits allows for 4096 combinations.

If you have a Catalyst 6000 family switch in your network and you have MAC address reduction enabled on it, you should also enable MAC address reduction on all your Catalyst 4000 series and 5000 family switches to avoid problems in the spanning-tree topology.

Figure 5-18 shows the Bridge ID when you do not enable MAC address reduction. The bridge identifier consists of the Bridge Priority and the MAC address.

Figure 5-18 The usual Bridge ID consists of an ordered pair—the Bridge Priority and a MAC address.

MAC address reduction is disabled by default on Catalyst 5000 and 6000 switches. To enable it on CatOS switches, use the command **set spantree macreduction enable**; you can then create VLANs numbered 1025 to 4094 on a Catalyst 6000. Figure 5-19 shows the Bridge Identifier when you enable MAC address reduction. The Bridge Identifier consists of the Bridge Priority, the System ID Extension, and the MAC address (an ordered triple). The Bridge Priority and the System ID Extension combined are known as the ***Bridge ID Priority***. The Bridge ID Priority is the unique identifier for the VLAN or the MISTP instance.

Figure 5-19 The Bridge Priority and the System ID Extension comprise the Bridge ID Priority.

Notice that the 4 high-order bits determine the Bridge Priority when MAC address reduction is enabled. This means that the 16 possible values for the Bridge Priority are 0, 4096, 8192, 12288, 16384, 20480, 24576, 28672, 32768, 36864, 40960, 45056, 49152, 53248, 57344, and 61440.

When you enter a **show spantree** command, you can see the Bridge ID Priority for a VLAN in PVST+ mode or for a MISTP instance in MISTP or MISTP-PVST+ mode. Example 5-17 shows the Bridge ID Priority for VLAN 1 when you enable MAC address reduction in PVST+ mode. The unique identifier for this VLAN is 4097.

Example 5-17 Bridge ID Priority

```
Console> (enable) set spantree macreduction enable
MAC address reduction enabled
Console> (enable) show spantree 1
VLAN 1
Spanning tree mode          PVST+
Spanning tree type          ieee
Spanning tree enabled
```

continues

Example 5-17 Bridge ID Priority (Continued)

```
Designated Root              00-02-fd-43-62-00
Designated Root Priority     4097
Designated Root Cost         0
Designated Root Port         1/0
Root Max Age   20 sec   Hello Time 2  sec   Forward Delay 15 sec

Bridge ID MAC ADDR           00-02-fd-43-62-00
Bridge ID Priority           4097  (bridge priority: 4096, sys ID ext: 1)
Bridge Max Age 20 sec   Hello Time 2  sec   Forward Delay 15 sec
<output omitted>
```

Now we are ready to talk about setting the Bridge Priority. To set the spanning-tree Bridge Priority for a VLAN, use the command **set spantree priority** *bridge_ID_priority* [*vlan*]. Example 5-18 illustrates this command when MAC address reduction is enabled and then disabled.

Example 5-18 Setting the Bridge ID Priority with MAC Address Reduction Enabled and Then Disabled

```
Console> (enable) set spantree priority 25000 1
Spantree priority must be one of these numbers: 0, 4096, 8192, 12288, 16384,
20480, 24576, 28672, 32768, 36864, 40960, 45056, 49152, 53248, 57344, 61440
Console> (enable) set spantree priority 28672 1
Spantree 1 bridge ID priority set to 28673
(bridge priority: 28672 + sys ID extension: 1)
Console> (enable) show spantree 1
VLAN 1
Spanning tree mode           PVST+
Spanning tree type           ieee
Spanning tree enabled

Designated Root              00-04-c0-98-e3-42
Designated Root Priority     7000
Designated Root Cost         100019
Designated Root Port         2/27-28 (agPort 13/46)
Root Max Age   20 sec   Hello Time 2  sec   Forward Delay 15 sec
```

Example 5-18 Setting the Bridge ID Priority with MAC Address Reduction Enabled and Then Disabled (Continued)

```
Bridge ID MAC ADDR          00-02-fd-43-62-00
Bridge ID Priority          28673  (bridge priority: 28672, sys ID ext: 1)
Bridge Max Age 20 sec   Hello Time 2  sec   Forward Delay 15 sec
<output omitted>
Console> (enable) set spantree macreduction disable
MAC address reduction disabled
Console> (enable) set spantree priority 25000 1
Spantree 1 bridge priority set to 25000.
Console> (enable) show spantree 1
VLAN 1
Spanning tree mode          PVST+
Spanning tree type          ieee
Spanning tree enabled

Designated Root             00-04-c0-98-e3-42
Designated Root Priority    7000
Designated Root Cost        100019
Designated Root Port        2/27-28 (agPort 13/46)
Root Max Age   20 sec   Hello Time 2  sec   Forward Delay 15 sec

Bridge ID MAC ADDR          00-02-fd-43-62-00
Bridge ID Priority          25000
Bridge Max Age 20 sec   Hello Time 2  sec   Forward Delay 15 sec
<output omitted>
```

In this example, you see that the switch does not allow setting the Bridge Priority to 25000, because that is not one of the acceptable values when MAC address reduction is enabled.

IOS-Based Switch

Here we describe the configuration of Root Bridges on IOS-based switches. First note that PVST+ is automatically enabled on 802.1Q trunks, so no user configuration is required to set the STP mode.

Use the **spanning-tree vlan** *vlan-id* **root** global configuration command to alter the Bridge Priority. When you enter this command on a switch, it checks the Bridge Priority of the current Root Switch for each VLAN and sets its own Bridge Priority for the specified VLAN to 8192 if this value causes this switch to become the Root Switch for the specified VLAN. If any Root Switch for the specified VLAN has a Bridge Priority lower than 8192, the switch sets its own priority for the specified VLAN to 1 less than the lowest Bridge Priority.

The syntax of this command is **spanning-tree vlan** *vlan-id* **root primary** [**diameter** *net-diameter* [**hello-time** *seconds*]].

Use the **diameter** keyword to specify the network diameter. When you specify the network diameter, the switch automatically sets an optimal Hello Time, Forward Delay, and Max Age for a network of that diameter, which can significantly reduce the convergence time. You can use the **hello** keyword to override the automatically calculated Hello Time. Use the **no spanning-tree vlan** *vlan-id* **root primary** command to return the switch to the default setting (32768).

Example 5-19 shows how to configure a switch as the Root Switch for VLAN 10 with a network diameter of 4.

Example 5-19 Configuring an IOS-Based Switch as a Root Switch

```
Switch(config)#spanning-tree vlan 10 root primary diameter 4
```

To configure a switch as a secondary root, use the preceding command, but replace **primary** with **secondary**. The STP Bridge Priority is then changed from the default value (32768) to 16384 so that the switch is likely to become the Root Switch for the specified VLAN if the primary Root Switch fails.

You can configure multiple backup Root Switches. Use the same network diameter and Hello Time you used when configuring the primary Root Switch. Again, the **no** form of this command returns the switch to the default STP values.

Example 5-20 shows how to configure a switch as a secondary Root Switch for VLAN 10 with a network diameter of 4.

Example 5-20 Configuring an IOS-Based Switch as a Secondary Root Switch

```
Switch(config)#spanning-tree vlan 10 root secondary diameter 4
```

For most situations, it is recommended that you use the global configuration commands **spanning-tree vlan** *vlan-id* **root primary** and **spanning-tree vlan** *vlan-id* **root secondary** to modify the Bridge Priority. However, you might find it necessary in some circumstances to manually configure Bridge Priorities to set a switch as root.

The second way to configure a Root Switch is to use the global configuration command **spanning-tree vlan** *vlan-id* **priority** *priority*. Example 5-21 illustrates the use of this command.

Example 5-21 Manually Configuring Bridge Priority on an IOS-Based Switch

```
Switch(config)#spanning-tree vlan 2 priority 8000
Switch(config)#exit
Switch#show spanning-tree vlan 2

Spanning tree 2 is executing the IEEE compatible Spanning Tree protocol
  Bridge Identifier has priority 8000, address 0004.c098.e344
  Configured hello time 2, max age 20, forward delay 15
  We are the root of the spanning tree
  Topology change flag set, detected flag set, changes 38
  Times:  hold 1, topology change 35, notification 2
          hello 2, max age 20, forward delay 15
  Timers: hello 0, topology change 23, notification 0

Interface Fa0/3 (port 1) in Spanning tree 2 is down
   Port path cost 19, Port priority 128
   Designated root has priority 8000, address 0004.c098.e344
   Designated bridge has priority 8000, address 0004.c098.e344
   Designated port is 1, path cost 0
   Timers: message age 0, forward delay 0, hold 0
   BPDU: sent 35, received 580144
```

Next, we describe the configuration of STP Path Cost.

Configuring Path Cost

After the Root Bridge has been elected, all switches determine the best loop-free path to the root. STP uses, in decreasing priority, the following STP parameters in determining the best path to the Root Bridge:

- Path Cost
- Bridge ID
- Port Priority

If you want to change the path of frames between a particular switch and the root, carefully calculate the current Path Cost, and then change the Port Costs of the desired path. Again, the Path Cost is prescribed by selectively configuring individual Port Costs along the STP paths. Ports with lower Port Costs are more likely to be chosen to forward frames.

Make sure that you calculate the sum of potential alternative paths in addition to the desired path before making changes. This ensures a proper assessment of Path Costs before you make Port Cost changes. It also ensures that frames are forwarded over the desired path.

You might want to refresh your memory regarding Path Cost by reviewing the earlier sections "Path Cost" and "Electing Root Ports." Also, Tables 5-1 and 5-2 summarize Port Cost values.

The following sections describe how to configure Port Cost on a CatOS switch and an IOS-based switch.

CatOS Switch

To configure the Port Cost for a port, use the command **set spantree portcost** {*mod/*port} *cost*. Also, if you want to change the Port Cost for each port in an EtherChannel, use the **set spantree channelcost** command, introduced in CatOS 7.1 (it is supported by Catalyst 4000s and 6000s). This command is actually a macro, entering individual **set spantree portcost** commands in the configuration file. EtherChannel is discussed in detail later in this chapter.

Example 5-22 shows how to configure the Port Cost on a port and verify the configuration.

Example 5-22 Configuring Port Cost on a CatOS Switch

```
Console> (enable) set spantree portcost 2/3 17
Spantree port 2/3 path cost set to 17.
Console> (enable) show spantree 2/3
Port                     Vlan Port-State    Cost      Prio Portfast Channel_id
------------------------ ---- ------------- --------- ---- -------- ----------
 2/3                     1    forwarding           17   32 disabled 0
Console> (enable) show port channel
Port  Status     Channel              Admin Ch
                 Mode                 Group Id
----- ---------- -------------------- ----- -----
 2/27 connected  on                      12   814
 2/28 connected  on                      12   814
----- ---------- -------------------- ----- -----
```

Example 5-22 Configuring Port Cost on a CatOS Switch (Continued)

```
Port  Device-ID                      Port-ID                 Platform
----- ------------------------------ ----------------------- ---------------
 2/27 Switch                         FastEthernet0/24       cisco WS-C2950-24
 2/28 Switch                         FastEthernet0/23       cisco WS-C2950-24
----- ------------------------------ ----------------------- ---------------
Console> (enable) set spantree channelcost 814 12

Port(s) 2/27-28 port path cost are updated to 19.
Channel 814 cost is set to 12.
Warning: channel cost may not be applicable if channel is broken.
Console> (enable)
```

You can configure the Port Cost for a port on a per-VLAN basis. Ports with a lower Port Cost in the VLAN are more likely to be chosen to forward frames.

To configure the Port Cost for a VLAN, use the command **set spantree portvlancost** {*mod/port*} [**cost** *cost*] [*vlan_list*]. Use the command **clear spantree portvlancost** *mod/port* [*vlans*] to restore the default cost to a VLAN on a port.

The **set spantree portvlancost** command is illustrated in Example 5-23. Following the output, details explaining the behavior of this command are provided.

Example 5-23 Machinations of set spantree portvlancost

```
Console> (enable) set spantree portcost 2/28 80
Spantree port 2/28 path cost set to 80.
Console> (enable) show spantree portvlancost 2/28
Port 2/28 VLANs 1-1005 have path cost 80.
Console> (enable) set spantree portvlancost 2/28 cost 15             (A)
Port 2/28 VLANs 1-1005 have path cost 80.
Console> (enable) set spantree portvlancost 2/28 cost 15 1-20        (A)
Port 2/28 VLANs 21-1005 have path cost 80.
Port 2/28 VLANs 1-20 have path cost 15.
Console> (enable) set spantree portvlancost 2/28 cost 15             (B)
Port 2/28 VLANs 21-1005 have path cost 80.
Port 2/28 VLANs 1-20 have path cost 15.
Console> (enable) set spantree portvlancost 2/28 cost 10             (B)
Port 2/28 VLANs 21-1005 have path cost 80.
Port 2/28 VLANs 1-20 have path cost 10.
```

continues

Example 5-23 Machinations of set spantree portvlancost (Continued)

```
Console> (enable) set spantree portvlancost 2/28 cost 10 1-10
Port 2/28 VLANs 21-1005 have path cost 80.
Port 2/28 VLANs 1-20 have path cost 10.
Console> (enable) set spantree portvlancost 2/28 cost 15 1-10
Port 2/28 VLANs 21-1005 have path cost 80.
Port 2/28 VLANs 1-20 have path cost 15.
Console> (enable) set spantree portvlancost 2/28 22
Port 2/28 VLANs 21,23-1005 have path cost 80.
Port 2/28 VLANs 1-20,22 have path cost 15.
Console> (enable) set spantree portvlancost 2/28 cost 10 22
Port 2/28 VLANs 21,23-1005 have path cost 80.
Port 2/28 VLANs 1-20,22 have path cost 10.
Console> (enable) set spantree portvlancost 2/28 cost 19            (B)
Port 2/28 VLANs 21,23-1005 have path cost 80.
Port 2/28 VLANs 1-20,22 have path cost 19.
Console> (enable) set spantree portvlancost 2/28 cost 10 21
Port 2/28 VLANs 23-1005 have path cost 80.
Port 2/28 VLANs 1-22 have path cost 10.
Console> (enable) clear spantree portvlancost 2/28 1-20
Port 2/28 VLAN 1-20,23-1005 have path cost 80.
Port 2/28 VLAN 21-22 have path cost 10.
Console> (enable) clear spantree portvlancost 2/28 1-22
Port 2/28 VLAN 1-1005 have path cost 80.
Console> (enable) set spantree portvlancost 2/28 cost 10            (A)
Port 2/28 VLANs 1-1005 have path cost 80.
Console> (enable) set spantree portvlancost 2/28 cost 10 1-1005     (A)
Port 2/28 VLANs 1-1005 have path cost 10.
Console> (enable) clear spantree portvlancost 2/28 1-1005
Port 2/28 VLAN 1-1005 have path cost 80.
Console> (enable) set spantree portvlancost 2/28 1-1005             (F)
Port 2/28 VLANs 1-1005 have path cost 79.
Console> (enable) set spantree portvlancost 2/28 cost 10            (B)
Port 2/28 VLANs 1-1005 have path cost 10.
Console> (enable) clear spantree portvlancost 2/28 1-1005
Port 2/28 VLAN 1-1005 have path cost 80.
Console> (enable) set spantree portcost 2/28 19
```

Example 5-23 Machinations of set spantree portvlancost (Continued)

```
Spantree port 2/28 path cost set to 19.
Console> (enable) show spantree portvlancost 2/28
Port 2/28 VLANs 1-1005 have path cost 19.
Console> (enable)
```

In Example 5-23, the command **clear spantree portvlancost 2/28 1-1005** resets the Port Cost for each VLAN to the value specified in the **set spantree portcost** command (in this case, 80). If the **set spantree portcost** command was never entered, the Port Cost defaults to the value associated with the link speed (here the link speed is 100 Mbps, so the default Port Cost is 19). Note that there are only two possible values for a VLAN's Port Cost: the Port Cost specified by the **set spantree portcost** command (or the default Port Cost if this command was never used) and a value effected by the **set spantree portvlancost** command.

We proceed to provide excruciating details of how to predict the output of the **set spantree portvlancost** command. If you don't like math, skip to the next section.

For convenience, let X denote the set of all VLANs with Port Cost 80, and let Y denote the set of all VLANs with Port Costs other than 80. Let c(Y) denote the cost of VLANs in Y. Note that there's nothing sacred about 80; we just use 80 to have a number to work with instead of a variable.

By analyzing the output of the commands in Example 5-23, you can deduce the following (the letters correspond to the letters in the output):

- (A)—If Y is empty, using the **cost** keyword alone has no effect. If Y is empty, using the **cost** keyword together with a set S of VLANs has the effect of expanding Y (empty) to S with the new **cost** value applied to these VLANs.
- (B)—If the **cost** keyword is used alone with Y nonempty and
  - — if the **cost** value equals c(Y), the **cost** value has no effect.
  - — if the **cost** value is different from c(Y), the value is applied to all VLANs in Y.
- (C)—if the **cost** value matches c(Y) and the VLANs listed form a subset of Y, specifying the **cost** value has no effect. If the **cost** value does not match c(Y) and the VLANs listed form a subset of Y, all VLANs in Y assume the new **cost** value.
- (D)—Using the command without the **cost** keyword and listing a set S of VLANs not forming a subset of Y has the effect of expanding Y to Y∪S (Y∪S is the set of all VLANs in Y or S). Recall that Y denotes the set of all VLANs with Port Costs other than 80.

- (E)—If Y is nonempty, a **cost** value is given that differs from c(Y), and a set S of VLANs not forming a subset of Y is listed. The effect is to expand Y to Y∪S with the new **cost** value.
- (F)—If Y is empty and the **cost** parameter is left off, all VLANs listed in the **set spantree portvlancost** command have their Port Cost decremented by 1 (unless the Port Cost was set to 1 with the **set spantree portcost** command, in which case you get the error "Failed to set spantree portvlancost").

NOTE

Beginning with CatOS 7.1, the command **clear spantree portcost** *mod/port* was introduced to clear the Port Cost for a port. This command is not used in the examples.

Note that the **clear spantree portvlancost 2/28 1-1005** command near the end of the output has the effect of returning all VLANs to the value specified by the **set spantree portcost** command (80).

IOS-Based Switch

To configure the Port Cost, proceed to interface configuration mode and enter the command **spanning-tree cost** *cost*. Example 5-24 illustrates this.

Example 5-24 Configuring Port Cost on an IOS-Based Switch

```
Switch(config)#interface fa0/23
Switch(config-if)#spanning-tree cost 8
Switch(config-if)#end
Switch#show spanning-tree interface fa0/23
Interface Fa0/23 (port 2) in Spanning tree 1 is FORWARDING
   Port path cost 8, Port priority 60
   Designated root has priority 8000, address 0004.c098.e342
   Designated bridge has priority 8192, address 0007.84f9.f680
   Designated port is 2, path cost 19
   Timers: message age 0, forward delay 0, hold 0
   BPDU: sent 2618, received 153642
Interface Fa0/23 (port 2) in Spanning tree 2 is FORWARDING
   Port path cost 10, Port priority 50
   Designated root has priority 8000, address 0004.c098.e344
   Designated bridge has priority 32768, address 0007.84f9.f681
   Designated port is 2, path cost 19
   Timers: message age 0, forward delay 0, hold 0
   BPDU: sent 149049, received 9405
Interface Fa0/23 (port 2) in Spanning tree 3 is FORWARDING
   Port path cost 10, Port priority 128
   Designated root has priority 8000, address 0004.c098.e345
```

Example 5-24 Configuring Port Cost on an IOS-Based Switch (Continued)

```
   Designated bridge has priority 32768, address 0007.84f9.f682
   Designated port is 2, path cost 19
   Timers: message age 0, forward delay 0, hold 0
   BPDU: sent 1927, received 154303
<output omitted>
```

Notice that the Port Cost for trunk interface Fa0/23 is changed only for VLAN 1 in Example 5-23. This command is normally applied to access ports. Here, only the Path Cost for the Native VLAN (VLAN 1) is affected by the command. Use the **no** form of this command to return the interface to its default setting.

Now, to configure the Port Cost for VLANs on a trunk port, use the command **spanning-tree vlan** *vlan-id* **cost** *cost*. For the *vlan-id*, the range is 1 to 1005. Use the **no** form of this command to return the interface to its default setting.

As with CatOS switches, this command allows you to tailor Path Costs to force intra-VLAN traffic to follow a prescribed path. Example 5-25 demonstrates configuring the Port Cost for VLANs 2 and 3 to be 15.

Example 5-25 Configuring Per-VLAN Port Cost on an IOS-Based Switch

```
Switch(config-if)#spanning-tree vlan 2 3 cost 15
Switch#show spanning-tree interface fa0/2
Interface Fa0/2 (port 8) in Spanning tree 1 is FORWARDING
   Port path cost 20, Port priority 128
   Designated root has priority 8000, address 0004.c098.e342
   Designated bridge has priority 8000, address 0004.c098.e342
   Designated port is 14, path cost 0
   Timers: message age 2, forward delay 0, hold 0
   BPDU: sent 162894, received 177589
Interface Fa0/2 (port 8) in Spanning tree 2 is FORWARDING
   Port path cost 15, Port priority 128
   Designated root has priority 8000, address 0004.c098.e344
   Designated bridge has priority 8000, address 0004.c098.e344
   Designated port is 14, path cost 0
   Timers: message age 2, forward delay 0, hold 0
   BPDU: sent 163255, received 155756
Interface Fa0/2 (port 8) in Spanning tree 3 is FORWARDING
```

continues

Example 5-25 Configuring Per-VLAN Port Cost on an IOS-Based Switch (Continued)

```
   Port path cost 15, Port priority 128
   Designated root has priority 8000, address 0004.c098.e345
   Designated bridge has priority 8000, address 0004.c098.e345
   Designated port is 14, path cost 0
   Timers: message age 2, forward delay 0, hold 0
   BPDU: sent 163522, received 153267
Interface Fa0/2 (port 8) in Spanning tree 4 is FORWARDING
   Port path cost 30, Port priority 128
   Designated root has priority 8000, address 0004.c098.e340
   Designated bridge has priority 8000, address 0004.c098.e340
   Designated port is 14, path cost 0
   Timers: message age 4, forward delay 0, hold 0
   BPDU: sent 163702, received 606704
<output omitted>
```

Configuring Port Priority

After prescribing primary and secondary Root Bridges and tailoring Path Costs, you might find it necessary to configure Port Priorities. Recall that STP uses, in decreasing priority, the following STP parameters in determining the best path to the Root Bridge:

- Path Cost
- Bridge ID
- Port Priority

The Port Priority is a 2-byte quantity consisting of the Port ID (high-order bits) and the Port Number (low-order bits). The Port ID can be configured, allowing you to fine-tune your STP configuration. This method is applicable when all other STP parameters are equal.

In the event of a loop, STP considers the Port Priority when selecting a port to put into Forwarding state.

The first thing to notice here is that, if the Path Costs and sender Bridge IDs coincide, there is only one possibility: The BPDUs are coming from the same switch. So, altering Port Priorities is relevant only if multiple trunks are connecting two switches.

In this case, you can use Port Priorities to load-balance data across the parallel trunks by setting distinct Port Priorities on the trunk ports for mutually exclusive sets of VLANs.

We have not discussed the use of Bridge Priorities beyond prescribing the Root Bridges. It is possible to configure Bridge IDs in your network to affect the STP topology. However, changing Bridge ID parameters is not recommended beyond doing so to configure Root Bridges.

Changing Bridge IDs to further define Layer 2 traffic flow is an indirect method and has the effect of an accident waiting to happen. This method relies on altering the sender's Bridge ID, upstream of the flow of BPDUs in the spanning-tree topology. Furthermore, this method is not scalable, because it is difficult to keep track of which Bridge IDs were altered for which VLANs and for what reason. In general, the preferred method is to use Port Costs and per-VLAN Port Costs, as described in the preceding section.

CatOS Switch

You can configure the Port Priority of switch ports with values ranging from 0 to 63. The default Port Priority value is 32. The port with the lowest-priority value forwards frames for all VLANs. If all ports have the same Port Priority, the port with the lowest Port Number forwards frames.

To configure Port Priority for a port, use the command **set spantree portpri** *mod/port priority*, as shown in Example 5-26.

Example 5-26 Configuring Port Priority on a CatOS Switch

```
Console> (enable) set spantree portpri 2/3 16
Bridge port  2/3 port priority set to 16.
Console> (enable) show spantree 2/3
Port                     Vlan Port-State    Cost      Prio Portfast Channel_id
------------------------ ---- ------------- --------- ---- -------- ----------
 2/3                     1    forwarding          100   16 disabled 0
```

You can also set the Port Priority on a per-VLAN basis on a trunk port. To do this, you use the **set spantree portvlanpri** *mod/port priority* [*vlans*] command.

Use the **clear spantree portvlanpri** *mod/port* [*vlans*] command to reset the spanning-tree VLAN Port Priority.

Example 5-27 illustrates the use of the **set spantree portvlanpri** command. Note that, disregarding the special case of VLAN 1005, again there are only two possible values for a VLAN's Port Priority: the Port Priority specified by the **set spantree portpri** command (or the default Port Priority if this command was never used) and a value effected by the **set spantree portvlanpri** command. With the caveat that the VLAN Port Priority for a VLAN must be lower than the Port Priority, the rules for predicting the

NOTE

The VLAN Port Priority for a VLAN must be lower than the Port Priority. The CatOS does not permit you to configure it to be higher than the Port Priority.

values of the per-VLAN Port Priority effected by the **set spantree portvlanpri** command mirror those for predicting the values of the per-VLAN Port Cost effected by the **set spantree portvlancost** command.

Example 5-27 Configuring Per-VLAN Port Priority on a CatOS Switch

```
Console> (enable) clear spantree portvlanpri 2/28 1-1005
Port 2/28 vlans 1-1004 using portpri 32.
Port 2/28 vlans 1005 using portpri 4.
Console> (enable) set spantree portvlanpri 2/28 20 1-5
Port 2/28 vlans 1-5 using portpri 20.
Port 2/28 vlans 6-1004 using portpri 32.
Port 2/28 vlans 1005 using portpri 4.
Console> (enable) set spantree portvlanpri 2/28 40 7
Portvlanpri must be less than portpri. Portpri for 2/28 is 32.
Console> (enable) set spantree portvlanpri 2/28 15 7
Port 2/28 vlans 1-5,7 using portpri 15.
Port 2/28 vlans 6,8-1004 using portpri 32.
Port 2/28 vlans 1005 using portpri 4.
Console> (enable) clear spantree portvlanpri 2/28 15 7
Port 2/28 vlans 1-5 using portpri 15.
Port 2/28 vlans 6-1004 using portpri 32.
Port 2/28 vlans 1005 using portpri 4.
Console> (enable)
```

IOS-Based Switch

> **NOTE**
> Beginning with CatOS 7.1, the command **clear spantree portpri** *mod/port* was introduced to clear a port's Port Priority. We do not use this command in the examples.

You can configure the Port Priority for switch ports with values ranging from 0 to 255. The default value is 128. If all interfaces have the same Port Priority, STP puts the interface with the lowest interface number in the Forwarding state and blocks other interfaces. Note that the interface with the lowest physical interface number is the same as the interface with the lowest Port Number.

To change the Port Priority, use the interface configuration mode command **spanning-tree port-priority** *priority*. This command is normally applied to *access ports*.

Example 5-28 illustrates this command.

Example 5-28 Configuring Port Priority on an IOS-Based Switch

```
Switch(config-if)#spanning-tree port-priority 100
Switch#show spanning-tree interface fa0/23
```

Example 5-28 Configuring Port Priority on an IOS-Based Switch (Continued)

```
Interface Fa0/23 (port 2) in Spanning tree 1 is FORWARDING
   Port path cost 80, Port priority 100
   Designated root has priority 8000, address 0004.c098.e342
   Designated bridge has priority 8192, address 0007.84f9.f680
   Designated port is 2, path cost 20
   Timers: message age 0, forward delay 0, hold 0
   BPDU: sent 32999, received 153694
Interface Fa0/23 (port 2) in Spanning tree 2 is FORWARDING
   Port path cost 80, Port priority 50
   Designated root has priority 8000, address 0004.c098.e344
   Designated bridge has priority 32768, address 0007.84f9.f681
   Designated port is 2, path cost 15
   Timers: message age 0, forward delay 0, hold 0
   BPDU: sent 179380, received 9412
<output omitted>
```

Notice that the Port Priority for interface Fa0/23 was changed only for VLAN 1 in Example 5-28. As just mentioned, this command is normally applied to access ports. Here, only the Path Cost for the Native VLAN (VLAN 1) is affected by the command. Use the **no** form of this command to return the interface to its default setting.

To configure the Port Priority for a VLAN on an interface serving as a trunk port, use the command **spanning-tree vlan** *vlan-id* **port-priority** *priority*, as demonstrated in Example 5-29.

Example 5-29 Configuring Per-VLAN Port Priority on an IOS-Based Switch

```
Switch(config-if)#spanning-tree vlan 2 port-priority 10
Switch(config-if)#end
Switch#show spanning-tree interface fa0/23
Interface Fa0/23 (port 2) in Spanning tree 1 is FORWARDING
   Port path cost 80, Port priority 128
   Designated root has priority 8000, address 0004.c098.e342
   Designated bridge has priority 8192, address 0007.84f9.f680
   Designated port is 2, path cost 20
   Timers: message age 0, forward delay 0, hold 0
   BPDU: sent 33171, received 153694
Interface Fa0/23 (port 2) in Spanning tree 2 is FORWARDING
```

continues

Example 5-29 Configuring Per-VLAN Port Priority on an IOS-Based Switch (Continued)

```
   Port path cost 80, Port priority 10
   Designated root has priority 8000, address 0004.c098.e344
   Designated bridge has priority 32768, address 0007.84f9.f681
   Designated port is 2, path cost 15
   Timers: message age 0, forward delay 0, hold 0
   BPDU: sent 179552, received 9412
<output omitted>
```

Enhancing STP Convergence

Fortunately, a variety of tools are available for tuning STP performance. Without exception, these tools are designed to speed up STP convergence when the conditions are right. These options include manually setting STP timers, using macros such as **set spantree root** on CatOS switches and the corresponding IOS-based macro, configuring PortFast, configuring UplinkFast, and configuring BackboneFast. Each of these is discussed in the following sections.

Configuring STP Timers

Earlier in this chapter, we described the three STP timers:

- **Hello Time**—Determines how often the Root Switch broadcasts Configuration BPDUs to other switches. The default is 2 seconds.
- **Forward Delay**—Monitors the time spent by a port in the Learning and Listening states. The default is 15 seconds.
- **Max Age**—Measures the age of the received BPDU information recorded for a port and ensures that this information is discarded when its age limit exceeds the value of the maximum age parameter recorded by the switch. The default is 20 seconds.

The STP timers are necessary to prevent bridge loops in your network. These timers are put into place to give the network enough time to get all the correct information about the topology and to determine if there are redundant links.

With the default STP timers, it can take up to 50 seconds after a link has failed for a backup link to take over. The length of time that it takes spanning tree to converge when a link has failed can have a negative impact on some protocols and applications, resulting in lost connections, sessions, or data. One example of this is when you're using Hot Standby Routing Protocol (HSRP) with two routers connected to a Catalyst switch. The default timers for STP in some cases are too long for HSRP, causing the wrong router to become the "active" router. (See Chapter 8 for more information on the interaction between STP and HSRP.)

Another case in which you might opt to lower STP timers is on the Root Switch. By lowering the values for the Hello Time, Forward Delay, and Max Age on the Root Switch, you can reduce STP convergence time. Reducing the timer parameter values is possible only if your network has LAN links of 10 Mbps or faster. This might appear to be a given, but if you have a router with a T1 link configured for transparent bridging, you see the potential problem. In a network with links of 10 Mbps or faster, the network diameter can reach the maximum value of 7.

To speed up convergence, an alternative is to use nondefault parameters permitted by the 802.1D standard: a network diameter of 2, Hello Time = 2 seconds, Forward Delay = 4 seconds, and Max Age = 6 seconds. This allows STP convergence in 14 seconds or less. You should use these alternative timers only after careful consideration. Just as with routing protocols, changing default timers can cause serious problems in a network if the alternative timers are not predicated on a thorough network analysis.

We now show how to change these timers on Catalyst switches. These parameters can be set for both PVST+ and MISTP modes, but all configurations in this chapter are restricted to PVST+ configuration. With PVST+, STP timers are set for each VLAN.

On Catalyst switches, the Hello Time can be configured between 1 and 10 seconds. The Forward Delay can be configured between 4 and 30 seconds. The Max Age can be configured between 6 and 40 seconds.

CatOS Switch

To change the Hello Time, use the command **set spantree hello** *interval* [*vlan*].

To change the Forward Delay, use the command **set spantree fwddelay** *delay* [*vlan*].

To change the Max Age, use the command **set spantree maxage** *agingtime* [*vlans*].

Example 5-30 illustrates these commands on a Root Switch, with the nondefault STP parameters permitted by the 802.1D standard.

Example 5-30 Setting STP Timers on a CatOS Switch

```
Console> (enable) set spantree hello 2 200
Spantree 200 hello time set to 2 seconds.
Console> (enable) set spantree fwddelay 4 200
Spantree 200 forward delay set to 4 seconds.
Console> (enable) set spantree maxage 6 200
Spantree 200 max aging time set to 6 seconds.
```

For most situations, Cisco recommends that you use the **set spantree root** and **set spantree root secondary** commands to modify spanning-tree performance parameters.

IOS-Based Switch

The global configuration command to change the Hello Time is **spanning-tree vlan** *vlan-id* **hello-time** *seconds*.

The global configuration command to change the Forward Delay is **spanning-tree vlan** *vlan-id* **forward-time** *seconds*.

The global configuration command to change the Max Age is **spanning-tree vlan** *vlan-id* **max-age** *seconds*.

Example 5-31 illustrates these commands.

Example 5-31 Changing the STP Timers on an IOS-Based Switch

```
Switch(config)#spanning-tree vlan 2 hello-time 2
Switch(config)#spanning-tree vlan 2 forward-time 4
Switch(config)#spanning-tree vlan 2 max-age 6
```

Again, for most situations, Cisco recommends that you use the **spanning-tree vlan** *vlan-id* **root primary** and the **spanning-tree vlan** *vlan-id* **root secondary** global configuration commands to modify the STP parameters.

PortFast

Spanning tree ***PortFast*** is a Catalyst feature that causes a switch or trunk port to enter the spanning-tree Forwarding state immediately, bypassing the Listening and Learning states. Although PortFast can be configured on trunk ports, this is generally reserved for the case in which you have a CatOS switch connected to a server with a trunk. IOS-based switches use PortFast only on access ports connected to end stations, as shown in Figure 5-20.

When a device is connected to a port, the port normally enters the spanning-tree Listening state. When the Forward Delay timer expires, the port enters the Learning state. When the Forward Delay timer expires a second time, the port is transitioned to the Forwarding or Blocking state. When you enable PortFast on a switch or trunk port, the port is immediately transitioned to the Forwarding state. As soon as the switch detects the link, the port is transitioned to the Forwarding state (less than 2 seconds after the cable is plugged in).

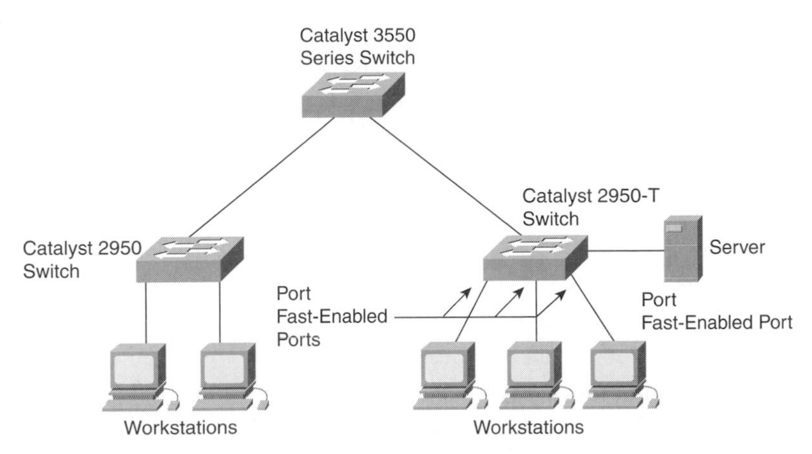

Figure 5-20 PortFast is normally enabled on access ports of access-layer Catalyst switches.

If a loop is detected and PortFast is enabled, the port is transitioned to the Blocking state. It is important to note that PortFast begins only when the port first initializes. If the port is forced into the Blocking state for some reason and later needs to return to the Forwarding state, the usual Listening and Learning processes are performed (note the difference here from the usual PortFast operation when a link first initializes).

Now, to help you understand the argument for enabling PortFast on an access port, we take a step back for a bit. Think for a moment about what happens when you boot your PC. You turn on the power, the monitor flickers, and the machine beeps and buzzes. Sometime during that process, your NIC initializes the Ethernet link, causing the switch port to jump from "not connected" to the STP Listening state. Thirty seconds later, the Catalyst switch puts your port into Forwarding mode, and you can access the network. Normally, this sequence goes on unnoticed, because it takes your PC at least 30 seconds to boot. However, in a couple of cases, this might not be true.

First, some NICs do not enable a link until the MAC-layer software driver is actually loaded. Because most operating systems try to use the network almost immediately after loading the driver, this can create a problem (the 30 seconds of STP delay begins right when the OS begins trying to access the network). This problem is fairly common with PC Card (PCMCIA) NICs used in laptop computers.

Second, there is a race between operating systems and CPU manufacturers. CPU manufacturers keep making the chips faster, while at the same time operating systems keep slowing down, but the chips are speeding up at a greater rate than the operating systems are slowing down. As a result, PCs are booting faster than ever. In fact, modern machines are often finished booting and need to use the network before the STP 30-second delay is over.

This problem motivates some network administrators to disable STP altogether. This certainly fixes any STP booting problems, but it can easily create other problems. If you employ this strategy, you must eliminate all physical loops (which means no redundancy). Also, keep in mind that you cannot disable STP for a single port. **set spantree disable** [*vlan*] is a per-VLAN command that disables STP for every port that participates in the specified VLAN. In short, rather than disabling STP, you should consider using the PortFast feature. This feature gives you the best of both worlds—immediate end-station access and the safety net of STP.

The following sections describe how to configure PortFast for CatOS and IOS-based switches.

CatOS Switch

To configure PortFast on an access port, use the command **set spantree portfast** *mod_num/port_num* **enable | disable**.

For a trunk port, use the command **set spantree portfast** *mod_num/port_num* **enable trunk**. If you leave off the **trunk** keyword, PortFast remains disabled for the trunk port.

Example 5-32 demonstrates these commands, first on access port 2/3, and then on trunk port 2/28.

Example 5-32 Enabling PortFast on a CatOS Switch

```
Console> (enable) set spantree portfast 2/3 enable

Warning: Spantree port fast start should only be enabled on ports connected
to a single host.  Connecting hubs, concentrators, switches, bridges, etc. to
a fast start port can cause temporary spanning tree loops.  Use with caution.

Spantree port  2/3 fast start enabled.
Console> (enable) show spantree 2/3
Port                     Vlan Port-State    Cost      Prio Portfast Channel_id
------------------------ ---- ------------- --------- ---- -------- ----------
 2/3                     1    forwarding          100   16 enabled  0

Console> (enable) set spantree portfast 2/28 enable trunk
```

Example 5-32 Enabling PortFast on a CatOS Switch (Continued)

```
Warning: Spantree port fast start should only be enabled on ports connected
to a single host.  Connecting hubs, concentrators, switches, bridges, etc. to
a fast start port can cause temporary spanning tree loops.  Use with caution.

Spantree port  2/28 fast start enabled.
Console> (enable) show spantree 2/28
Port                     Vlan Port-State    Cost      Prio Portfast Channel_id
------------------------ ---- ------------- --------- ---- -------- ----------
 2/28                    1    forwarding            1   15 enabled  0
 2/28                    2    forwarding            1   15 enabled  0
 2/28                    3    forwarding            1   15 enabled  0
 2/28                    4    forwarding            1   15 enabled  0
 2/28                    5    forwarding            1   15 enabled  0
 2/28                    6    forwarding            1   32 enabled  0
 2/28                    7    forwarding            1   15 enabled  0
 2/28                    8    forwarding            1   32 enabled  0
 2/28                    200  forwarding            2   32 enabled  0
```

IOS-Based Switch

To configure PortFast, enter the interface configuration mode command **spanning-tree portfast.** This is illustrated in Example 5-33.

Example 5-33 Enabling PortFast on an IOS-Based Switch

```
Switch(config-if)#spanning-tree portfast
%Warning: portfast enabled on FastEthernet0/2.
 Usually portfast should be enabled on ports connected to a single host.
 When portfast is enabled, connecting hubs, concentrators, switches, bridges,
 etc. to this interface may cause temporary spanning tree loops.
 Use with CAUTION.
Switch(config-if)#end
Switch#show spanning-tree interface fa0/2
Interface Fa0/2 (port 8) in Spanning tree 1 is FORWARDING
   Port path cost 20, Port priority 128
   Designated root has priority 8000, address 0004.c098.e342
   Designated bridge has priority 8000, address 0004.c098.e342
   Designated port is 14, path cost 0
```

continues

Example 5-33 Enabling PortFast on an IOS-Based Switch (Continued)

```
Timers: message age 2, forward delay 0, hold 0
BPDU: sent 162897, received 190994
The port is in the portfast mode
```

UplinkFast

During the time it takes for STP to converge, some end stations might become inaccessible, depending on the STP state of the switch port to which the station is attached. This disrupts network connectivity, so the point is to decrease STP convergence time and reduce the length of the disruption. *UplinkFast* was developed to facilitate fast STP convergence. UplinkFast is a Catalyst feature that accelerates the choice of a new Root Port when a link or switch fails.

Switches in hierarchical networks can be grouped into core layer switches, distribution layer switches, and access layer switches. Figure 5-21 shows a portion of a campus network in which distribution switches and access switches each have at least one redundant link that STP blocks to prevent loops.

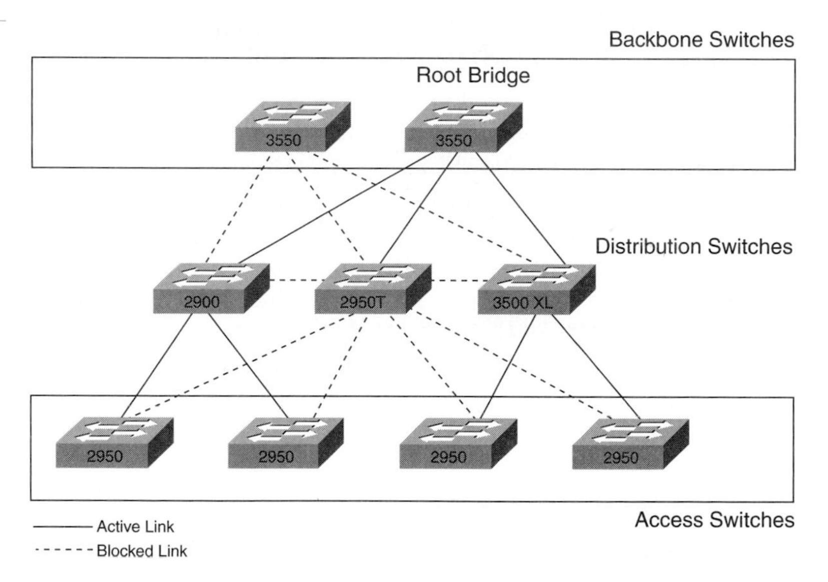

Figure 5-21 Switches in a hierarchical network provide an ideal setting for UplinkFast.

If a switch loses connectivity on a link, it begins using the alternative paths as soon as STP selects a new Root Port. When STP reconfigures the new Root Port, other interfaces flood the network with multicast packets, at least one for each address that was learned on the interface.

By using STP UplinkFast, you can accelerate the choice of a new Root Port when a link or switch fails or when STP reconfigures itself. The Root Port transitions to the Forwarding state immediately without going through the Listening and Learning states, as it would with the usual STP process. UplinkFast also limits the burst of multicast traffic by reducing the *max-update-rate* (IOS) or *station_update_rate* (CatOS) parameter. The default for these parameters is 150 packets per second.

UplinkFast is most useful in wiring-closet switches at the edge of the network. It is not appropriate for backbone devices.

UplinkFast provides fast convergence after a direct link failure and achieves load balancing between redundant links using uplink groups. An *uplink group* is a set of interfaces (per VLAN), only one of which is forwarding at any given time. Specifically, an uplink group consists of the Root Port (which is forwarding) and a set of blocked ports. The uplink group provides an alternative path in case the currently forwarding link fails.

Figure 5-22 shows a sample topology with no link failures. Switch A, the Root Switch, is connected directly to Switch B over link L1 and to Switch C over link L2. The interface on Switch C that is connected directly to Switch B is in a blocking state. Switch C is configured with UplinkFast.

Figure 5-22 UplinkFast is enabled, waiting for an uplink to fail.

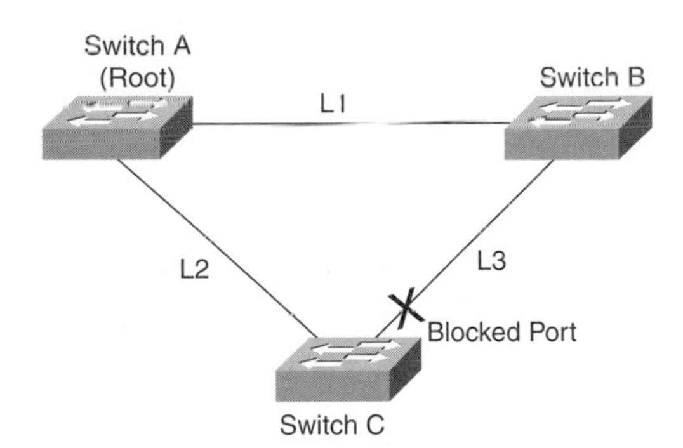

If Switch C detects a link failure on the currently active link L2 on the Root Port (a direct link failure), UplinkFast unblocks the blocked port on Switch C and transitions it to the Forwarding state without going through the Listening and Learning states, as shown in Figure 5-23. This change takes approximately 1 to 5 seconds.

As soon as a switch transitions an alternative port to the Forwarding state, the switch begins transmitting dummy multicast frames on that port, one for each entry in the local bridge table (except entries that are associated with the failed Root Port). By default, approximately 15 dummy multicast frames are transmitted per 100 milliseconds.

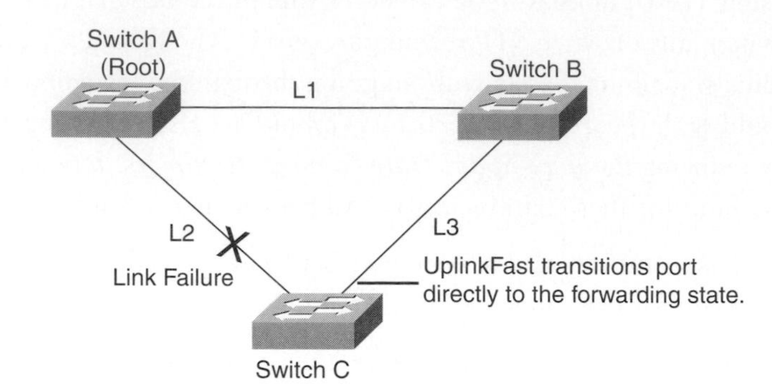

Figure 5-23 UplinkFast unblocks the blocked port on link L3 less than 5 seconds after link L2 fails.

Each dummy multicast frame uses the station address in the bridge table entry as its source MAC address and a dummy multicast address (01-00-0C-CD-CD-CD) as the destination MAC address.

Switches receiving these dummy multicast frames immediately update their bridge table entries for each source MAC address to use the new port, allowing the switches to begin using the new path almost immediately.

In the event that connectivity on the original Root Port is restored, the switch waits for a period equal to twice the Forward Delay time plus 5 seconds before transitioning the port to the Forwarding state in order to allow the neighbor port time to transition through the Listening and Learning states to the Forwarding state.

Configuring UplinkFast on Catalyst switches is straightforward.

CatOS Switch

To enable UplinkFast, use the command **set spantree uplinkfast enable**. This command increases the Path Cost of all ports on the switch, making it unlikely that the switch will become the Root Switch. *station_update_rate* represents the number of multicast packets transmitted per 100 milliseconds (the default is 15 packets per 100 milliseconds, or 150 packets per second).

When you enable the **set spantree uplinkfast** command, it affects all VLANs on the switch. You cannot configure UplinkFast on an individual VLAN.

The syntax for this command is **set spantree uplinkfast enable** [**rate** *station_update_rate*]. You verify UplinkFast settings with the **show spantree uplinkfast** [*vlans*] command.

Example 5-34 illustrates how to enable UplinkFast with a station-update rate of 40 packets per 100 milliseconds and how to verify that UplinkFast is enabled.

Example 5-34 Enabling UplinkFast on a CatOS Switch

```
Console> (enable) set spantree uplinkfast enable rate 40
VLANs 1-1005 bridge priority set to 49152.
The port cost and portvlancost of all ports set to above 3000.
Station update rate set to 40 packets/100ms.
uplinkfast all-protocols field set to off.
uplinkfast enabled for bridge.
Console> (enable) show spantree uplinkfast 1,2,4,200
Station update rate set to 40 packets/100ms.
uplinkfast all-protocols field set to off.

VLAN          port list
-----------------------------------------------
1             2/28(fwd)
2             2/28(fwd)
4             2/28(fwd)
200           2/28(fwd)
Console> (enable)
```

The **set spantree uplinkfast disable** command disables UplinkFast on the switch, but the switch priority and port cost values are not reset to the factory defaults. When you enter the **set spantree uplinkfast disable** command, it affects all VLANs on the switch. You cannot disable UplinkFast on an individual VLAN. Example 5-35 shows how to disable UplinkFast and return STP parameters to their default values.

Example 5-35 Disabling UplinkFast on a CatOS Switch

```
Console> (enable) set spantree uplinkfast disable
uplinkfast disabled for bridge.
Use clear spantree uplinkfast to return stp parameters to default.
Console> (enable) clear spantree uplinkfast
This command will cause all portcosts, portvlancosts, and the
bridge priority on all vlans to be set to default.
Do you want to continue (y/n) [n]? y
VLANs 1-1005 bridge priority set to 32768.
The port cost of all bridge ports set to default value.
The portvlancost of all bridge ports set to default value.
```

continues

Example 5-35 Disabling UplinkFast on a CatOS Switch (Continued)

```
uplinkfast all-protocols field set to off.
uplinkfast disabled for bridge.
```

IOS-Based Switch

UplinkFast cannot be enabled on VLANs that have been configured for Bridge Priority. To enable UplinkFast on a VLAN with Bridge Priority configured, first restore the switch priority on the VLAN to the default value by using the **no spanning-tree vlan** *vlan-id* **priority** global configuration command.

When UplinkFast is enabled, the Bridge Priority of all VLANs is set to 49152, and the Path Cost of all interfaces and VLAN trunks is increased by 3000 if you did not modify the Port Cost from its default setting. This change reduces the chance that the switch will become the Root Bridge. Also, as with CatOS switches, when you enable UplinkFast, it affects all VLANs on the switch. You cannot configure UplinkFast on an individual VLAN.

To configure UplinkFast, use the command **spanning-tree uplinkfast** [**max-update-rate** *pkts-per-second*]. For *pkts-per-second,* the range is 0 to 65535 packets per second. The default is 150; this setting is normally adequate. To verify the UplinkFast configuration, you can use the **show spanning-tree summary** command. Cisco IOS Software Release 12.1 is the first release with the command **show spanning-tree uplinkfast**.

When UplinkFast is disabled, the Bridge Priorities of all VLANs and Port Costs of all interfaces are set to default values if you did not modify them from their defaults. To disable UplinkFast, use the **no spanning-tree uplinkfast** command.

Example 5-36 illustrates these commands.

Example 5-36 Configuring UplinkFast on an IOS-Based Switch

```
Switch(config)#spanning-tree uplinkfast max-update-rate 40
Switch#show spanning-tree summary
UplinkFast is enabled


Name                 Blocking Listening Learning Forwarding STP Active
-------------------- -------- --------- -------- ---------- ----------
VLAN1                17       0         0        2          19
<output omitted>
Switch(config)#no spanning-tree uplinkfast max-update-rate 40
Switch(config)#no spanning-tree uplinkfast
```

BackboneFast

BackboneFast is a Catalyst feature that is initiated when a Root Port or blocked port on a switch receives inferior BPDUs from its Designated Bridge. An inferior BPDU identifies one switch as both the Root Bridge and the Designated Bridge. When a switch receives an inferior BPDU, it means that a link to which the switch is not directly connected (an indirect link) has failed; that is, the Designated Bridge has lost its connection to the Root Bridge. Under STP rules, the switch ignores inferior BPDUs for the configured Max Age (the default is 20 seconds). The role of BackboneFast is essentially to cheat this 20-second delay. When the switch receives the inferior BPDU, the switch tries to determine if it has an alternative path to the Root Bridge.

There are two cases to consider. First, if the inferior BPDU arrives on a blocked port, the Root Port and other blocked ports on the switch become alternative paths to the Root Bridge.

Second, if the inferior BPDU arrives on the Root Port, all blocked ports become potential alternative paths to the Root Bridge. If the inferior BPDU arrives on the Root Port and there are no blocked ports (for example, if the switch has two trunks—one connecting to the Root Bridge and one connecting downstream in the STP topology), the switch assumes that it has lost connectivity to the Root Bridge, causes the Max Age on the root to expire, and becomes the Root Bridge according to normal STP rules. If the switch has alternative paths to the Root Bridge (that is, the switch has blocked ports), it uses these alternative paths to transmit a new kind of Protocol Data Unit called the Root Link Query PDU. The switch sends the Root Link Query PDU on all potential alternative paths to the Root Bridge. If the switch determines that it still has an alternative path to the root, it causes the Max Age on the ports on which it received the inferior BPDU to expire. The switch then makes all ports on which it received an inferior BPDU its Designated Ports and moves them out of the Blocking state (if they were in the Blocking state), through the Listening and Learning states, and into the Forwarding state. On the other hand, if the switch learns via the Root Link Query process that all the alternative paths to the Root Bridge have lost connectivity to the root, the switch causes the Max Age on the ports on which it received inferior BPDUs to expire, and a new STP topology is calculated.

To illustrate this process, Figure 5-24 shows a sample topology with no link failures. Switch A, the Root Bridge, connects directly to Switch B over link L1 and to Switch C over link L2. The interface on Switch C that connects directly to Switch B is in the Blocking state.

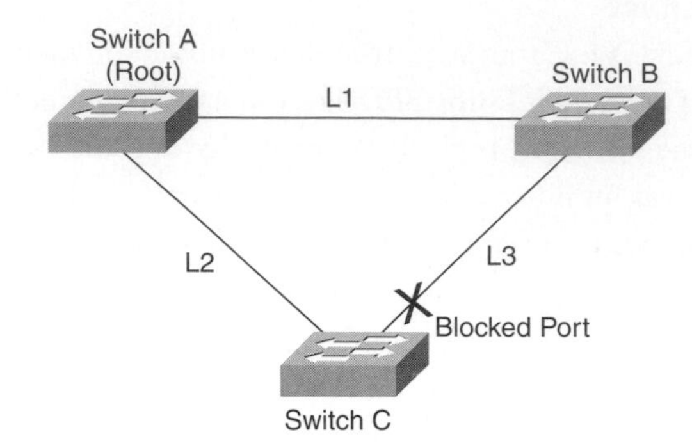

Figure 5-24 BackboneFast goes into effect when an indirect link failure occurs.

If link L1 fails, Switch C cannot directly detect this failure, because it is not connected directly to link L1. However, because Switch B is directly connected to the Root Bridge over L1, it detects the failure, elects itself the root, and begins sending BPDUs to Switch C, identifying itself as the root. When Switch C receives the inferior BPDUs from Switch B, Switch C assumes that an indirect failure has occurred. At that point, BackboneFast allows the blocked port on Switch C to move immediately to the Listening state without waiting for Max Age on the port to expire. BackboneFast then transitions the interface on Switch C to the Forwarding state, providing a path from Switch B to Switch A. This switchover takes approximately 30 seconds, twice the Forward Delay time if the default Forward Delay time of 15 seconds is set; this saves up to 20 seconds. Figure 5-25 shows how BackboneFast reconfigures the topology to account for the failure of link L1.

NOTE

If you use BackboneFast, you must enable it on all switches in the network.

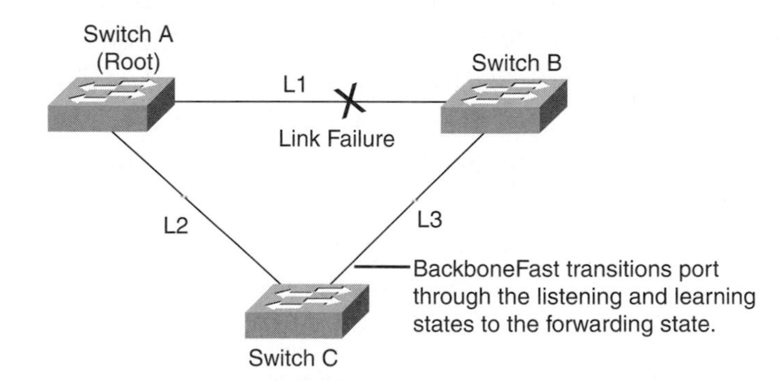

Figure 5-25 When L1 goes down, BackboneFast transitions the blocked port on Switch C to Forwarding state in less than 30 seconds.

CatOS Switch

To enable BackboneFast, use the command **set spantree backbonefast enable.** To verify the configuration, use the command **show spantree backbonefast.** BackboneFast statistics can be viewed with the command **set spantree summary.**

Example 5-37 illustrates the BackboneFast commands.

Example 5-37 Configuring BackboneFast on a CatOS Switch

```
Console> (enable) set spantree backbonefast enable
Backbonefast enabled for all VLANs.
Console> (enable) show spantree backbonefast
Backbonefast is enabled.
Console> (enable) show spantree summary
MAC address reduction: disabled
Root switch for vlans: 1-8,200.
BPDU skewing detection disabled for the bridge
BPDU skewed for vlans:  none.
Portfast bpdu-guard disabled for bridge.
Portfast bpdu-filter disabled for bridge.
Uplinkfast disabled for bridge.
Backbonefast enabled for bridge.

Summary of connected spanning tree ports by vlan

VLAN  Blocking Listening Learning Forwarding STP Active
----- -------- --------- -------- ---------- ----------
   1         0         0        0         15         15
   2         0         0        0          3          3
   3         0         0        0          3          3
   4         0         0        0          3          3
   5         0         0        0          4          4
   6         0         0        0          3          3
   7         0         0        0          3          3
   8         0         0        0          3          3
 200         0         0        0          3          3

      Blocking Listening Learning Forwarding STP Active
----- -------- --------- -------- ---------- ----------
Total        0         0        0         40         40

BackboneFast statistics
-----------------------
```

continues

Example 5-37 Configuring BackboneFast on a CatOS Switch (Continued)

```
Number of transitions via backboneFast (all VLANS) : 0
Number of inferior BPDUs received (all VLANs)      : 0
Number of RLQ req PDUs received (all VLANs)        : 0
Number of RLQ res PDUs received (all VLANs)        : 0
Number of RLQ req PDUs transmitted (all VLANs)     : 0
Number of RLQ res PDUs transmitted (all VLANs)     : 0
```

To disable BackboneFast, use the command **set spantree backbonefast disable.**

IOS-Based Switch

To enable BackboneFast, use the command **spanning-tree backbonefast** in global configuration mode, and verify the configuration with the **show spanning-tree summary** command, as shown in Example 5-38.

Example 5-38 Configuring BackboneFast on an IOS-Based Switch

```
Switch(config)#spanning-tree backbonefast
Switch#show spanning-tree summary

Root bridge for: none.
PortFast BPDU Guard is disabled
UplinkFast is enabled
BackboneFast is enabled
Spanning tree default pathcost method used is short

Name                 Blocking Listening Learning Forwarding STP Active
-------------------- -------- --------- -------- ---------- ----------
VLAN1                13       0         0        1          14
VLAN2                1        0         0        1          2
VLAN3                1        0         0        1          2
<output omitted>
```

NOTE

The **spanning-tree backbonefast** command was introduced to IOS-based switches in Cisco IOS Software Release 12.1. This was also the first release to support the command **show spanning-tree backbonefast**, used to verify BackboneFast configuration.

To disable BackboneFast, use the **no spanning-tree backbonefast** command.

This brings us to the end of our study of spanning-tree theory and application. One of the arguments for using STP is to maintain redundancy and reliability at Layer 2 within a switched network, including Layer 2 load balancing. Layer 2 load balancing is enabled with STP by configuring distinct Root Bridges for distinct sets of VLANs, as well as by tweaking Port Costs on a per-VLAN basis. Another technology that enables Layer 2 load balancing (or load distribution) is EtherChannel. Our last journey in this chapter is to learn how EtherChannel works and to see how to take advantage of this Cisco technology in your network. The remaining chapters of this book are dedicated to considerations relating to OSI Layer 3 and above.

EtherChannel

EtherChannel is a Cisco-proprietary technology that, by aggregating links into a single logical link, provides incremental trunk speeds ranging from 10 Mbps to 160 Gbps (full duplex). EtherChannel comes in four flavors: standard EtherChannel (for history's sake), Fast EtherChannel (FEC), Gigabit EtherChannel (GEC), and 10-Gigabit EtherChannel. The term *EtherChannel* encompasses all these technologies. EtherChannel combines up to eight standard Ethernet links (160 Mbps on a Catalyst 3000—more history), up to eight Fast Ethernet links (1.6 Gbps), up to eight Gigabit Ethernet links (16 Gbps), and up to eight 10-Gigabit Ethernet links (160 Gbps)! EtherChannel provides fault-tolerant, high-speed links between switches, routers, and servers. Figure 5-26 illustrates how EtherChannel works.

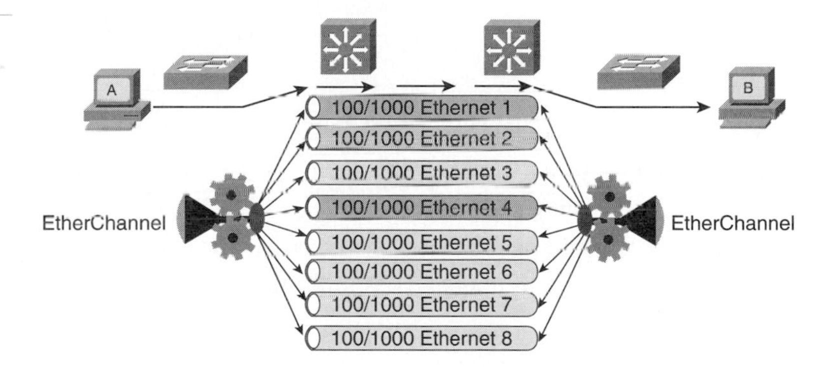

Figure 5-26 EtherChannel aggregates Ethernet links into a single logical link.

The term you probably hear most often in the context of EtherChannel is *resiliency*. Cisco defines resiliency as "the ability to recover from any network failure or issue, whether it is related to a disaster, link, hardware, design, or network services." This concept might remind you of STP and how it handles link failures within a Layer 2 topology. With EtherChannel, resiliency refers to the ability of the EtherChannel to continue to operate when one of the links in a "bundle" goes down and also the ability to restore the full bundle should the downed link come back into service.

Chapter 4 discussed trunking technologies; in particular, we looked at the IEEE 802.1Q and ISL Ethernet trunking technologies. What we might not have made clear at that time is that Ethernet trunks frequently go hand in hand with EtherChannel links. The reason for this should be clear: Trunk links carry more traffic and require more bandwidth than access links.

EtherChannel bundles segments into groups of two to eight links. This lets you scale links at rates between 100 Mbps and a healthy fraction of 1 Tbps. Do you remember when T1 or 1.544 Mbps was the fastest WAN link available? Now it is at least .16 Tbps for up to a 50-kilometer stretch with single-mode fiber (using the Catalyst 6500 10GBaseEX4 Metro 10 Gigabit Ethernet module). EtherChannel operates as either an access or trunk link. For example, you can deploy EtherChannel between the wiring closet and the data center, as shown in Figure 5-27. In the data center you can deploy EtherChannel links between servers and the network backbone to provide scalable incremental bandwidth.

Figure 5-27 EtherChannel is commonly deployed between wiring closets and the data center.

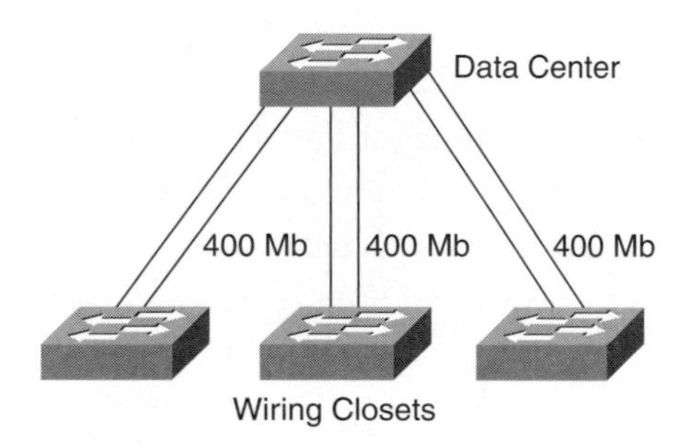

Here are some of the benefits of EtherChannel:

- **Transparent to network applications**—EtherChannel does not require any changes to networked applications. For support of EtherChannel on enterprise-class servers and network interface cards, smart software drivers can coordinate distribution of loads across multiple network interfaces.
- **Load balancing**—When EtherChannel is used within the campus, switches and routers provide load balancing transparently across multiple links to network users. Unicast, multicast, and broadcast traffic is distributed across the links in the channel.
- **Resiliency and fast convergence**—EtherChannel provides automatic recovery for loss of a link by redistributing loads across remaining links. If a link does fail, EtherChannel redirects traffic from the failed link to the remaining links in *less than a second*. This convergence is transparent to the end user. No host protocol timers expire, so no sessions are dropped.

Figure 5-28 shows a typical EtherChannel network configuration. Now that we have introduced EtherChannel, we can investigate how it works and how to configure it in your network.

Next, we explain how EtherChannel utilizes frame distribution.

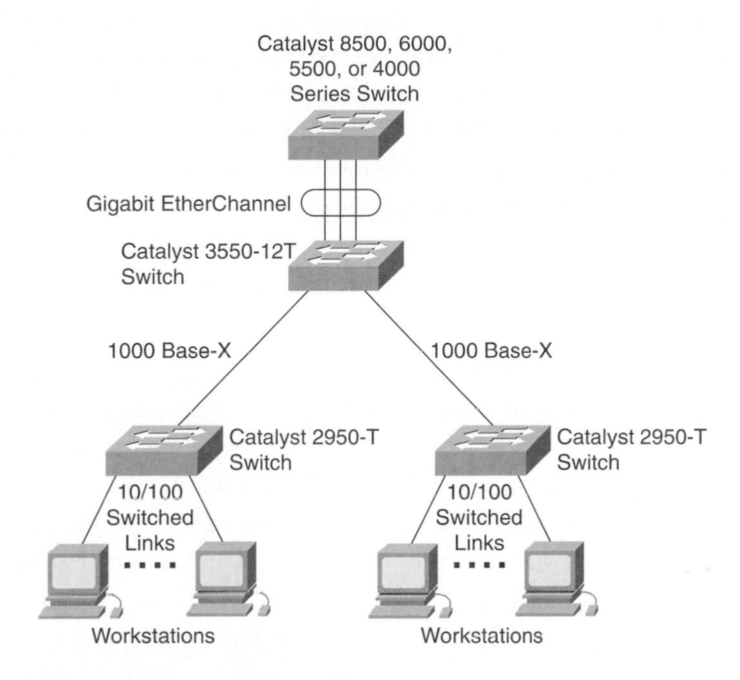

Figure 5-28 A typical network configuration utilizes EtherChannel between the core and distribution layers.

EtherChannel Operation: Frame Distribution

EtherChannel balances the traffic load across the links in a channel by applying an algorithm to a portion of the binary pattern extracted from addresses in the Ethernet frame or encapsulated data packets. The algorithm output is a numerical value that is then associated with one of the links in the channel.

On IOS-based switches, Catalyst 4000 series switches, and Catalyst 5000/5500 family switches, EtherChannel load balancing can be based on either source MAC or destination MAC addresses. On Catalyst 6000 family switches, the load-balancing policy (frame distribution) can be based on a MAC address, IP address, or port number.

With source MAC address forwarding, when packets are forwarded to an EtherChannel, they are distributed across the ports in the channel based on the incoming packet's source MAC address. Therefore, to provide load balancing, packets from different hosts use different ports in the channel, but packets from the same host use the same port in the channel.

With destination MAC address forwarding, when packets are forwarded to an EtherChannel, they are distributed across the ports in the channel based on the destination host's MAC address in the incoming packet. Therefore, packets to the same destination are forwarded over the same port, and packets to a different destination are sent on a different port in the channel.

NOTE

The most recent EtherChannel frame distribution process is based on a Cisco-proprietary hashing algorithm. The algorithm is deterministic: Given the same addresses and session information, you always hash to the same port in the channel, preventing out-of-order packet delivery.

Although EtherChannel frame distribution based on MAC address, IP address, or Layer 4 port number is possible, the configurable options allow for distribution based on source only (MAC, IP, or port), destination only (MAC, IP, or port), or both source and destination (MAC, IP, or port). The mode you select applies to all EtherChannels configured on the switch. Use the option that provides the greatest variety in your configuration. For example, if the traffic on a channel is going to a single MAC address only, using source addresses, IP addresses, or Layer 4 port numbers as the basis for frame distribution might provide better frame distribution than selecting MAC addresses as the basis.

NOTE

Cisco 7000, 7200, 7500, and 8500 routers support Fast EtherChannels via logical port channel interfaces binding Fast Ethernet interfaces. Cisco 8500 routers also support Gigabit EtherChannel.

In Figure 5-29, an EtherChannel connects a Catalyst 2950 to a router. The switch is also connected to four workstations. Because the router is referenced by the switch with a single MAC address, source-based forwarding on the switch EtherChannel ensures that the switch uses all available bandwidth to the router. The router is configured for destination-based forwarding because the large number of workstations ensures that the traffic is evenly distributed from the router EtherChannel.

Figure 5-29 EtherChannels can be configured between Catalyst switches and some Cisco routers.

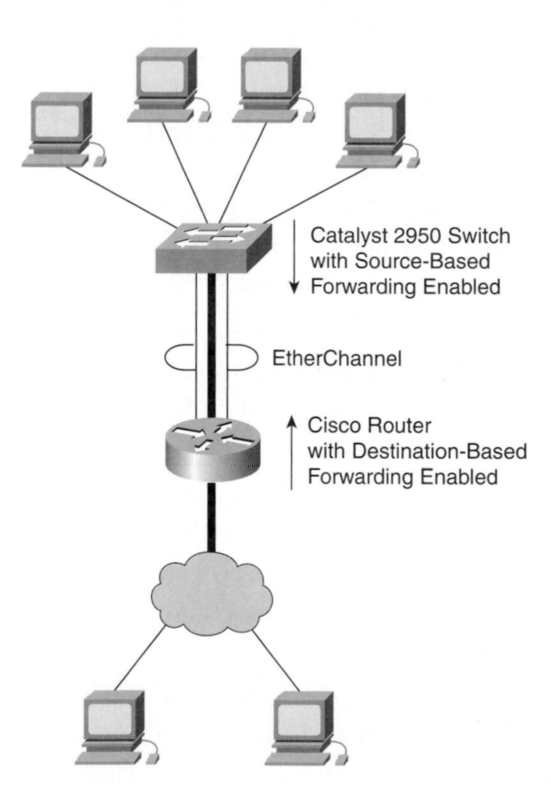

EtherChannel Operation: PAgP and LACP

Port Aggregation Protocol (PAgP) is a Cisco-proprietary technology that facilitates the automatic creation of EtherChannels by exchanging packets between Ethernet interfaces. By using PAgP, the switch learns the identity of partners that can support PAgP and learns the capabilities of each interface. It then dynamically groups similarly configured interfaces into a single logical link (channel or aggregate port). These interfaces are grouped based on hardware, administrative, and port parameter constraints. For example, PAgP groups the interfaces with the same speed, duplex mode, native VLAN, VLAN range, and trunking status and type. After grouping the links into an EtherChannel, PAgP adds the group to the spanning tree as a single switch port.

The various PAgP modes and their definitions are spelled out in Table 5-5. EtherChannel includes four user-configurable modes: on, off, auto, and desirable. Only auto and desirable are PAgP modes. You can modify the auto and desirable modes with the **silent** and **non-silent** keywords. By default, ports are in auto silent mode.

Table 5-5 EtherChannel Modes for PAgP

Mode or Keyword	Description
On	The mode that forces the port to channel without PAgP. With the on mode, a usable EtherChannel exists only when a port group in on mode is connected to another port group in on mode.
Off	The mode that prevents the port from channeling.
Auto	The mode that places a port into a passive negotiating state, in which the port responds to PAgP packets it receives but does not initiate PAgP packet negotiation. This is the default setting.
Desirable	The mode that places a port into an active negotiating state, in which the port initiates negotiations with other ports by sending PAgP packets.
silent	The keyword that is used with auto or desirable mode when no traffic is expected from the other device. This option prevents the link from being reported to the Spanning Tree Protocol as down. This is the default, secondary PAgP setting.
non-silent	The keyword that is used with auto or desirable mode when traffic is expected from the other device.

Both the auto and desirable modes allow ports to negotiate with connected ports to determine if they can form an EtherChannel, based on criteria such as port speed, trunking state, and VLAN numbers.

Ports can form an EtherChannel when they are in different PAgP modes as long as the modes are compatible. For example:

- A port in desirable mode can form an EtherChannel successfully with another port that is in desirable or auto mode.
- A port in auto mode can form an EtherChannel with another port in desirable mode.
- A port in auto mode cannot form an EtherChannel with another port that is also in auto mode, because neither port initiates negotiation.

On CatOS switches, configuring an EtherChannel automatically creates an administrative group, designated by an integer between 1 and 1024, to which the EtherChannel belongs. If you prefer, you can specify the administrative group number. Forming a channel without specifying an administrative group number creates a new automatically numbered administrative group. An administrative group can contain a maximum of eight ports. In addition, on CatOS switches, each EtherChannel is automatically assigned a unique EtherChannel ID.

Now, we change gears and introduce the open-standard equivalent to PAgP: the ***Link Aggregation Control Protocol (LACP)***. Defined in IEEE 802.3ad, LACP allows Cisco switches to manage Ethernet channels with non-Cisco devices conforming to the 802.3ad specification. It is likely that LACP will become the norm for Catalyst switches, just as Cisco migrated from ISL to IEEE 802.1Q. (Cisco can't always wait for the IEEE to complete its standards process before incorporating new features on its devices.)

If you want LACP to handle channeling, use the active and passive channel modes. To start automatic EtherChannel configuration with LACP, you need to configure at least one end of the link to active mode to initiate channeling, because ports in passive mode passively respond to initiation and never initiate the sending of LACP packets. The various LACP modes and their definitions are detailed in Table 5-6.

Table 5-6 EtherChannel Modes for LACP

Mode	Description
On	Forces the port to channel without LACP. With the on mode, a usable EtherChannel exists only when a port group in on mode is connected to another port group in on mode.
Off	Prevents the port from channeling.
Passive	Similar to PAgP's auto mode, places a port into a passive negotiating state, in which the port responds to LACP packets it receives but does not initiate LACP packet negotiation. This is the default.

Table 5-6 EtherChannel Modes for LACP (Continued)

Mode	Description
Active	Similar to PAgP's desirable mode, places a port into an active negotiating state, in which the port initiates negotiations with other ports by sending LACP packets.

The parameters used in configuring LACP are as follows:

- **System priority**—Each switch running LACP must be assigned a system priority. The system priority is used with the switch MAC address to form the system ID. It is also used during negotiation with other systems.
- **Port priority**—Each port in the switch must be assigned a port priority. The port priority is used with the port number to form the port identifier. The port priority decides which ports should be put in standby mode when a hardware limitation prevents all compatible ports from aggregating.
- **Administrative key**—Each port in the switch must be assigned an administrative key value. A port's ability to aggregate with other ports is defined with the administrative key. An administrative key is meaningful only in the context of the switch that allocates it; there is no global significance to administrative key values. A port's ability to aggregate with other ports is determined by these factors:
  - — Port physical characteristics, such as data rate, duplex capability, and point-to point or shared medium
  - — Configuration constraints that you establish

When enabled, LACP always tries to configure the maximum number of compatible ports in a channel, up to the maximum allowed by the hardware (eight ports for Catalyst switches). If LACP cannot aggregate all the ports that are compatible (for example, if the remote system has more restrictive hardware limitations), all the ports that cannot be actively included in the channel are put in hot standby state and are used only if one of the channeled ports fails.

EtherChannel Configuration Guidelines

There are a host of guidelines for configuring EtherChannel (see Figure 5-30). These relate to trunks, dynamic VLANs, STP, and otherwise. If improperly configured, some EtherChannel interfaces are automatically disabled to avoid network loops and other problems.

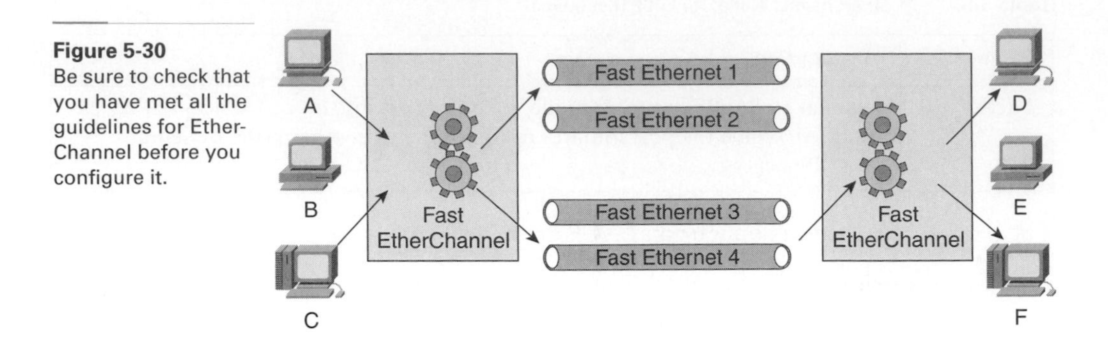

Figure 5-30
Be sure to check that you have met all the guidelines for EtherChannel before you configure it.

To minimize your time spent troubleshooting EtherChannel, you should follow these recommendations:

- Assign all ports in an EtherChannel to the same VLAN, or configure them as trunk ports.
- If you configure the EtherChannel as a trunk, configure the same trunk mode on all the ports in the EtherChannel. Configuring ports in an EtherChannel in different trunk modes can have unexpected results.
- An EtherChannel supports the same allowed range of VLANs on all the ports in a trunking EtherChannel. If the allowed range of VLANs is not the same for a port list, the ports do not form an EtherChannel even when set to auto or desirable mode.
- Do not configure the ports in an EtherChannel as dynamic VLAN ports. Doing so can adversely affect switch performance.
- An EtherChannel will not form with ports that have different GARP, GVRP, or QoS configurations. For more information on GARP and GVRP, see Chapter 4. See Chapter 10 for more information on QoS.
- If you configure a broadcast limit on the ports (discussed in Chapter 10), configure the broadcast limit as a percentage limit for the channeled ports. With a packets-per-second broadcast limit, unicast packets might get dropped for 1 second when the broadcast limit is exceeded.
- An EtherChannel will not form with ports that have the port security feature enabled (see Chapter 10). You cannot enable the port security feature for ports in an EtherChannel.
- If one port in an EtherChannel is used by IGMP multicast filtering (discussed in Chapter 9), you must set the EtherChannel mode for both PAgP and LACP to off. No other mode might be used.

- An EtherChannel will not form if one of the ports is a Switched Port Analyzer (SPAN) destination port (see Chapter 10).
- An EtherChannel will not form if protocol filtering is set differently on any of the ports. For more information on protocol filtering, see Chapter 10.
- Each EtherChannel can have up to eight compatibly configured Ethernet interfaces.
- Configure all interfaces in an EtherChannel to operate at the same speed and duplex modes.
- Enable all interfaces in an EtherChannel. If you shut down an interface in an EtherChannel, it is treated as a link failure, and its traffic is transferred to one of the remaining interfaces in the EtherChannel.
- Interfaces with different STP Port Costs can form an EtherChannel as long they are otherwise compatibly configured. Setting different STP Port Costs does not, by itself, make interfaces incompatible for the formation of an EtherChannel. But it is preferable to set STP port costs to be equal for all ports in an EtherChannel.

Beyond the restrictions listed here, the only remaining concern is what ports you are allowed to use when configuring an EtherChannel. We address this question next.

Catalyst 4000 and 6000 line cards have newer chipsets that allow you to use an even or odd number of links in your EtherChannel. The ports do not have to be contiguous or even on the same line card, as is true with some Catalyst devices and line modules. The newer chips on the Catalyst 4000s and 6000s are not available on all Catalyst hardware.

Older Catalyst switches use an Ethernet Bundle Controller (EBC) to manage aggregated EtherChannel ports (for example, performing frame distribution). The EBC communicates with the Enhanced Address Recognition Logic (EARL) Application-Specific Integrated Circuit (ASIC). The EARL is responsible for, among other things, learning MAC addresses. For example, the EARL ages out all addresses learned on a link in an EtherChannel bundle when it fails. After a link failure, the EBC and the EARL work together to recalculate in hardware the "lost" source-destination address pairs on a different link.

On the older switches, when selecting ports to group for an EtherChannel process, you must select ports that belong to the same EBC. On a 24-port module, there are three groups of eight ports. On a 12-port module, there are three groups of four ports. See Figure 5-31.

NOTE

Be sure to check your hardware documentation before attempting to create EtherChannel bundles. Older Catalyst switches restrict which ports are used in a given bundle. The remainder of the material in this section mostly pertains to Catalyst 5000 line modules. A few 5000s are still in action, so we include the details here. You might want to skip ahead to the next section on EtherChannel configuration.

Figure 5-31 Older Catalyst switches require you to use contiguous ports associated with Ethernet Bundle Controllers when configuring an EtherChannel.

1	2	3	4	5	6	7	8	9	10	11	12	13	14	15	16	17	18	19	20	21	22	23	24
1	1	1	1	1	1	1	1	2	2	2	2	2	2	2	2	3	3	3	3	3	3	3	3
1	1	1	1	2	2	2	2	3	3	3	3												

For example, in a 12-port module, you can create up to two dual-segment EtherChannel configurations within each group of four ports, as illustrated in Example A of Figure 5-32. Or, you can create one dual-segment EtherChannel configuration within each group, as in Example B of Figure 5-32. Example C illustrates a four-segment and a two-segment EtherChannel configuration.

Figure 5-32 Some examples of valid and invalid EtherChannel port configurations for older Catalyst switches help you get it right the first time.

Port		1	2	3	4	5	6	7	8	9	10	11	12
Example A	OK	1	1	2	2	3	3	4	4	5	5	6	6
Example B	OK	1	1			2	2			3	3		
Example C	OK	1	1	1	1	2	2						
Example D	Not OK			1	1								
Example E	Not OK	1	1	2	2	2	2						
Example F	Not OK	1		1									
Example G	Not OK		1	1									

You must avoid some EtherChannel configurations on older Catalyst devices, including most Catalyst 5000 line cards. Example D in Figure 5-32 illustrates an invalid two-segment EtherChannel configuration using Ports 3 and 4 of a group. The EBC must start its bundling with the first ports of a group. This does not mean that you have to use the first group. In contrast, a valid dual-segment EtherChannel configuration can use Ports 5 and 6 with no EtherChannel segment on the first group.

Example E illustrates another invalid configuration. In this example, two EtherChannel segments are formed. One is a dual-segment EtherChannel configuration, and the other is a four-segment EtherChannel configuration. The dual-segment EtherChannel configuration is valid. The four-segment EtherChannel configuration, however, violates the rule that all ports must belong to the same group (EBC). This EtherChannel configuration uses two ports from the first group and two ports from the second group.

Example F shows an invalid configuration in which an EtherChannel configuration is formed with discontiguous segments. You must use adjacent ports to form an Ether-Channel configuration.

Finally, Example G is an invalid EtherChannel configuration because it does not use the first ports on the module to start the EtherChannel process (like Example D). You cannot start the EtherChannel process with middle ports on the line module.

All the examples in Figure 5-32 apply to 24-port modules as well. The only difference between a 12- and 24-port module is the number of EtherChannel segments that can be formed within a group. The 12-port module allows only two (two-port) Ether-Channel segments in a group, whereas the 24-port module supports up to four (two-port) EtherChannel segments per group, as shown in Figure 5-31.

EtherChannel Configuration

With a bit of EtherChannel information behind us, we proceed to detail EtherChannel configuration. As with STP, there tends to be a lot more theory than configuration. You will find the configuration to be somewhat anticlimactic.

NOTE

We cover only PAgP configuration. LACP configuration is documented in the Configuration Guides for the various Catalyst switches at www.cisco.com.

CatOS Switch

The elements of EtherChannel configuration and monitoring on a CatOS switch include the following:

- Creating the EtherChannel
- Defining the administrative group
- Verifying EtherChannel configuration
- Setting the EtherChannel STP Port Cost
- Setting the EtherChannel STP Port VLAN Cost
- Removing the EtherChannel bundles
- Displaying EtherChannel information
- Displaying EtherChannel traffic statistics
- Displaying EtherChannel PAgP statistics

NOTE

You can configure EtherChannels as trunks. After a channel is formed, configuring any port in the channel as a trunk applies the configuration to all ports in the channel.

To determine what ports are capable of participating in an EtherChannel, use the command **show port capabilities** [*mod_num[/port_num*]], as shown in Example 5-39. Here, ports 1 and 2 on module 1 form an EtherChannel.

Example 5-39 Determining if EtherChannel is Supported on a Port

```
Console> (enable) show port capabilities 1/2
Model                    WS-X4013
Port                     1/2
Type                     1000BaseSX
Speed                    1000
Duplex                   full
Trunk encap type         802.1Q
Trunk mode               on,off,desirable,auto,nonegotiate
Channel                  1/1-2
Flow control             receive-(off,on,desired),send-(off,on,desired)
Security                 yes
Dot1x                    yes
Membership               static,dynamic
Fast start               yes
QOS scheduling           rx-(none),tx-(2q1t)
CoS rewrite              no
ToS rewrite              no
Rewrite                  no
UDLD                     yes
Inline power             no
AuxiliaryVlan            no
SPAN                     source,destination
```

To create an EtherChannel, use the command **set port channel** *port_list* [*admin_group*] **mode** {**on** | **off** | **desirable** | **auto**} [**silent** | **non-silent**].

To verify the configuration, enter the command **show port channel** [port_list]. With this command, entering a single port in the port channel is sufficient. Notice that the administrative group and EtherChannel ID are displayed.

Example 5-40 illustrates these commands.

Example 5-40 Creating an EtherChannel on a CatOS Switch

```
Console> (enable) set port channel 2/29-30 mode on
Port(s) 2/29-30 channel mode set to on.
Console> (enable) show port channel 2/29
Port  Status     Channel              Admin Ch
                 Mode                 Group Id
```

Example 5-40 Creating an EtherChannel on a CatOS Switch (Continued)

```
----- ---------- ------------------- ----- -----
 2/29 connected  on                      90   815
 2/30 connected  on                      90   815
----- ---------- ------------------- ----- -----

Port  Device-ID                       Port-ID                   Platform
----- ------------------------------- ------------------------- ---------------
 2/29 Switch                         FastEthernet0/3           cisco WS-C2912-XL
 2/30 Switch                         FastEthernet0/4           cisco WS-C2912-XL
----- ------------------------------- ------------------------- ---------------
```

Another command to view the EtherChannel ID (if you know the administrative group number) is **show channel group** *admin_group*. To change the administrative group for an EtherChannel, type the command **set port channel** *port_list* [*admin_group*]. These commands are illustrated in Example 5-41.

Example 5-41 Changing the Administrative Group for an EtherChannel

```
Console> (enable) show channel group 90
Admin Port  Status     Channel                Channel
group                  Mode                   id
----- ----- ---------- ------------------- --------
   90  2/29 connected  on                       815
   90  2/30 connected  on                       815

Admin Port  Device-ID                       Port-ID                   Platform
group
----- ----- ------------------------------- ------------------------- ---------
   90  2/29 Switch                         FastEthernet0/3           cisco WS-C2912-XL
   90  2/30 Switch                         FastEthernet0/4           cisco WS-C2912-XL
Console> (enable) set port channel 2/29-30 9
Port(s) 2/29-30 are assigned to admin group 9.
Console> (enable) 2002 Feb 03 18:39:04 %PAGP-5-PORTFROMSTP:Port 2/29 left bridge
  port 2/29-30
2002 Feb 03 18:39:04 %PAGP-5-PORTFROMSTP:Port 2/30 left bridge port 2/29-30
2002 Feb 03 18:39:06 %PAGP-5-PORTTOSTP:Port 2/29 joined bridge port 2/29-30
2002 Feb 03 18:39:06 %PAGP-5-PORTTOSTP:Port 2/30 joined bridge port 2/29-30
```

continues

Example 5-41 Changing the Administrative Group for an EtherChannel (Continued)

```
Console> (enable) show port channel 2/29
Port  Status     Channel              Admin Ch
                 Mode                 Group Id
----- ---------- -------------------- ----- -----
 2/29 connected  on                       9   815
 2/30 connected  on                       9   815
----- ---------- -------------------- ----- -----

Port  Device-ID                       Port-ID                   Platform
----- ------------------------------- ------------------------- ----------------
 2/29 Switch                         FastEthernet0/3           cisco WS-C2912-XL
 2/30 Switch                         FastEthernet0/4           cisco WS-C2912-XL
----- ------------------------------- ------------------------- ----------------
```

At this point, we stop to review some of the differences with the EtherChannel verification commands. Notice the difference between the outputs of the following commands, based on mod/port versus EtherChannel ID versus administrative group (see Example 5-42). Pay special attention to the leftmost column of the outputs.

Example 5-42 EtherChannel Verification Output Display Options

```
Console> (enable) show port channel ?
  info                          Show port channel information
  statistics                    Show port channel statistics
  <mod>                         Module number
  <mod/port>                    Module number and Port number
  <cr>
Console> (enable) show port channel 2/29
Port  Status     Channel              Admin Ch
                 Mode                 Group Id
----- ---------- -------------------- ----- -----
 2/29 connected  on                       9   815
 2/30 connected  on                       9   815
----- ---------- -------------------- ----- -----

Port  Device-ID                       Port-ID                   Platform
----- ------------------------------- ------------------------- ----------------
```

Example 5-42 EtherChannel Verification Output Display Options (Continued)

```
 2/29 Switch                          FastEthernet0/3          cisco WS-C2912-XL
 2/30 Switch                          FastEthernet0/4          cisco WS-C2912-XL
----- ------------------------------ ----------------------- ---------------
Console> (enable) show channel ?
  <channel_id>                Show a particular channel info
  group                       Show channel group status
  mac                         Show mac info in the channel
  traffic                     Show channel traffic
  <cr>
Console> (enable) show channel
Channel Id   Ports
-----------  -----------------------------------------------
814          2/27-28
815          2/29-30
Console> (enable) show channel 815
Channel Ports                                         Status    Channel
id                                                              Mode
------- ---------------------------------------------- --------- -------------
    815 2/29-30                                       connected on
Console> (enable) show channel group ?
  info                        Show Admin group info
  statistics                  Show Admin group statistics
  <admin_group>               Admin group (1..1024)
  <cr>
Console> (enable) show channel group 9
Admin Port  Status     Channel              Channel
group                  Mode                 id
----- ----- ---------- -------------------- --------
    9  2/29 connected  on                        815
    9  2/30 connected  on                        815

Admin Port  Device-ID                      Port-ID                   Platform
group
----- ----- ------------------------------ ------------------------- ---------
    9  2/29 Switch                          FastEthernet0/3          cisco WS-C2912-XL
    9  2/30 Switch                          FastEthernet0/4          cisco WS-C2912-XL
```

To set the EtherChannel Port Cost, enter the command **set channel cost** {*channel_id* | **all**} *cost*. If you do not specify the *cost*, the spanning tree Port Cost is updated based on the current Port Costs of the channeling ports. However, if you change the channel Port Cost, the Port Costs of member ports in the channel are modified to reflect the new cost.

To set the EtherChannel Port VLAN Cost, use the command **set channel vlancost** {*channel_id* | **all**} *cost*. If you change the channel Port VLAN Cost, the Port VLAN Costs of member ports in the channel are modified to reflect the new cost.

The latter command provides an alternative cost for some of the VLANs in a trunk channel. This enables load balancing of VLAN traffic across multiple channels configured with trunking, because some VLANs in the channel have Port VLAN Cost values and the others have Port Cost values. This might not sound useful (or meaningful), but these commands allow you, for example, to configure VLAN traffic to be load-balanced between one EtherChannel connected to one core switch and another EtherChannel connected to another core switch.

If you have access to a lab with three Catalyst switches, you need to practice these commands to understand how the Port Costs, Port VLAN Costs, channel costs, and channel VLAN costs interrelate. The two commands are illustrated in Example 5-43.

Example 5-43 EtherChannel Port Cost and EtherChannel Port VLAN Cost Configuration

```
Console> (enable) set channel cost 814 10
Port(s) 2/27-28 port path cost are updated to 16.
Channel 814 cost is set to 10.
Warning: channel cost may not be applicable if channel is broken.
Console> (enable) set channel vlancost 814 20
Port(s) 2/27-28 vlan cost are updated to 32.
Channel 814 vlancost is set to 20.
```

If you have tried to remove configurations on CatOS switches, you have probably experienced some frustration at some point. Sometimes you use **clear** commands and sometimes you use **set...disable** commands. And, sometimes after doing this, you view your configuration file (**show config**) only to find that unwanted vestigial commands remain. If only we lived in an ideal world!

Anyway, one way to remove an EtherChannel bundle is to perform the command **set port channel** *port_list* **mode auto** on both ends of the channel. See Example 5-44.

Example 5-44 Removing an EtherChannel

```
Console> (enable) set port channel 2/27-28 mode auto
Port(s) 2/27-28 channel mode set to auto.
2002 Feb 03 20:17:51 %PAGP-5-PORTFROMSTP:Port 2/27 left bridge port 2/27-28
2002 Feb 03 20:17:51 %PAGP-5-PORTFROMSTP:Port 2/28 left bridge port 2/28
2002 Feb 03 20:17:51 %PAGP-5-PORTTOSTP:Port 2/28 joined bridge port 2/28
2002 Feb 03 20:17:51 %PAGP-5-PORTFROMSTP:Port 2/28 left bridge port 2/28
2002 Feb 03 20:18:02 %PAGP-5-PORTTOSTP:Port 2/27 joined bridge port 2/27
2002 Feb 03 20:18:02 %PAGP-5-PORTTOSTP:Port 2/28 joined bridge port 2/28
```

In Example 5-44, you see the trunk ports 2/27 and 2/28 resuming their independence after being bound together.

Next, to display EtherChannel information, you use one of the following commands:

- **show port channel** [*mod_num*[/*port_num*]] **info**
- **show channel group** [*admin_group*] **info**
- **show channel** [*channel_id*] **info**

Due to the length of the output of these commands, we content ourselves with the output of the third command, as shown in Example 5-45.

Example 5-45 Displaying EtherChannel Information

```
Console> (enable) show channel 815 info
Chan Port  Status     Channel             Admin Speed Duplex Vlan
id                    mode                group
---- ----- ---------- ------------------- ----- ----- ------ ----
 815  2/29 connected  on                    815 a-100 a-full    1
 815  2/30 connected  on                    815 a-100 a-full    1
Chan Port  if-   Oper-group Neighbor   Chan  Oper-Distribution PortSecurity/
id         Index            Oper-group cost  Method            Dynamic Port
---- ----- ----- ---------- ---------- ----- -----------------
 815  2/29 48             2                20 mac both
 815  2/30 48             2                20 mac both

Chan Port  Device-ID                      Port-ID                   Platform
id
---- ----- ------------------------------ ------------------------- ---------
```

continues

Example 5-45 Displaying EtherChannel Information (Continued)

```
 815  2/29 Switch                                  FastEthernet0/3          cisco WS-C2912-XL
 815  2/30 Switch                                  FastEthernet0/4          cisco WS-C2912-XL

Chan Port  Trunk-status Trunk-type    Trunk-vlans
id
----- ----- ----------- ------------ ----------------------------------------
 815  2/29 trunking     dot1q         1-1005
 815  2/30 trunking     dot1q         1-1005

Chan Port  Portvlancost-vlans
id
---- ----- --------------------------------------------------------------------
 815  2/29
 815  2/30

Chan Port  Port      Portfast Port    Port
id         priority           vlanpri vlanpri-vlans
---- ----- -------- -------- ------- ----------------------------------------
 815  2/29       32 disabled       0
 815  2/30       32 disabled       0

Chan Port  IP       IPX      Group
id
---- ----- -------- -------- --------
 815  2/29 on       auto-on  auto-on
 815  2/30 on       auto-on  auto-on

Chan Port  GMRP     GMRP         GMRP
id         status   registration forwardAll
---- ----- -------- ------------ ----------
 815  2/29 enabled  normal       disabled
 815  2/30 enabled  normal       disabled

Chan Port  GVRP     GVRP         GVRP
id         status   registration applicant
---- ----- -------- ------------ --------
```

Example 5-45 Displaying EtherChannel Information (Continued)

```
 815  2/29 disabled normal       normal
 815  2/30 disabled normal       normal

Chan Port  Auxiliaryvlan
id
---- ----- -------------
 815  2/29 none
 815  2/30 none
```

The highlighted output shows that the EtherChannel frame distribution method in effect is based on both destination and source MAC addresses.

To display EtherChannel traffic statistics, use the command **show channel** [*channel_id*] **mac.** As seen in Example 5-46, this command gives a breakdown of traffic in terms of unicast, multicast, and broadcast.

Example 5-46 Displaying EtherChannel Traffic Statistics

```
Console> (enable) show channel 815 mac
Channel  Rcv-Unicast          Rcv-Multicast        Rcv-Broadcast
-------- -------------------- -------------------- --------------------
815                        15                 2447                    0

Channel  Xmit-Unicast         Xmit-Multicast       Xmit-Broadcast
-------- -------------------- -------------------- --------------------
815                        14                20619                  637

Channel  Rcv-Octet            Xmit-Octet
-------- -------------------- --------------------
815                    313686              1538058

Channel  Dely-Exced MTU-Exced  In-Discard Lrn-Discrd In-Lost    Out-Lost
-------- ---------- ---------- ---------- ---------- ---------- ----------
815               0          -          0          0          0          0
```

Finally, to display EtherChannel PAgP statistics, use one of the following commands:

- **show port channel** [*mod_num*[*/port_num*]] **statistics**

- **show channel group** [*admin_group*] **statistics**
- **show channel** [*channel_id*] **statistics**

Interestingly, the output of these commands is identical. Sample output is shown in Example 5-47.

Example 5-47 Displaying EtherChannel PAgP Statistics

```
Console> (enable) show channel 815 stat
Port  Channel PAgP Pkts   PAgP Pkts PAgP Pkts PAgP Pkts PAgP Pkts PAgP Pkts
      id      Transmitted Received  InFlush   RetnFlush OutFlush  InError
----- ------- ----------- --------- --------- --------- --------- ---------
 2/29     815         187        67         0         0         0         0
 2/30     815         198        93         0         0         0         0
```

NOTE

The **channel-group** command replaces the **port group** command, which was used prior to Cisco IOS Software Release 12.1.

IOS-Based Switch

To configure EtherChannel on an IOS-based switch, on each of the participating interfaces (up to eight), enter the command **channel-group** *channel-group-number* **mode** {**auto** [**non-silent**] | **desirable** [**non-silent**] | **on**}.

To remove an interface from the EtherChannel group, use the **no channel-group** interface configuration command.

Example 5-48 illustrates the configuration of EtherChannel on an IOS-based switch.

Example 5-48 Configuring an EtherChannel on an IOS-Based Switch

```
Switch(config)#interface range gigabitethernet0/1 - 2
Switch(config-if)#channel-group 1 mode desirable
```

NOTE

The **show etherchannel** command replaces the **show port group** command, which was used prior to Cisco IOS Software Release 12.1.

A shortcut is now available, beginning with Cisco IOS Software Release 12.1(5)T: Identical commands can be entered once for a range of interfaces, rather than being entered separately for each interface, using the **interface range** command, as shown in Example 5-48. The space before and after the dash is required.

To verify your EtherChannel configuration and view EtherChannel information, use the command **show etherchannel** [*channel-group-number*] {**brief** | **detail** | **load-balance** | **port** | **port-channel** | **summary**}.

The **summary** option is illustrated in Example 5-49.

Example 5-49 Verifying the EtherChannel Configuration on an IOS-Based Switch

```
Switch#show etherchannel 1 summary
Flags:  D - down         P - in port-channel
        I - stand-alone s - suspended
        R - Layer3       S - Layer2
        U - port-channel in use
Group Port-channel  Ports
-----+------------+-----------------------------------------------------------
1     Po1(SU)      Gi0/1(P)   Gi0/2(P)
```

To specify the technique for load balancing (frame distribution) among links comprising an EtherChannel, use the command **port-channel load-balance** {**dst-mac** | **src-mac**}, as shown in Example 5-50.

Example 5-50 Configuring the EtherChannel Frame Distribution Method

```
Switch(config)#port-channel load-balance dst-mac
```

You can verify your configuration with the command **show etherchannel load-balance**.

Finally, to view PAgP status information, use the command **show pagp** [*channel-group number*] {**counters** | **internal** | **neighbor**}, as illustrated in Example 5-51.

Example 5-51 Viewing PAgP Status

```
Switch#show pagp 1 counters


           Information        Flush
Port       Sent   Recv      Sent   Recv
--------------------------------------
Channel group: 1
  Gi0/1    45     42        0      0
  Gi0/2    45     41        0      0
```

EtherChannel Example

Now that you have seen all the elements of EtherChannel configuration, this is a good time to work through a complete example that demonstrates all the steps in the configuration. You will configure a two-port Gigabit EtherChannel link between two CatOS switches. A fiber link connects 1000BaseSX Gigabit Ethernet ports 2/1-2 on one side and 3/1-2 on the other (see Figure 5-33).

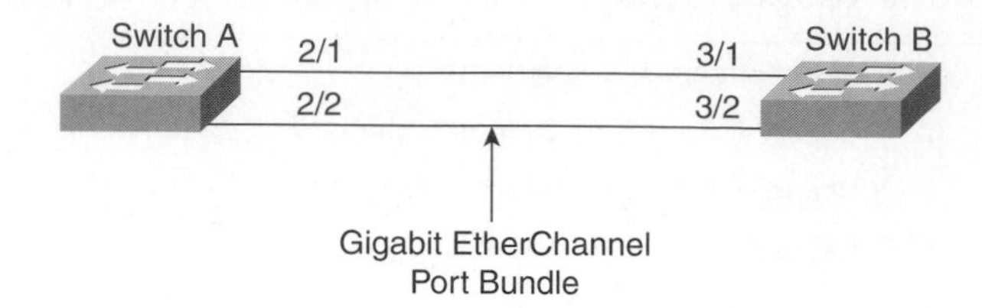

Figure 5-33 An EtherChannel is configured with the **set port channel** command on CatOS switches.

Step 1 Make sure that all ports on Switch A and Switch B have the same port configuration, such as VLAN membership:

```
Switch_A> (enable) set vlan 100 2/1-2

VLAN 100 modified.
VLAN 1 modified.
VLAN  Mod/Ports
---- -----------------------
100   2/1-2

Switch_A> (enable)

Switch_B> (enable) set vlan 100 3/1-2

VLAN 100 modified.
VLAN 1 modified.
VLAN  Mod/Ports
---- -----------------------
100   3/1-2

Switch_B> (enable)
```

Step 2 Confirm the channeling status of the switches using the **show port channel** command:

```
Switch_A> (enable) show port channel

No ports channelling
Switch_A> (enable)
Switch_B> (enable) show port channel

No ports channelling
Switch_B> (enable)
```

Step 3 In this example, EtherChannel will be configured as **on** for all ports. In this case, you have to configure the ports on both ends of the EtherChannel bundle as **on** (see Table 5-5). The system logging messages provide information about the formation of the EtherChannel bundle:

```
Switch_A> (enable) set port channel 2/1-2 on
Port(s) 2/1-2 channel mode set to on.
```

```
Switch_A> (enable) %PAGP-5-PORTFROMSTP:Port 2/1 left bridge port 2/1
%PAGP-5-PORTFROMSTP:Port 2/2 left bridge port 2/2
%PAGP-5-PORTTOSTP:Port 2/1 joined bridge port 2/1-2
%PAGP-5-PORTTOSTP:Port 2/2 joined bridge port 2/1-2
Switch_B> (enable) set port channel 3/1-2 on

Port(s) 3/1-2 channel mode set to on.
Switch_B> (enable) %PAGP-5-PORTFROMSTP:Port 3/1 left bridge port 3/1
%PAGP-5-PORTFROMSTP:Port 3/2 left bridge port 3/2
%PAGP-5-PORTTOSTP:Port 3/1 joined bridge port 3/1-2
%PAGP-5-PORTTOSTP:Port 3/2 joined bridge port 3/1-2
```

Step 4 After the EtherChannel bundle is negotiated, enter the **show port channel** command to verify the configuration. If you configure only the ports on one side of the link as **on**, the **show port channel** command will show that the ports are channeling, but no traffic will pass over the EtherChannel. In this event, spanning-tree loops can occur, and the switch will eventually disable the EtherChannel:

```
Switch_A> (enable) show port channel

Port  Status     Channel   Channel     Neighbor          Neighbor
        mode      status     device                       port
----- ---------- --------- ----------- ------------------------ -----
-----
2/1  connected  on        channel    WS-C4003    JAB023806LN(  3/1
2/2  connected  on        channel    WS-C4003    JAB023806LN(  3/2
----- ---------- --------- ----------- ------------------------ -----
-----
Switch_A> (enable)
Switch_B> (enable) show port channel

Port  Status     Channel   Channel     Neighbor          Neighbor
        mode      status     device                       port
----- ---------- --------- ----------- ------------------------ -----
-----
3/1  connected  on        channel    WS-C4003    JAB023806JR(  2/1
 3/2  connected  on        channel    WS-C4003    JAB023806JR(  2/2
----- ---------- --------- ----------- ------------------------ -----
-----
Switch_B> (enable)
```

Summary

STP enables Layer 2 reliability, resiliency, and redundancy in a campus network. STP dynamically configures a Layer 2 topology that blocks physical loops in a switched network while at the same time providing a number of configuration options for tailoring the flow of traffic in your network.

A Root Bridge serves as the focal point of the topology for each VLAN configured in PVST+ STP mode. With Path Cost, Port Priority, Bridge Priority, STP timers, PortFast, UplinkFast, and BackboneFast, you have the tools to design a robust Layer 2 topology.

EtherChannel provides the finishing touch to your Layer 2 switched infrastructure. EtherChannel provides for incremental bandwidth Ethernet link options, ranging from 10 Mbps (half-duplex) 100-meter links to 160 Gbps (full-duplex) 50-kilometer links. EtherChannel relies on Port Aggregation Protocol (PAgP) to dynamically manage bundles of Ethernet links.

Several protocols are responsible for the wide variety of tools available in Layer 2 switched design and deployment. Chapter 4 introduced IEEE 802.1Q, ISL, DISL, DTP, and VTP. This chapter focused on STP and PAgP.

The remaining chapters of this book explore LAN switching technologies that involve OSI Layers 3 to 7. The next step is to move beyond intra-VLAN considerations and begin taking advantage of the inter-VLAN routing functionality of Catalyst switches used in the core and distribution layers of a campus network. Along the way, you will learn how to configure Cisco routers to interoperate with Catalyst switches.

Check Your Understanding

Test your understanding of the concepts covered in this chapter by answering these review questions. Answers are listed in Appendix A, "Check Your Understanding Answer Key."

1. When you're creating a loop-free logical topology, which STP parameter takes precedence in the four-step decision sequence *following the election of a Root Bridge?*

   A. Sender Bridge ID

   B. Port ID

   C. Root Path Cost

   D. VLAN ID

2. What is the default Max Age?

   A. 10 seconds

   B. 15 seconds

   C. 20 seconds

   D. 30 seconds

3. What is the default Forward Delay?

   A. 10 seconds

   B. 15 seconds

   C. 20 seconds

   D. 30 seconds

4. What is the default Hello Time?

   A. 1 second

   B. 2 seconds

   C. 4 seconds

   D. 5 seconds

5. BackboneFast allows for STP to reconverge after a trunk link failure. Which of the following best represents the time saved in this reconvergence relative to the time it would take without BackboneFast?

   A. Hello Time

   B. Forward Delay

C. Max Age

D. None of the above

6. Where would you likely never configure UplinkFast?

A. Core layer

B. Distribution layer

C. Access layer

7. EtherChannel frame distribution on Catalyst 6000 switches can be configured to be a function of which of the following parameters?

A. Source MAC address

B. Destination IP address

C. Destination MAC address

D. Source port number

8. Among all the Catalyst switches, what is the greatest number of Fast Ethernet links that can be bundled to form a Fast EtherChannel?

A. 2

B. 4

C. 6

D. 8

9. Which VLANs have a Path Cost of 15 after you enter the command **set spantree portvlancost 2/28 22**, given that:

Port 2/28 VLANs 21 to 1005 have a Path Cost of 80.

Port 2/28 VLANs 1 to 20 have a Path Cost of 15.

A. 1 to 20

B. 1 to 1005

C. 1 to 20 and 22

D. None

10. What is the CatOS command to bind ports 2/3 and 2/4 into an EtherChannel?

A. **set port group 2/3-4 desirable**

B. **channel-group 2/3-4 nonegotiate**

C. **set channel 2/3-4 auto**

D. **set port channel 2/3-4 on**

Key Terms

BackboneFast A Catalyst feature that is initiated when a root port or blocked port on a switch receives inferior BPDUs from its designated bridge. Max Age is skipped in order to allow appropriate blocked ports to transition quickly to the Forwarding state in the case of an indirect link failure.

BID (Bridge ID) An 8-byte field consisting of an ordered pair of numbers. The first is a 2-byte decimal number called the bridge priority, and the second is a 6-byte (hexadecimal) MAC address.

Blocking state The STP state in which a bridge listens for BPDUs. This state follows the Disabled state. Forwarding does not occur in the Blocking state.

BPDU (Bridge Protocol Data Unit) Frame passed between switches to communicate STP information used in the STP decision-making process.

Bridge ID Priority The Bridge Priority and the system ID extension combined.

Bridge Priority The decimal number used to measure the preference of a bridge in the Spanning-Tree Algorithm. The possible values range between 0 and 65,535.

CST (Common Spanning Tree) The spanning-tree implementation specified in the IEEE 802.1Q standard. CST defines a single instance of spanning tree for all VLANs with BPDUs transmitted over VLAN 1.

cost For switches with active 1 Gbps links but no active 10 Gbps links, a link's STP cost is defined according to the following table:

Bandwidth	STP Cost
4 Mbps	250
10 Mbps	100
16 Mbps	62
45 Mbps	39
100 Mbps	19
155 Mbps	14
622 Mbps	6
1 Gbps	4
10 Gbps	2

Designated Bridge The bridge containing the designated port for that segment.

Designated Port The bridge port connected to that segment that both sends traffic toward the root bridge and receives traffic from the root bridge over that segment.

Disabled state The administratively shut-down STP state.

EtherChannel A Cisco-proprietary technology that, by aggregating links into a single logical link, provides incremental trunk speeds ranging from 10 Mbps to 160 Gbps.

Forward Delay The time that the bridge spends in the Listening and Learning states.

Forwarding state The STP state in which data traffic is both sent and received on a port. It is the "last" STP state, following the Learning state.

Hello Time The time interval between the sending of configuration BPDUs.

LACP (Link Aggregation Control Protocol) Defined in IEEE 802.3ad. Allows Cisco switches to manage Ethernet channels with non-Cisco devices conforming to the 802.3ad specification.

Learning state The STP state in which the bridge does not pass user data frames but builds the bridging table and gathers information, such as the source VLANs of data frames. This state follows the Listening state.

Listening state The STP state in which no user data is passed, but the port sends and receives BPDUs in an effort to determine the active topology. It is during the Listening state that the three initial convergence steps take place—elect a Root Bridge, elect Root Ports, and elect Designated Ports. This state follows the Blocking state.

Max Age An STP timer that controls how long a bridge stores a BPDU before discarding it. The default is 20 seconds.

MISTP (Multiple Instances of Spanning Tree) Allows you to group multiple VLANs under a single instance of spanning tree. MISTP combines the Layer 2 load-balancing benefits of PVST+ with the lower CPU load of IEEE 802.1Q.

MST (Mono Spanning Tree) The spanning-tree implementation used by non-Cisco 802.1Q switches. One instance of STP is responsible for all VLAN traffic.

MST (Multiple Spanning Tree) Extends the IEEE 802.1w Rapid Spanning Tree (RST) algorithm to multiple spanning trees, as opposed to the single CST of the original IEEE 802.1Q specification. This extension provides for both rapid convergence and load balancing in a VLAN environment.

PAgP (Port Aggregation Protocol) A Cisco-proprietary technology that facilitates the automatic creation of EtherChannels by exchanging packets between Ethernet interfaces.

Path Cost An STP measure of how close bridges are to each other. Path Cost is the sum of the costs of the links in a path between two bridges.

Port ID A 2-byte STP parameter consisting of an ordered pair of numbers. The first is the port priority, and the second is the port number.

Port Number A numerical identifier used by Catalyst switches to enumerate a port.

Port Priority A configurable STP parameter with values ranging from 0 to 63 on a CatOS switch (the default is 32) and from 0 to 255 on an IOS-based switch (the default is 128). Port Priority influences Root Bridge selection when all other STP parameters are equal.

PortFast A Catalyst feature that, when enabled, causes an access or trunk port to enter the spanning tree Forwarding state immediately, bypassing the Listening and Learning states.

PVST (Per-VLAN Spanning Tree) A Cisco-proprietary spanning tree implementation requiring ISL trunk encapsulation. PVST runs a separate instance of STP for each VLAN.

PVST+ A Cisco-proprietary implementation that allows CST and PVST to exist on the same network.

Root Path Cost The cumulative cost of all links to the Root Bridge.

Root Port The port on a bridge closest to the Root Bridge in terms of Path Cost.

STP (Spanning-Tree Protocol) A Layer 2 protocol that utilizes a special-purpose algorithm to discover physical loops in a network and effect a logical loop-free topology.

System ID Extension The 12-bit number of the VLAN or the MISTP instance on Catalyst 6000 family switches that support 4096 VLANs.

tunnel A mechanism that enables the propagation of packets or frames through a series of devices in such a way that the contents of the packets or frames are not changed in transit.

UplinkFast A Catalyst feature that accelerates the choice of a new Root Port when a link or switch fails.

After completing this chapter, you will be able to perform tasks related to the following:

- The role of the distribution layer in inter-VLAN routing
- The router-on-a-stick solution to inter-VLAN routing
- Identify the combinations of Catalyst chassis and line cards needed for inter-VLAN routing with an internal route processor
- Configure Catalyst switches and Cisco routers to enable inter-VLAN routing

Chapter 6

Inter-VLAN Routing

Up to this point, we focused primarily on Layer 2 considerations, such as Spanning-Tree Protocol (STP), VLAN Trunking Protocol (VTP), and IEEE 802.1Q. If we stopped here, our campus networks would consist only of independent Layer 3 broadcast domains that could not communicate with each other. We need mechanisms to allow for inter-VLAN communication. This is where routing functionality enters the picture.

We now investigate Catalyst switches with Layer 3+ functionality, including the Catalyst 4000, 5000, 6000, and 8500 families of switches. These devices, if populated with the appropriate line cards and/or Supervisor daughter cards, permit the configuration of IP networks, as well as Internetwork Packet Exchange (IPX) networks and AppleTalk cable ranges, on logical and physical interfaces. In this case, IP subnets normally correspond with campus VLANs in a one-to-one fashion.

After exploring the important role of inter-VLAN routing in campus networks, we will look at the various configuration options for inter-VLAN routing. These options include external routers trunked to Catalyst switches, 3550 series switches, 4000 series switches with an L3 module, 4000 series switches with the Supervisor III module, 5000 family switches with a *Route Switch Module (RSM)*, 5000 family switches with a *Route Switch Feature Card (RSFC)*, 6000 family switches with a *Multilayer Switch Module (MSM)*, and 6000 family switches with a *Multilayer Switch Feature Card (MSFC)*.

Role of Routers in a Campus Network

Configuring VLANs helps control the size of broadcast domains and localizes traffic. VLANs are associated with individual networks or subnetworks. Network devices in different VLANs cannot communicate with one another without an external router or an internal *route processor*.

When an end station in one VLAN needs to communicate with an end station in another VLAN, inter-VLAN communication is required. Routing is required to support communication between VLANs. You configure one or more route processors or routers to route traffic between VLANs.

NOTE

From a hardware-engineering standpoint, an internal route processor (or simply route processor) is a set of hardware components that provide routing functionality. When we use the term *route processor* in this chapter, we take the liberty of an "abuse of language" by equating route processors to line cards or daughter cards containing route processors that are inserted into the appropriate switches. The route processors we discuss in this chapter are the RSM, RSFC, MSM, and MSFC.

Routers are defined by their capability to facilitate communication between networks. Because routers prevent broadcast propagation and use more intelligent forwarding algorithms than bridges and switches, routers enable more efficient use of bandwidth. Routers also enable multiple redundant paths from source to destination and optimal path selection. With routers, it is relatively easy to implement load balancing across multiple paths; on the other hand, Layer 2 load balancing can be difficult to design, implement, and maintain. In addition, routers play a key role when multicast traffic is in use.

Routers provide additional benefits that reach beyond data forwarding. If Layer 3 addressing is designed hierarchically, routers can be used to implement designs that utilize route summarization. Summarization reduces routing protocol overhead, increases table lookup performance, and improves network stability—allowing you to scale your network. Cisco routers also provide extensive access list capabilities that can be used to provide policy controls. Finally, Cisco routers provide important features such as Dynamic Host Configuration Protocol (DHCP) relay, proxy Address Resolution Protocol (ARP) in IP networks, Get Nearest Server (GNS) functions in IPX networks, and quality of service (QoS).

When VLANs are interconnected, several technical issues arise. Table 6-1 lists some of the most common issues and solutions when you're trying to route between VLANs.

Table 6-1 Routers Solve the Issues of Interconnecting VLANs

Issue	Solution
Isolated broadcast domains	Routers
End stations send nonlocal packets	Default gateways
Support for multiple VLAN traffic across VLAN boundaries	IEEE 802.1Q

The issues addressed in Table 6-1 are discussed in the next three sections. After that, we discuss the role of the distribution layer, external routers, and internal route processors.

Isolated Broadcast Domains

VLANs are designed to control the size of broadcast domains and to keep local traffic within its respective VLAN, as shown in Figure 6-1. Because a VLAN isolates traffic to one broadcast domain, normally that broadcast domain is associated with a single Layer 3 subnet. VLANs cannot communicate with other VLANs without an intervening Layer 3 device.

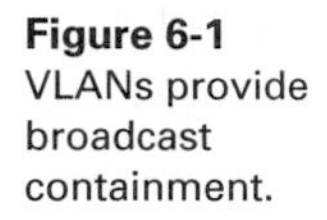

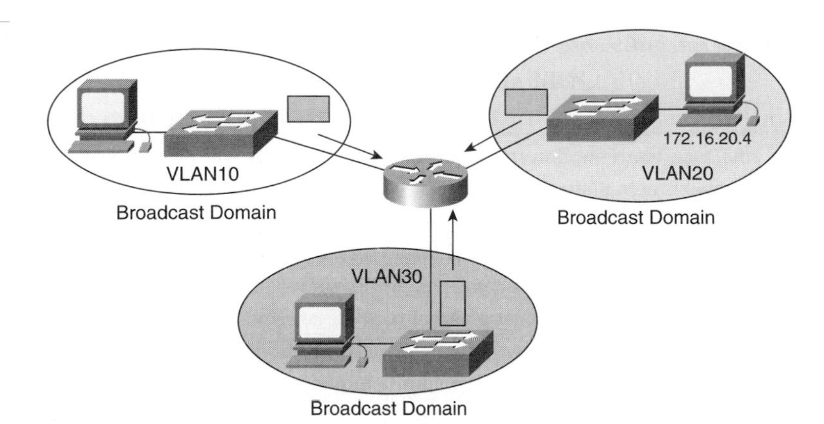

Figure 6-1
VLANs provide broadcast containment.

A *route processor* is the main system processor in a Layer 3 networking device. It is responsible for managing tables and caches, for sending and receiving routing protocol updates, and for routing between networks. The term *route processor* is used in this chapter to indicate a switch module on which the main system processor resides. Within switched networks, route processors provide communication between VLANs. Route processors also provide VLAN access to the server block in the campus network. In addition, route processors provide either direct or indirect access to WAN links connecting the campus network to other intranets, extranets, or the Internet.

Before you can configure routing between VLANs, you must have defined the VLANs on the switches in your network. Issues related to network design and VLAN definition need to be addressed during the network design phase. The issues that you need to consider include the following:

- Sharing resources between VLANs
- Load balancing
- Redundant links
- Logical addressing
- How to segment the network using VLANs

These issues are discussed in context. Next, we consider the role of default gateways.

NOTE

We use the term *external router* to indicate a self-contained standalone device that provides routing functionality. Although an external router contains a route processor, we differentiate between route processors and external routers. So in this chapter, we use the term *route processor* for a Layer 3 card inserted into a switch to provide routing functionality, and we use the term *external router* for a self-contained standalone routing device.

We use the term *router* to encompass both internal route processors and external routers. If there is a chance of confusion in a given context, we clarify.

How Does Host Traffic Move Beyond the Local VLAN?

Connecting the separate subnets through a route processor introduces the issue of how end stations can communicate with other devices through multiple LAN segments. Some network devices use routing tables to identify where to deliver packets outside the local network segment. Even though it is not the responsibility of end stations to route data, end stations still must be able to send data to addresses on subnets other than their own.

So that each end device does not have to manage its own routing tables, most devices are configured with the IP address of a designated route processor. This designated route processor is the default router to which all nonlocal network packets are sent. The route processor then forwards the packets toward the appropriate destination. A network device's default router IP address depends on which IP subnet contains that network device. Each VLAN has its own default gateway on the route processor.

Similarly, when configuring inter-VLAN routing, the standard operating procedure is to configure an IP default gateway (sometimes called a default route in switch-speak) on a switch, pointing to the IP address configured for the management VLAN on the route processor. This enables connectivity via the management VLAN to other devices in the campus network.

How Do You Support Traffic for Multiple VLANs Across VLAN Boundaries?

As the number of VLANs in a network increases, you must determine whether you need an individual router interface per VLAN, or if VLAN trunking is needed to assign multiple VLANs to a single routed interface.

One solution is to dedicate one interface on a router for each VLAN, as shown in Figure 6-2. However, as the number of VLANs per switch increases, so does the requirement for the number of interfaces on the router. (Recall that we use the term router to include external routers and internal route processors.) With the expense of a Layer 3 interface, it might not be feasible to use separate interfaces as a solution for inter-VLAN routing. In addition, some VLANs might not require inter-VLAN routing on a regular basis, creating a situation in which interfaces on the router are underutilized.

Another solution is to carry multiple VLAN traffic over a single link using ISL or IEEE 802.1Q trunking (see Chapter 4 for more information on trunking). In Figure 6-3, the clients on VLANs 10, 20, and 30 all need to establish sessions with a server that is attached to a port in VLAN 60.

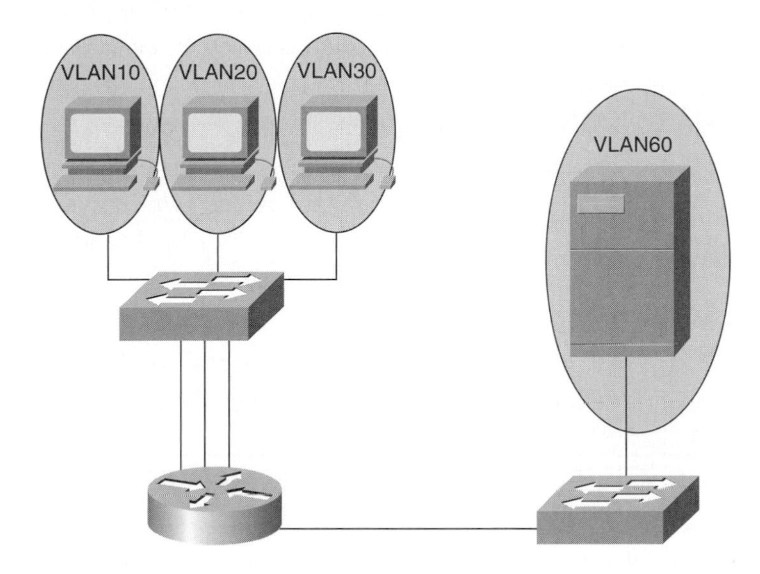

Figure 6-2
One router link per VLAN is an option.

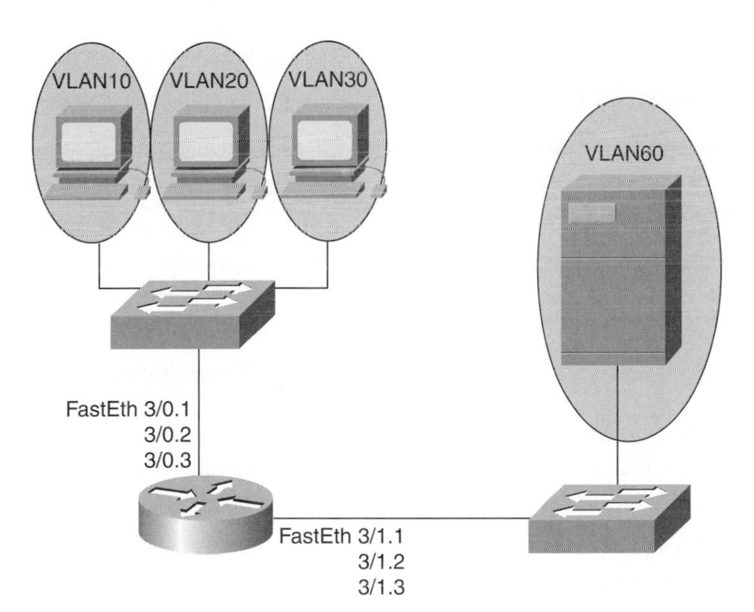

Figure 6-3
ISL or 802.1Q trunks are another option.

Because the file server resides in a different VLAN than any of the clients, you need to configure inter-VLAN routing. The route processor performs this function as follows:

Step 1 The route processor accepts the frames sent from VLANs 10, 20, and 30 on interface 3/0.

Step 2 The route processor classifies the packet based on the destination IP address. The route processor then prepends the original frame with an

ISL header or tags the frame with an 802.1Q header. In either case, the destination VLAN ID of 60 is coded in the frame before it is sent out the appropriate port.

Step 3 The router switches the frame to the appropriate interface based on the destination IP address—in this case, FastEthernet 3/1. Note that prior to sending the frame out the interface, the router also rewrites the Ethernet frame header with FastEthernet 3/1's MAC address as source and the server's MAC address as destination.

Step 4 When the switch attached to the server receives the frame, it strips the ISL or 802.1Q header and forwards the frame out the port associated with the destination MAC address (the port connected to the server).

INTEGRATED ROUTING AND BRIDGING

In Figure 6-3, if the members of VLAN 60 reside on segments connected to FastEthernet 3/0 and 3/1, respectively, you have to configure Integrated Routing and Bridging (IRB) on the router in order to preserve both the MAC and ISL/802.1Q headers of an intra-VLAN frame moving between these interfaces. The IP addresses of the source and destination reside within the same network in this case; also, the source and destination VLAN are both 60. IRB is the only way to allow a VLAN to "span" a router so that the ISL/802.1Q header is not terminated upon arrival at the ingress router port.

The complexities of IRB can be avoided if the two switches and router pictured are replaced with a single multilayer Catalyst switch, such as a Catalyst 3550 or a Catalyst 4006 with a Supervisor Engine III.

Another option is to connect the pictured switches with a trunk or VLAN 60 access link to circumvent the router in order to propagate VLAN 60 traffic "around" the router.

Next, we consider the role of routing in the distribution layer in a campus network.

Role of the Distribution Layer

The distribution layer provides boundaries for different types of traffic, such as broadcast and multicast traffic. The distribution layer is also where inter-VLAN routing is normally configured, as shown in Figure 6-4. In the case of a Layer 3 core, the routing normally takes place independent of VLANs, which are terminated at the distribution layer.

The distribution layer often consists of a combination of relatively high-end switches and route processors (internal to the switches). Because of its Layer 3 capabilities, the distribution layer becomes the demarcation point between networks in the access layer and networks in the core. In modern campus design, each wiring-closet switch corresponds to one or two logical subnets connecting to the route processor, thus allowing the route processors in the distribution layer to provide broadcast control.

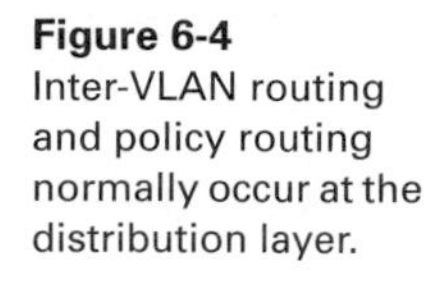

Figure 6-4
Inter-VLAN routing and policy routing normally occur at the distribution layer.

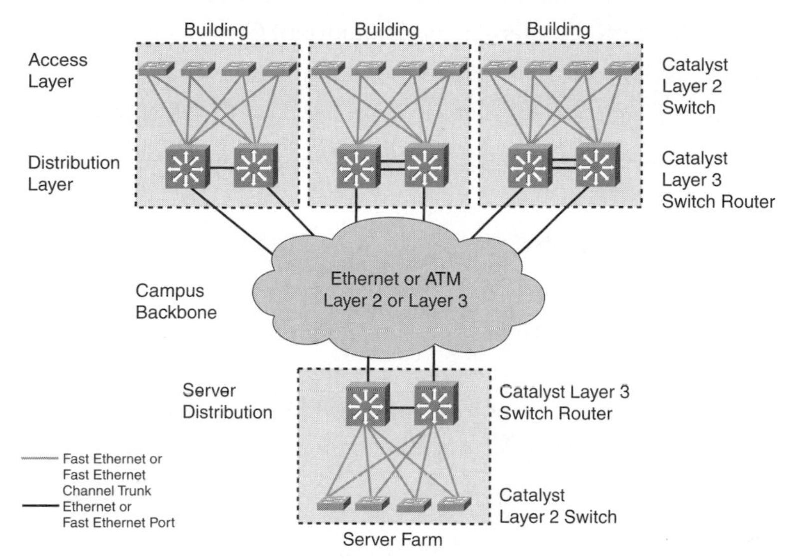

The switch/route processor model is straightforward to configure and maintain because of its inherent modularity. Each route processor in the distribution layer is programmed with the same features and functionality. Common configuration elements can be copied across the layer, allowing for predictable behavior and ease of troubleshooting.

External Routers

Inter-VLAN routing can be performed with a router and a switch working in tandem. The configuration consists of a router connected to a switch via a trunk link, as shown in Figure 6-5.

Figure 6-5
Inter-VLAN routing is possible with an external router connected to a Catalyst switch.

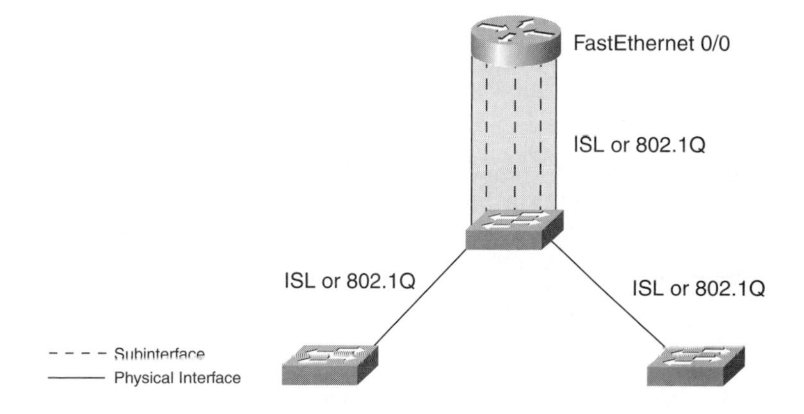

A number of combinations with Cisco products support this solution. In fact, this tandem configuration is possible with almost all Cisco routers and Catalyst switches that support both of the following:

- ISL or 802.1Q (both devices must support the same trunking technology)
- Fast Ethernet or Gigabit Ethernet (both devices must support the same speed setting)

The Fast Ethernet interfaces on Cisco 1720 and 1750 routers are exceptions to this tandem configuration option; trunking is not supported on these interfaces.

> **NOTE**
>
> Although it is possible to route between VLANs with switch access ports (corresponding to single VLANs) connecting to individual router interfaces, there is not much sense in doing this unless you have fewer VLANs than you have router interfaces and you have no foreseeable use for the other router interfaces. Even so, it's probably not advisable because you might add more VLANs later, beyond the capacity of your router's port density, forcing you to resort to using a trunk.
>
> When multiple router interfaces are used for inter-VLAN routing, this is normally done with an EtherChannel—a single logical connection referred to in this context as a ***port channel***. Subinterfaces are then configured on the port channel interface, one subinterface per VLAN.

The primary advantage of using a trunk link is a reduction in the number of required router and switch ports. Not only can this save money, but it can also reduce configuration complexity. Consequently, the trunk-connected router approach can scale to a much larger number of VLANs than a one-link-per-VLAN design.

The trunk-connected router configuration does have some disadvantages:

- There is a possibility of inadequate bandwidth for each VLAN.
- Additional overhead on the router can occur.
- Older versions of the Cisco IOS Software support only a limited set of features on ISL interfaces.

With regard to inadequate bandwidth for each VLAN, consider, for example, the use of a Fast Ethernet link in which all VLANs must share 100 Mbps of bandwidth. A single VLAN could easily consume the entire capacity of the router or the link (especially in the event of a broadcast storm).

With regard to the additional overhead on the router caused by using a trunk-connected router, not only must the router perform normal routing and data-forwarding duties, but it must also handle the additional encapsulation used by the trunking protocol. Consider ISL running on a Cisco 7505 router as an example. These routers have many different switching modes (process, fast, optimum, and so on), a term that Cisco uses to refer to the particular internal process used to forward packets. Note that fast switching, in particular, cannot handle all types of traffic (such as many types of IBM Systems Network Architecture [SNA] traffic). Getting back to the example of an ISL interface on a 7505, these routers normally use techniques such as optimum switching and distributed switching to achieve data-forwarding rates from 300,000 to more than 1,000,000 packets per second (pps). When you run ISL on an interface, that interface is limited to fast switching. Because of this restriction, ISL routing is limited to approximately 50,000 to 100,000 pps on this router (and considerably less on lower-end platforms).

The third disadvantage of the trunk-connected router design is that older versions of the Cisco IOS Software support only a limited set of features on ISL interfaces.

Although most limitations were removed in 11.3 and some later 11.2 images, networks using older images need to have their inter-VLAN routing carefully planned. Some of the more significant limitations prior to 11.3 include the following:

- Earlier versions support only IP and IPX. All other protocols (including AppleTalk and DECnet) must be bridged. Inter-VLAN bridging is almost always a bad idea, because IPX supports only the **novell-ether** encapsulation (Novell calls this Ethernet_802.3).
- Hot Standby Router Protocol (HSRP) is not supported. This can make it difficult or impossible to provide default gateway redundancy.
- Secondary IP addresses are not supported.

Before we get into configuring inter-VLAN routing, we must consider the role of internal route processors.

Internal Route Processors

Another option for inter-VLAN routing that is much more common nowadays is that of a switch with an integrated route processor. In this case, the route processor resides on one of the line cards in the switch chassis. The trend is for modern distribution-layer switches to incorporate the route processor on the same module as the switch engine itself. The switch backplane (the high-speed switching path used inside the switch chassis) provides the communication path between the switch engine and the route processor.

In Figure 6-6, traffic flows from the source VLAN to the router, where it is routed to the destination VLAN. This creates an out-and-back flow to the router.

Figure 6-6 Internal route processors can manage inter-VLAN traffic.

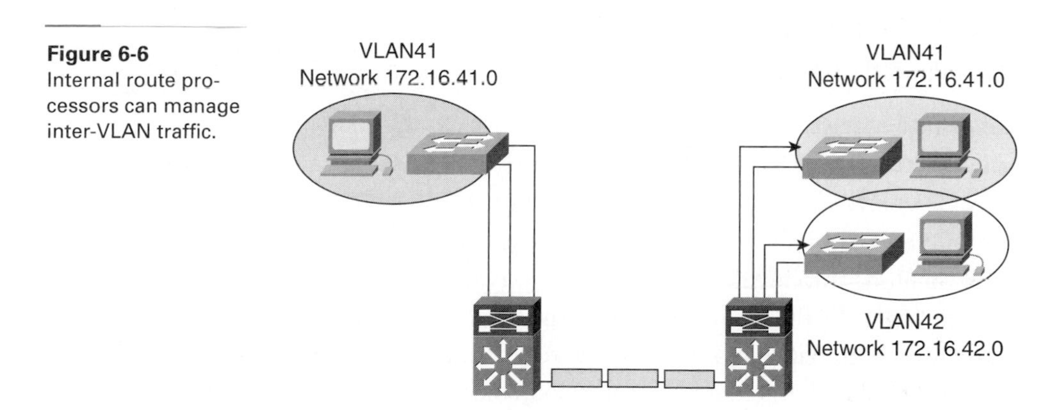

Technically, the switch's internal route processor uses a flow that is similar to that utilized with an external router working in tandem with a switch. However, there is one important difference: The trunk connecting the switch to the route processor exists on the switch backplane. This difference provides two key benefits:

- Speed
- Integration

Because the route processor connects directly to the switch's backplane, it allows the router to be much more tightly integrated with the switching process. Not only can this ease configuration tasks, but it can also provide intelligent communication between the Layer 2 and Layer 3 portions of the network. Also, because it provides a faster link than is possible with an external trunk link, performance is improved. Two examples of switches utilizing internal route processors are the Catalyst 5000 with an RSM, providing inter-VLAN switching of up to .175 Mpps (million packets per second), and the Catalyst 6500 with an MSFC2, providing inter-VLAN switching of up to 210 Mpps.

Configuring Inter-VLAN Routing

The procedures for configuring inter-VLAN routing fall into four categories:

- External router (router-on-a-stick)
- Internal route processor with the Catalyst 4000 L3 module or Catalyst 6000 with an MSM
- Internal route processor with the Catalyst 4000 Supervisor Engine III, Catalyst 3550, and Catalyst 6000 with MSFC or MSFC2 running Native IOS
- Catalyst 5000 family with RSM or RSFC and Catalyst 6000 family with MSFC or MSFC2 running CatOS

The configuration of each of these categories is discussed in the following sections.

Router-on-a-Stick

The *router-on-a-stick* method of inter-VLAN routing consists of an external router with a Fast Ethernet, Gigabit Ethernet, or EtherChannel trunk connecting to a switch, utilizing ISL or 802.1Q. Subinterfaces on the trunk are created to correspond to VLANs in a one-to-one fashion. Figure 6-5 illustrates the router-on-a-stick configuration. The "stick" is the trunk connecting the switch and the router. With this design, all inter-VLAN traffic has to travel to the router and back. This "out-to-the-router-and-back" flow is characteristic of router-on-a-stick design.

To define subinterfaces on a physical or logical interface for the purposes of inter-VLAN routing with IP, perform the following tasks:

- Identify the interface.
- Define the trunk encapsulation.
- Assign an IP address to the subinterface.

To identify the interface, create a Fast Ethernet, Gigabit Ethernet, or port-channel subinterface in global configuration mode or interface configuration mode. A *port-channel* interface is a logical interface into which physical interfaces are grouped to form a single logical link.

To define the trunk encapsulation, enter either the command **encapsulation isl** *vlan-number* or **encapsulation dot1Q** *vlan-number* in subinterface configuration mode, where *vlan-number* identifies the VLAN for which the subinterface will carry traffic. A VLAN ID is added to the frame only when the frame is destined for a nonlocal network.

To assign the IP address, enter the command **ip address** *ip-address subnet-mask* in subinterface configuration mode, where *ip-address* and *subnet-mask* are the 32-bit network address and mask of the specific interface.

Example 6-1 illustrates this procedure on a Catalyst 8510 Campus Switch Router (CSR), where Fast Ethernet interfaces Fa0/0/0 and Fa0/0/1 comprise a port-channel interface (EtherChannel).

Example 6-1 Configuring a Port Channel and Subinterfaces for Inter-VLAN Routing

```
8510CSR#configure terminal
8510CSR(config)#interface port-channel1
8510CSR(config-if)#exit
8510CSR(config-if)#interface fa0/0/0
8510CSR(config-if)#channel-group 1
FastEthernet0/0/0 added as member-1 to port-channel1
00:20:20: %LINK-3-UPDOWN: Interface Port-channel1, changed state to up
00:20:21: %LINEPROTO-5-UPDOWN: Line protocol on Interface Port-channel1,
changed state to up
8510CSR(config-if)#exit
8510CSR(config)#interface fa0/0/1
8510CSR(config-if)#channel-group 1
```

continues

Example 6-1 Configuring a Port Channel and Subinterfaces for Inter-VLAN Routing (Continued)

```
FastEthernet0/0/1 added as member-2 to port-channel1
8510CSR(config-if)#interface port-channel 1.100
8510CSR(config-subif)#encapsulation isl 100
8510CSR(config-subif)#ip address 172.20.52.33 255.255.255.224
8510CSR(config-subif)#interface port-channel 1.200
8510CSR(config-subif)#encapsulation isl 200
8510CSR(config-subif)#ip address 172.20.52.65 255.255.255.224
```

NOTE

The fact that a Catalyst 8510 CSR was used in Example 6-1 has little relevance to the technique employed. The same type of configuration could have been performed with a Cisco 2621 router with interface Fa0/0 substituting for the port channel; subinterfaces of Fa0/0 would substitute for subinterfaces of the port channel interface used on the 8510. You don't need the **channel-group** commands on the 2621. They are used to form an EtherChannel, which is not supported on 2621 routers.

The configuration of the switch on the other end of the trunk in Example 6-1 is carried out as described in the section "Configuring Ethernet Trunks" in Chapter 4. The **nonegotiate** option is used on the switch end because Cisco routers don't support Dynamic Trunking Protocol (DTP).

The next section explores the router-on-a-stick option in more detail.

802.1Q Trunking Between a Catalyst 3500 XL Switch and a Cisco 2621 Router

The scenario in this section explains how to create an IEEE 802.1Q trunk that carries traffic from two VLANs (VLAN 1 and VLAN 2) across a single link between a Catalyst 3512 XL and a Cisco 2621 router. This scenario uses the Cisco 2621 router to do the inter-VLAN routing between VLAN 1 and VLAN 2. Catalyst 2900 XL, 3500 XL, and 2950 series switches are Layer 2 switches and cannot route between VLANs.

This example also demonstrates the configuration on the 2600 series for Cisco IOS Software releases before and after Cisco IOS Software Release 12.1(3)T. The **encapsulation dot1Q 1 native** command was added in Cisco IOS Software Release 12.1(3)T. For details on native VLANs and 802.1Q trunking, see the Tech Note titled "Native VLANs on 802.1Q Trunks" in Chapter 4. In this example, the native VLAN is VLAN 1 (by default) on both the Cisco 2621 router and the Catalyst 3512 XL switch. Depending on your network needs, you might have to use a native VLAN other than the default native VLAN. The commands to change the native VLAN are explained in Chapter 4.

Figure 6-7 illustrates this scenario. We begin with the configuration of the Catalyst 3512 XL in Example 6-2. You've seen this type of configuration in earlier chapters—specifically, Chapters 3 through 5.

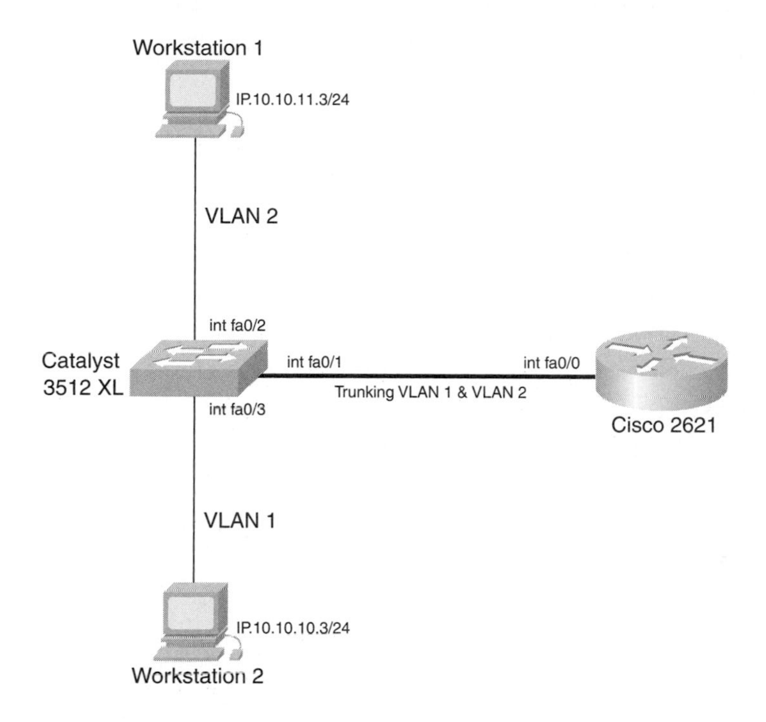

Figure 6-7 Configuring router-on-a-stick is an option for inter-VLAN routing.

NOTE

In Catalyst 2900 XL, 3500 XL, and 2950 switches, a VLAN interface—for example, interface vlan 1, interface vlan 2, or interface vlan *x*—can be created for every VLAN that is configured on the switch. However, only one VLAN at a time can be used as a management VLAN. The IP address is assigned to the VLAN interface of the management VLAN only. If the IP address is assigned to another VLAN interface whose VLAN is not used as management VLAN, that interface does not come up. It is preferable to create the VLAN interface only for the management VLAN.

Example 6-2 The Catalyst 3512 XL Is Configured with a Trunk and Two Access Ports

```
3512xl(config)#interface vlan 1
3512xl(config-if)#ip address 10.10.10.2 255.255.255.0
3512xl(config-if)#exit
3512xl(config)#ip default-gateway 10.10.10.1
3512xl(config)#end
3512xl#vlan database
3512xl(vlan)#vtp transparent
Setting device to VTP TRANSPARENT mode.
3512xl(vlan)#vlan 2
VLAN 2 added:
Name: VLAN0002
3512xl(vlan)#exit
```

continues

Example 6-2 The Catalyst 3512 XL Is Configured with a Trunk and Two Access Ports (Continued)

```
APPLY completed.
Exiting....
3512xl#configure terminal
Enter configuration commands, one per line. End with CNTL/Z.
3512xl(config)#interface fastEthernet 0/1
3512xl(config-if)#switchport mode trunk
3512xl(config-if)#switchport trunk encapsulation dot1q
3512xl(config-if)#switchport trunk native vlan 1  <------- Note: just to be sure.
3512xl(config-if)#switchport trunk allowed vlan all  <-- Note: just to be sure.
3512xl(config-if)#exit
3512xl(config)#interface fastEthernet 0/2
3512xl(config-if)#switchport access vlan 2
3512xl(config-if)#spanning-tree portfast
3512xl(config-if)#exit
3512xl(config)#interface fastEthernet 0/3     <------- In VLAN 1 by default.
3512xl(config-if)#spanning-tree portfast
3512xl(config-if)#end
3512xl#wr
Building configuration...
```

The first, third, and fourth instances of the **exit** command are not necessary in Example 6-2. The configuration works just as well without them, but it's a little more difficult to read a sequence of commands without using **exit** as a separator.

Next, we configure the 2621 router. This is detailed in Example 6-3 for Cisco IOS Software Releases 12.1(3)T and later. Refer to Figure 6-7 to put the configuration into context.

Example 6-3 This Cisco 2621 Router Is Configured with a Trunk Port with a Cisco IOS Software Release of 12.1(3)T

```
c2600(config)#interface fastEthernet 0/0
c2600(config-if)#no shutdown        <--------- Make sure the interface is up.
c2600(config-if)#exit
c2600(config)#interface fastEthernet 0/0.1
c2600(config-subif)#encapsulation dot1Q 1 native   <-- Need IOS 12.1(3)T or greater.
c2600(config-subif)#ip address 10.10.10.1 255.255.255.0
c2600(config-subif)#exit
```

Example 6-3 This Cisco 2621 Router Is Configured with a Trunk Port with a Cisco IOS Software Release of 12.1(3)T (Continued)

```
c2600(config)#interface fastEthernet 0/0.2
c2600(config-subif)#encapsulation dot1Q 2
c2600(config-subif)#ip address 10.10.11.1 255.255.255.0
c2600(config-subif)#end
c2600#wr
Building configuration...
[OK]
c2600#
```

In the earlier Cisco IOS Software releases, it is important not to configure the VLAN 1 interface as a subinterface. In Cisco IOS Software releases earlier than 12.1(3)T, you cannot define the native VLAN explicitly. You cannot use the **encapsulation dot1Q 1 native** command under the subinterface. If you configure VLAN 1 as a subinterface, the router expects a tagged dot1q frame on VLAN 1, and the switch does not expect a tag on VLAN 1. (Recall that on a Catalyst switch supporting 802.1Q, if a frame has a VLAN ID that is the same as the outgoing port native VLAN ID, the packet is transmitted untagged. Otherwise, the switch transmits the packet with a tag.) As a result, no traffic passes between VLAN 1 on the switch and the router. So, if the 2621 router has a Cisco IOS Software release prior to 12.1(3)T, you configure it as shown in Example 6-4.

Example 6-4 This Cisco 2621 Router Is Configured with a Trunk Port with a Cisco IOS Software Release of 12.1(1)T

```
c2600#configure terminal
Enter configuration commands, one per line. End with CNTL/Z.
c2600(config)#interface fastEthernet 0/0
c2600(config-if)#no shutdown
c2600(config-if)#ip address 10.10.10.1 255.255.255.0
c2600(config-if)#exit
c2600(config)#interface fastEthernet 0/0.2
c2600(config-subif)#encapsulation dot1Q 2
c2600(config-subif)#ip address 10.10.11.1 255.255.255.0
c2600(config-subif)#end
c2600#wr
```

This completes our look at the router-on-a-stick configuration option.

> **NOTE**
>
> The 2/1 and 2/2 ports are internal ports on the L3 module that provide Layer 2 functionality. The L3 module actually has four Gigabit Ethernet ports. Two ports are external (G1 and G2), and two are internal (G3 and G4). These four ports provide Layer 3 functionality. You configure the external ports using Layer 3 Cisco IOS Software only and the internal ports using Layer 3 Cisco IOS Software and the Layer 2 Supervisor. You must use the Cisco IOS Software (the OS on the Layer 3 module) to configure G3 and G4 and use the Supervisor to configure ports 2/1 and 2/2.

Catalyst 4000 with an L3 Module and Catalyst 6000 with an MSM

Of the four categories listed for inter-VLAN configuration, this category is probably the most difficult to configure. On the Catalyst 4000 L3 module, the way the ports are designed to communicate with the Supervisor Engine is unusual. Assuming that the Layer 3 Services module resides in slot 2, the port numbering is 2/1 and 2/2 for the internal Gigabit Ethernet ports, and 2/3 through 2/34 for the external 10/100 ports. Ports 2/3 through 2/34 are directly accessible via the external RJ-45 connectors. Ports 2/1 and 2/2 are internal ports and have no external interfaces. Figure 6-8 shows the Catalyst 4000 L3 Services module.

Figure 6-8
The Catalyst 4000 L3 module is commonly used for inter-VLAN routing.

To the Catalyst 4000, the Layer 3 Services module appears to be an external router connected to the switch through two internal full-duplex Gigabit Ethernet ports.

The port numbering for the external Gigabit Ethernet interfaces on the front panel is Gigabit Ethernet1 (G1) and Gigabit Ethernet2 (G2). The port numbering for the internal Gigabit Ethernet interfaces is Gigabit Ethernet3 (G3) and Gigabit Ethernet4 (G4).

If the Layer 3 Services module is installed in slot 2, port 2/1 on the Catalyst switch side is connected to interface Gigabit Ethernet3 on the Catalyst 4000 Layer 3 Services module side, and port 2/2 is connected to interface Gigabit Ethernet4.

Figure 6-9 shows the internal interface connections when the Layer 3 Services module is installed in slot 2 in a Catalyst 4000 switch.

You have the option to configure the Catalyst 4000 Layer 3 Services module ports as trunks, as EtherChannels, or as independent links (access links).

To connect to the Layer 3 Services module (assuming that it resides in slot 2) via the Supervisor Engine II, enter the command **session 2**. As soon as you are connected to the module via the backplane, you enter commands much as you do on any Cisco router. To access G3 from global configuration mode, enter the command **interface GigabitEthernet 3** (or **int g3** for short).

NOTE

The G3 and G4 interfaces on the L3 module only support IEEE 802.1Q trunking, even though the option for ISL appears. You can see why 802.1Q is the only viable option. If you enter the command **show port capabilities 2/1** on the Supervisor Engine II, you can see that 802.1Q is the only trunk encapsulation type supported. Because G3 is internally connected to 2/1, the limitation to 802.1Q on G3 (and G4) makes sense.

Interestingly, the Supervisor Engine III on the Catalyst 4006 supports 802.1Q and ISL encapsulation but does not support the Layer 3 Services module.

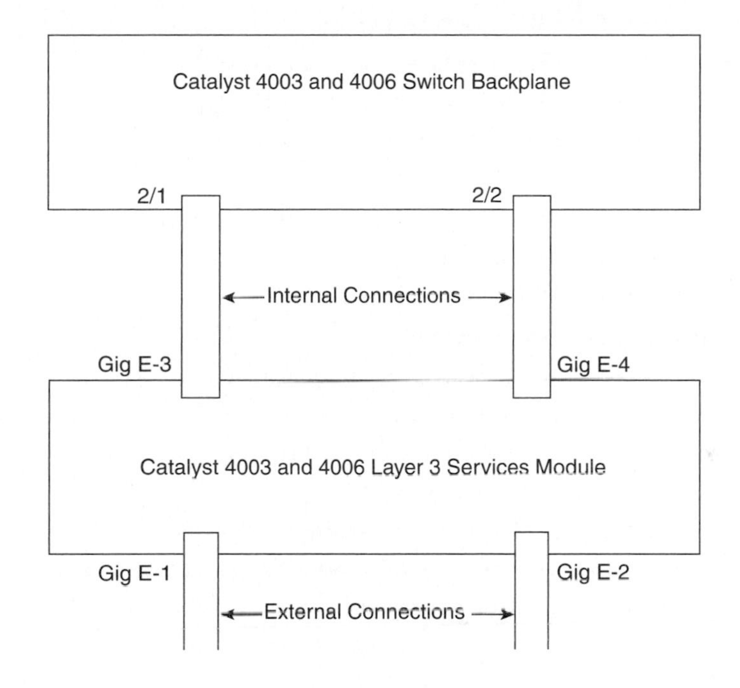

Figure 6-9
The Catalyst 4000 L3 module has two internal Layer 2 ports (2/1 and 2/2) that connect via the backplane to two internal Layer 3 ports (G3 and G4).

Example 6-5 demonstrates the procedure for configuring inter-VLAN routing on the Catalyst 4000 L3 module. The following data is used:

- VLAN 1 is the management VLAN with default gateway 10.1.1.1/24 on the L3 module.
- VLAN 1 is the native VLAN (it must match on both ends of an 802.1Q trunk!).
- The default gateway for VLAN 10 is 10.1.10.1/24.
- The default gateway for VLAN 20 is 10.1.20.1/24.
- The default gateway for VLAN 30 is 10.1.30.1/24.
- The VTP domain is cisco.
- The Catalyst is set as a VTP server.
- The sc0 interface address on the Catalyst is 10.1.1.11/24 and resides in VLAN 1.
- The internal ports 2/1 and 2/2 form a Gigabit EtherChannel with the internal ports G3 and G4, and 802.1Q trunking is enabled on this channel.
- Four subinterfaces are configured on the port channel on the L3 module, corresponding to VLANs 1, 10, 20, and 30.

Example 6-5 Configuring Inter-VLAN Routing on the Catalyst 4000 L3 Module

```
Switch>(enable) set vtp domain cisco
Switch>(enable) set vtp mode server
Switch (enable) set interface sc0 up
Switch (enable) set interface sc0 1 10.1.1.11/24
Switch (enable) set ip route default 10.1.1.1
Switch (enable) set port channel 2/1-2 mode on
Switch (enable) set vlan 1 2/1-2
Switch (enable) set trunk 2/1 nonegotiate dot1q 1-1005
Switch> (enable) session 2
Router(config)#interface Port-channel1
Router(config-if)#no shutdown
Router(config-if)#interface GigabitEthernet3
Router(config-if)#channel-group 1
Router(config-if)#interface GigabitEthernet4
Router(config-if)#channel-group 1
Router(config-if)#interface Port-channel1.1
Router(config-subif)#ip address encapsulation dot1Q 1 native
Router(config-subif)#ip address 10.1.1.1 255.255.255.0
Router(config-subif)#interface Port-channel1.10
Router(config-subif)#encapsulation dot1Q 10
Router(config-subif)#ip address 10.1.10.1 255.255.255.0
Router(config-subif)#interface Port-channel1.20
Router(config-subif)#encapsulation dot1Q 20
Router(config-subif)#ip address 10.1.20.1 255.255.255.0
Router(config-subif)#interface Port-channel1.30
Router (config-subif)#encapsulation dot1Q 30
Router (config-subif)#ip address 10.1.30.1 255.255.255.0
```

The **set trunk** command does not reference port 2/2. Because ports 2/1 and 2/2 form a channel, it is only necessary to reference port 2/1 in the **set trunk** command.

For the Catalyst 6000 with a Multilayer Switch Module (MSM) installed, the setup and procedure are almost exactly the same. The MSM is a line card that runs Cisco IOS Software router software and directly interfaces to the Catalyst 6000 backplane to provide Layer 3 switching.

The MSM connects to the Supervisor Engine through four internal full-duplex Gigabit Ethernet interfaces. The Catalyst switch sees the MSM as an external router connected to the switch through the four interfaces. You can group the four Gigabit interfaces into a single Gigabit EtherChannel or configure them as independent interfaces (links), just as with the Catalyst 4000/L3 module combination. If channeled, the channel supports trunking through 802.1Q or ISL. After you configure a channel and specify a trunk type, the port-channel interface on the MSM is configured with one subinterface for every VLAN on the switch, providing inter-VLAN routing. Alternatively, you can configure each Gigabit interface (link) independently as a separate VLAN trunk or nontrunked routed interface.

The Supervisor Engine software sees each Gigabit interface as a configurable port. For example, if the MSM is installed in slot 4 and you enter the **show module 4** command, you see ports 4/1, 4/2, 4/3, and 4/4. Similarly, the MSM software sees each Gigabit interface as a configurable interface. For example, if you do a **show interface** from the MSM, you see interfaces g0/0/0, g1/0/0, g3/0/0, and g4/0/0 (there is no g2/0/0). Port 1 on the "switch side" is connected to interface g0/0/0 on the *MSM side*, port 2 is connected to interface g1/0/0, port 3 is connected to interface g3/0/0, and port 4 is connected to interface g4/0/0, as shown in Figure 6-10.

Figure 6-10 The Catalyst 6000 Multilayer Switch Module internal interfaces are similar to those of the Catalyst 4000 L3 module.

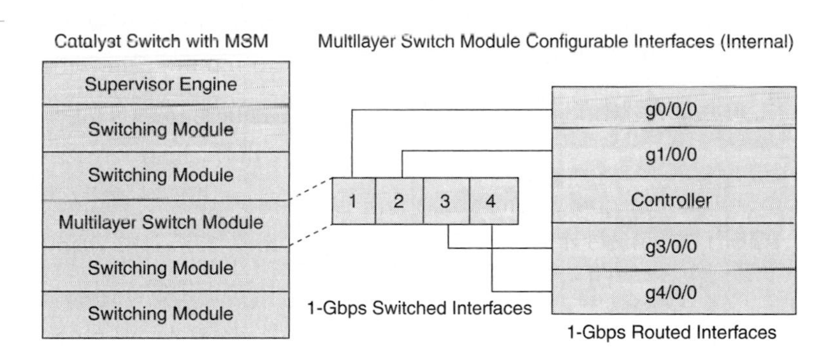

Configuring inter-VLAN routing is performed as shown in Example 6-5, with the following exceptions:

- ISL is also an option for the trunk.
- Up to four Gigabit interfaces (instead of two) can be included in the port channel (EtherChannel).

This completes our look at configuring a Catalyst 4000 Series switch with a Layer 3 Services module and a Catalyst 6000 family switch with an MSM.

Catalyst 4000 with Supervisor Engine III, Catalyst 3550 and Catalyst 6000 with MSFC Running Native IOS

This section shows you how to configure inter-VLAN routing for Catalyst 4000s with the Supervisor Engine III, Catalyst 3550s, and or Catalyst 6000s with the Multilayer Switch Feature Card 1 or 2 (MSFC1 or MSFC2) running Native-mode IOS. The MSFC1 and MSFC2 are daughter cards to the Cat6000 Supervisor Engine, which provides multilayer switching functionality and routing services between VLANs. Note that these configurations do not involve an independent line card in the Cat4000 or Cat6000 chassis (as was the case with the L3 module and the MSM).

Both of these configurations make use of what is called Native IOS or Catalyst IOS (CatIOS). This operating system consists of a single image, rather than a separate CatOS image for the Switching Engine and an IOS image for the route processor. The combined image has the advantage that you only have to learn the commands for a single operating system for all your Cisco switch and router tasks. It appears that this is the trend for future Cisco devices. If you compare the command syntax used in these CatOS-free devices, you see that they are very similar, so it is apparent that somewhat of a convergence is taking place with the operating systems. The following hardware/software combinations reflect this trend:

- CatIOS used on the Supervisor Engine III for the Catalyst 4006 and Native mode for the Catalyst 6000
- 12.1+ versions of Cisco IOS Software used on "IOS-based" switches (such as Cisco IOS Release 12.1(9) for the Catalyst 2950 and 3550)
- January 2002-launched 16-Port Ethernet Switch Module for Cisco 2600/3600 series routers (requires Cisco IOS Software 12.2(2) or higher)

NOTE

You cannot configure subinterfaces or use the **encapsulation** keyword on LAN ports on a Catalyst running Native mode. Catalyst 6000 switches running in Native mode support Layer 2 trunks and Layer 3 VLAN interfaces, which together provide inter-VLAN routing capability. With the CatIOS, you configure the Layer 2 and Layer 3 interfaces using the same command line.

Inter-VLAN routing is configured from a single command-line interface in the case of a multilayer Catalyst switch running CatIOS. This means that you won't be configuring any internal trunks or internal EtherChannels. All trunks and EtherChannels (port channels) will be external to the router. So how do you configure inter-VLAN routing? That is, how do you configure the Cat6000 to internally route between VLANs? You no longer have an internal trunk or EtherChannel to configure. You no longer have internal Layer 2 ports and internal Layer 3 interfaces connecting via the switch's backplane.

You already know how to configure Layer 2 trunks. The only remaining question is how to configure Layer 3 VLAN interfaces. To configure VLANs, first you create them, and then you add Layer 2 ports (LAN interfaces) to them. You create VLANs exactly the same way as you create VLANs on an IOS-based Layer 2 switch in VTP server or transparent mode (by way of VLAN configuration mode, which is discussed in Chapter 4). You add ports to the VLANs in almost exactly the same way you do on an IOS-based Layer 2 switch:

Step 1 Enter the command **switchport** in interface configuration mode. This configures the LAN port as a Layer 2 port (this is not required on IOS-based Layer 2 switches). You might want to do a **shutdown** before this command and a **no shutdown** after the configuration to prevent traffic flow during the configuration process.

Step 2 Enter the command **switchport access vlan** to configure the port as a Layer 2 access port.

Step 3 Enter the command **switchport access vlan** *vlan_ID* to place the port in the associated VLAN.

Finally, you need to create Layer 3 VLAN interfaces to enable inter-VLAN routing. To do this, create the logical VLAN interface (also known as a switch virtual interface) with the command **interface vlan** *vlan_ID*. The switch internally associates this Layer 3 construct with the Layer 2 construct created in VLAN configuration mode (and recorded in the vlan.dat file). Next, configure an IP address on the interface (as usual), and enter **no shutdown** on the switch virtual interface. If you have multiple routing devices configured, you also need to configure the appropriate routing protocols on the switch.

Catalyst 5000 with RSM or RSFC and Catalyst 6000 with MSFC or MSFC2 Running CatOS

Probably the easiest scenario for configuring inter-VLAN routing with Cisco devices is the one described in the title of this section. The Route Switch Module (RSM) is a separate line card that interfaces with the Supervisor Engine of a Catalyst 5000/5500 switch via the backplane to provide multilayer switching and inter-VLAN routing functionality. The RSM runs Cisco IOS Software. The Route Switch Feature Card (RSFC) is a daughter card to the Catalyst 5000/5500 Supervisor Engine IIG and IIIG that provides inter-VLAN routing and multilayer switching functionality. The RSFC runs the Cisco IOS router software and directly interfaces with the Catalyst switch backplane.

Each of these feature card/module/chassis combinations permits inter-VLAN routing in a fairly transparent, intuitive fashion.

The procedure is as follows:

Step 1 Create VLANs on the Supervisor Engine (with CatOS), and add the appropriate ports to the VLANs.

Step 2 "Session" to the route processor card and create logical/virtual VLAN interfaces.

Step 3 Configure IP addresses on the VLAN interfaces.

The next section illustrates this process.

Inter-VLAN Routing Scenario: Catalyst 5509 with RSFC

The equipment combination in this scenario is somewhat historical in nature, but the point is that the exact same commands are used with the Catalyst 5000/RSM (even more historical) and Catalyst 6000/MSFC combinations. We point out the superficial differences at the end of this section. We begin with the topology shown in Figure 6-11.

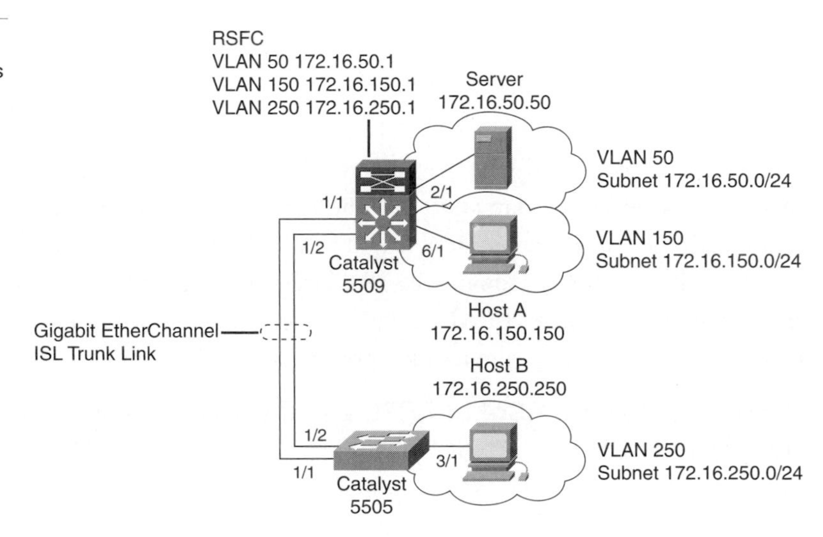

Figure 6-11 Inter-VLAN routing is as easy as it gets with the RSFC on a Catalyst 5000 family switch.

The network is configured as follows:

- There are three VLANs (IP subnets):
  - — VLAN 50 (172.16.50.0/24)
  - — VLAN 150 (172.16.150.0/24)
  - — VLAN 250 (172.16.250.0/24)
- Three VLAN interfaces are configured on the RSFC:
  - — Interface vlan50 (172.16.50.1)
  - — Interface vlan150 (172.16.150.1)
  - — Interface vlan250 (172.16.250.1)
- The Catalyst 5509 has the following hardware:
  - — Supervisor Engine IIIG with the RSFC in slot 1
  - — 12-port 100-Mbps Fast Ethernet module in slot 2
  - — Two-slot 48-port 10-Mbps Ethernet module in slot 6

- The Catalyst 5505 has the following hardware:
  - — Supervisor Engine III with Gigabit Ethernet uplink ports in slot 1
  - — Two-slot 48-port 10-Mbps Ethernet module in slot 3
- The Catalyst 5509 and the Catalyst 5505 are connected through a Gigabit EtherChannel ISL trunk link on ports 1/1 and 1/2.
- The switches are in the VTP Corporate domain.
- The Catalyst 5509 is the VTP server, and the Catalyst 5505 is a VTP client.

The following configuration tasks must be performed to configure the network in this example:

Step 1 Configure the Catalyst 5509 as a VTP server, and assign a VTP domain name.

Step 2 Configure the Catalyst 5505 as a VTP client in the same VTP domain.

Step 3 Create the VLANs on the Catalyst 5509. The VLANs propagate to the 5505 through VTP.

Step 4 Configure the Gigabit EtherChannel ISL trunk link between the switches.

Step 5 Assign the end station switch ports to the appropriate VLANs.

Step 6 On the RSFC, create VLAN interfaces, and assign IP addresses to them, one for each VLAN configured on the switch.

After you successfully configure the network, all end stations should be able to communicate with one another. Whenever an end station in one VLAN transmits to an end station in another VLAN, the traffic travels to the Catalyst 5509 and is passed to the RSFC on the appropriate VLAN interface. The RSFC checks the routing table, determines the correct outgoing VLAN interface, and sends the traffic out that interface to the Catalyst 5509. The Catalyst 5509 forwards the traffic out the appropriate switch port to the destination.

For example, if Host A transmits to the server, the Catalyst 5509 receives the traffic on port 6/1 and passes it to the RSFC on the VLAN 150 interface. The RSFC performs a routing table lookup and forwards the traffic out the VLAN 50 interface. The Catalyst 5509 forwards the traffic to the server out port 2/1.

Similarly, if Host B transmits to the server, the Catalyst 5505 receives the traffic on port 3/1 and passes it over the Gigabit EtherChannel ISL trunk link to the Catalyst 5509. The Catalyst 5509 passes the traffic to the RSFC over the VLAN 250 interface. The RSFC routes the traffic out the VLAN 50 interface, and the Catalyst 5509 forwards the traffic to the server.

We begin with the Catalyst 5509 configuration:

Catalyst 5509 Configuration

```
Cat5509> (enable) set vtp domain Corporate mode server

VTP domain Corporate modified
Cat5509> (enable) set vlan 50
Vlan 50 configuration successful
Cat5509> (enable) set vlan 150

Vlan 150 configuration successful
Cat5509> (enable) set vlan 250

Vlan 250 configuration successful
Cat5509> (enable) set port channel 1/1-2 desirable

Port(s) 1/1-2 channel mode set to desirable.
Cat5509> (enable) set trunk 1/1 desirable isl

Port(s) 1/1 trunk mode set to desirable.
Port(s) 1/1 trunk type set to isl.
Cat5509> (enable) set port duplex 2/1 full

Port 2/1 set to full-duplex.
Cat5509> (enable) set vlan 50 2/1

VLAN 50 modified.
VLAN 1 modified.
VLAN  Mod/Ports
---- -----------------------
50    2/1

Cat5509> (enable) set port duplex 6/1 full

Port 6/1 set to full-duplex.
Cat5509> (enable) set vlan 150 6/1

VLAN 150 modified.
VLAN 1 modified.
VLAN  Mod/Ports
---- -----------------------
150   6/1

Cat5509> (enable)
```

Catalyst 5505 Configuration

```
Cat5505> (enable) set vtp domain Corporate mode client

VTP domain Corporate modified
Cat5509> (enable) set port duplex 3/1 full
```

```
Port 3/1 set to full-duplex.
Cat5505> (enable) set vlan 250 3/1

VLAN 250 modified.
VLAN 1 modified.
VLAN  Mod/Ports
---- -----------------------
250   3/1

Cat5505> (enable)
```

RSFC Configuration

```
Console> (enable) session 15

Trying Router-15...
Connected to Router-15.
Escape character is '^]'.

RSFC>enable
RSFC#configure terminal
Enter configuration commands, one per line.  End with CNTL/Z.
RSFC(config)#interface vlan50
RSFC(config-if)#ip address 172.16.50.1 255.255.255.0
RSFC(config-if)#no shutdown
RSFC(config-if)#interface vlan150
RSFC(config-if)#ip address 172.16.150.1 255.255.255.0
RSFC(config-if)#no shutdown
RSFC(config-if)#interface vlan250
RSFC(config-if)#ip address 172.16.250.1 255.255.255.0
RSFC(config-if)#no shutdown
RSFC(config-if)#end
RSFC#
```

If the 5509/RSFC combo were replaced with a 6509/MSFC combination, the command syntax would be identical. The only difference would be the line cards referenced that contain the switch ports. Interestingly, the command **session 15** on the Supervisor Engine is the same in each case (to access the route processor).

If the 5509/RSFC combination were replaced with a 5509/RSM combination, the only difference would be the module referenced by the **session** command.

Summary

Inter-VLAN routing is a mechanism to switch packets between VLANs with a Layer 3 device, whether an external router or an internal route processor. In the case of an external router, the configuration is called router-on-a-stick. In the case of an internal route processor, you are dealing with a multilayer Catalyst switch with a daughter card or line card or built-in route processor that interfaces with the switching engine via the switch backplane.

With router-on-a-stick, you configure an ISL or 802.1Q trunk between a switch and a router and configure subinterfaces on the router end of the trunk, configuring one subinterface per VLAN. The trunk port on the router can be a Fast Ethernet, Gigabit Ethernet, or port channel (EtherChannel) interface.

With an internal route processor, there are several configuration options, depending on the Catalyst chassis, the Supervisor Engine, the route processor module or feature card, and whether you are using CatOS or CatIOS.

You have seen how to configure inter-VLAN routing in each of the various scenarios. The next chapter explores how multilayer switching (MLS and CEF) works, what devices support it, and how to configure it. The configuration requirements are practically nonexistent for CEF. CEF is the Layer 3+ switching method used primarily by newer Catalyst switches and Cisco routers.

Check Your Understanding

Test your understanding of the concepts covered in this chapter by answering these review questions. The answers are listed in Appendix A, "Check Your Understanding Answer Key."

1. What layer of the Cisco hierarchical switching model is normally responsible for inter-VLAN routing?
   - **A.** Core
   - **B.** Distribution
   - **C.** Access
2. What two entities are involved in all inter-VLAN routing scenarios?
   - **A.** Switching engine
   - **B.** Routing protocol
   - **C.** Route processor
   - **D.** Catalyst OS
3. What three daughter cards or line cards used on Catalyst 6000 family switches can be used for inter-VLAN routing?
   - **A.** RSM
   - **B.** MSFC1
   - **C.** MSFC2
   - **D.** MSM
4. What two daughter cards or line cards used on Catalyst 5000 family switches can be used for inter-VLAN routing?
   - **A.** RSM
   - **B.** MSM
   - **C.** MSFC
   - **D.** RSFC
5. What combination(s) available on a Catalyst 4000 series switch can be used for inter-VLAN routing?
   - **A.** Supervisor Engine III
   - **B.** Supervisor Engine II
   - **C.** Supervisor Engine I or II together with an L3 module
   - **D.** Any Supervisor Engine

6. What Cisco IOS Software command is used on a Catalyst 5000 RSM to create a logical connection with a "Layer 2" VLAN created on the Supervisor Engine?

   A. set vlan

   B. interface port-channel

   C. interface vlan

   D. set port channel

7. What CatOS command can be used on a Catalyst 4000 Supervisor Engine II to link internally with the L3 module?

   A. set port channel

   B. set channel

   C. set vlan channel

   D. set interface vlan

8. What Cisco IOS Software command can be used on a Catalyst 8510 CSR router to create a port-channel subinterface corresponding to VLAN 10 in a router-on-a-stick configuration?

   A. interface port-channel 1.10

   B. interface port-channel 10.1

   C. interface etherchannel 1.10

   D. interface etherchannel 10.1

9. What CatOS command is used on a Catalyst 4000 Supervisor Engine II to direct all management VLAN traffic to IP address 10.1.1.1/24 on the L3 module?

   A. ip route 0.0.0.0 0.0.0.0 10.1.1.1

   B. set ip default-gateway 10.1.1.1

   C. ip default-gateway 10.1.1.1

   D. set ip route default 10.1.1.1

10. Is a trunk required to enable inter-VLAN routing?

   A. No

   B. Yes

Key Terms

MSFC (Multilayer Switch Feature Card) MSFC1 and MSFC2 are daughter cards to the Catalyst 6000 Supervisor Engine that provide multilayer switching functionality and routing services between VLANs.

MSM (Multilayer Switch Module) A line card for the Catalyst 6000 family of switches that runs the Cisco IOS router software and directly interfaces with the Catalyst 6000 backplane to provide Layer 3 switching.

port channel A logical interface on a switch or router into which physical interfaces are grouped to form a single logical link.

route processor The main system processor in a Layer 3 networking device. Responsible for managing tables and caches and for sending and receiving routing protocol updates.

router-on-a-stick A method for inter-VLAN routing consisting of an external router with a Fast Ethernet, Gigabit Ethernet, or EtherChannel trunk connecting to a switch, utilizing ISL or 802.1Q. Subinterfaces on the trunk are created to correspond to VLANs in a one-to-one fashion.

RSFC (Route Switch Feature Card) A daughter card to the Catalyst 5000/5500 Supervisor Engine IIG and IIIG that provides inter-VLAN routing and multilayer switching functionality. The RSFC runs Cisco IOS router software and directly interfaces with the Catalyst switch backplane.

RSM (Route Switch Module) A line card that interfaces with the Supervisor Engine of a Catalyst 5000/5500 switch to provide inter-VLAN routing functionality. The RSM runs the Cisco IOS.

After completing this chapter, you will be able to perform tasks related to the following:

- Familiarity with MLS's hardware and software requirements
- Understanding the key steps in MLS operation
- Function of the MLS Switching Engine (MLS-SE) and the MLS Route Processor (MLS-RP)
- Configuring the MLS-SE and the MLS-RP
- Verifying MLS configuration
- Key advantages of Cisco Express Forwarding (CEF) over MLS
- Understanding the key principles of CEF operation
- Familiarity with CEF's hardware requirements

Chapter 7

MLS and CEF

The desire for high-speed networking provides the impetus to develop new technologies. One of the bottlenecks in high-speed networking is the routers' decision-making process. Until recently, routers forwarded traffic without the aid of specialized hardware designed to rewrite PDU headers to speed up the process. In recent years, technologies and accompanying hardware have enhanced the process of rewriting PDU headers to move traffic to its ultimate destination.

A number of software and hardware combinations enable hardware-based PDU header rewrites and forwarding. Two of the methods used by Cisco devices are Multilayer Switching (MLS) and Cisco Express Forwarding (CEF). The hardware requirements for each of these technologies are strict, so it is important to do some research before you make any purchases. This chapter clarifies the hardware and software requirements for supporting MLS and CEF.

In general, MLS is a technology used by a small number of older Catalyst switches to provide wire-speed routing (with the exception of IP Multicast MLS, which is still supported on the Catalyst 6000 family). CEF is the technology used by most newer Cisco devices to provide wire-speed routing. In a nutshell, MLS looks at the first packet in a flow and caches the requisite information to permit subsequent packets to be switched independent of the route processor. CEF enables packet switching to circumvent the route processor altogether by employing specialized data structures that are dynamically updated by a communication process between the route processor and the switch processor. Therefore, the traditional routing of the "first" packet required by MLS is avoided.

This chapter shows you how MLS and CEF work and how you can configure these technologies. You also learn how to interpret the output of various **show** commands relating to MLS and CEF, for verification and troubleshooting.

MLS

The first thing to note about MLS is its name. The term *multilayer switching (MLS)* is used for two distinct things. On one hand, it is a general network design term that refers to hardware-based PDU header rewriting and forwarding based on information specific to one or more OSI layers. The term is used in this context in Chapter 1. Basically, multilayer switching used in this way refers to Chapter 1's subject matter.

On the other hand, in this chapter, MLS refers to a specific technology employed by various Cisco devices to perform wire-speed PDU header rewrites and forwarding. To help differentiate between the two uses of the term multilayer switching, when the acronym MLS is used (as opposed to the expanded term), it refers to the Cisco-specific technology.

MLS was the first technological step in solving the router-as-a-bottleneck issue. It is also responsible for blurring the lines we have taken for granted with respect to using the OSI model as a means of delineating per-OSI-layer functions. With MLS, Layer 3 switching speeds approximate those of Layer 2 switching. This has a dramatic impact on network design, as discussed in Chapter 1. The trend is to incorporate more and more Layer 3 functionality into what are traditionally Layer 2 devices. (For example, the Catalyst 3550 access-layer switch is now a true multilayer switch.)

Something to speculate about is that, with Layer 3 switching as fast as Layer 2 switching, and assuming that garden-variety access-layer switches will support Layer 3 switching in the near future, VLANs will play a more cosmetic role as time progresses. With Layer 3 switching as fast as Layer 2 switching, and assuming Layer 3 switching functionality in access-layer switches, data would move around the campus network just as efficiently without VLANs in place, so the argument for functional grouping of users by VLAN is much less relevant. Users could be artificially labeled part of a workgroup, similar to an AppleTalk zone, independent of their IP network. The only clear advantage remaining for VLANs would be ease of filtering between functional groups. So, you see that multilayer switching is changing the dynamics of network design. MLS was Cisco's first step in this (r)evolution.

We begin our discussion of MLS by detailing the hardware and software requirements. This is followed by an explanation of MLS components and operations. Last, we detail the configuration of MLS and the verification and troubleshooting tools.

MLS Hardware and Software Requirements

The requirements for running MLS in your network are specific. Many technologies are widely available on Cisco products; for example, most Cisco router platforms have Cisco IOS Software releases that support IPX. With MLS, widespread support is not the case—the options are limited.

Catalyst 5000 series switches are normally the first devices that come to mind when a network engineer thinks of MLS. These devices comprise the majority of MLS deployments. For Catalyst 5000 switches, the MLS requirements are as follows:

- Supervisor Engine III with NetFlow Feature Card (NFFC) or NFFC II, a Supervisor Engine IIG, or a Supervisor Engine IIIG (the IIG and IIIG have integrated NFFCs). The only NFFCs still available for purchase are the NFFC-A and NFFC II-A. The other NFFCs were last sold on April 30, 2000.
- Catalyst 5000 Route Switch Module (RSM), Catalyst 5000 Route Switch Feature Card (RSFC), Catalyst 6000 MSFC, or Catalyst 6000 MSM.
- CatOS 4.1(1) or later.
- Cisco IOS Software Release 12.0(3c)W5(8a) or later on the RSFC.
- Cisco IOS Software Release 11.3(2)WA4(4) or later on the RSM.

Catalyst 2926G switches can also work in tandem with external Cisco routers to support MLS. 2926G switches are fixed-configuration switches. They were last sold in May 2000 and have been replaced by the 2948G switches.

Here are the hardware and software guidelines for external routers:

- Cisco 3600, 4500, 4700, 7200, or 7500 router or Catalyst 8500 switch router externally attached to the Catalyst 5000 series switch.
- The connection between the external router and the Catalyst 5000 series switch must be a Fast or Gigabit Ethernet link, an ISL or IEEE 802.1Q trunk, or a Fast or Gigabit EtherChannel.
- Cisco IOS Software Release 11.3(2)WA4(4) or later on Cisco 4500, 4700, 7200, or 7500 routers. Cisco IOS Software Release 12.0 is the earliest release available on 8500 series switch routers.
- Cisco IOS Software Release 12.0(2) or later on Cisco 3600 series routers.
- Catalyst 6000 MSFC on a Supervisor Engine with CatOS 6.3.
- Catalyst 6000 MSM.

For Catalyst 6000 switches, the MLS hardware and software guidelines are as follows:

- Catalyst 6000 family switches do *not* support an external Multilayer Switching Route Processor (MLS-RP). The MLS-RP must be a Multilayer Switch Module (MSM) or Multilayer Switch Feature Card (MSFC). As of July 6, 2000, the MSM is no longer available for purchase and has been replaced by the MSFC.
- MSFC, Policy Feature Card (PFC), and CatOS 5.1CSX or later.
- MSM, CatOS 5.2(1)CSX or later, and Cisco MSM IOS Software Release 12.0(1a)WX5(6d) or later.

NOTE

MLS support for IPX is available in various hardware/software combinations. MLS also supports IP multicast traffic in most of the hardware/software combinations just listed. This book doesn't discuss MLS support of IPX. Chapter 9 discusses configuring multicast technologies on Cisco routers and switches outside of the multilayer switching context.

MLS, in the generic sense, is implemented with CEF in newer Catalyst 6000 products. In particular, Supervisor Engine 2, Policy Feature Card 2 (PFC2), and Multilayer Switch Feature Card 2 (MSFC2) comprise a standard configuration for newly purchased Catalyst 6000s as of the year 2002; this configuration relies natively on CEF. The Catalyst 4000 switch with the Layer 3 Services Module or the Supervisor Engine III also natively supports CEF.

With the hardware and software requirements clarified, you are ready to see how MLS works. The following sections detail the concepts of MLS flow, MLS-SE, and MLS-RP.

MLS Flows

The idea of MLS is simple. MLS looks at the first packet in a flow of data and caches some header information describing the flow. Subsequent packets in the flow circumvent the router because the switch has cached the data necessary to rewrite the packet header, which it does. Thus, you have wire-speed switching.

Specifically, IP MLS switches unicast IP data packet flows between IP subnets using advanced ASIC switching hardware, offloading processor-intensive packet routing from network routers. The packet forwarding function is moved onto Layer 3 switches whenever a partial or complete switched path exists between two hosts. Packets that do not have a partial or complete switched path to reach their destinations are still forwarded in software by routers. Standard routing protocols, such as OSPF, EIGRP, RIP, and IS-IS are used for route determination.

IP is a connectionless protocol—every packet is delivered independently of every other packet. However, actual network traffic consists of many end-to-end conversations, or flows, between users or applications.

A *flow* is a unidirectional sequence of packets between a particular source and destination that share the same Layer 3 and Layer 4 information. For example, Telnet traffic transferred from a particular source to a particular destination is a separate flow from FTP packets between the same source and destination. Communication from a client to a server and from the server to the client comprise distinct flows.

If a flow is defined administratively by destination IP address alone, traffic from multiple users or applications to that destination is included in that flow.

The *MLS Switching Engine (MLS-SE)* is the set of hardware components on a Catalyst switch, excluding the route processor, necessary to support MLS. For example, a Catalyst 5000 with a Supervisor Engine IIIG is an MLS-SE. These requirements are detailed in the preceding section. The *MLS Route Processor (MLS-RP)* is a Cisco device with a route processor that supports MLS. For example, a Cisco 3620 router is an MLS-RP. The devices meeting these criteria are detailed in the preceding section. Only certain combinations of MLS-SEs and MLS-RPs can run MLS.

There are several cases in networking in which a term or acronym has multiple meanings. The term *virtual circuit* is one example. It is used in the context of TCP and internal Layer 2 switch operations and with technologies such as Frame Relay and ATM. These uses connote a logical connection between two entities. The acronym RP is another such case. RP can mean both route processor and rendezvous point. It helps to be aware of these terms and acronyms so that when you hear them, the first thing you think is, "In what context is this term being used?" If you are talking to someone else about an RP, it is helpful to expand the acronym so that your audience knows what you are talking about. This helps you avoid some confusion.

An MLS-SE maintains a Layer 3 switching table (the Layer 3 MLS cache) for Layer 3 switched flows. After the MLS cache is created, packets identified as belonging to an existing flow can be switched based on the cached information. The MLS cache maintains flow information for all active flows. The maximum size of the MLS cache is 128,000 entries. The larger the active cache, the greater the likelihood that a flow will not be switched by the MLS-SE and will get forwarded to the router.

An MLS cache entry is created for each flow's initial packet. Upon receipt of a packet that does not match any flow currently in the MLS cache, a new IP MLS entry is created. This first packet has special significance and is detailed more in the next section.

The flow's state and identity are maintained while packet traffic is active; when traffic for a flow ceases, the entry ages out. You can configure the aging time for MLS entries kept in the MLS cache. If an entry is not used for the specified period of time, the entry ages out.

The MLS-SE uses flow masks to determine how MLS entries are created. The *flow mask* is a set of criteria, based on a combination of source IP address, destination IP address, protocol, and protocol ports that describes the flow's characteristics. The MLS-SE learns the flow mask through *Multilayer Switching Protocol (MLSP)* messages from each MLS-RP for which the MLS-SE is performing Layer 3 switching. MLSP is the protocol used to communicate MLS information between the MLS-SE and MLS-RP. In particular, the MLS-SE populates its Layer 2 Content-Addressable Memory (CAM) table with updates received from MLSP packets. When the MLS-SE flow mask changes, the entire MLS cache is purged.

The three flow masks are as follows:

- **destination-ip**—The least specific flow mask. The MLS-SE maintains one MLS entry for each destination IP address. All flows to a given destination IP address use this MLS entry. This mode is used if no access lists are configured on any of the MLS-RP interfaces.
- **source-destination-ip**—The MLS-SE maintains one MLS entry for each source and destination IP address pair. All flows between a given source and destination use this MLS entry, regardless of the IP protocol ports. This mode is used if any of the MLS-RP interfaces has a standard access list.
- **ip-flow**—The most-specific flow mask. The MLS-SE creates and maintains a separate MLS cache entry for every IP flow. An ip-flow entry includes the source IP address, destination IP address, protocol, and protocol ports. This mode is used if any of the MLS-RP interfaces has an extended access list.

When a packet is Layer 3 switched from a source host to a destination host, the MLS-SE performs a packet rewrite based on information learned from the MLS-RP and stored in the MLS cache.

If Host A and Host B are on different VLANs, and Host A sends a packet to the MLS-RP to be routed to Host B, the MLS-SE recognizes that the packet was sent to the MLS-RP's MAC address. The MLS-SE checks the MLS cache and finds the entry matching the flow in question.

When the MLS-SE receives the packet, it is formatted as shown in Table 7-1.

Table 7-1 MLS Flow Packet Received by the MLS-SE

Frame Header		**IP Header**				**Payload**	
Destination	Source	Destination	Source	TTL	Checksum	Data	Checksum
MLS-RP MAC	*Host A MAC*	*Host B IP*	*Host A IP*	*n*	*Calculation1*		

The MLS-SE rewrites the Layer 2 frame header, changing the destination MAC address to Host B's MAC address and the source MAC address to the MLS-RP's MAC address. (These MAC addresses are stored in the MLS cache entry for this flow.) The Layer 3 IP addresses remain the same, but the IP header Time To Live (TTL) is decremented, and the IP checksum is recomputed. The MLS-SE rewrites the switched Layer 3 packets so that they appear to have been routed by a router. The MLS-SE forwards the rewritten packet to Host B's VLAN (the destination VLAN is stored in the MLS cache entry), and Host B receives the packet.

After the MLS-SE performs the packet rewrite, the packet is formatted as shown in Table 7-2.

Table 7-2 MLS Flow Packet After a Rewrite by the MLS-SE

Frame Header		IP Header				Payload	
Destination	Source	Destination	Source	TTL	Checksum	Data	Checksum
Host B MAC	*MLS-RP MAC*	*Host B IP*	*Host A IP*	*n–1*	*Calcu-lation2*		

An IP MLS network topology is shown in Figure 7-1. In this example, Host A is on the Sales VLAN (IP subnet 171.59.1.0), Host B is on the Marketing VLAN (IP subnet 171.59.3.0), and Host C is on the Engineering VLAN (IP subnet 171.59.2.0).

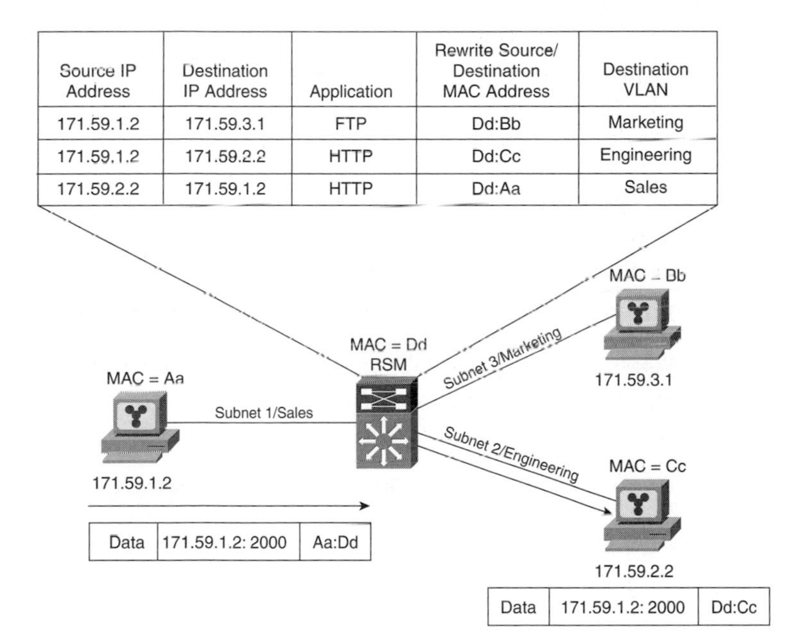

Figure 7-1 Catalyst 5000 switches with Supervisor Engine III, NFFC, and RSM support MLS. IP addresses, MAC addresses, protocol ports, and destination VLANs are stored in the MLS cache.

When Host A initiates an FTP file transfer to Host B, an MLS entry for this flow is created (this entry is the first item in the MLS cache shown in Figure 7-1). The MLS-SE stores the MAC addresses of the MLS-RP and Host B in the MLS entry when the MLS-RP forwards the first packet from Host A through the switch to Host B. The MLS-SE uses this information to rewrite subsequent packets from Host A to Host B.

Similarly, a separate MLS entry is created in the MLS cache for the HTTP traffic from Host A to Host C and for the HTTP traffic from Host C to Host A. The destination VLAN is stored as part of each MLS entry so that the correct VLAN identifier is used when encapsulating traffic on trunk links.

MLS creates flows based on access lists configured on the MLS-RP. IP MLS allows you to enforce access lists on every packet of the flow without compromising IP MLS performance. When you enable IP MLS, the MLS-SE handles standard and extended access list permit traffic at wire speed. Access list deny traffic is always handled by the MLS-RP, not the MLS-SE. Route topology changes and the addition or modification of access lists are reflected in the IP MLS switching path automatically on the MLS-SE.

For example, when Host A wants to communicate with Host B, it sends the first packet to the MLS-RP. If an access list is configured on the MLS-RP to deny access from Host A to Host B, the MLS-RP receives the packet, checks the access list to see if the packet flow is permitted, and discards the packet based on the access list. Because the first packet for this flow does not return from the MLS-RP, an MLS cache entry is not established by the MLS-SE.

If a flow is already being Layer 3 switched by the MLS-SE, and the access list is created on the MLS-RP, the MLS-SE learns of the change through MLSP and immediately enforces security for the affected flow by purging it from the MLS cache. New flows are created based on the restrictions imposed by the access list.

Similarly, when the MLS-RP detects a routing topology change, the appropriate MLS cache entries are deleted in the MLS-SE. New flows are created based on the new topology.

Before we discuss MLS configuration, we need to detail the four steps of MLS operation.

MLS Operation

Now that we have the MLS concepts in place, we are ready to describe the four steps of MLS operation:

Step 1 MLSP multicast hello packets are sent every 15 seconds to inform the MLS-SE(s) of the MLS-RP MAC addresses used on different VLANs and the MLS-RP's routing and access-list information (when the router boots or after changes in access lists). When an MLS-SE hears the MLSP hello message indicating an IP MLS initialization, the MLS-SE is programmed with a locally unique *XTAG* for the MLS-RP and with the MLS-RP MAC addresses associated with each VLAN. The XTAG is a 1-byte value that the MLS-SE attaches to each VLAN to all MAC addresses learned from the same MLS-RP via MLSP. See Figure 7-2.

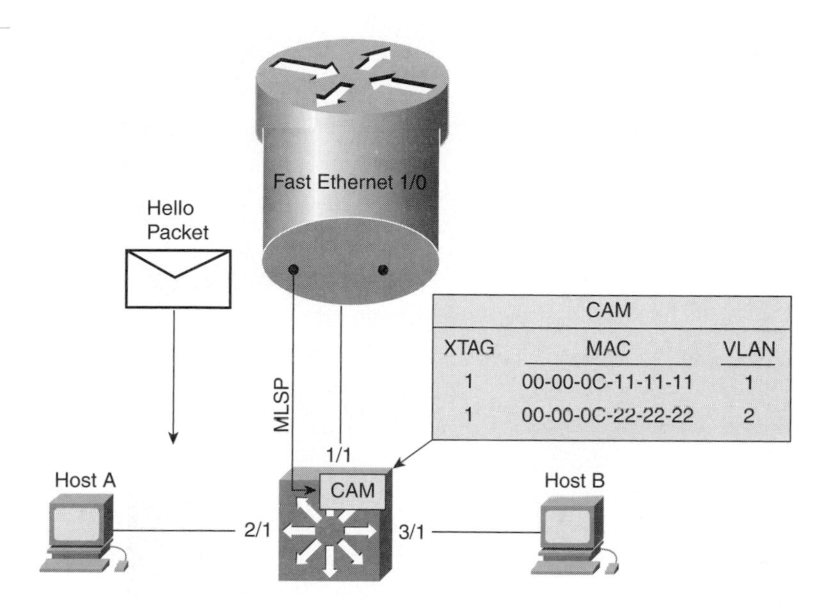

Figure 7-2 The first step of MLS operation is the exchange of hello packets identifying the MLS-RP to the MLS-SE.

Step 2 In Figure 7-3, Host A and Host B are located on different VLANs. Host A initiates a data transfer to Host B. When Host A sends the first packet to the MLS-RP, the MLS-SE recognizes this packet as a *candidate packet* for Layer 3 switching because the MLS-SE has learned the MLS-RP's destination MAC addresses and VLANs through MLSP. The MLS-SE learns the Layer 3 flow information (such as the destination IP address, source IP address, and protocol port numbers) and forwards the first packet to the MLS-RP. A partial MLS entry for this Layer 3 flow is created in the MLS cache.

The MLS-RP receives the packet, looks at its route table to determine how to forward the packet, and applies services such as access control lists and class of service (CoS) policy.

The MLS-RP rewrites the MAC header, adding a new destination MAC address (Host B's) and its own MAC address, associated with Host B's VLAN, as the source.

Step 3 The MLS-RP routes the packet to the destination host. When the switch receives the packet, the MLS-SE recognizes that the source MAC address belongs to the MLS-RP, that the XTAG matches that of the candidate packet, and that the packet's flow information matches the flow for which the candidate entry was created. The MLS-SE considers this packet an *enable packet* and completes the MLS entry in the MLS cache. See Figure 7-4.

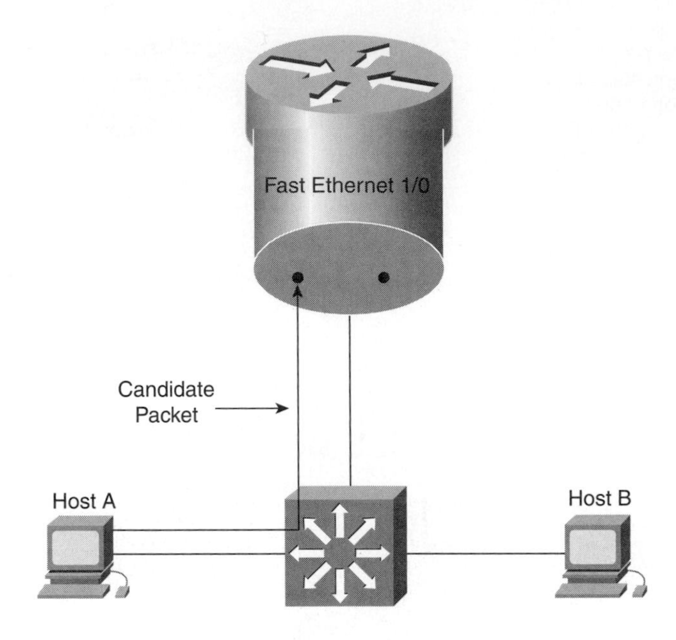

Figure 7-3
The second step of MLS operation is the creation of a partial MLS cache entry based on a candidate packet.

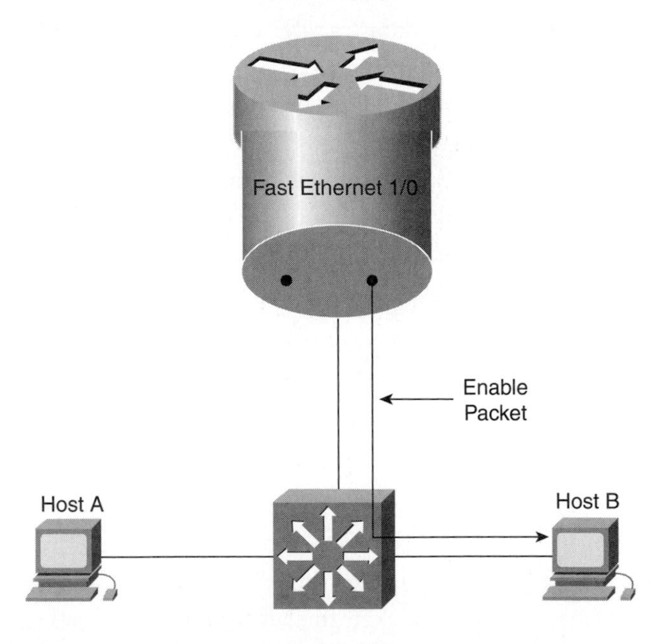

Figure 7-4
The third step of MLS operation is the completion of the MLS cache entry based on an enable packet.

Step 4 After the MLS entry has been completed, all Layer 3 packets in the same flow from the source host to the destination host are Layer 3/4 switched directly by the switch, bypassing the router, as shown in Figure 7-5. IP MLS is unidirectional: A separate Layer 3 switched path is created for traffic from Host B to Host A.

After the Layer 3 switched path is established, the MLS-SE rewrites the PDU headers from the source host before it is forwarded to the destination host. The rewritten information includes the MAC addresses, encapsulations (if applicable), protocol port information, and Layer 3 information.

The resultant packet format and protocol behavior are identical to that of a packet routed by the MLS-RP (RSM, RSFC, MSM, MSFC, or external router).

Figure 7-5
The fourth step of MLS operation is the Layer 3 switching of subsequent packets in the same flow.

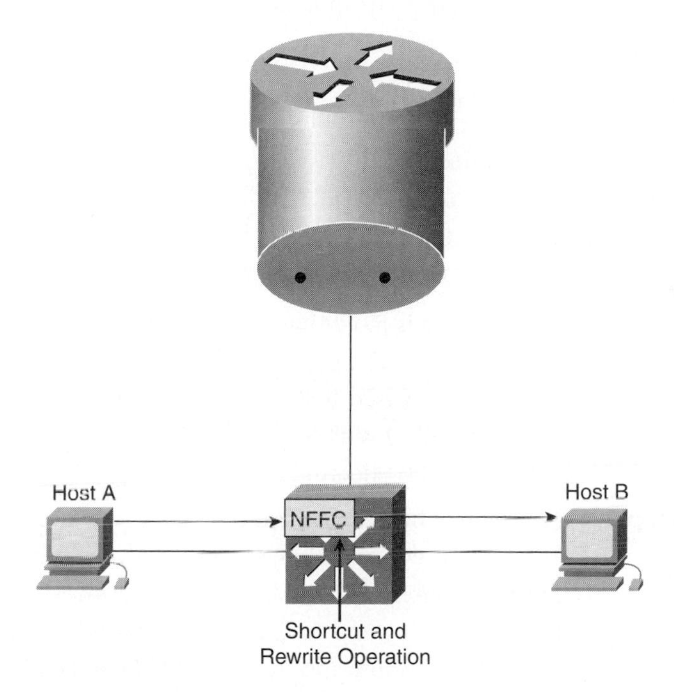

Finally, some Catalyst 5000 family switching line cards have onboard hardware that performs the rewrite (as opposed to rewrites on the Supervisor Engine), maximizing IP MLS performance. This performance enhancement is also used on the Catalyst 2926G series switch ports. When the line cards or individual ports perform the PDU header rewrites, this is called *inline rewrite*. With inline rewrite, frames traverse the switch bus only once.

The Catalyst 5000 *NetFlow Feature Card (NFFC)* is a daughter card for the Supervisor module that enables intelligent network services, such as high-performance multilayer switching and accounting and traffic management. The NFFC contains *central rewrite engines* (one per bus) to handle PDU header rewrites when inline rewrite is not an option. If the central rewrite engines are used, this means that a frame must traverse the bus twice—first with a VLAN tag for the source (on its way to the NFFC), and second with a VLAN tag for the destination (on its way to the egress port). To determine whether a port supports packet rewrite, use the **show port capabilities** command. If the port does not support inline rewrite, the packet rewrite is done in the Supervisor Engine.

This completes the theory portion of the MLS discussion. You should now have a good idea of how MLS works. The basic ingredients are the MLS-RP, MLS-SE, and MLSP. The MLS-SE creates flows in the MLS cache based on candidate and enable packets and access lists configured on the MLS-RP. The result is wire-speed routing of packets based on criteria such as destination IP address, source IP address, and protocol port numbers.

What remains to be seen is the MLS configuration methodology. The next sections describe MLS configuration and the commands that verify and troubleshoot it.

MLS-RP Configuration

Configuring MLS support on an MLS-RP has five basic steps, which are detailed in the following list. After detailing the configuration steps for an MLS-RP and an MLS-SE, we provide a complete example that covers the entire process:

> **NOTE**
> Perform Step 2 before you enter any other IP MLS interface commands on the IP MLS interface (specifically, the **mls rp ip** or **mls rp management-interface** interface commands in Steps 4 and 5). Entering IP MLS interface commands on an interface prior to putting the interface into a VTP domain places the interface in the null domain. Then, to put the IP MLS interface into a domain other than the null domain, you must clear the IP MLS interface configuration before you can add it to another VTP domain.

Step 1 This first step is to configure the MLS-RP to globally enable MLS. This is analogous to enabling IPX or multicast routing on a router. The global configuration Cisco IOS Software command to enable MLS is **mls rp ip**. A Catalyst 6000 with Supervisor Engine I and an MSFC has IP MLS globally enabled by default.

Step 2 Determine which router interfaces you will use as IP MLS interfaces, and add those interfaces to the same VTP domain as the MLS-SE. Remember that router interfaces will be VLAN interfaces (not physical interfaces) on the RSM, RSFC, and MSFC. The complete syntax for the command to add an IP MLS interface to a VTP domain is **mls rp vtp-domain** [*domain_name*].

On ISL or 802.1Q trunks, enter the **mls rp vtp-domain** command on the primary interface (not on the individual subinterfaces). All subinterfaces on the primary interface inherit the VTP domain assigned to the primary interface. The command **mls rp vtp-domain** is also used on physical interfaces configured as access ports (ports with only one VLAN association). On the RSFC, RSM, and MSFC, this command is entered on each VLAN interface participating in MLS.

Step 3 For Layer 3 interfaces that are not trunk ports and for which you want to configure IP MLS support, first double-check that the interface has been assigned to a VLAN (as an access port). Then assign a VLAN ID to the IP MLS interface with the interface command **mls rp** *vlan-id* [*vlan_id_num*].

Step 4 Enable IP MLS on the interfaces participating in IP MLS with the command **mls rp ip**. This is the same command used in global configuration mode, but here it is applied at the interface level. In the case of an ISL or 802.1Q trunk, this command must be configured on all subinterfaces (it is not entered on the interface itself). On the MSFC, RSM, and RSFC, this command is entered on VLAN interfaces.

Step 5 Specify an MLS-RP interface as an *MLS-RP management interface*. MLSP runs on the management interface, sending hello messages, advertising routing changes, and announcing VLANs and MAC addresses of interfaces participating in MLS. You must specify at least one router interface as a management interface (more than one is permitted, but this introduces a proportional increase in overhead). *If you do not specify a management interface, IP MLS will not function.*

Also, the MLS-SE must have an active port in at least one VLAN that has a corresponding router interface configured as a management interface.

The interface command to specify an MLS-RP interface as a management interface is **mls rp management-interface**. In the case of an ISL or 802.1Q trunk, this command is applied to a particular subinterface (not to the interface itself).

Finally, configuring output access lists on MLS-RP interfaces works seamlessly with MLS, meaning that MLS cache entries are created that reflect the desired behavior of the access list while still providing wire-speed switching of packets. On the other hand, input access lists on MLS-RP interfaces force every packet to be routed by the MLS-RP. To enable MLS to cooperate with input access list, enter the global configuration command **mls rp ip input-acl**.

MLS-SE Configuration

The configuration is platform-dependent for the MLS-SE.

IP MLS is enabled by default on Catalyst 5000 family and 2926G series switches. If the MLS-RP is an RSM or RSFC installed in the Catalyst 5000 chassis, you do not need to configure the switch. You only need to configure the switch in these circumstances:

- When an external router is the MLS-RP (this is always the case with Catalyst 2926G series switches)
- When you want to change the IP MLS aging time parameters or packet threshold values
- When you want to enable NetFlow Data Export (NDE) to monitor all IP MLS intersubnet traffic through the NFFC (NDE configuration and monitoring are not explored in this book)

Similarly, IP MLS is enabled by default on Catalyst 6000 family switches. Recall that the Catalyst 6000 does not support an external MLS-RP; the only options for MLS-RP are the MSM and the MSFC. You only need to configure the switch in the following circumstances:

- When you want to specify MLS aging time parameters or packet threshold values
- When you want to set the minimum IP MLS flow mask

Assuming that you have a Catalyst 5000 or a 2926G switch connecting to an external MLS-RP, the configuration proceeds as follows:

Step 1 Enable MLS on the switch with the command **set mls enable**. When you enable IP MLS on the switch, the switch (MLS-SE) starts to process MLSP messages from the MLS-RPs and starts Layer 3 switching. IP MLS is enabled by default on the MLS-SE, so this step is necessary only if MLS has been disabled with the command **set mls disable**.

Step 2 Specify the external router(s) participating in MLS. Before specifying a router to participate in IP MLS, enter the **show mls rp** command on the router to identify the MLS-RP IP address. Use the displayed address when you enter the **set mls include** *ip_addr* command on the switch, the command that specifies the external router(s) participating in MLS. The MLS-SE does not process MLSP messages from external routers that have not been included as MLS-RPs. Also, when you configure the MLS-SE in a scenario where the MLS-RP is not an external router, you don't need the **set mls include** command.

In addition to the MLS-RP and MLS-SE commands you have seen up to this point, there are a few optional MLS parameters, such as MLS aging time and minimum flow mask, that you might want to configure on the switch. These are discussed in the following sections.

MLS Aging Time

The IP MLS *aging time* determines the time before an MLS entry is aged out. The default is 256 seconds. You can configure the aging time in the range of 8 to 2032 seconds in 8-second increments. Any aging-time value that is not a multiple of 8 seconds is adjusted to the closest one. For example, a value of 65 is adjusted to 64, and a value of 127 is adjusted to 128. Other events might cause MLS entries to be purged, such as routing changes or a change in link state (MLS-SE link down).

To specify the IP MLS aging time, use the command **set mls agingtime** [*agingtime*] in privileged mode.

Another useful MLS cache aging parameter is the fast aging time. Cisco recommends that you keep the number of MLS entries in the MLS cache below 32,000 entries. If the number of MLS entries amounts to more than 32,000, some flows are sent to the router.

To help keep the MLS cache size below 32,000, enable the IP MLS *fast aging time*. The IP MLS fast aging time applies to MLS entries that have no more than *pkt_threshold* packets switched within fastagingtime seconds after it is created. A typical cache entry that is removed is the entry for flows to and from a Domain Name Server or TFTP server; the entry might never be used again after it is created. Detecting and aging out these entries saves space in the MLS cache for other data traffic.

The default *fastagingtime* value is 0 (no fast aging). You can configure the *fastagingtime* value to 32, 64, 96, or 128 seconds. Any *fastagingtime* value that is not configured exactly as the indicated values is adjusted to the closest one. You can configure the *pkt_threshold* value to 0, 1, 3, 7, 15, 31, or 63 packets.

If you need to enable IP MLS fast aging time, initially set the value to 128 seconds. If the size of the MLS cache continues to grow over 32 KB, decrease the setting until the cache size stays below 32 KB. If the cache continues to grow over 32 KB, decrease the normal IP MLS aging time.

Typical values for *fastagingtime* and *pkt_threshold* are 32 seconds and 0 packets (no packets switched within 32 seconds after the entry is created).

To specify the IP MLS fast aging time and packet threshold, use the command **set mls agingtime fast** [*fastagingtime*] [*pkt_threshold*] in privileged mode.

The remaining configurable MLS parameter for an MLS-SE is the minimum IP MLS flow mask, which is discussed in the next section.

Minimum IP MLS Flow Mask

You can set the minimum granularity of the flow mask for the MLS cache on the MLS-SE. The actual flow mask used consists of a granularity specified by this command. If you configure a more-specific flow mask, the number of active flow entries increases. To limit the number of active flow entries, you might need to decrease the MLS aging time.

If you do not configure access lists on any MLS-RP, the IP MLS flow mask on the MLS-SE is **destination-ip** by default (as discussed in the earlier section "MLS Flows"). However, you can force the MLS-SE to use a particular flow mask granularity by setting the minimum IP MLS flow mask using the **set mls flow** {**destination** | **destination-source** | **full**} command in privileged mode. Depending on the MLS-RP configuration, the actual flow mask used might be more specific than the specified minimum flow mask. For example, if you configure the minimum flow mask to **destination-source**, but an MLS-RP interface is configured with IP extended access lists, the actual flow mask used is **full** (**ip-flow**).

The **set mls flow** command purges all existing MLS cache entries and affects the number of active entries on the MLS-SE, so this should factor into your decision of whether to use this command.

MLS Configuration Example

This example details the configuration of IP MLS for the topology displayed in Figure 7-6. A Cisco 7505 router serves as the MLS-RP, and a Catalyst 5509 with a Supervisor Engine III and an NFFC serves as the MLS-SE. An IEEE 802.1Q trunk connects the two devices participating in MLS.

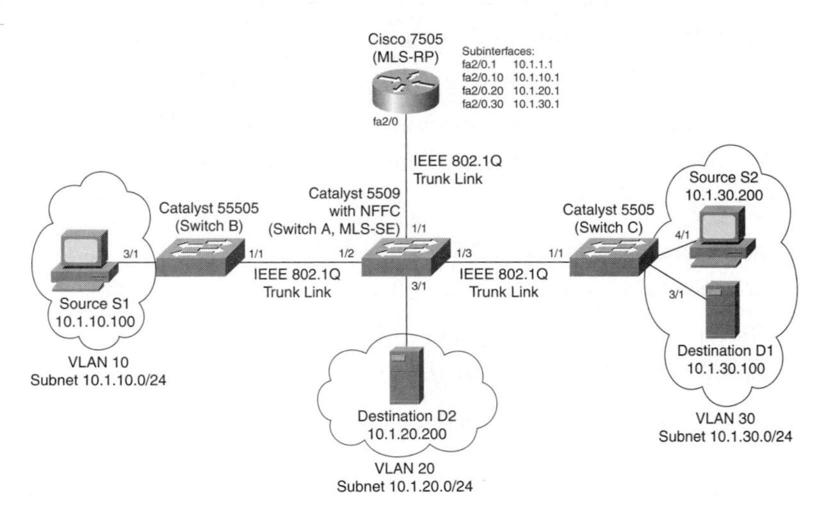

Figure 7-6
A 7505 router (MLS-RP) connects via an 802.1Q trunk to an MLS-SE. Here, the MLS-SE is a Catalyst 5000 with a Supervisor Engine III and an NFFC.

The IP MLS network topology shown in Figure 7-6 includes two Catalyst 5505 switches connecting via 802.1Q trunks to the MLS-SE. The 5505s do not play a role in MLS. The network is configured as follows:

- There are four VLANs (IP subnetworks):
  — VLAN 1 (the management VLAN), subnet 10.1.1.0/24
  — VLAN 10, subnet 10.1.10.0/24
  — VLAN 20, subnet 10.1.20.0/24
  — VLAN 30, subnet 10.1.30.0/24
- The MLS-RP is a Cisco 7505 router with a Fast Ethernet interface (interface FastEthernet 2/0)

- The subinterfaces on the router interface have these IP addresses:
  - FastEthernet 2/0.1—10.1.1.1 255.255.255.0
  - FastEthernet 2/0.10—10.1.10.1 255.255.255.0
  - FastEthernet 2/0.20—10.1.20.1 255.255.255.0
  - FastEthernet 2/0.30—10.1.30.1 255.255.255.0
- A standard output access list is configured on subinterface FastEthernet 2/0.20 (the interface in VLAN 20) on the MLS-RP. The ACL denies all traffic from VLAN 30.
- Switch A, the MLS-SE, is a Catalyst 5509 switch with Supervisor Engine III and the NFFC.
- Switch B and Switch C are Catalyst 5505 switches.
- Switch A is the VTP server in domain "Corporate."
- Switch B and Switch C are VTP clients.

Next, we detail the inter-VLAN switching path both before and after IP MLS operation.

Operation Before IP MLS

Before IP MLS is implemented, when the source host S1 (on VLAN 10) transmits traffic destined for server D1 (on VLAN 30), Switch B forwards the traffic (based on the Layer 2 forwarding table) to Switch A over the 802.1Q trunk link. Switch A forwards the packet to the router over the 802.1Q trunk. The router receives the packet on the VLAN 10 subinterface, checks the destination IP address, and routes the packet to the VLAN 30 subinterface. Switch A receives the routed packet and forwards it to Switch C. Switch C receives the packet and forwards it to destination server D1. This process is repeated for each packet in the flow between source host S1 and destination server D1.

When source host S2 sends traffic to destination server D2, Switch C forwards the packets over the 802.1Q trunk to Switch A. Switch A forwards the packet to the MLS-RP, which receives it on the VLAN 30 subinterface. Because the standard access list configured on the outgoing VLAN 20 subinterface denies all traffic from VLAN 30, the router drops the traffic to Destination D2 from Source S2. Any subsequent traffic from Source S2 for Destination D2 also reaches the router and is dropped.

Operation After IP MLS

After IP MLS is implemented, when the source host S1 transmits traffic destined for destination server D1, Switch B forwards the traffic (based on the Layer 2 forwarding table) to Switch A (the MLS-SE) over the 802.1Q trunk link. When the first packet enters Switch A, a *candidate* flow entry is established in the MLS cache. Switch A forwards the packet to the MLS-RP over the 802.1Q trunk. The MLS-RP receives the packet on the VLAN 10 subinterface, checks the destination IP address, and routes the packet to the VLAN 30 subinterface. Switch A receives the routed packet (the *enable* packet) and completes the flow entry in the MLS cache for destination IP address 10.1.30.100. Switch A forwards the packet to Switch C, where it is forwarded to destination server D1.

Subsequent packets destined for IP address 10.1.30.100 are multilayer-switched by the MLS-SE based on the flow entry in the MLS cache. For example, subsequent packets in the flow from source host S1 are forwarded by Switch B to Switch A (the MLS-SE). The MLS-SE determines that the packets are part of the established flow, rewrites the packet headers, and switches the packets directly to Switch C, bypassing the router.

Because a standard access list is applied on subinterface FastEthernet 2/0.20, the MLS-SE must use the **source-destination-ip** flow mask for all MLS cache entries. When source host S2 sends traffic to destination server D2, Switch C forwards the packets over the 802.1Q trunk to Switch A. Switch A forwards the candidate packet to the MLS-RP, which receives it on the VLAN 30 subinterface. Because the standard access list configured on the outgoing VLAN 20 subinterface denies all traffic from VLAN 30, the router drops the traffic to Destination D2 from Source S2. Switch A never receives the enable packet for the flow on VLAN 20, and no MLS cache entry is completed for the flow. Any subsequent traffic from Source S2 for Destination D2 also reaches the router and is dropped.

The configurations for each of the devices are shown in the next sections. The steps outlined in the previous sections for MLS-RP (five steps) and MLS-SE (two steps) configuration are indicated next to the commands. MLS-RP Step 3 isn't used because a trunk is connecting the MLS-SE and MLS-RP.

MLS-RP Configuration. The following is the MLS-RP configuration:

```
Cisco7505(config)#mls rp ip                         <--------------------MLS-RP step 1
Cisco7505(config)#access-list 1 deny 10.1.30.0 0.0.0.255
Cisco7505(config)1#access-list 1 permit any
Cisco7505(config)#interface fastethernet 2/0
Cisco7505(config-if)#speed 100
Cisco7505(config-if)#full-duplex
Cisco7505(config-if)#mls rp vtp-domain Corporate    <-------------MLS-RP step 2
Cisco7505(config-if)#interface fastethernet2/0.1
```

```
Cisco7505(config-subif)#encapsulation dot1q 1

Cisco7505(config-subif)#ip address 10.1.1.1 255.255.255.0

Cisco7505(config-subif)#mls rp ip           <--------------------MLS-RP step 4

Cisco7505(config-subif)#mls rp management-interface  <------------MLS-RP step 5

Cisco7505(config-subif)#interface fastethernet2/0.10

Cisco7505(config-subif)#encapsulation dot1q 10

Cisco7505(config-subif)#ip address 10.1.10.1 255.255.255.0

Cisco7505(config-subif)#mls rp ip           <--------------------MLS-RP step 4

Cisco7505(config-subif)#interface fastethernet2/0.20

Cisco7505(config-subif)#encapsulation dot1q 20

Cisco7505(config-subif)#ip address 10.1.20.1 255.255.255.0

Cisco7505(config-subif)#ip access-group 1 out

Cisco7505(config-subif)#mls rp ip           <--------------------MLS-RP step 4

Cisco7505(config-subif)#interface fastethernet2/0.30

Cisco7505(config-subif)#encapsulation dot1q 30

Cisco7505(config-subif)#ip address 10.1.30.1 255.255.255.0

Cisco7505(config-subif)#mls rp ip           <--------------------MLS-RP step 4
```

Switch A Configuration. In some Cisco IOS Software releases, traffic on the IEEE 802.1Q native VLAN is not routed. The default native VLAN on Catalyst switches is VLAN 1. If your Cisco IOS Software release does not route traffic on the native VLAN and you want to route traffic on VLAN 1, change the native VLAN on the switch-to-router trunk link to an unused VLAN. In the Switch A, Switch B, and Switch C configuration examples, the native VLAN on all the 802.1Q trunk links is set to an unused VLAN, VLAN 5. The following is the configuration for Switch A:

```
SwitchA> (enable) set vtp domain Corporate mode server

VTP domain Corporate modified
SwitchA> (enable) set vlan 5

Vlan 5 configuration successful
SwitchA> (enable) set vlan 10

Vlan 10 configuration successful
SwitchA> (enable) set vlan 20
```

continues

```
Vlan 20 configuration successful
SwitchA> (enable) set vlan 30

Vlan 30 configuration successful
SwitchA> (enable) set port name 1/1 Router Link

Port 1/1 name set.
SwitchA> (enable) set trunk 1/1 on dot1q

Port(s) 1/1 trunk mode set to on.
Port(s) 1/1 trunk type set to dot1q.
SwitchA> (enable) set port name 1/2 SwitchB Link

Port 1/2 name set.
SwitchA> (enable) set trunk 1/2 desirable dot1q

Port(s) 1/2 trunk mode set to desirable.
Port(s) 1/2 trunk type set to dot1q.
SwitchA> (enable) set port name 1/3 SwitchC Link

Port 1/3 name set.
SwitchA> (enable) set trunk 1/3 desirable dot1q

Port(s) 1/3 trunk mode set to desirable.
Port(s) 1/3 trunk type set to dot1q.
SwitchA> (enable) set vlan 5 1/1-3

VLAN 5 modified.
VLAN 1 modified.
VLAN  Mod/Ports
---- -----------------------
5     1/1-3

SwitchA> (enable) set mls enable             <---------------------MLS-SE step 1

IP Multilayer switching is enabled.
SwitchA> (enable) set mls include 10.1.1.1   <---------------------MLS-SE step 2

IP Multilayer switching enabled for router 10.1.1.1.
SwitchA> (enable) set port name 3/1 Destination D2

Port 3/1 name set.
SwitchA> (enable) set vlan 20 3/1

VLAN 20 modified.
VLAN 1 modified.
VLAN  Mod/Ports
---- -----------------------
20    3/1
```

Switch B Configuration. The following is the Switch B configuration example:

```
SwitchB> (enable) set port name 1/1 SwitchA Link

Port 1/1 name set.
SwitchB> (enable) set vlan 5 1/1
```

```
VLAN 5 modified.
VLAN 1 modified.
VLAN  Mod/Ports
---- -----------------------
5     1/1

SwitchB> (enable) set port name 3/1 Source S1

Port 3/1 name set.
SwitchB> (enable) set vlan 10 3/1

VLAN 10 modified.
VLAN 1 modified.
VLAN  Mod/Ports
---- -----------------------
10    3/1
```

Switch C Configuration. The following is the Switch C configuration example:

```
SwitchC> (enable) set port name 1/1 SwitchA Link

Port 1/1 name set.
SwitchC> (enable) set vlan 5 1/1

VLAN 5 modified.
VLAN 1 modified.
VLAN  Mod/Ports
---- -----------------------
5     1/1

SwitchC> (enable) set port name 3/1 Destination D1

Port 3/1 name set.
SwitchC> (enable) set vlan 30 3/1

VLAN 30 modified.
VLAN 1 modified.
VLAN  Mod/Ports
---- -----------------------
30    3/1

SwitchC> (enable) set port name 4/1 Source S2

Port 4/1 name set.
SwitchC> (enable) set vlan 30 4/1

VLAN 30 modified.
VLAN 1 modified.
VLAN  Mod/Ports
---- -----------------------
30    3/1
      4/1
```

This example illustrates a complete IP MLS configuration for an MLS-SE and MLS-RP. As you can see, the steps are not complicated. What remains to be seen is how to monitor an MLS configuration.

Verifying MLS Configuration

We begin by analyzing the most frequently used commands to verify the MLS configuration on an MLS-RP. These commands are all variations of the **show mls rp** privileged-mode command. (This command is not available on a Catalyst 6000 MSFC.) Example 7-1 displays the output of the **show mls rp** command.

Example 7-1 Verifying the MLS Configuration on the MLS-RP with the show mls rp Command

```
Router#show mls rp
Multilayer switching is globally enabled
mls id is 0010.f6b3.d000       <------- This MAC address appears in the MLS cache
mls ip address 172.16.1.142           <------- The IP Address given to the MLS-SE
mls flow mask is destination-ip
number of domains configured for mls 1
vlan domain name: bcmsn             <------- The domain name must match the MLS-SE
  current flow mask: destination-ip
  current sequence number: 779898001
  current/maximum retry count: 0/10
  current domain state: no-change
  current/next global purge: false/false
  current/next purge count: 0/0
  domain uptime: 00:21:40
  keepalive timer expires in 6 seconds
  retry timer not running
  change timer not running
1 management interface(s) currently defined:
vlan 1 on Vlan1                         <------- The interface sending MLSP messages
  2 mac-vlan(s) configured for multi-layer switching:
        mac 0010.f6b3.d000
vlan id(s)
 1 41 42
router currently aware of following 0 switch(es):  <------- The number of switches
  for which the MLS-RP is routing
```

The **show mls rp** command displays the following information:

- Whether MLS switching is globally enabled or disabled
- The MLS ID for this MLS-RP
- The MLS IP address for this MLS-RP

- The MLS flow mask
- The name of the VTP domain in which the MLS-RP interfaces reside
- Statistical information for each VTP domain
- The number of management interfaces defined for the MLS-RP
- The number of VLANs configured for MLS
- The ID of each VLAN configured for this MAC address
- The number of MLS-SEs to which the MLS-RP is connected
- The MAC address of each switch

Each MLS-RP is identified to the switch by both the MLS ID and the MLS IP address of the route processor. The MLS ID is the route processor's MAC address. The MLS-RP automatically selects the IP address of one of its interfaces from MLS-enabled interfaces and uses that IP address as its MLS IP address.

The MLS-SE uses the MLS ID in the process of populating the MLS cache.

This MLS IP address is used in the following situations:

- By the MLS-RP and the MLS-SE when sending MLS statistics to a data-collection application
- In the included MLS RP list on the switch

To display IP MLS information about a specific interface, use the command **show mls rp interface** *type number*. Example 7-2 shows the output of this command for interface VLAN 10.

Example 7-2 Displaying IP MLS Information Specific to an Interface

```
Router#show mls rp interface vlan 10
mls active on Vlan10, domain Corporate
Router#
```

The **show mls rp vtp-domain** *domain-name* command displays IP MLS information for a specific VTP domain on the MLS-RP. Example 7-3 demonstrates this command.

Example 7-3 Displaying IP MLS Information for a Particular VTP Domain

```
Router#show mls rp vtp-domain Corporate

vlan domain name: Corporate
   current flow mask: ip-flow
   current sequence number: 80709115
```

continues

Example 7-3 Displaying IP MLS Information for a Particular VTP Domain (Continued)

```
current/maximum retry count: 0/10
current domain state: no-change
current/next global purge: false/false
current/next purge count: 0/0
domain uptime: 13:07:36
keepalive timer expires in 8 seconds
retry timer not running
change timer not running
fcp subblock count = 7

1 management interface(s) currently defined:
   vlan 1 on Vlan1

7 mac-vlan(s) configured for multi-layer switching:

   mac 00e0.fefc.6000
      vlan id(s)
      1    10   91   92   93   95   100

router currently aware of following 1 switch(es):
   switch id 0010.1192.b5ff
```

On the Catalyst 6000 MSFC, the **show mls status** command displays basic MLS information, as shown in Example 7-4.

Example 7-4 Displaying IP MLS Information on a Catalyst 6000 MSFC

```
Router#show mls status
MLS global configuration status:
global mls ip:                      enabled
global mls ipx:                     enabled
global mls ip multicast:            disabled
current ip flowmask for unicast:    destination only
current ipx flowmask for unicast:   destination only
Router#
```

Next, we look at the MLS-SE commands that verify IP MLS configuration. To display MLS information on an MLS-SE, enter the command **show mls** in privileged mode. The display that results from this command is shown in Example 7-5.

Example 7-5 Displaying IP MLS Information on an MLS-SE

```
Console> (enable) show mls

Multilayer switching enabled
Multilayer switching aging time = 256 seconds
Multilayer switching fast aging time = 0 seconds, packet threshold = 1
Destination-ip flow
Total packets switched = 101892
Active entries = 2153
Netflow data export enabled
Netflow data export configured for port 8010 on host 10.0.2.15
Total packets exported = 20

MLS-RP IP   MLS-RP ID       Xtag   MLS-RP            MAC-Vlans
----------- -----------     ----   ---------------------
172.20.25.2 0000808cece0    2      00-00-80-8c-ec-e0 1-20
                                   00-00-80-8c-ec-e1 21-30
                                   00-00-80-8c-ec-e2 31-40
                                   00-00-80-8c-ec-e3 41-50
                                   00-00-80-8c-ec-e4 51-60

172.20.27.1 0000808c1214    3      00-00-80-8c-12-14 1-20,31-40
                                   00-00-80-8c-12-15 21-30
                                   00-00-80-8c-12-16 41-50
```

The **show mls** output provides the following information:

- Whether MLS is enabled on the switch
- The aging time, in seconds, for an MLS cache entry
- The fast aging time, in seconds, and the packet threshold for a flow
- The flow mask

- Total packets switched
- The number of active MLS entries in the cache
- Whether NetFlow data export is enabled and, if so, for which port and host
- The MLS-RP IP address, MAC address, XTAG, and supported VLANs

To display information about a specific MLS-RP, enter the **show mls rp** [*ip_addr*] command. Note the difference between the **show mls rp** command for the MLS-SE (refer to Example 7-1) and the **show mls rp** command on the MLS-RP (see Example 7-6). The MLS-SE command requires a specific MLS-RP IP address. It displays information used by the switch with regard to Layer 3 switched packets, as shown in Example 7-6.

Example 7-6 Displaying IP MLS Information on an MLS-SE Specific to an NLS-RP

```
Console> (enable) show mls rp 166.122.20.2

MLS-RP IP   MLS-RP ID        Xtag   MLS-RP MAC-Vlans
----------- -----------      ----   ---------------------
166.122.20.2 0000808cece0     2      00-00-80-8c-ec-e0 1-20
                                    00-00-80-8c-ec-e1 21-30
                                    00-00-80-8c-ec-e2 31-40
                                    00-00-80-8c-ec-e3 41-50
                                    00-00-80-8c-ec-e4 51-60
```

Arguably, the most useful **show** commands for MLS are those that display the MLS cache entries or that use the **show mls entry** command or one of the four variations on this command. To display all MLS entries on the switch, simply enter the command **show mls entry**, as shown in Example 7-7. The flow mask is **ip-flow.** An extended access list is configured on one of the MLS-RP interfaces.

Example 7-7 Displaying All IP MLS Cache Entries on an MLS-SE

```
Console> (enable) show mls entry

                 Last Used          Last    Used
Destination IP  Source IP        Port DstPrt SrcPrt Destination Mac   Vlan Port
--------------  --------------   ---- ------ ------ ----------------- ---- -----
MLS-RP 10.20.6.161:
```

Example 7-7 Displaying All IP MLS Cache Entries on an MLS-SE (Continued)

```
10.19.6.2       10.19.26.9      UDP  6009   69     00-10-0b-16-98-00 250  1/1-2
10.19.26.9      10.19.6.2       UDP  6002   69     00-00-00-00-00-09 26   4/7
MLS-RP 132.68.9.10:
10.19.86.12     10.19.85.7      TCP  6007   SMTP   00-00-00-00-00-12 86   4/10
10.19.85.7      10.19.86.12     TCP  6012   WWW    00-00-00-00-00-07 85   4/5
MLS-RP 10.20.6.82:
10.19.63.13     10.19.73.14     TCP  6014   Telnet 00-00-00-00-00-13 63   4/11
10.19.73.14     10.19.63.13     TCP  6013   FTP    00-00-00-00-00-14 73   4/12
```

You can narrow down the **show mls entry** command output to get more-specific information by using four different keywords to show MLS cache entries that are specific to that keyword:

- For a specific destination IP address, use the command **show mls entry destination** *ip_addr_spec,* as shown in Example 7-8. *ip_addr_spec* can be a full IP address or a subnet address in the format *ip_subnet_addr*, *ip_addr/subnet_mask*, or *ip_addr/subnet_mask_bits*.

Example 7-8 Displaying IP MLS Entries Based on the Destination IP Address

```
Console> (enable) show mls entry destination 172.20.22.14/24

Destination IP  Source IP       Port DstPrt SrcPrt Destination Mac   Vlan Port
--------------- --------------- ---- ------ ------ ----------------- ---- ----
MLS-RP 172.20.25.1:
172.20.22.14    172.20.25.10    TCP  6001   Telnet 00-60-70-6c-fc-22 4    2/1
MLS-RP 172.20.27.1:
172.20.22.16    172.20.27.139   TCP  6008   Telnet 00-60-70-6c-fc-24 4    2/3
```

- For a specific source IP address, use the command **show mls entry source** *ip_addr_spec*, as shown in Example 7-9. *ip_addr_spec* includes the syntax option of simply the source IP address.

Example 7-9 Displaying IP MLS Entries Based on the Source IP Address

```
Console> (enable) show mls entry source 10.0.2.15

Destination IP  Source IP       Port DstPrt SrcPrt Destination Mac   Vlan Port
--------------- --------------- ---- ------ ------ ----------------- ---- ----
MLS-RP 51.0.0.3:
51.0.0.2        10.0.2.15       TCP  Telnet 37819  00-e0-4f-15-49-ff 51   1/9
51.0.0.2        10.0.2.15       ICMP               00-e0-4f-15-49-ff 51   1/9
```

- For a specific MLS-RP ID, use the command **show mls entry rp** *ip-address*, as shown in Example 7-10.

Example 7-10 Displaying IP MLS Entries for a Particular MLS-RP

```
Console> (enable) show mls entry rp 172.20.27.1

Destination IP  Source IP       Port DstPrt SrcPrt Destination Mac   Vlan Port
--------------- --------------- ---- ------ ------ ----------------- ---- -----
MLS-RP 172.20.27.1:
172.20.22.16    172.20.27.139   TCP  DNS    DNS    00-60-70-6c-fc-24 4    2/3
172.20.21.17    172.20.27.138   TCP  7001   7003   00-60-70-6c-fc-25 3    2/4
```

- For a specific IP flow, use the command **show mls entry flow** *protocol source-port destination-port*, as shown in Example 7-11.

Example 7-11 Displaying IP MLS Entries for a Particular Flow

```
Console> (enable) show mls entry flow tcp 23 37819

Destination IP  Source IP       Port DstPrt SrcPrt Destination Mac   Vlan Port
--------------- --------------- ---- ------ ------ ----------------- ---- -----
MLS-RP 51.0.0.3:
10.0.2.15       51.0.0.2        TCP  37819  Telnet 08-00-20-7a-07-75 10   3/1
```

Now you have seen all the major **show** commands for both an MLS-RP and an MLS-SE. Finally, just as you would occasionally want to clear ARP cache entries, content-addressable memory (CAM), or the routing table, sometimes you need to remove entries from the MLS cache. To do so, enter the **clear mls entry** command. The full syntax for this command when used for IP flows is **clear mls entry ip** [**destination** *ip_addr_spec*] [**source** *ip_addr_spec*] [**flow** *protocol src_port dst_port*] [**all**]. The **all** keyword clears all MLS entries. Other than the **all** keyword option, the syntax parallels that of the **show mls entry** command. Example 7-12 shows the output of a prescriptive use of the **clear mls entry ip** command.

Example 7-12 Clearing IP MLS Entries for a Particular Flow

```
Console> (enable) clear mls entry destination 172.20.26.22 source 172.20.22.113
  flow tcp 1652 23


MLS IP entry cleared
```

This completes our discussion of MLS. The remainder of this chapter is devoted to Cisco Express Forwarding (CEF), the multilayer switching technology that is replacing MLS on Cisco products.

Cisco Express Forwarding

Cisco Express Forwarding (CEF) is a multilayer-switching technology that allows for increased scalability and performance to meet the requirements of large enterprise networks. CEF has evolved to accommodate the traffic patterns realized by modern networks. These networks are characterized by an increasing number of short-duration flows. Shorter flows are common in environments with a high degree of web-based or other highly interactive types of traffic.

As these types of applications continue to grow in popularity, a higher-performing and more scalable forwarding methodology is required to meet the needs of these environments. The following sections compare and contrast MLS and CEF and provide a detailed description of CEF operation.

CEF Versus MLS

MLS is a flow-based model that entails software path handling of the first packet in a given flow. This, in turn, creates a cache entry that can be used by subsequent packets in the flow for accelerated performance. Although this model works extremely well, it has been shown that in environments such as those found in the core of large enterprise networks, this model might not always prove optimal. Specifically, three areas have been identified in which a flow-based model is suboptimal:

- Anytime a large number of routers are controlled by a large number of independent organizations, it is possible to encounter situations in which some devices negatively affect the stability of other devices. In this case (a peering relationship), router instability or flapping leads to the purging of previously cached data. In a flow-based model, the loss of cache entries requires that the first packet of a given flow be handled within the software path. In a peering situation, such as that required to support the core of an enterprise network, this means that potentially tens of thousands of flows would need to be relearned or cached by the software path to enable further hardware acceleration. However, it is difficult to scale software-based forwarding to simultaneously handle tens of thousands of concurrent flows per second. So stability, and the ability to withstand route flaps, is critical. The CEF-based forwarding model is designed to handle these situations gracefully and to minimize the impact of these events on network stability.

- Another area in which a flow-based model encounters scalability challenges is with respect to the actual number of flows that could be cached in the necessary lookup tables. Recall that a flow entry can be expressed in one of several ways, including **destination-ip**, **source-destination-ip**, and **ip-flow**. The greater the granularity selected, the more specific the flow information will be, but the lower the access and storage efficiency. For example, multiple entries are likely to map to a given entry for a **destination-ip** flow (for example, many clients all accessing a single server). When **ip-flow** mode is selected, full protocol and port information is stored in addition to the full source and destination IP addresses. This, in turn, reduces the likelihood that any two entries will be able to share a given flow entry. Although in a large percentage of the enterprise networks in existence today this is not cause for concern, as networks grow and the number of flows that must be supported increases (such as in a service-provider environment), this can lead to a situation in which the number of entries exceeds the available flow cache table space. This leads to suboptimal packet handling, in that packets not matched in the flow cache must be processed in software. One option is to change the flow mask to specify a less-specific mask, such as **destination-ip**. This increases cache efficiency, but it too has its limitations, beyond which software processing will be required.

NOTE

Recall that the flow cache on MLS-SE devices is limited to 128,000 flow entries. Although, in theory, as many as 128,000 entries can be supported in the MLS flow cache table hardware, in practice (based on the efficiency of hashing algorithms and the flow mask in use) this number ranges from 32,000 to 120,000 entries.

Although the simple answer to the problem of exceeding the size of the flow cache would appear to be increasing the number of possible entries available, in practice it is not so simple. There is a direct relationship between the speed of hardware-based forwarding and the size of the flow cache. Simply increasing the size of the flow cache does not guarantee that the appropriate lookup engine can parse the table sufficiently fast enough to forward flows in hardware at high speed. It would then seem logical to increase the speed of the lookup engine to work against the increased flow cache table. To a certain extent this approach could be made to work, but network administrators would have to continually upgrade equipment, an extremely costly and inefficient solution.

- A cache-based forwarding model relies on actual traffic flows in order to establish cache entries for expedited forwarding. In a classical enterprise environment, this model has proven highly effective, because traffic patterns associate a large number of sources with a much smaller number of destinations. An example is a traditional centralized server farm, in which a large number of clients send traffic to and receive traffic from a much smaller number of servers. When you compare these traffic patterns to those of the Internet, it is far less likely to achieve the many-to-one ratio typified by the enterprise data center. Because the Internet is a global entity, it becomes increasingly difficult to maintain the flow cache and to ensure the integrity of the data therein. This means that a certain amount of cache flux must be expected in terms of the maintenance required to keep the state of the flow cache current. This is reflected in terms of CPU utilization.

With a CEF-based mechanism, the challenges presented by flow caching models are greatly reduced, and hence scalability is greatly increased. With flow-cached models, a complete forwarding table must be maintained for the proper handling of first packets by the CPU or packets that for some other reason are incapable of being processed in hardware. After the CPU makes a forwarding decision (part of which involves a lookup against the routing table), an entry is made in the flow cache table, and subsequent packets are handled in hardware.

With a CEF-based forwarding model, all packets, including the first packet in a given flow, are handled in hardware. A routing table is still maintained by the router CPU, but two additional tables are created in the CEF-based model. These tables are populated before any actual user traffic is present in the network, such as would be the case with a cache-based model. The first of these tables is a copy of the forwarding information from the routing table. It is called the *Forwarding Information Base (FIB)* table. The FIB table contains the minimum information from the routing table necessary to forward packets; in particular, it does not contain any routing protocol information. The second table is called the *adjacency table*. The adjacency table maintains a database of node adjacencies (two nodes are said to be adjacent if they can reach each other via a single Layer 2 hop) and their associated Layer 2 MAC rewrite or next-hop information. The third table is the *NetFlow table*, which provides network accounting data. The NetFlow table is updated in parallel with the CEF-based forwarding mechanism provided by the FIB and adjacency tables.

By performing a high-speed lookup against the FIB and adjacency, a forwarding decision along with the appropriate rewrite information can be accessed in a highly efficient and consistent manner, while also providing a mechanism offering a high degree of scalability.

Although high-end Cisco routers have offered support for CEF for several years now, some differences between the CEF implementation on Catalyst switches and Cisco routers should be pointed out. Specifically, both implementations provide the following:

- High-speed forwarding based on a "longest address match" lookup
- Equal-cost path load balancing
- Reverse Path Forwarding (RPF) checks (this is discussed in Chapter 9)
- Invalidation of less-specific routes

The CEF implementation on the Catalyst 6000 family also adds the following:

- Support for IP multicast
- Support for IPX forwarding
- Hardware-based forwarding of rates up to 210 Mpps in distributed CEF mode

Unsupported features of the CEF implementation on the Catalyst switches include the following:

- Turning off CEF-based forwarding
- CEF per-prefix and per-prefix length accounting
- Accounting data from load-sharing paths
- Per-packet load balancing

CEF is supported on Catalyst 8500 switch routers, Catalyst 3550 switches, Catalyst 2948G-L3 switches, Catalyst 4000 switches, and Catalyst 6000 switches. Cisco routers running Cisco IOS Software Release 12.2 or later also support CEF. Some routers, such as the 7500 series, supported CEF beginning with Cisco IOS Software Release 12.0.

The Catalyst 4000 switch supports CEF in the following combinations: with the Layer 3 Services Module, with the Access Gateway Module, and with Supervisor Engine III.

The remainder of this chapter focuses on CEF technology in the context of Catalyst 6000 family switches. In particular, the Catalyst 4000 Supervisor Engine III multilayer-switching mechanism does rely on CEF, but the architecture and Layer 2/3/4 packet flow methodology are similar, but not identical, to that of the Catalyst 6000.

TECH NOTE: CATALYST 4000 SUPERVISOR ENGINE III ARCHITECTURE

Supervisor Engine III is based on a low-latency, centralized, shared-memory switching fabric architecture that delivers leading-edge capabilities and eliminates any possibility of head-of-line blocking. This architecture delivers wire-speed Layer 2/3/4 switching on all ports. For every incoming packet, the packet data is written to the shared memory, and a packet descriptor is generated. Only the packet descriptor passes through a series of intermediate processing and modification points, while the original packet data resides in the packet buffer memory.

Supervisor III has an integrated switching engine that is implemented by two main ASICs—the Packet Processing Engine (PPE) and the Fast Forwarding Engine (FFE), as shown in Figure 7-7.

The PPE manages the switch's shared-packet buffer memory. It handles packet header parsing, inline rewrite of MAC headers, and VLAN tagging and untagging. It also performs output queuing of data packets. PPE buffers packets only once—the Supervisor III never copies the data portion of the packet. The PPE has 32 nonblocking gigabit ports (the switch can handle independent packets simultaneously between these ports) and provides a 6 Gbps connection to each of the five slots. This 6 Gbps line card connection can be further divided into blocking 10/100 or 10/100/1000 ports.

The FFE makes the Layer 2/3/4 forwarding decision while handling ACL and QoS processing. The FFE looks only at the packet descriptors and also implements an interface to the CPU.

The PPE and FFE ASICs work in conjunction with each other to achieve nonblocking and line-rate Layer 2/3/4 switching for 32 Gigabit ports or 240 10/100/1000 ports. Both ASICs have on-chip memory as well as external memory to save the state information required for switching the packets.

On the software side, the Supervisor Engine III is the first chassis-based Catalyst switching engine that does not have a CatOS option. The only OS available for the Supervisor Engine III is the Cat IOS, similar to the Catalyst 3550 OS and Native mode on a Catalyst 6000. This likely reflects a trend for future Catalyst devices.

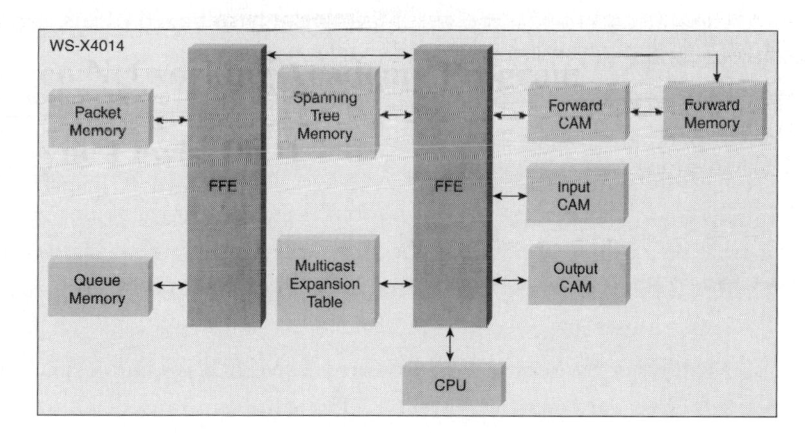

Figure 7-7 The Catalyst 4000 Supervisor Engine III integrated switching engine delivers state-of-the-art MLS.

For Catalyst 6000 switches, the CEF hardware requirements are Supervisor Engine 2, MSFC2, and PFC2.

CEF Operation

Understanding how CEF functions on a Catalyst 6000 Supervisor Engine 2 requires a basic understanding of the architecture of the Supervisor Engine 2. Figure 7-8 details the architecture.

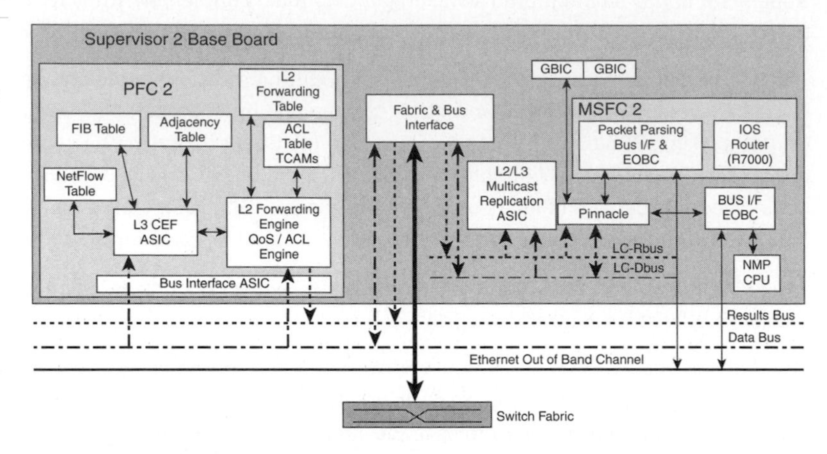

Figure 7-8
The Catalyst 6000 Supervisor Engine 2 with MSFC2 and PFC2 supports CEF technology.

Supervisor Engine 2 consists of three primary components:

- **Supervisor base board**—The Supervisor base board provides the connectivity to the switching fabric, 32-Gbps bus, multicast replication ASIC, Network Management Processor (NMP), and two Gigabit Interface Converter (GBIC)-based Gigabit Ethernet ports. In addition, the base board offers connectors for attaching the PFC2 and MSFC2 daughter boards.
- **Policy Feature Card 2 (PFC2)**—PFC2 is a factory-installed daughter card of a Supervisor Engine 2. The PFC2 provides an array of ASICs to enable all hardware-based forwarding. The Layer 3 forwarding engine ASIC provides the actual CEF function via its inherent logic. It also provides access to the various Layer 3 tables (the FIB, adjacency table, and NetFlow table). In addition, the PFC2 daughter card also contains the ASICs that deliver hardware-based access control lists and QoS mechanisms.
- **Multilayer Switching Feature Card 2 (MSFC2)**—The MSFC2 is an optional daughter card that provides a CPU for the handling of all Layer 3 control-plane activities. The *control plane* is the portion of the hardware architecture that handles route calculations. The MSFC2 is responsible for handling any functions that cannot be handled in the hardware elements of the PFC2, as well as the processing of all routing protocol activities such as OSPF and BGP routing updates. The MSFC2 is also responsible for populating the IP routing table, FIB table, and adjacency table.

Of these three elements, the Supervisor base board and PFC2 are mandatory components, and the MSFC2 element is optional (although it is effectively required for Layer 3 switching because it contains the CPU for the population of the CEF tables).

The CPU on the MSFC2 daughter card runs all instances of whatever configured routing protocols are required. In addition, it handles packets that cannot be processed in hardware. The MSFC2 is a Cisco IOS Software-based router, and it is configured as such. With the Supervisor Engine 2, CEF is enabled by default. In fact, it cannot be disabled. For most common CEF functions, no specific configuration is required to enable CEF, beyond the standard configuration of routing protocols, network interface addressing, and so on.

When the router is initialized, a routing table is constructed based on information in the router software configuration (such as static routes, directly connected routes, and dynamically learned routes via routing protocol exchanges). After the routing table is constructed, the CPU creates the FIB and adjacency tables automatically. The FIB and adjacency tables represent the data present within the routing table in a manner in which optimal forwarding can be performed.

Unlike a flow cache, which is based on traffic flow, the CEF table is based on the network topology. When a packet enters the switch, the switch's Layer 3 forwarding engine ASIC performs a longest-match lookup based on the destination network and the most-specific netmask. For example, instead of switching based on a destination address of 172.34.10.3, the PFC-2 looks for the network 172.34.10.0/24 and switches to the interface connecting to that network. This scheme is highly efficient and does not involve the software for anything other than the routing table and prepopulation of the FIB table. In addition, cache invalidation because of a route flap does not occur; as soon as a change in the routing table is made, the CEF is updated immediately. This makes the CEF table more adaptable to changes in the network topology.

The next section describes the role of the FIB in more detail.

Forwarding Information Base

The FIB table consists of a four-level hierarchical tree, as shown in Figure 7-9. The four levels are derived from the fact that IPv4 uses a 32-bit address: Each level of the hierarchy is based on 8 of the possible 32 bits. CEF relies on a longest-match forwarding algorithm, meaning that the tree is searched in descending order until the "longest match," or greatest number of bits, is matched. The FIB tree is represented hierarchically, with the least-specific address at the top of the tree and the most-specific address at the bottom. Each leaf is based on an 8-bit boundary, with more-specific entries in descending order. This tree (commonly known as a 256-way radix tree) provides a highly efficient mechanism for rapid lookups. This ensures that minimal latency is incurred during the lookup process. It also provides for a highly scalable architecture in that IPv4 addressing can be completely accounted for with a minimal trade-off of performance versus table efficiency. Each leaf offers a pointer to the appropriate next-hop entry in the adjacency table.

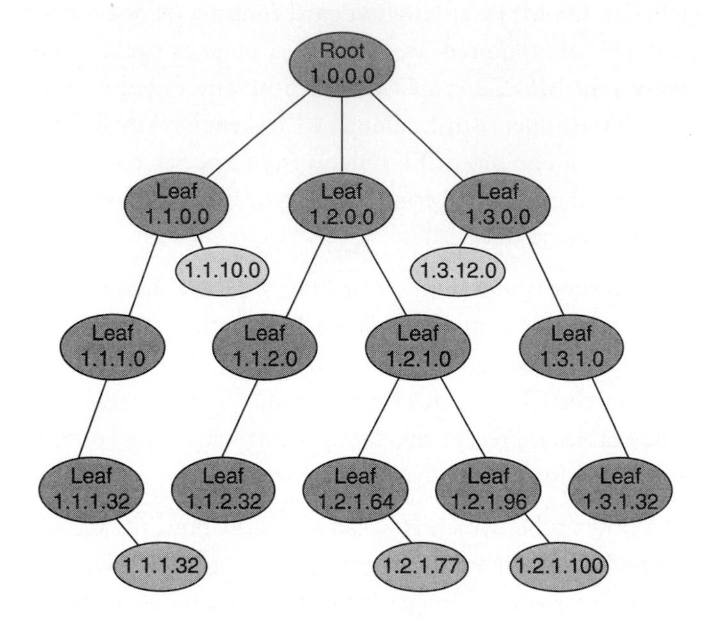

Figure 7-9
The FIB table is a four-level hierarchical tree with 256 branch options per level.

The CPU of the MSFC2 builds the FIB table from the IP routing table. The appropriate IP routing protocols first resolve the IP routing table, at which point the CEF process is invoked and the corresponding FIB and adjacency tables are constructed. The FIB table is maintained separately from the routing table in that the routing table contains additional information that is important for the purposes of routing protocols, but not for the actual forwarding of packets. The routing protocol that created the route cannot be determined by the FIB. The FIB and adjacency tables are optimized to provide only the information that is necessary for a forwarding decision to be made, and nothing more. It is also important to note that recursive routes are resolved when the FIB table is built. This removes what has classically been a CPU-intensive event in non-CEF implementations.

If the FIB table becomes full, subsequent entries are compared to the existing entries, and the more-specific entries are maintained at the expense of less-specific entries. This method is necessary to ensure that proper forwarding is maintained. For example, the less-specific route of 1.2.0.0 is not specific enough to ensure proper routing to 1.2.1.0. You can view the contents of the FIB table via the command **show ip cef detail** from the MSFC2. An example of the output of the **show ip cef detail** command is shown in Example 7-13.

Example 7-13 Catalyst 6000 MSFC2 show ip cef detail Output Displays the Contents of the FIB Table

```
MSFC2#show ip cef detail
IP CEF with switching (Table Version 13), flags=0x0
11 routes, 0 reresolved, 0 unresolved (0 old, 0 new)
14 leaves, 16 nodes, 18376 bytes, 16 inserts, 2 invalidations
0 load sharing elements, 0 bytes, 0 references
2 CEF resets, 0 revisions of existing leaves
refcounts: 2083 leaf, 2076 node
Adjacency Table has 2 adjacencies
0.0.0.0/32, version 0, receive
10.0.0.0/24, version 11, attached, connected
0 packets, 0 bytes
via Vlan1, 0 dependencies
valid glean adjacency
10.0.0.0/32, version 9, receive
10.0.0.1/32, version 12, connected, cached adjacency 10.0.0.1
0 packets, 0 bytes
via 10.0.0.1, Vlan1, 0 dependencies
next hop 10.0.0.1, Vlan1
valid cached adjacency
10.0.0.5/32, version 8, receive
10.0.0.255/32, version 10, receive
127.0.0.0/8, version 6, attached, connected
0 packets, 0 bytes
via EOBC0/0, 0 dependencies
valid glean adjacency
127.0.0.0/32, version 4, receive
127.0.0.11/32, version 7, connected, cached adjacency 127.0.0.11
0 packets, 0 bytes
via 127.0.0.11, EOBC0/0, 0 dependencies
next hop 127.0.0.11, EOBC0/0
valid cached adjacency
```

continues

Example 7-13 Catalyst 6000 MSFC2 show ip cef detail Output Displays the Contents of the FIB Table (Continued)

```
127.0.0.12/32, version 3, receive
127.255.255.255/32, version 5, receive
224.0.0.0/4, version 2
0 packets, 0 bytes, Precedence routine (0)
via 0.0.0.0, 0 dependencies
next hop 0.0.0.0
valid drop adjacency
224.0.0.0/24, version 2, receive
```

The second table used by CEF is the adjacency table, which is discussed in the next section.

Adjacency Table

The adjacency table contents are fundamentally a function of the ARP process, whereby Layer 2 addresses are mapped to corresponding Layer 3 addresses. When the router issues an ARP request, a corresponding reply is received, and a host entry is added to the adjacency table to reflect this. In addition, the router can also glean next-hop routers from routing updates and make entries in the adjacency table to reflect this. This lets the router build the next-hop rewrite information necessary for Layer 3 packet forwarding. By having this data already stored in a table, CEF can perform highly efficient and consistent forwarding, because no discovery process is required. You can use the command **show ip cef** to view the contents of the CEF adjacency table from the MSFC2. The command **show ip cef summary** gives you a brief overview of the CEF process. It shows information such as the total number of adjacencies and routes. Sample output for both these commands is displayed in Example 7-14.

Example 7-14 Output of the show ip cef and show ip cef summary Commands Displays Adjacency Table Information

```
MSFC2#show ip cef summary
IP CEF with switching (Table Version 17498), flags=0x0, 2676 routes, 0
reresolve, 0 unresolved (0 old, 0 new)
2679 leaves, 126 nodes, 463236 bytes, 17501 inserts, 14822 invalidations, 0
load sharing elements, 0 bytes, 0 references 2 CEF resets, 0 revisions of
existing leaves, refcounts: 33678 leaf, 28486 node
adjacency Table has 2522 adjacencies
MSFC2#show ip cef
```

Example 7-14 Output of the show ip cef and show ip cef summary Commands Displays Adjacency Table Information (Continued)

```
Prefix              Next Hop            Interface
0.0.0.0/32          receive
11.0.0.0/8          attached            Vlan11
11.0.0.0/32         receive
11.0.0.1/32         receive
11.0.0.2/32         11.0.0.2            Vlan11
11.0.0.4/32         receive
11.255.255.255/32   receive
12.0.0.0/8          attached            Vlan12
12.0.0.0/32         receive
12.0.0.1/32         receive
12.0.0.11/32        12.0.0.11           Vlan12
12.0.0.12/32        12.0.0.12           Vlan12
12.0.0.13/32        12.0.0.13           Vlan12
12.0.0.14/32        12.0.0.14           Vlan12
12.0.0.15/32        12.0.0.15           Vlan12
12.0.0.16/32        12.0.0.16           Vlan12
12.0.0.17/32        12.0.0.17           Vlan12
12.0.0.18/32        12.0.0.18           Vlan12
12.0.0.19/32        12.0.0.19           Vlan12
12.0.0.20/32        12.0.0.20           Vlan12
12.0.0.21/32        12.0.0.21           Vlan12
12.0.0.22/32        12.0.0.22           Vlan12
<output omitted>
```

The third table used by CEF is the NetFlow table. Because this table compiles network accounting data and does not play a role in CEF's PDU header rewrite mechanism, it isn't discussed in this book.

Packet Flow for Layer 2 and Layer 3 Forwarding Decisions

This section details the data flow through the Layer 2 and Layer 3 forwarding process on a Catalyst 6000 Supervisor Engine 2. The process is fundamentally unchanged between a Layer 2 and Layer 3 decision; it is a function of whether the Layer 3 data exists. In other words, if the MSFC2 is present, the Layer 3 path is valid; otherwise, forwarding decisions are made based purely on Layer 2 forwarding information. This discussion details the data path for Layer 2 and Layer 3 lookups.

Figure 7-10 details the CEF forwarding process. The steps are as follows:

Step 1 The Layer 3 forwarding engine and Layer 2 forwarding engine ASICs receive the packet headers from the data bus.

Step 2 The following operations occur in parallel:

- The Layer 2 forwarding engine performs a lookup on the destination MAC address.
- The Layer 2 forwarding engine ASIC performs any necessary input security ACL lookups.
- The Layer 2 forwarding engine ASIC performs any necessary input QoS/ACL lookups.
- The Layer 3 forwarding engine ASIC performs a FIB table lookup.
- The Layer 3 forwarding engine ASIC performs a NetFlow table lookup.

Step 3 The following operations occur in parallel:

- The input ACL and QoS results from the Layer 2 forwarding engine ASIC are forwarded to the Layer 3 forwarding engine ASIC.
- The Layer 3 forwarding engine ASIC sends the destination VLAN information for the packet to the Layer 2 forwarding engine ASIC.

Step 4 The following operations occur in parallel:

- The Layer 3 forwarding engine ASIC performs the adjacency lookup.
- The Layer 2 forwarding engine ASIC performs the outbound security ACL lookup.
- The Layer 2 forwarding engine ASIC performs the outbound QoS ACL lookup.

Step 5 The Layer 2 forwarding engine sends the results of the security and QoS ACL lookups to the Layer 3 forwarding engine ASIC.

Step 6 The following operations occur in parallel:

- The Layer 3 forwarding engine ASIC generates the rewrite result and sends it to the Layer 2 forwarding engine ASIC.
- The Layer 3 forwarding engine ASIC updates the adjacency table statistics as necessary.
- The Layer 3 forwarding engine ASIC updates the NetFlow table statistics as necessary.

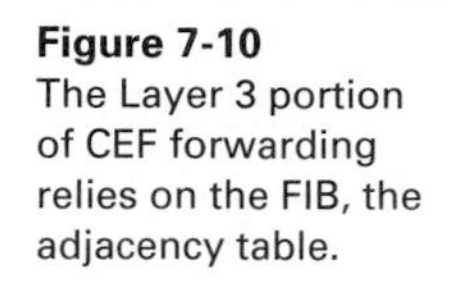

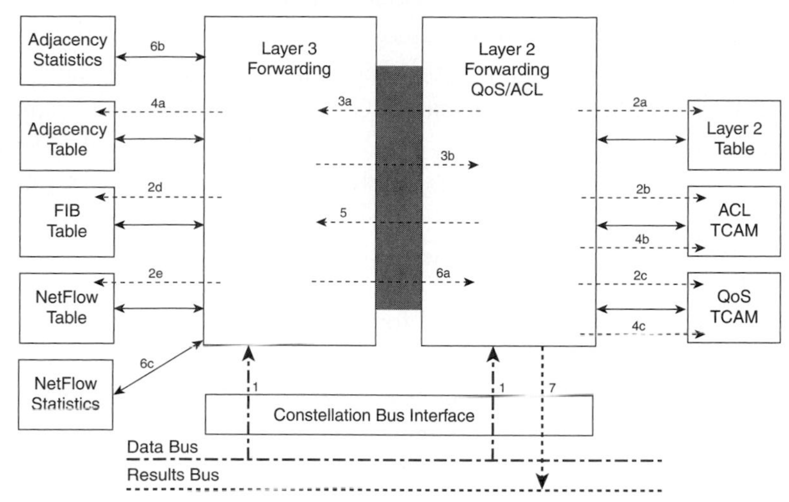

Figure 7-10 The Layer 3 portion of CEF forwarding relies on the FIB, the adjacency table.

The Layer 2 forwarding engine looks up the destination MAC address received from the Layer 3 forwarding engine ASIC. It then chooses between a Layer 2 and a Layer 3 result and sends the result onto the results bus, an out-of-band control-plane mechanism used for this purpose.

Figure 7-11 summarizes the CEF process with a traffic flow from a source workstation (on the right side of the diagram) to a destination workstation (on the left side of the diagram). The critical elements of the diagram are the FIB and adjacency entries of the intermediate routers along the path. All the IP addresses and MAC addresses referenced by the FIB and adjacency tables are indicated as the packet traverses the network.

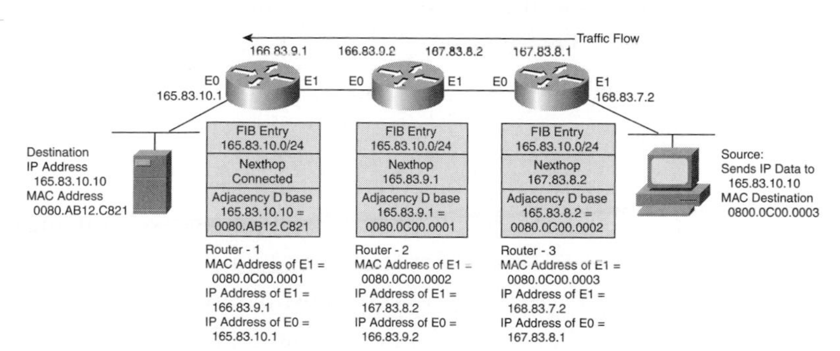

Figure 7-11 The Layer 3 portion of CEF forwarding relies on the FIB, the adjacency table, and the NetFlow table.

Additional Benefits of CEF-Based Forwarding

In addition to the benefits described in the section, "CEF Versus MLS," CEF-based forwarding provides several other benefits. These benefits include enhanced scalability, network stability, load balancing, ACL processing, and multicasting. Many of these benefits are supported only on the Catalyst 6500 series (a subset of the Catalyst 6000 family). The following sections briefly describe each of these additional benefits.

Scalability

CEF increases scalability not only by the raw number of available FIB entries, but also by the replication of CEF-based forwarding technology on a per-slot basis. This "distributed" capability (dCEF) allows the Catalyst 6500 series to provide sufficient forwarding capabilities for the largest of networks. The Catalyst 6500 series, when equipped with a Supervisor Engine 2, can provide as many as 256,000 entries in the FIB table and 256,000 entries in the adjacency table. These entries are not stored via a traditional hashing algorithm, as is the case with MLS.

Availability

Although the Catalyst 6000 family offers a breadth of features to support high availability, the addition of CEF further increases the availability and stability demanded in today's enterprise networks. As networks increase in size, a natural side effect is the increased chance of instability and change. Network instability, whether due to failures, configuration changes, or bursty traffic patterns, can have a tremendous impact on routing implementations, which rely on heavy CPU computations for routing-table maintenance. Because CEF employs a mechanism by which the forwarding information is constructed based on the network topology rather than a representation of traffic within the network, the CPU is no longer burdened by having to set up large numbers of entries. This also means that network availability is not directly linked to the network's actual size.

Load Balancing

CEF delivers some additional functionality with respect to load balancing of traffic across multiple equal-cost parallel paths. Traditionally, routing protocols such as OSPF and EIGRP have limited equal-cost load-balancing paths to four paths. With CEF, this has been increased to six equal-cost parallel paths. With routing protocols, more than four parallel paths might exist, but a maximum of four can be installed in the actual routing table. Routing protocols such as EIGRP and OSPF support the concept of a Routing Information Base (RIB). If a route cannot be installed into the routing table, it can be maintained in the RIB. Because CEF relies on a FIB and an adjacency table for forwarding decisions, it is not limited to what is actually present in the routing table. In fact, it can reference the RIB, which lets CEF install a greater number of parallel paths (with entries gleaned from the RIB). It is important to note that the Catalyst 6000 family offers the ability to load-balance across parallel paths on a per-flow basis only. This means that per-packet load balancing is not supported on the Catalyst 6000 family of switches when you utilize CEF-based forwarding.

Access Control Lists

Another manner in which various CEF implementations vary is with respect to their support of ACLs. In the case of the Catalyst 6500 series, ACLs can be handled in hardware for most common configurations. CEF-based forwarding does not have any impact on this functionality. The most notable ACL options that disable hardware processing for a given ACL are the use of the **log** keyword or enabling "ip unreachables" on a given interface (which enable the sending of ICMP unreachable messages).

Multicasting

The CEF implementation on the Catalyst 6000 family also includes support for IP multicast. The FIB table can hold as many as 16,000 entries for IP multicast, including both (S,G) and (*,G) entries. (S,G) and (*,G) entries, associated with dense-mode and sparse-mode Protocol-Independent Multicast, respectively, are discussed in Chapter 9.

In conclusion, CEF provides high performance and high scalability for the largest and most demanding networks. The Supervisor Engine 2 offers the scalability and performance requirements necessary for both large enterprise and service-provider environments. Cisco has also managed to maintain the rich accounting support of the MLS model (utilizing NetFlow features) while introducing CEF-based forwarding.

Finally, CEF's most-appealing feature might be its configuration. CEF for PFC2 is permanently enabled on Supervisor Engine 2 with the PFC2 and the MSFC2. No configuration is required.

Summary

MLS and CEF are Cisco's multilayer-switching technologies. CEF is newer than MLS and is supported by all the newer devices, such as the Catalyst 4000 Supervisor Engine III, the Catalyst 6000 with MSFC2 and PFC2, and Cisco routers with Cisco IOS Software Release 12.2 and later. CEF has been supported on some routers since Cisco IOS Software Release 12.0.

MLS is a flow-based switching mechanism whereby PDU header information from a candidate packet is cached for subsequent MLS-RP-independent wire-speed packet switching by the MLS-SE. MLS relies on two components, the MLS-SE and the MLS-RP, communicating header rewrite and forwarding information via MLSP. This technology was originally used on Catalyst 5000 switches, but it is also supported by Catalyst 6000 switches with Supervisor Engine 1 and MSFC1.

CEF is an improved multilayer switching mechanism that utilizes three tables: the Forwarding Information Base, the adjacency table, and the NetFlow table. The FIB and the adjacency table are responsible for the high-speed PDU header rewrites that facilitate multilayer switching. Catalyst 6000 switches with Supervisor Engine 2, MSFC2, and PFC2 use CEF by default (no configuration is required). This is also true of the Catalyst 4000 switches with Supervisor Engine III.

Check Your Understanding

Test your understanding of the concepts covered in this chapter by answering these review questions. Answers are listed in Appendix A, "Check Your Understanding Answer Key."

1. What three components does MLS use?

   A. MLS-SE

   B. MLS-RP

   C. Catalyst 5000

   D. MLSP

2. What MLS component stores the MLS cache entries?

   A. MLS-SE

   B. MLS-RP

   C. Catalyst 5000

   D. MLSP

3. What is the term for the first packet in an MLS flow prior to reaching the MLS-RP?

   A. Enable

   B. Priority

   C. Candidate

   D. Initiator

4. What is the maximum number of entries in the MLS cache?

   A. 16,000

   B. 32,000

   C. 64,000

   D. 128,000

5. What global configuration command is used on the MLS-RP to enable IP MLS?

   A. **set mls ip**

   B. **mls ip rp**

   C. **mls rp ip**

   D. **set mls enable**

6. What interface configuration command is used on all MLS-RP interfaces participating in IP MLS?

   A. **set mls ip**

   B. **mls ip rp**

   C. **mls rp ip**

   D. **set mls enable**

7. What command is used on an MLS-SE to view all the MLS cache entries based on Layer 4 information?

   A. **show mls entry rp**

   B. **show mls entry flow**

   C. **show mls entry source**

   D. **show mls entry destination**

8. What command clears IP MLS entries?

   A. **clear mls**

   B. **clear mls ip**

   C. **clear mls ip entry**

   D. **clear mls entry ip**

9. What three tables are used by CEF?

   A. Adjacency

   B. Candidate

   C. NetFlow

   D. FIB

10. What three Catalyst 6000 hardware components are required for CEF operation?

    A. Supervisor Engine 2

    B. MSFC2

    C. PFC2

    D. FIB

Key Terms

adjacency table One of three tables used by CEF. The adjacency table is a database of node adjacencies (two nodes are said to be adjacent if they can reach each other via a single Layer 2 hop) and their associated Layer 2 MAC rewrite or next-hop information. Each leaf of the FIB tree offers a pointer to the appropriate next-hop entry in the adjacency table.

aging time Determines the time before an MLS entry is aged out. The default is 256 seconds. You can configure the aging time in the range of 8 to 2032 seconds in 8-second increments.

CAM (Content-Addressable Memory) Memory that is accessed based on its contents, not on its memory address. Sometimes called associative memory.

candidate packet When a source initiates a data transfer to a destination, it sends the first packet to the MLS-RP through the MLS-SE. The MLS-SE recognizes the packet as a candidate packet for Layer 3 switching, because the MLS-SE has learned the MLS-RP's destination MAC addresses and VLANs through MLSP. The MLS-SE learns the candidate packet's Layer 3 flow information (such as the destination address, source address, and protocol port numbers) and forwards the candidate packet to the MLS-RP. A partial MLS entry for this Layer 3 flow is created in the MLS cache.

CEF (Cisco Express Forwarding) A Cisco multilayer-switching technology that allows for increased scalability and performance to meet the requirements for large enterprise networks. CEF has evolved to accommodate the traffic patterns realized by modern networks, characterized by an increasing number of short-duration flows.

central rewrite engine The Catalyst 5000 NFFC contains central rewrite engines (one per bus) to handle PDU header rewrites when inline rewrite is not an option. If central rewrite engines are used, a frame must traverse the bus twice—first with a VLAN tag for the source (on its way to the NFFC), and second with a VLAN tag for the destination (on its way to the egress port).

control plane The portion of hardware and software on a Cisco device that handles Layer 3 traffic forwarding.

enable packet When an MLS-SE receives a packet from the MLS-RP that originated as a candidate packet, it recognizes that the source MAC address belongs to the MLS-RP, that the XTAG matches that of the candidate packet, and that the packet's flow information matches the flow for which the candidate entry was created. The MLS-SE considers this packet an enable packet and completes the MLS entry created by the candidate packet in the MLS cache.

fast aging time The amount of time before the purging of MLS entries that have no more than *pkt_threshold* packets switched within fastagingtime seconds after they are created.

FIB (Forwarding Information Base) One of three tables used by CEF. The FIB table contains the minimum information necessary to forward packets; in particular, it does not contain any routing protocol information. This table consists of a four-level hierarchical tree, with 256 branch options per level (reflecting four octets in an IP address).

flow A unidirectional sequence of packets between a particular source and destination that share the same Layer 3 and Layer 4 PDU header information.

flow mask A set of criteria, based on a combination of source IP address, destination IP address, protocol, and protocol ports, that describes a flow's characteristics.

inline rewrite Some Catalyst 5000 family switching line cards have onboard hardware that performs PDU header rewrites and forwarding, maximizing IP MLS performance. When the line cards perform the PDU header rewrites, this is called inline rewrite. With inline rewrite, frames traverse the switch bus only once.

MSFC (Multilayer Switch Feature) MSFC1 and MSFC2 are daughter cards to the Catalyst 6000 Supervisor Engine that provide multilayer switching functionality and routing services between VLANs.

MLS (multilayer switching) A specific multilayer switching technology employed by various Cisco devices to perform wire-speed PDU header rewrites. The first packet in a flow is routed as normal, and subsequent packets are switched by the MLS-SE based on cached information.

MLS-RP (MLS Route Processor) A Cisco device with a route processor that supports MLS. For example, a Catalyst 3620 router is an MLS-RP.

MLS-RP management interface Sends hello messages, advertises routing changes, and announces VLANs and MAC addresses of interfaces participating in MLS.

MLS-SE (MLS Switching Engine) The set of hardware components on a Catalyst switch, excluding the route processor, that are necessary to support MLS. For example, a Catalyst 5000 with a Supervisor Engine IIIG is an MLS-SE.

MLSP (Multilayer Switching Protocol) The protocol used to communicate MLS information between the MLS-SE and MLS-RP. In particular, the MLS-SE populates its Layer 2 CAM table with updates received from MLSP packets.

multilayer switching A general network design term used to refer to hardware-based PDU header rewrites and the forwarding of PDUs based on information specific to one or more OSI layers.

NetFlow table One of three tables used by CEF. The NetFlow table provides network accounting data. It is updated in parallel with the CEF-based forwarding mechanism provided by the FIB and adjacency tables.

NFFC (NetFlow Feature Card) The daughter card for a Catalyst 5000 Supervisor module that enables intelligent network services, such as high-performance multilayer switching and accounting and traffic management.

XTAG A 1-byte value that the MLS-SE attaches to each VLAN for all MAC addresses learned from the same MLS-RP via MLSP.

After completing this chapter, you will be able to perform tasks related to the following:

- Router discovery methods
- Hot Standby Router Protocol operation
- HSRP preempt and interface tracking
- HSRP configuration
- HSRP in campus network design

Chapter 8

Hot Standby Router Protocol

Businesses and consumers relying on intranet and Internet services for their mission-critical communications require their networks and applications to be continuously available. Customers can satisfy their demands for near-100 percent network uptime by leveraging the Hot Standby Router Protocol (HSRP) in Cisco IOS Software. HSRP, unique to Cisco platforms, provides network redundancy for IP networks in a manner that ensures that user traffic will quickly and transparently recover from first-hop failures in network edge devices or access circuits.

By sharing an IP address and a MAC address, two or more routers on a LAN segment can act as a single "virtual" router. The members of the virtual router group continually exchange status messages. This way, one router can assume the routing responsibility of another should one of them go out of commission. Hosts on the LAN segment are configured with a single default gateway and continue to forward IP packets to a consistent IP and MAC address, and the changeover of devices performing the routing is transparent to the hosts. The vast majority of hosts has relatively primitive routing tables and uses the default gateway as a single next-hop IP and MAC address. HSRP provides default-gateway redundancy to hosts in lieu of their maintaining routing tables.

This chapter details the operation, configuration, and campus network design issues relating to HSRP. We begin with a discussion of the various methods available for hosts to discover routers.

Router Discovery

IP hosts need to forward packets to a default gateway so that communication can take place beyond the immediate LAN segment. *Router discovery* is the process of a host's determining a default gateway for sending data beyond the local LAN segment. Ideally, router discovery provides a mechanism for quick and efficient redundancy or failover in case a default gateway becomes unavailable (assuming that more than one candidate default gateway resides on the local segment). This situation is pictured in Figure 8-1. Without a router-discovery mechanism, the failure of a single gateway will likely isolate hosts.

> **NOTE**
> In the context of router discovery, routers with IP routing disabled are classified as *hosts* unless indicated otherwise. You disable IP routing on a Cisco router in global configuration mode with the command **no ip routing**. Catalyst switches are also classified as hosts in the context of router discovery. Catalyst switches are configured with a default gateway, as detailed in Chapter 3. The term *end station* includes workstations and servers, but does *not* include routers with IP routing disabled or Catalyst switches.

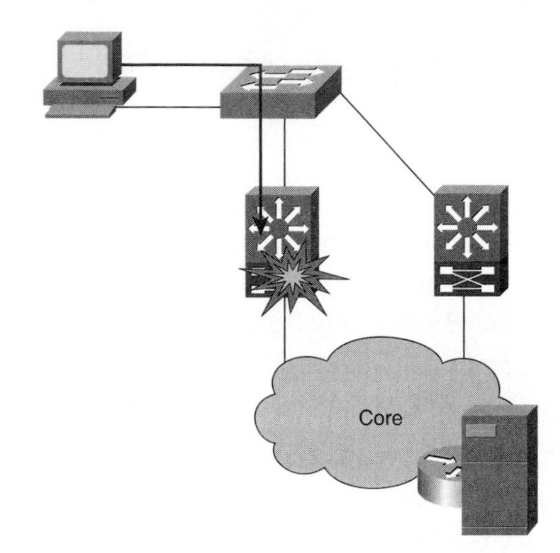

Figure 8-1
Router discovery enables hosts to fail over to an alternative default gateway.

Here are the ways in which IP hosts can discover where to forward packets:

- **Proxy ARP**—An end station can discover a router by using ARP to find the MAC addresses of end stations that are not on its directly connected LAN. A router that is configured to support Proxy ARP answers ARP requests with its own MAC address when it has a specific route for these addresses. Unfortunately, many end stations have long timeout values (or no timeout values at all) on their ARP caches. As a result, if a router becomes unavailable, the host continues trying to send traffic for these hosts to the router that originally sent the proxy ARP reply.
- **Default gateway**—This is the simplest and most frequently used method of forwarding packets. Using this method, a host is statically configured to know the IP address of its default router. However, if that router becomes unavailable, the host can no longer communicate with devices off the local LAN segment, even if another router is available.
- **ICMP Router Discovery Protocol (IRDP)**—IP end stations can use IRDP to listen to router hellos. This allows an end station to adapt to changes in network topology. However, this solution requires that all end stations have an IRDP implementation.

These three methods provide a means by which an IP host can access hosts on other networks. Each method is described in more detail in the following three sections.

Proxy ARP

One common way for hosts to access remote networks is proxy ARP. *Proxy ARP*, defined in RFC 1027, allows an Ethernet host with no knowledge of routing to communicate with hosts on other networks or subnets. Such a host assumes that all hosts are on the same local segment and that it can use ARP to determine their hardware addresses.

Under proxy ARP, if a router receives an ARP request for a host that is not on the same network as the ARP request sender, the Cisco IOS Software evaluates whether it has the best route to that host. If it does, the device sends an ARP reply packet, giving its own Ethernet hardware address. The host that sent the ARP request then sends its packets to the router, which forwards them to the intended host. The host software treats all networks as if they are local and performs ARP requests for every IP address. Proxy ARP is enabled by default on Ethernet interfaces on Cisco routers.

Proxy ARP works as long as other routers on the local segment support it. Cisco routers support proxy ARP by default on Ethernet interfaces. However, many other routers, especially servers configured as routers, do not support it.

Default Gateways

A *default gateway* is also called a default router. The default gateway provides a definitive location for IP packets to be sent in case the source device has no IP routing functionality built into it. This is the method most end stations use to access nonlocal networks. The default gateway is set manually or is learned from a DHCP server.

When IP routing is disabled on a Cisco router, the default gateway feature and the router discovery client are enabled, and proxy ARP is disabled. When IP routing is enabled, the default gateway feature is disabled, and you can configure router discovery servers.

For Cisco routers with IP routing disabled, the Cisco IOS Software sends all nonlocal packets to the default gateway (default router), which either routes them appropriately or sends back an ICMP redirect message, telling the source router of a better route. The ICMP redirect message also indicates which local router the host should use. The software caches the redirect messages and routes each packet thereafter as efficiently as possible. The limitations of this method are that there is no means of detecting when the default router has gone down or is unavailable, and there is no method of picking another device if one of these events should occur.

To set up a default gateway for a Cisco router, use the command **ip default-gateway** *ip-address* in global configuration mode. To display the address of the default gateway, use the **show ip redirects** command in privileged mode.

Setting up a router with a default gateway and IP routing disabled is often useful in testing configurations on a remote "pod," where a set of routers and switches comprise a closed network topology and are remotely accessed via reverse Telnet from an intermediary access server (such as a Cisco 2511 router). With the default-gateway option on a router, the router effectively acts as a host to be used in testing IP connectivity.

End stations running Microsoft operating systems can be configured for multiple default gateways. However, these multiple default gateways are not dynamic. The operating system uses only a single default gateway at a time. The system selects an additional configured default gateway at boot time only when the first configured default gateway is determined unreachable by ICMP.

ICMP Router Discovery Protocol

With *ICMP Router Discovery Protocol (IRDP)*, the IRDP-enabled hosts dynamically discover routers in order to access nonlocal networks. IRDP allows hosts to locate routers. Router discovery packets are exchanged between hosts (IRDP servers) and Cisco routers (IRDP clients). The Cisco IRDP implementation fully conforms to the router discovery protocol outlined in RFC 1256.

The software also can wire-tap RIP and IGRP routing updates and infer the location of routers from those updates. (This feature is available on some end-station operating systems as well as on Cisco routers that have IP routing disabled.) The client/server implementation of router discovery does not actually examine or store the full routing tables sent by routing devices; it merely keeps track of which systems are sending such data.

You can configure the three protocols (RIP, IGRP, and IRDP) in any combination on a set of Cisco routers populating a LAN segment. Cisco recommends that you use IRDP when possible, because it allows each router to specify *both* a priority and the time after which a device should be assumed to be down if no further packets are received. Devices discovered using IGRP are assigned an arbitrary priority of 60. Devices discovered through RIP are assigned a priority of 50. For IGRP and RIP, the software attempts to measure the time between updates. It assumes that the device is down if no updates are received for 2.5 times that interval.

Each device discovered becomes a candidate for the default router. The list of candidates is scanned, and a new highest-priority router is selected when any of the following events occurs:

- When a higher-priority router is discovered (the list of routers is polled at 5-minute intervals).
- When the current default router is declared down.
- When a TCP connection is about to time out because of excessive retransmissions. In this case, the IRDP server flushes the ARP cache and the ICMP redirect cache. It also picks a new default router in an attempt to find a successful route to the destination.

With IRDP, dead-gateway detection is much faster than simply relying on proxy ARP or multiple gateways on a network interface card (NIC).

Windows 98, 2000, and XP hosts natively support IRDP. When a host running a Microsoft OS initializes, it joins the all-devices IP multicast group (224.0.0.1) and then listens for the router advertisements that IRDP clients send to that group. Hosts can also send router solicitation messages to the all-routers IP multicast address (224.0.0.2) when an interface initializes to avoid any delay in being configured. The router responds by sending a router advertisement. This ensures that the host receives a default gateway immediately without waiting for a periodic router advertisement. Windows hosts send a maximum of three solicitation messages at intervals of approximately 600 milliseconds.

To enable IRDP on an Ethernet interface on a Cisco router (IRDP client), use the interface command **ip irdp**.

IRDP's default behavior is to broadcast advertisements. To force IRDP to use multicasts to 224.0.0.1 (all-devices) instead of broadcasts, use the interface command **ip irdp multicast**. This command is required for compatibility with Sun Microsystems' Solaris OS.

To set the router's IRDP preference level, use the command **ip irdp preference** *number*. The default preference is 0. A higher value increases the router's preference level. You can modify a particular router so that it is the preferred router to serve as a default gateway. To optimize traffic flow in an IRDP-enabled network, it is good to get in the habit of configuring the preferences on IRDP client routers to provide a more deterministic failover mechanism.

GATEWAY DISCOVERY PROTOCOL

Another method used for router discovery by Cisco routers with IP routing disabled is Gateway Discovery Protocol (GDP). Support for this protocol ended with Cisco IOS Software Release 11.2. On pre-11.3 Cisco IOS Software releases, the Cisco IOS Software also can wire-tap RIP and IGRP routing updates and infer the location of routers from those updates. This server/client implementation of router discovery does not actually examine or store the full routing tables sent by routing devices; it merely keeps track of which systems are sending such data. Each device discovered becomes a candidate for the default router.

GDP functionality was replaced by IRDP. Pre-11.3 Cisco IOS Software releases had four options for router discovery: GDP, IRDP, RIP, and IGRP. These options can be configured only on routers that have IP routing disabled. For these Cisco IOS Software releases, router discovery is configured with the associated discovery mechanism via the respective interface configuration commands **ip gdp gdp**, **ip gdp irdp**, **ip gdp rip**, and **ip gdp igrp**. These options are used in any combination on a given LAN segment (assuming that your routers are running pre-11.3 images).

Interestingly, IOS-based switches (such as the Catalyst 2900XL, 3500XL, 2950, and 3550) support router discovery. On these devices, use the global configuration commands **ip gdp irdp**, **ip gdp rip**, and **ip gdp igrp** in Cisco IOS Software Release 12.0 and later.

You have now seen the three methods of router discovery: proxy ARP, multiple manually configured default gateways, and IRDP. Because none of these three mechanisms are satisfactory in the majority of networking situations, Cisco developed HSRP to allow hosts to adapt to network topology changes almost immediately without requiring hosts to run any special software. HSRP is used in conjunction with the configuration of a default gateway in the host devices. This makes the protocol easy to use in any networking environment and provides redundancy for the critical first hop, or default gateway. HSRP also includes a number of additional features that are not available with the three router discovery methods. The following section defines and discusses HSRP in more depth.

HSRP Operation

Hot Standby Router Protocol (HSRP) lets a set of routers on a LAN segment work together to present the appearance of a single virtual router or default gateway to the hosts on the segment, as shown in Figure 8-2. In so doing, HSRP lets fast rerouting alternate default gateways in case one of them fails. HSRP is particularly useful in environments where critical applications are running and fault-tolerant networks have been designed. By sharing an IP address and a MAC address, two or more routers acting as one virtual router can seamlessly assume the routing responsibility in the case of planned downtime or unexpected failure. This lets hosts on a LAN continue forwarding IP packets to a consistent IP and MAC address, with the routers' failover being transparent to the hosts and their sessions. Existing TCP sessions can survive the failover.

Figure 8-2
HSRP creates a virtual router with its own MAC address and IP address.

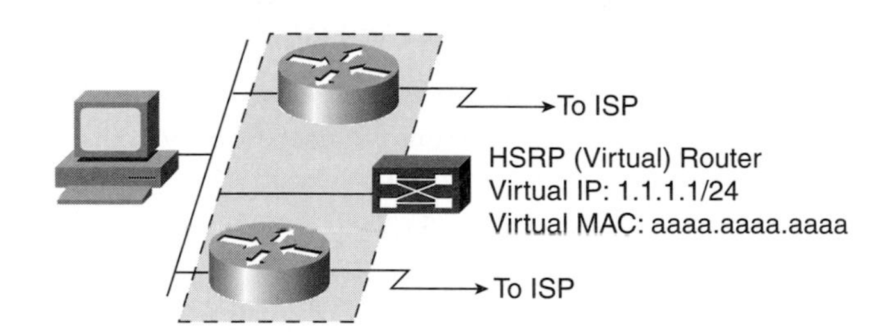

HSRP works by allowing the configuration of *HSRP groups*, in which each group shares a virtual router IP address (or virtual IP address for short). The groups can be composed of independent subnets/VLANs or subsets of a single subnet, as shown in Figure 8-3.

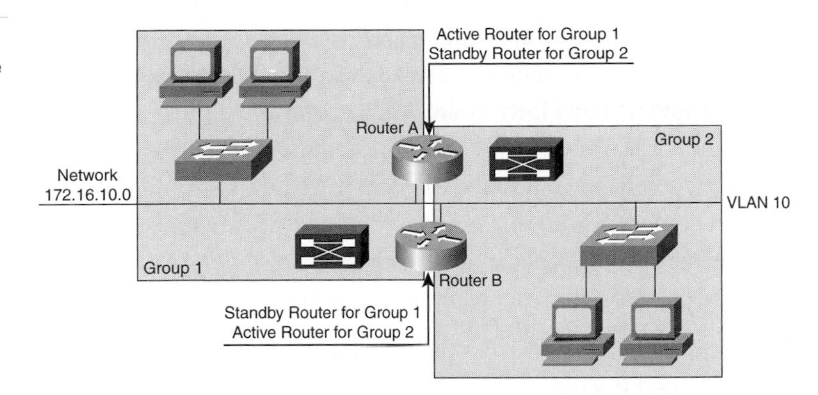

Figure 8-3
HSRP groups enable independent default gateways for different subnets or portions of subnets.

Each router in an HSRP group is given a *priority* to weigh the prioritization of routers for active router selection, as shown in Figure 8-4. The default priority is 100. One of the routers in each group is elected the active forwarder, and one is the standby router, waiting to take over this functionality if necessary. This is done according to the router's configured priorities (see Figure 8-4). The router with the highest priority wins. In the case of a tie in priority, the greater value of the configured IP addresses breaks the tie. For example, if two routers each have an HSRP priority of 100, one has IP address 1.1.1.2 for its interface on the LAN segment, and the other has IP address 1.1.1.3 for its interface on the segment, the router with IP address 1.1.1.3 becomes the active HSRP router for that segment. If there are other routers in the HSRP group besides the active and standby routers, they monitor the active and standby routers' status to enable further fault tolerance.

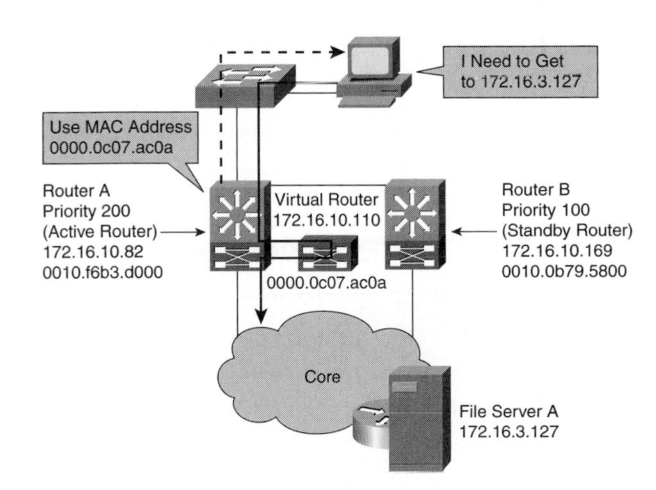

Figure 8-4
The active HSRP router is determined by the priority settings of the HSRP group members.

NOTE

HSRP was introduced in Cisco IOS Software Release 10.0. Since Cisco IOS Software Release 10.0, HSRP functionality has been available on Ethernet, Token Ring, and FDDI. We restrict our discussion to Ethernet. HSRP is also supported on ISL and 802.1Q trunks. Cisco recommends Cisco IOS Software Release 12.0(7) or later for ISL trunks and 12.0(8.1)T or later for 802.1Q trunks.

HSRP routers participating in a standby group watch for HSRP *hello packets* from both the active and standby routers. Devices that are running HSRP send and receive multicast UDP-based hello packets to detect router failure and to determine active and standby routers. The HSRP routers in a group learn the hello interval and hold time, as well as the virtual IP address, from the active router in the HSRP group, assuming that these parameters are not explicitly configured on each individual router. If the active router becomes unavailable due to scheduled maintenance, power failure, or other reasons, the standby router can assume this functionality transparently within a few seconds. This transfer of responsibility takes place only after the hold time has expired.

The following sections discuss the HSRP packet format, HSRP states, HSRP addresses, HSRP timers, HSRP preempt, HSRP interface tracking, and HSRP authentication.

HSRP Packet Format

Table 8-1 shows the format of the data portion of the HSRP frame, encapsulated by UDP.

Table 8-1 HSRP Packet Format

Version	Op Code	State	Hello Interval
Hold Time	Priority	Group	Reserved
Authentication Data			
Authentication Data			
Virtual IP Address			

The fields of the HSRP packet are as follows:

- **Version**—RFC 2281, "Cisco Hot Standby Router Protocol," lists the current version as 0.
- **Op Code**—This field describes the type of message contained in the packet. Possible values are as follows:
  - **0: hello**—Hello messages.
  - **1: coup**—By default, an HSRP router sends hello messages every 3 seconds. When a standby router assumes the function of the active router, it sends a coup message.
  - **2: resign**—A router that is the active router sends a resign message when it is about to shut down or when a router that has a higher priority sends a hello message.
- **State**—Each router in the standby group implements a state machine. Hello messages convey the HSRP state. The State field describes the current state of the router sending the message. The individual states are 0—initial, 1—learn, 2—listen, 4—speak, 8—standby, 16—active.

- **Hello Interval**—This field is meaningful only in hello messages. It specifies the approximate period of time between the hello messages that the router sends. The time is given in seconds. The default is 5 seconds.
- **Hold Time**—This field is meaningful only in hello messages. It specifies how long the routers wait for a hello message before initiating a state change. The default is 10 seconds.
- **Priority**—This field is used to elect the active and standby routers. This parameter is conveyed by HSRP hello messages. When comparing priorities of two routers, the router with the higher value becomes the active router. The tie-breaker is the higher IP address of the respective router interfaces.
- **Group**—This field identifies the standby group (also known as an HSRP group). The default is 0.
- **Reserved**—This field is not presently used.
- **Authentication Data**—This field contains a clear-text, eight-character password.
- **Virtual IP Address**—If the virtual IP address is not configured on a router, it can be learned from the hello message from the active router. An address is learned only if no HSRP standby IP address was configured and the hello message is authenticated (if authentication is configured). This is also called the standby IP address.

The packet fields within HSRP packets are used to exchange HSRP parameters. UDP is the transport-layer protocol used by HSRP. The destination address of HSRP messages is the all-routers multicast address, 224.0.0.2, on UDP port 1985. The source address is the router's primary IP address assigned to the interface.

HSRP States

Each router in an HSRP group participates in the protocol by implementing a simple state machine. *HSRP state* is a term describing the current HSRP condition for a particular router interface and a particular HSRP group. At any given time, HSRP-configured routers are in one of the states listed in Table 8-2.

Table 8-2 HSRP States

State	Definition
Initial	The starting state, which indicates that HSRP is not running. This state is entered via a configuration change or when an interface first comes up.
Learn	The router has not determined the virtual IP address and has not yet seen an authenticated hello message from the active router. In this state, the router is still waiting to hear from the active router.

continues

Table 8-2 HSRP States (Continued)

State	Definition
Listen	The router knows the virtual IP address but is neither the active router nor the standby router. It listens for hello messages from the active and standby routers.
Speak	The router sends periodic hello messages and actively participates in the election of the active and/or standby router. A router cannot enter speak state unless it has the virtual IP address.
Standby	The router is a candidate to become the next active router. It sends periodic hello messages. Excluding transient conditions, at most, only one router in the group is in standby state.
Active	The router is currently forwarding packets that are sent to the group's virtual MAC address. The router sends periodic hello messages. Excluding transient conditions, there must be at most one router in active state in the group.

As you can see, the state machine used by HSRP is relatively simple. HSRP runs on only a single LAN segment, so it does not require the creation and maintenance of a specialized table, as is the case with routing protocols such as RIP and OSPF.

HSRP Addresses

In most cases, when routers are configured to be part of an HSRP group, they recognize their own native MAC address plus the HSRP group MAC address, unless they can utilize only one MAC address. Routers whose Ethernet controllers recognize only a single MAC address use the HSRP MAC address when they are the active router. They use their burned-in address (BIA) when they are not in the active router, unless they are configured with the **standby use-bia** command.

The Cisco 2500 series, 3000 series, 4000 series, and 4500 series routers that use Lance Ethernet hardware do not support multiple HSRP groups, or multiple hot standby groups, on a single Ethernet interface. The Cisco 800 series, 1000 series, and 1600 series that use PQUICC Ethernet hardware do not support multiple HSRP groups on a single Ethernet interface. Multigroup HSRP (MHSRP) further enables redundancy and load sharing within networks, allowing redundant routers to be more fully utilized. While a router is actively forwarding traffic for one HSRP group, it can be in standby or can listen for another group. This enables load sharing of LAN-originated traffic rather than having a standby router unutilized. You can configure a workaround solution for devices that do not support MHSRP by using the **standby use-bia** interface configuration command. It uses the interface's BIA as the virtual MAC address instead of the preassigned MAC address.

The MAC addresses used by HSRP on media that support HSRP, except Token Ring, are 0000.0c07.ac**. The asterisks (**) denote the HSRP group number.

The MAC address used by the virtual router is made up of the following three components:

- **Vendor ID**—Comprises the first 3 bytes of the MAC address.
- **HSRP code**—These 2 bytes indicate that the MAC address is for an HSRP virtual router. The HSRP code is always 0x07.ac.
- **Group ID**—The last byte of the MAC address is the group ID number. In Figure 8-5, decimal group number 10 is being converted to hexadecimal 0a.

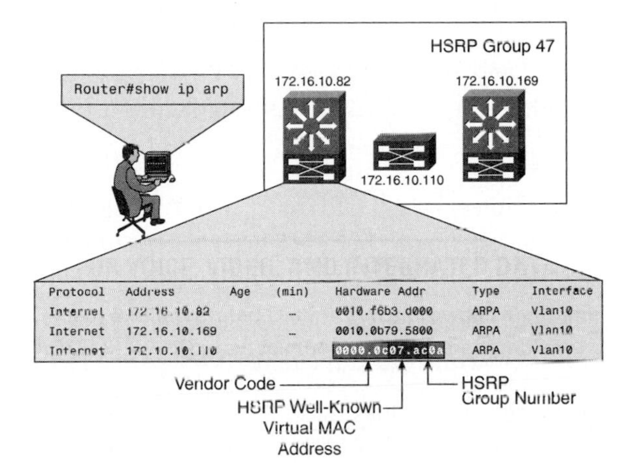

Figure 8-5
The HSRP MAC address is distinguished by the telltale 07.ac in the fifth and sixth bytes.

The **show ip arp** privileged-mode command displays the virtual MAC address associated with the virtual IP address. The three components of the HSRP MAC address are displayed in the **show ip arp** output of Figure 8-5.

To display the virtual IP and MAC addresses for each HSRP group, use the command **show standby** in privileged mode, as shown in Example 8-1.

Example 8-1 show standby Command Displays the Virtual IP and MAC Addresses

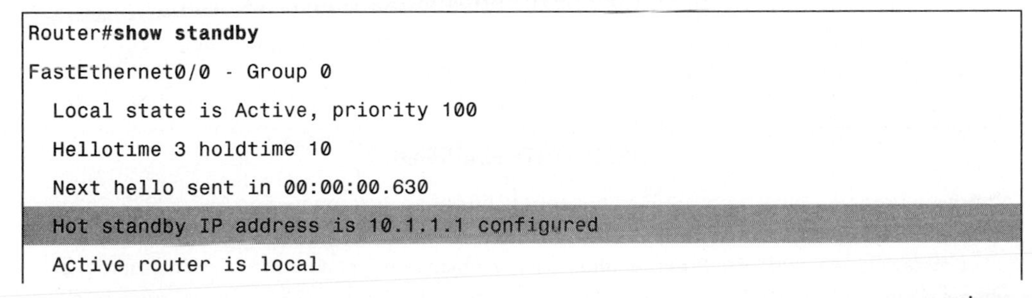

```
Router#show standby
FastEthernet0/0 - Group 0
  Local state is Active, priority 100
  Hellotime 3 holdtime 10
  Next hello sent in 00:00:00.630
  Hot standby IP address is 10.1.1.1 configured
  Active router is local
```

continues

Example 8-1 show standby Command Displays the Virtual IP and MAC Addresses (Continued)

```
Standby router is 10.1.1.3 expires in 00:00:09
Standby virtual mac address is 0000.0c07.ac00FastEthernet0/1 - Group 0
Local state is Standby, priority 100
Hellotime 3 holdtime 10
Next hello sent in 00:00:01.542
Hot standby IP address is 10.1.2.1 configured
Active router is 10.1.2.3 expires in 00:00:08
Standby router is local
Standby virtual mac address is 0000.0c07.ac00
```

HSRP virtual IP and MAC addresses let HSRP route traffic just as actual IP and MAC addresses would for a Cisco router Ethernet interface.

HSRP Timers

HSRP uses two timers—the hello interval and the hold time. The *hello interval* is the time between the sending of hello packets. The *hold time* is the time it takes before HSRP routers in an HSRP group declare the active router down. HSRP routers in any HSRP state should generate a hello packet when the hello timer expires. The default hello interval is 3 seconds, and the default hold time is 10 seconds.

Routers on which timer values that are not configured can learn timer values from the active or standby router. The timers configured on the active router always override any other timer settings. The routers in an HSRP group should use the same timer values. Normally, the hold time is greater than or equal to three times the value of the hello interval. The range of values for the hold time forces the hold time to be greater than the hello interval.

The active router is monitored by the other HSRP routers according to hold time: Anytime a hello packet is received from any active router, a timer is set to expire by the hold time value in the corresponding field of the HSRP hello message.

Also, the standby router is monitored by the other HSRP routers according to the hold time: Anytime a hello packet is received from the standby router, a timer is set to expire by the hold time value in the respective hello packet.

HSRP STATE FLAPPING

A fairly common problem with HSRP state flapping has been documented on Cisco 2620, 2621, and 3600 series routers with FastEthernet interfaces connected to a Catalyst switch running STP. This problem is seen with FastEthernet interfaces in which network connectivity is disrupted, or when a higher-priority HSRP router is added to a network. When the HSRP state changes from active to speaking, the router resets the interface. Only specific hardware used on the FastEthernet interfaces for Cisco 2600 and 3600 routers have this issue. The router interface reset causes a link state change on FastEthernet interfaces, which is detected by the connected switch. If the switch is running STP, an STP transition takes place. The STP takes 30 seconds to transition the port to the forwarding state. At the same time, the speaking router transitions to the standby state after 10 seconds (the HSRP hold time). STP is not forwarding yet, so no HSRP hello messages are received from the active router. This causes the standby router to become active after about 10 seconds. Both routers are now active. When the STP ports enter forwarding state, the lower-priority router changes from active to speaking, and the whole process repeats.

The workaround for this problem is to enable PortFast on the connected switch ports. An alternative workaround is to adjust the HSRP timers so that the STP forward delay (the default is 15 seconds) is less than half the default HSRP hold time, which is 10 seconds.

This issue is resolved in Cisco IOS Software Release 12.1.3 and later.

Normally, the default HSRP timers are suitable for most LAN segments. Be careful to use the same timers for all the routers in an HSRP group.

HSRP Preempt and HSRP Interface Tracking

The HSRP *preempt* feature allows the router with the highest priority to immediately assume the active role at any time. Otherwise, HSRP's default behavior is for the standby HSRP router to wait for the hold time to expire before it takes over active functionality, regardless of the respective priorities. This does not apply to the higher IP address values used to break a tie when two routers originally vie for the active router role when they boot up. Only a higher priority allows a router to utilize the preempt mechanism and become the active router.

The HSRP *tracking* feature allows you to specify another interface on the router for the HSRP process to monitor. If the tracked interface goes down, the router takes over as the active router. This process is facilitated by a decrement to the priority resulting from the tracked interface line protocol going down. The decrement is configured to lower the router's HSRP priority below that of the currently active router's HSRP priority; the default decrement value is 10. In concert with the preempt feature, this enables an inactive HSRP router to immediately assume the role of active router should the tracked interface go down. As with other HSRP features, the tracking feature is configured per HSRP group. This feature is useful in the following network scenario.

Consider that a router with a T1 line is normally the preferred path from a LAN. If this router's serial interface should become inactive for any reason, HSRP can automatically change the active router by decrementing the priority of the currently active router to be less than that of the HSRP peer. This can be used to ensure that the active router has connectivity to the rest of the network, regardless of line problems. The preempt and tracking features also enable the usually preferred (T1-connected) router to regain its active HSRP role if connectivity is regained on its serial interface. This occurs automatically, because the priority of the T1-connected router increases with the change in the serial interface's line protocol. This allows traffic to be rerouted through this preferred route transparently.

As a general rule, setting the preempt option on HSRP interfaces tends to provide the best failover results.

HSRP Authentication

The authentication option built into HSRP consists of the addition of a shared clear-text key within the HSRP packets. This feature prevents the lower-priority router from learning the standby IP address and standby timer values from the higher-priority router. The purpose of this password is to disallow misconfigured routers from participating in an HSRP group they were not intended to participate in. Although this feature allows for a certain level of security, in practice it is possible to cause more issues than this authentication can prevent. Because the key is clear-text on the wire, it would be relatively simple for an attacker forging an HSRP packet to sniff and replay this key.

The behavior of an HSRP peer, in the event that it receives a packet that contains a bad authentication value, is to ignore it. This means that, in the case of a misconfigured router meant to participate in an HSRP group, it is possible that two or more routers will see no other valid hellos and decide to become active at the same time. In this event, they would be using the same IP and MAC address concurrently. This could create problems in the network. If the authentication feature is used on a network, it should be used cautiously. It is advisable to explicitly configure the standby IP address for which a router may assume responsibility on each device participating in an HSRP group, as opposed to learning the standby IP address from the active router (when a router is not explicitly configured with a standby IP address).

> **NOTE**
>
> When HSRP is configured on an interface, it used to be that ICMP redirect messages were automatically disabled. ICMP is a network-layer protocol that provides message packets to report errors and other information relevant to IP processing. ICMP provides many diagnostic functions and can send and redirect error packets to the host. When running HSRP, it is important to prevent hosts from discovering the actual interface MAC addresses of routers in the HSRP group. If a host is redirected by ICMP to a router's real MAC address, and that router later fails, packets from the host are lost. With Cisco IOS Software Release 12.1(3)T and later, ICMP redirect messages are automatically enabled on interfaces configured with HSRP. A sophisticated mechanism is built into these later Cisco IOS Software releases to accurately handle all the intricacies of ICMP redirects in conjunction with HSRP functionality.
>
> Also, when the **standby ip** command is enabled on an interface, the handling of proxy ARP requests is changed (unless proxy ARP is disabled). If the interface's HSRP state is active, proxy ARP requests are answered using the standby group's MAC address. If the interface is in a different state, proxy ARP responses are suppressed.

HSRP Configuration

A basic HSRP configuration on a Cisco router simply requires configuring the standby IP address and the HSRP priority on an Ethernet interface. If more than one HSRP group is needed, the standby IP address and other HSRP parameters are configured for each group.

To activate HSRP on an interface and configure an interface as a member of an HSRP standby group, enter the command **standby** [*group-number*] **ip** [*ip-address* [**secondary**]] in interface configuration mode.

If the *group-number* is omitted, the default group number is 0. Be sure to include the *group-number* when configuring HSRP groups. It is a common mistake to forget this part of the configuration.

The *ip-address* parameter is the virtual or standby IP address. If an IP address is specified, that address is used as the HSRP address for the group. If no IP address is specified, the virtual address is learned via HSRP hello messages. For HSRP to elect an active router, at least one router on the cable must have been configured with, or have learned, the virtual IP address. Configuring the virtual IP address on the active router always overrides a designated address that is currently in use.

The **secondary** option is used when associating a virtual IP address with a secondary IP network already configured on the router interface.

When the **standby ip** command is issued, the interface changes to the appropriate HSRP state. Example 8-2 shows the state message that is generated after HSRP is enabled. This message is automatically generated upon successful execution of the command.

Example 8-2 State Changes Are Reflected Upon Configuration of HSRP

```
Router(config-if)#ip address 10.1.1.2 255.255.255.0
Router(config-if)#standby 50 ip 10.1.1.1
Router(config-if)#exit
Router(config)#
1w0d: %STANDBY-6-STATECHANGE: Standby: 50: FastEthernet0/0 state
Speak    -> Standby
1w0d: %STANDBY-6-STATECHANGE: Standby: 50: FastEthernet0/0 state
Standby -> Active
```

To remove an interface from an HSRP group, enter the **no standby** *group* **ip** command.

The next sections discuss the various HSRP parameters that can be configured, including HSRP priority, preempt, authentication, timers, and tracking. This is followed by a comprehensive HSRP configuration example.

Configuring HSRP Priority and HSRP Preempt

To configure HSRP priorities and the preempt feature, use the interface command **standby** [*group-number*] **priority** *priority* [**preempt** [**delay** *delay*]]. This allows you to configure both priority and preempt with a single command. You also have the option of configuring the preempt feature separately with the interface command **standby** [*group-number*] **preempt.**

Here are some key points to consider with this command:

- Configure group numbers if you are configuring MHSRP. The default is 0, which is actually reflected in NVRAM as no group number to provide backward compatibility. The configurable range is 0 to 255.

- The configurable range for *priority* is 1 to 255. The default priority is 100.
- The assigned priority has a higher priority than the authentication string specified in the **standby authentication** command (discussed in the next section). A router with a higher HSRP priority ignores the authentication string.
- When a router first comes up, it does not have a complete routing table. If it is configured to preempt, it becomes the active router, yet it is unable to provide adequate routing services. You can solve this problem by configuring a delay before the preempting router actually preempts the currently active router. The default delay is 0 seconds: If the router wants to preempt, it does so immediately. The *delay* argument causes the local router to postpone taking over the active role for *delay* seconds since that router was last restarted. The range is from 0 to 3600 seconds.

Next, we consider the configuration of HSRP authentication.

Configuring HSRP Authentication

To configure HSRP authentication on an interface, use the interface command **standby** [*group-number*] **authentication** *string*. The *string* parameter can have up to eight characters. The default string is **cisco**.

The authentication string is sent unencrypted in all HSRP messages. The same authentication string must be configured on all routers on a LAN segment to ensure interoperation. Authentication mismatch prevents a device from learning the designated virtual IP address and the HSRP timer values from other routers configured with HSRP. Authentication mismatch does not prevent protocol events such as one router taking over as the active router.

HSRP authentication prevents a lower-priority router from learning the standby IP address and standby timer values from the higher-priority router. A router with a higher HSRP priority ignores the authentication string.

Configuring HSRP Timers

An HSRP-enabled router sends hello messages to indicate that the router is running and can become either the active or standby router. The hello message contains the router's priority, as well as *hellotime* and *holdtime* values. The *hellotime* value indicates the interval between the hello messages that the router sends. The *holdtime* value specifies how long the current hello message is considered valid. The *holdtime* should be at least three times the value of the *hellotime*. The range of values for holdtime forces the holdtime to be greater than the *hellotime*.

Both the hellotime and the *holdtime* parameters can be configured. To configure the time between hello messages and the time before other group routers declare the active or standby router down, enter the interface command **standby** [*group-number*] **timers** [**msec**] *hellotime* [**msec**] *holdtime*.

The timers configured on the active router always override any other timer settings. All routers in an HSRP group should use the same timer values. The default *hellotime* is 3 seconds. *hellotime* can be configured between 1 and 254 seconds. If the **msec** option is specified, the hello interval is configured as a number (in milliseconds) between 50 and 999. Millisecond timers allow for faster failover.

The default *holdtime* is 10 seconds. The minimum value possible for *holdtime* depends on the configured hellotime value. The maximum *holdtime* is 255 seconds. If the **msec** option is specified, *holdtime* is in milliseconds, with a maximum value of 3000.

Our last configuration task to consider for HSRP is interface tracking, which is discussed in the next section.

Configuring HSRP Interface Tracking

In many situations, an interface's status directly affects which router needs to become the active router, particularly when each router in an HSRP group has a different path to resources within the campus network.

In the network pictured in Figure 8-6, Routers A and B each support a FastEthernet link to the backbone. Router A has the higher priority and is the active forwarding router for standby group 50. Router B is the standby router for that group. Routers A and B are exchanging hello messages through their FastEthernet 0/0 interfaces.

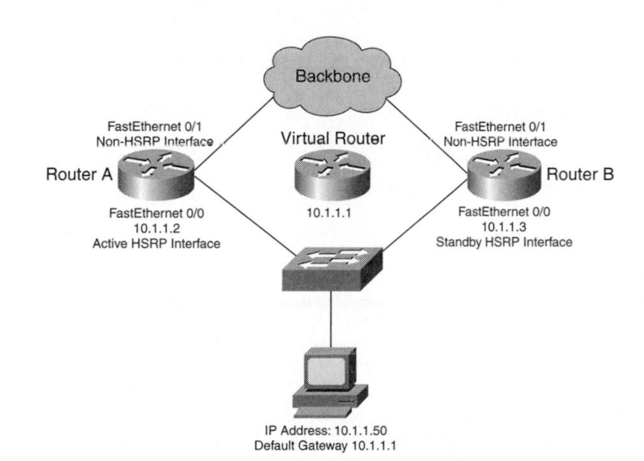

Figure 8-6 HSRP interface tracking extends HSRP's functionality to determine the active router as a function of line protocol status on non-HSRP interfaces.

If the Router A interface FastEthernet 0/1 goes down, Router A loses its direct connection to the backbone. Although this prevents traffic from reaching the backbone (Router A could send everything through Router B), it leads to inefficient traffic flow.

Router A's FastEthernet 0/0 interface is still active, so packets destined for the core would still be sent to the active HSRP router, Router A, just to be forwarded in turn to Router B (regardless of HSRP). The standby track option can be used to prevent this situation, as shown in Figure 8-6.

To configure HSRP tracking, enter the interface command **standby** [*group-number*] **track** *interface-type interface-number* [*interface-priority*].

The *interface-priority* is the amount by which the router's priority is decremented (or incremented) when the interface goes down (or comes back up). The default value is 10.

NOTE

When multiple tracked interfaces are down and interface-priority values are configured, these configured priority decrements are cumulative. If tracked interfaces are down, but none of them is configured with priority decrements, the default decrement is 10, and it is noncumulative.

When a tracked interface goes down, the priority decreases by the value specified as *interface-priority*. If an interface is not tracked, its state changes do not affect the priority. For each interface configured for HSRP, you can configure a separate list of interfaces to be tracked.

The HSRP interface-tracking feature is almost always used in conjunction with the preempt feature to permit quick failover.

HSRP Configuration Scenario

Figure 8-6 displays a typical network topology employing HSRP. Router A is initially the active router with a priority of 150, and Router B is the standby router with a priority of 100. Tracking is configured on Router A's FastEthernet 0/1 interface. When Router A's FastEthernet 0/1 interface goes down, you can see that Router A's priority goes down to 95. At that point, Router B becomes the active router, and Router A becomes the standby router. This sequence of events is illustrated in Example 8-3.

Here are the relevant portions of the configurations for Router A and Router B:

Router A Configuration

```
interface FastEthernet0/0
 ip address 10.1.1.2 255.255.255.0
 no ip redirects
 no ip directed-broadcast
 duplex auto
 speed auto
 standby 50 priority 150 preempt
 standby 50 ip 10.1.1.1
 standby 50 track FastEthernet0/1 55
!
interface FastEthernet0/1
 ip address 10.1.2.1 255.255.255.0
 no ip directed-broadcast
 duplex auto
 speed auto
```

Router B Configuration

```
interface FastEthernet0/0
 ip address 10.1.1.3 255.255.255.0
 no ip redirects
 no ip directed-broadcast
 standby 50 preempt
 standby 50 ip 10.1.1.1
!
interface FastEthernet0/1
 ip address 10.1.3.1 255.255.255.0
 no ip directed-broadcast
 duplex auto
 speed auto
```

Example 8-3 shows the **show standby brief** command outputs for Router A and Router B.

Example 8-3 show standby brief Command Normally Provides the Information Necessary to Verify Your HSRP Configuration

```
Router-A#show standby brief
                     P indicates configured to preempt.
                     |
Interface   Grp Prio P State    Active addr  Standby addr  Group addr
Fa0/0       50  150  P Active   local        10.1.1.3      10.1.1.1
Router-A#
1w0d: %LINEPROTO-5-UPDOWN: Line protocol on Interface FastEthernet0/1,
changed state to down

1w0d: %STANDBY-6-STATECHANGE: Standby: 50: FastEthernet0/0 state Active
-> Speak
1w0d: %STANDBY-6-STATECHANGE: Standby: 50: FastEthernet0/0 state Speak
-> Standby

Router-A#show standby brief
                     P indicates configured to preempt.
                     |
Interface   Grp Prio P State    Active addr  Standby addr  Group addr
Fa0/0       50  95   P Standby  10.1.1.3     local         10.1.1.1
```

continues

Example 8-3 show standby brief Command Normally Provides the Information Necessary to Verify Your HSRP Configuration (Continued)

```
Router-B#show standby brief
                     P indicates configured to preempt.
                     |
Interface   Grp Prio P State    Active addr     Standby addr    Group addr
Fa0/0       50  100  P Standby  10.1.1.2        local           10.1.1.1
Router-B#
1w0d: %STANDBY-6-STATECHANGE: Standby: 50: FastEthernet0/0 state Standby
 -> Active
Router-B#show standby brief
                     P indicates configured to preempt.
                     |
Interface   Grp Prio P State    Active addr     Standby addr    Group addr
Fa0/0       50  100  P Active   local           10.1.1.2        10.1.1.1
```

This completes the HSRP configuration scenario. Next, we look at HSRP configuration on trunk links.

HSRP Configuration on Trunk Links

This example illustrates the configuration of two HSRP-enabled routers, each with an ISL trunk supporting two VLANs. Running HSRP over ISL allows users to configure redundant connections to multiple routers configured as VLAN gateways. By configuring HSRP over ISL, users can eliminate situations in which a single point of failure causes traffic interruptions.

To configure HSRP over an ISL trunk, follow these steps for each Ethernet subinterface:

Step 1 Define the encapsulation format and VLAN.

Step 2 Define an IP address.

Step 3 Enable HSRP.

For Step 3, you configure HSRP using the same steps detailed in the previous sections. Figure 8-7 shows a sample configuration using HSRP groups 1 and 2. HSRP group 1 uses Router A as the active router. HSRP group 2 uses Router B as the active router.

HSRP is also supported over 802.1Q trunks. Cisco recommends Cisco IOS Software Release 12.0(8.1)T or later for 802.1Q trunks and 12.0(7) or later for ISL trunks. The next section explores HSRP's verification and troubleshooting options.

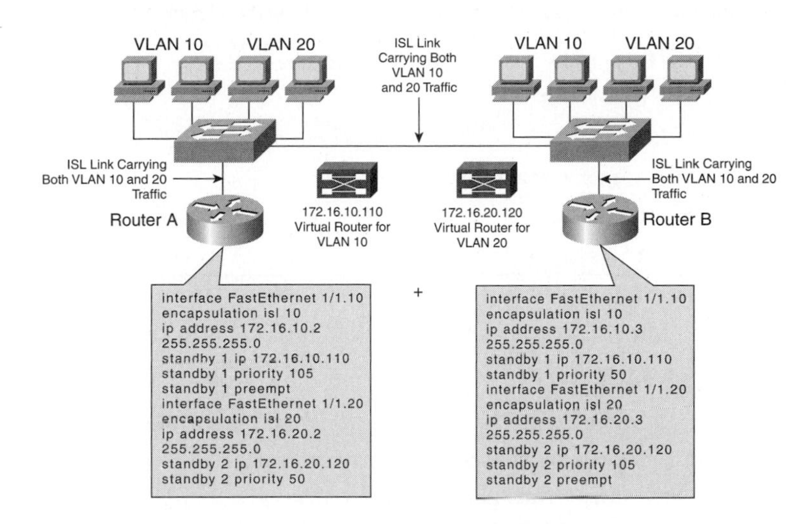

Figure 8-7 Configuring HSRP on trunks is done per subinterface (per VLAN), with subinterfaces corresponding to independent HSRP groups.

Verifying and Troubleshooting HSRP Configuration

To display an HSRP router's status, use the privileged-mode command **show standby** [*type number* [*group*]] [**active** | **init** | **listen** | **standby**] [**brief**].

The **active**, **init**, **listen**, and **standby** options display output regarding HSRP groups in the active, init, listen or learn, and standby or speak states, respectively. These four options require Cisco IOS Software Release 12.1(3)T or later. The **brief** option provides a single line of output summarizing each standby group. For most purposes, the **show standby brief** command is sufficient.

If the optional interface parameters are not included, the **show standby** command displays HSRP information for all interfaces and groups. Example 8-4 displays output for the **show standby** and **show standby init brief** commands.

Example 8-4 show standby Command Provides Interface Tracking Information

```
Router#show standby
Ethernet0 - Group 0
  Local state is Active, priority 100, may preempt
  Hellotime 3 holdtime 10
  Next hello sent in 0:00:00
  Hot standby IP address is 198.92.72.29 configured
  Active router is local
  Standby router is 198.92.72.21 expires in 0:00:07
  Standby virtual mac address is 0000.0c07.ac00
```

continues

Example 8-4 show standby Command Provides Interface Tracking Information (Continued)

```
  Tracking interface states for 2 interfaces, 2 up:
    Up      Ethernet0
    Up     Serial0
Router#show standby ethernet0 init brief
Interface    Grp Prio P State    Active addr     Standby addr    Group addr
Et0          0   120    Init     20.0.0.1        unknown         20.0.0.12
```

Notice the information regarding tracked interfaces in the **show standby** output. This is useful for a quick check of the status of tracked interfaces.

With Cisco IOS Software Release 12.1(0.2) and later, HSRP debugging is greatly improved over previous Cisco IOS Software releases. In particular, the output is not cluttered with noise from periodic hello messages. HSRP debugging is further enhanced in Cisco IOS Software Release 12.1(1.3), based on improvements made to the HSRP state machine. HSRP debugging is useful for troubleshooting HSRP operation. Enhanced HSRP debugging options are outlined in Table 8-3.

Table 8-3 Enhanced HSRP Debugging

Command	Description
debug standby	Displays all HSRP errors, events, and packets.
debug standby terse	Displays all HSRP errors, events, and packets, except hello and advertisement.
debug standby errors	Displays HSRP errors.
debug standby events [[*all* | *terse*] | [*icmp* | *protocol* | *redundancy* | *track*]] [**detail**]	Displays HSRP events.
debug standby packets [[*all* | *terse*] | [*advertise* | *coup* | *hello* | *resign*]] [**detail**]	Displays HSRP packets.

You can filter the **debug** output using interface and HSRP group *conditional debugging* (introduced in Cisco IOS Software Release 12.0). To enable interface conditional debugging, use the **debug condition interface** *interface* command. This prevents debug output for all interfaces except the specified interface. To enable HSRP conditional debugging, use the **debug condition standby** *interface group* command. This limits HSRP debug output to HSRP state changes for the specified interface and group. An interface debug condition applies only when you have not set any standby debug conditions. See Example 8-5.

Example 8-5 debug standby Command Is Useful for Displaying Detailed Information Related to HSRP State Changes

```
Router#debug standby
SB: Ethernet0 state Virgin -> Listen
SB: Starting up hot standby process
SB:Ethernet0 Hello in 192.168.72.21 Active pri 90 hel 3 hol 10 ip 192.168.72.29
SB:Ethernet0 Hello in 192.168.72.21 Active pri 90 hel 3 hol 10 ip 192.168.72.29
SB:Ethernet0 Hello in 192.168.72.21 Active pri 90 hel 3 hol 10 ip 192.168.72.29
SB:Ethernet0 Hello in 192.168.72.21 Active pri 90 hel 3 hol 10 ip 192.168.72.29
SB: Ethernet0 state Listen -> Speak
SB:Ethernet0 Hello out 192.168.72.20 Speak pri 100 hel 3 hol 10 ip 192.168.72.29
SB:Ethernet0 Hello in 192.168.72.21 Active pri 90 hel 3 hol 10 ip 192.168.72.29
SB:Ethernet0 Hello out 192.168.72.20 Speak pri 100 hel 3 hol 10 ip 192.168.72.29
SB:Ethernet0 Hello in 192.168.72.21 Active pri 90 hel 3 hol 10 ip 192.168.72.29
SB:Ethernet0 Hello out 192.168.72.20 Speak pri 100 hel 3 hol 10 ip 192.168.72.29
SB:Ethernet0 Hello in 192.168.72.21 Active pri 90 hel 3 hol 10 ip 192.168.72.29
SB: Ethernet0 state Speak -> Standby
SB:Ethernet0 Hello out 192.168.72.20 Standby pri 100 hel 3 hol 10 ip 192.168.72.29
SB:Ethernet0 Hello in 192.168.72.21 Active pri 90 hel 3 hol 10 ip 192.168.72.29
SB:Ethernet0 Hello out 192.168.72.20 Standby pri 100 hel 3 hol 10 ip 192.168.72.29
SB:Ethernet0 Hello in 192.168.72.21 Active pri 90 hel 3 hol 10 ip 192.168.72.29
SB:Ethernet0 Hello out 192.168.72.20 Standby pri 100 hel 3 hol 10 ip 192.168.72.29
SB:Ethernet0 Hello in 192.168.72.21 Active pri 90 hel 3 hol 10 ip 192.168.72.29
SB: Ethernet0 Coup out 192.168.72.20 Standby pri 100 hel 3 hol 10 ip 192.168.72.29
SB: Ethernet0 state Standby -> Active
SB:Ethernet0 Hello out 192.168.72.20 Active pri 100 hel 3 hol 10 ip 192.168.72.29
SB:Ethernet0 Hello in 192.168.72.21 Speak pri 90 hel 3 hol 10 ip 192.168.72.29
SB:Ethernet0 Hello out 192.168.72.20 Active pri 100 hel 3 hol 10 ip 192.168.72.29
SB:Ethernet0 Hello in 192.168.72.21 Speak pri 90 hel 3 hol 10 ip 192.168.72.29
SB:Ethernet0 Hello out 192.168.72.20 Active pri 100 hel 3 hol 10 ip 192.168.72.29
```

This completes the analysis of HSRP theory, configuration, and verification. The rest of this chapter discusses the role of HSRP in campus network design.

HSRP in Campus Network Design

Several campus design models employing HSRP are recommended by Cisco, depending on the type of switches used in the network core and whether Layer 2 or Layer 3 switching is utilized in the core. Our focus is on two particular options involving a dual-path Layer 3 core:

- Standard building block
- VLAN building block

We consider these two designs within a single campus network, as shown in Figure 8-8.

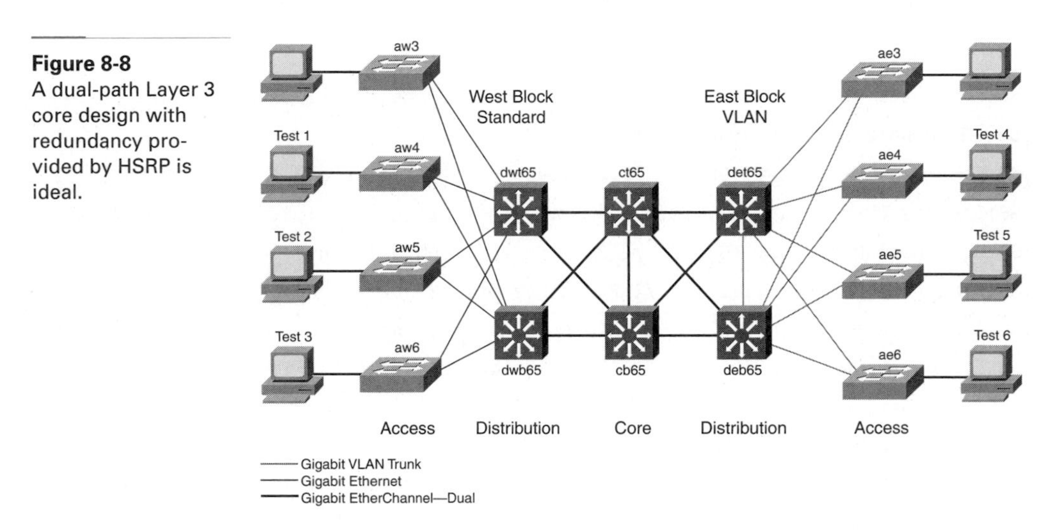

Figure 8-8
A dual-path Layer 3 core design with redundancy provided by HSRP is ideal.

Each distribution switch has redundant two-port Gigabit EtherChannel connections to both Layer 3 switches in the core. Wherever possible, Gigabit EtherChannel is configured with ports on two different line cards to increase availability. HSRP interface tracking is configured on the distribution switches so that fast HSRP recovery takes place if both links to the core (in an EtherChannel) are broken.

Redundant Gigabit Ethernet uplinks are used to connect the access switches to the distribution switches. Catalyst 6000s are used in the wiring closets. Catalyst 6500s with MSFCs are used in the distribution and core layers. HSRP is tuned for fast convergence with a 1-second hello time and a 3-second hold time.

Every router and switch in the network has an out-of-band Ethernet management interface on VLAN 99 with IP subnet 172.29.196.0/23. One Ethernet port on each device is configured in VLAN 99 and is wired outside the network to a separate switched management network. The network's IP addressing is shown in Figures 8-9 and 8-10.

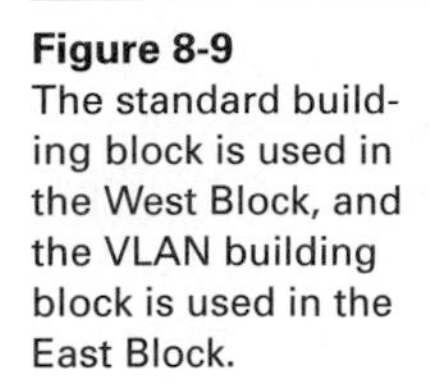

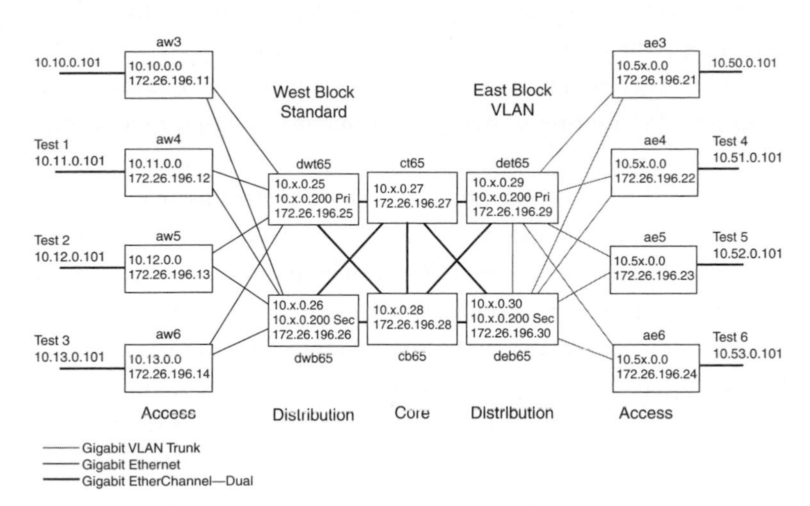

Figure 8-9
The standard building block is used in the West Block, and the VLAN building block is used in the East Block.

The standard building block appears on the left side (West Block) of Figure 8-8. Configuring the standard building block is very simple, because all spanning-tree loops are eliminated. Hence, there is no Layer 2 tuning, such as selecting the best root switch for a given VLAN. No VLAN trunks are used, and uplinks are connected to native routed interfaces on the Layer 3 switches in the distribution layer. Channel negotiation on the uplinks is disabled to accelerate connectivity after a failure has been restored.

The core's IP subnet and VLAN layout are detailed in Figure 8-10.

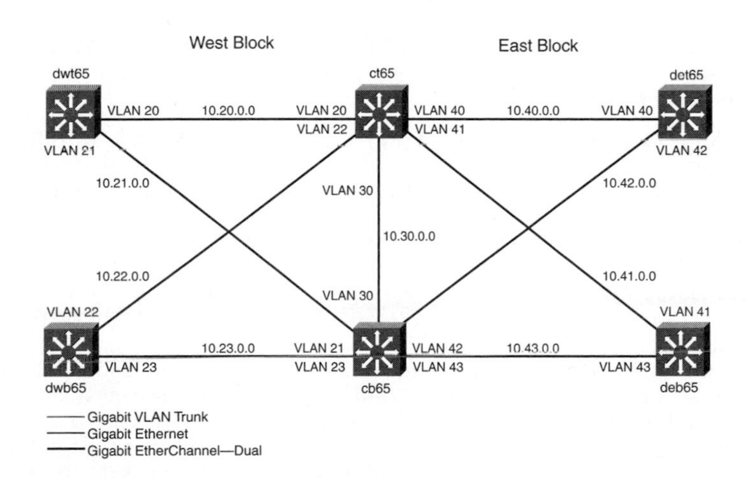

Figure 8-10
The links between core and distribution switches are all routed links.

The VLAN building block appears on the right side (East Block) of Figure 8-8. The VLAN building block is typically used in a server farm to provide for redundant server connections with dual NICs. If dual-attached servers are used, a VLAN trunking configuration is required within the server distribution block. The two distribution switches are the root bridges for the even- and odd-numbered VLANs, respectively. UplinkFast is configured on the access switches. BackboneFast is configured on the access and distribution switches for faster spanning-tree recovery. Because the number of VLANs is relatively small, VTP transparent mode is used, and all VLANs are configured explicitly. VLAN traffic for all four of the East Block VLANs is supported on each access switch. HSRP at Layer 3 is configured to match the Layer 2 spanning-tree configuration. This way, the HSRP primary gateway router for even-numbered subnets is also the spanning-tree root for even-numbered VLANs. The HSRP primary gateway router for odd-numbered subnets is also the spanning-tree root for odd-numbered VLANs. A trunk connects the two distribution switches in the East Block. This completes the characteristic triangle prevalent in the access-distribution switch block design. This triangle forms a Layer 2 loop that is required for Layer 2 load balancing.

The configurations for representative devices are as follows:

Catalyst 6000 Access Switch: Standard Building Block

```
Set prompt aw3         (access layer, west block, Catalyst 6000)
Set vtp domain west
Set vtp mode transparent (no VLAN trunks, use transparent mode)
Set vlan 99             (used for out-of-band management)
Set vlan 99 3/48       (last physical port on switch used for out-of-band management)
Set int sc0 99 172.29.196.51 255.255.254.0 (logical console port for out-of-band
  management)
Set ip route default 172.29.196.1 (gateway router in out-of-band management network)
Set port channel 1/1-2 mode off (not using EtherChannel in this configuration)
Set trunk 1/1-2 off      (turn off VLAN trunking on uplinks)
Set vlan 10             (VLAN 10 corresponds to subnet "10", i.e., 10.10.0.0)
Set vlan 10 1/1-2      (all other ports are part of VLAN 10)
Set vlan 10 3/1-47
```

Configuration of the Catalyst 6000 Access Switch: VLAN Building Block

```
Set prompt ae6         (access layer, east block, Catalyst 6000)
Set vtp domain east
Set vtp mode transparent (use transparent mode, configure all VLANs explicitly)
Set vlan 99            (used for out-of-band management)
Set vlan 99 4/48       (last physical port on switch used for out-of-band management)
Set int sc0 99 172.29.196.52 255.255.254.0   (logical console port for out-of-band
  management)
Set ip route default 172.29.196.1 (gateway router in oob management network)
set spantree uplinkfast enable (set uplinkfast on the access switch only)
set spantree backbonefast enable (enable BackboneFast on all switches in block)
Set port channel 1/1-2 mode off (not using EtherChannel on this switch)
set trunk 1/1-2 on 50,51,52,53 dot1q (dot1q VLAN trunking on uplinks, state VLANs
  explicitly)
set vlan 50 2/1-48   (VLAN 50 corresponds to subnet "50", i.e., 10.50.0.0)
```

Configuration of the Catalyst 6500 Distribution Switch

```
Set prompt det65 (distribution layer, east block, top, Catalyst 6500)
Set vtp domain east
Set vtp mode transparent (use transparent mode, configure all VLANs explicitly)
Set vlan 99             (used for out-of-band management)
Set vlan 99 3/48      (last physical port on switch used for out-of-band management)
Set int sc0 99 172.29.196.53 255.255.254.0 (logical console port for out-of-band
  management)
Set ip route default 172.29.196.1 (gateway router in out-of-band management network)
set spantree backbonefast enable (enable BackboneFast on all switches in block)
Set port channel 1/1-2 mode off (turn off channel negotiation on non-EtherChannel
  links)
set port channel 3/7-8 mode off
set port channel 4/7-8 mode off
set port channel 3/1-2 mode on (EtherChannel used for routed links to the core)
set port channel 4/1-2 mode on (EtherChannel used for routed links to the core)
set vlan 40 3/1,4/1 (routed EtherChannel link VLAN40=10.40.0.0 spans two cards)
set vlan 42 3/2,4/2 (routed EtherChannel link VLAN42=10.42.0.0 spans two cards)
set trunk 3/7 on 50,51,52,53 dot1q
(set VLANs and dot1q trunking explicitly)
(3/7-8 and 4/7-8 are uplinks to wiring closet switches)
set trunk 3/8 on 50,51,52,53 dot1q
set trunk 4/7 on 50,51,52,53 dot1q
set trunk 4/8 on 50,51,52,53 dot1q
set trunk 1/1 on 50,51,52,53 dot1q (this is the backup trunk to other distribution
  switch)
set spantree root 50,52 (make this root bridge for even VLANs)
set spantree root secondary 51,53 (make this backup root bridge odd VLANs)
set trunk 3/1-2 off (no VLAN trunking on routed links to core)
set trunk 4/1-2 off (no VLAN trunking on routed links to core)
```

Configuration of MSFC: Client-Side Interface on the Distribution Switch—HSRP Primary

```
interface Vlan10
 ip address 10.10.0.81 255.255.0.0
 no ip redirects
 no ip directed-broadcast
 standby 10 timers 1 3 (HSRP hellotime 1, holdtime 3, HSRP group number 10 matches
   VLAN number)
 standby 10 priority 200 preempt delay 60 (this is the primary gateway router for
   subnet 10)
(preempt delay 60 seconds allows EIGRP to stabilize before HSRP switches back upon
  power recovery)
 standby 10 ip 10.10.0.200 (10.10.0.200 is the HSRP gateway router address)
 standby 10 track Vlan20 75
 standby 10 track Vlan21 75 (if you lose both links to the backbone, drop priority
   by 150 to initiate HSRP recovery)
```

Configuration of MSFC: Client-Side Interface on the Distribution Switch—HSRP Secondary

```
interface Vlan10
 ip address 10.10.0.82 255.255.0.0
 no ip redirects
 no ip directed-broadcast
 standby 10 timers 1 3
 standby 10 priority 100 preempt delay 60 (HSRP secondary or backup gateway router
   for subnet 10)
 standby 10 ip 10.10.0.200
 standby 10 track Vlan22 25
 standby 10 track Vlan23 25   (track both links to the backbone)
```

Additional MSFC Configuration: Any Routed Interface to the Core

```
interface Vlan21
 ip address 10.21.0.81 255.255.0.0
 no ip directed-broadcast
```

Additional MSFC Configuration: Interface on the Management VLAN

```
interface Vlan99
 ip address 172.26.196.81 255.255.254.0
 no ip directed-broadcast
```

Additional MSFC Configuration: EIGRP with Passive Interfaces to Wiring Closets

```
router eigrp 1
 passive-interface Vlan10
 passive-interface Vlan11
 passive-interface Vlan12
 passive-interface Vlan13
 passive-interface Vlan99
 network 10.0.0.0
```

This section provided some design options for a network engineer to consider when deploying HSRP in a campus network. The two options discussed were deploying HSRP with the standard building block and deploying HSRP with the VLAN building block. The standard building block option uses routed interfaces, thus avoiding STP issues. The VLAN building block is more traditional, segmenting the network logically by VLAN. This approach has the advantage of permitting spanning-tree load balancing.

Summary

HSRP is a Layer 3 protocol supported by Cisco routers that enables redundant default gateways for a LAN segment. HSRP is preferable to the various router-discovery mechanisms, such as IRDP and proxy ARP, because it does not rely on any host operating systems, and it provides faster failover.

HSRP uses hello messages to communicate virtual IP addresses and other useful information between HSRP routers. HSRP permits the use of several HSRP groups, allowing a set of routers on a LAN segment to support multiple alternative gateways on a per-VLAN basis. HSRP groups also allow for dividing traffic within a given subnet so that separate gateways are used for independent subsets of a given subnet.

The active router forwards traffic for an HSRP group. The standby router takes over for the active router if the active router fails or if the HSRP interface on the active router fails.

HSRP interface tracking is an HSRP feature that results in HSRP state changes according to the line protocol status on a tracked interface. This allows a change in the active HSRP router as a result of the failure of a non-HSRP interface that links to critical network resources.

HSRP **show** and enhanced debugging commands verify and troubleshoot HSRP configurations.

The Layer 3 redundancy provided by HSRP enhances campus network design. HSRP is easily configured within switch blocks composed of access and distribution devices with purely routed links or employing trunks and VLANs. HSRP tracking of distribution-core links improves the configuration.

Check Your Understanding

Use the following review questions to test your understanding of the concepts covered in this chapter. Answers are listed in Appendix A, "Check Your Understanding Answer Key."

1. What is the default HSRP hello interval?
   - A. 3 seconds
   - B. 5 seconds
   - C. 10 seconds
   - D. 15 seconds
2. What is the default HSRP hold time?
   - A. 3 seconds
   - B. 5 seconds
   - C. 10 seconds
   - D. 15 seconds
3. Which HSRP command option allows immediate failover to a previously active HSRP router as soon as its interface comes back up?
   - A. **priority**
   - B. **track**
   - C. **tracking**
   - D. **preempt**
4. Which command displays current HSRP status?
   - A. **show backup**
   - B. **show HSRP**
   - C. **show debug**
   - D. **show standby**
5. Which of the following are not HSRP states?
   - A. Listen
   - B. Exstart
   - C. Speak
   - D. Wait

6. What is required for HSRP to enable multiple virtual gateways for a single subnet?
   - A. Trunks
   - B. VLANs
   - C. HSRP groups
   - D. Secondary addresses
7. What do the characters 07.ac in the fifth and sixth bytes of a MAC address indicate?
   - A. OUI
   - B. Multicast MAC address
   - C. Virtual MAC address
   - D. Vendor code
8. What is the default decrement in priority for an active HSRP router when the line protocol of one of its tracked interfaces goes down?
   - A. 5
   - B. 10
   - C. 15
   - D. 20
9. Which protocol is natively supported in Windows 2000 workstations?
   - A. Proxy ARP
   - B. IRDP
   - C. Default gateway
   - D. HSRP
10. What are the three types of HSRP messages?
    - A. Coup
    - B. Resign
    - C. Join
    - D. Hello

Key Terms

default Provides a definitive location for IP packets to be sent in case the source device has no IP routing functionality built into it. This is the method most end stations use to access nonlocal networks. The default gateway can be set manually or learned from a DHCP server.

hello interval The amount of time that elapses between the sending of HSRP hello packets. The default is 3 seconds.

hello packet An HSRP protocol data unit used to communicate HSRP information such as the virtual IP address, the hello time, and the hold time. This is also called a hello message.

hold time The amount of time it takes before HSRP routers in an HSRP group declare the active router down. The default is 10 seconds.

HSRP (Hot Standby Router Protocol) The Layer 3 Cisco-proprietary protocol that lets a set of routers on a LAN segment work together to present the appearance of a single virtual router or default gateway to the hosts on the segment. HSRP enables fast rerouting to alternate default gateways should one of them fail.

HSRP group Routers on a subnet, VLAN, or a subset of a subnet participating in an HSRP process. Each group shares a virtual IP address. The group is defined on the HSRP routers.

HSRP state The descriptor for the current HSRP condition for a particular router interface and a particular HSRP group. The possible HSRP states are initial, learn, listen, speak, standby, and active.

IRDP (ICMP Router Discovery Protocol) Hosts supporting IRDP dynamically discover routers in order to access nonlocal networks. IRDP allows hosts to locate routers (default gateways). Router discovery packets are exchanged between hosts (IRDP servers) and routers (IRDP clients).

preempt An HSRP feature that lets the router with the highest priority immediately assume the active role at any time.

priority An HSRP parameter used to facilitate the election of an active HSRP router for an HSRP group on a LAN segment. The default priority is 100. The router with the greatest priority for each group is elected the active forwarder for that group.

proxy ARP Lets an Ethernet host with no knowledge of routing communicate with hosts on other networks or subnets. Such a host assumes that all hosts are on the same local segment and that it can use ARP to determine their hardware addresses. Routers handle the proxy ARP function.

router discovery The process in which a host determines a default gateway in order to send data beyond the local LAN segment.

tracking An HSRP feature that allows you to specify other interfaces on the router for the HSRP process to monitor. If the tracked interface goes down, the HSRP standby router takes over as the active router. This process is facilitated by a decrement to the HSRP priority resulting from the tracked interface line protocol going down.

After completing this chapter, you will be able to perform tasks related to the following:

- Layer 2 and Layer 3 multicast addressing
- Internet Group Management Protocol (IGMP) operation and configuration
- Cisco Group Management Protocol (CGMP) operation and configuration
- IGMP snooping operation and configuration
- Router-Port Group Management Protocol (RGMP) operation and configuration
- IP multicast routing protocols
- Protocol-Independent Multicast (PIM) configuration

Chapter 9

Multicasting

Multicasting is an exciting set of network technologies at the forefront of research and development in the field of networking. New multicast technologies are being created each year to enable greater functionality or improve efficiency. To some extent, multicast technologies are driving the development of faster and more efficient networking devices, but it's probably more accurate to say that multicast technologies are evolving to enable and optimize multicast traffic in the existing Internet infrastructure.

Multicast technologies, such as security, multiservice, and QoS technologies, are changing rapidly. What was standard practice five years ago is presently passé. That's the joy of networking. The fields of endeavor in which the rate of change is greatest are the same fields in which opportunity is greatest. The upside is that you have to keep learning to stay on top of the latest technologies. If this seems like a downside, you might want to consider another career. Suffice it to say that multicasting is a thriving area of research that is pushing the envelope of campus network deployments.

In practical terms, IP multicast is a bandwidth-conserving technology that reduces traffic by simultaneously delivering a single stream of information to potentially thousands of corporate recipients and home users. Applications that take advantage of multicast include videoconferencing, corporate communications, distance learning, and distribution of software, stock quotes, and news.

IP multicast delivers source traffic to multiple receivers without adding any burden on the source or the receivers (multicast destinations) while minimizing network bandwidth. Multicast packets are replicated in the network by Cisco routers enabled with Protocol-Independent Multicast (PIM) and other supporting multicast protocols, resulting in the most efficient delivery of data to multiple receivers.

All alternatives to IP multicast require the source to send more than one copy of the data. Some even require the source to send an individual copy to each receiver. If there are thousands of receivers, even low-bandwidth applications benefit from using IP multicast.

High-bandwidth applications, such as MPEG video, might require a large portion of the available network bandwidth for a single stream. In these applications, the only way to send to more than one receiver simultaneously (without crippling the network) is by using IP multicast. Figure 9-1 shows how IP multicast is used to deliver data from one source to many interested recipients.

Figure 9-1 Multicasting conserves bandwidth.

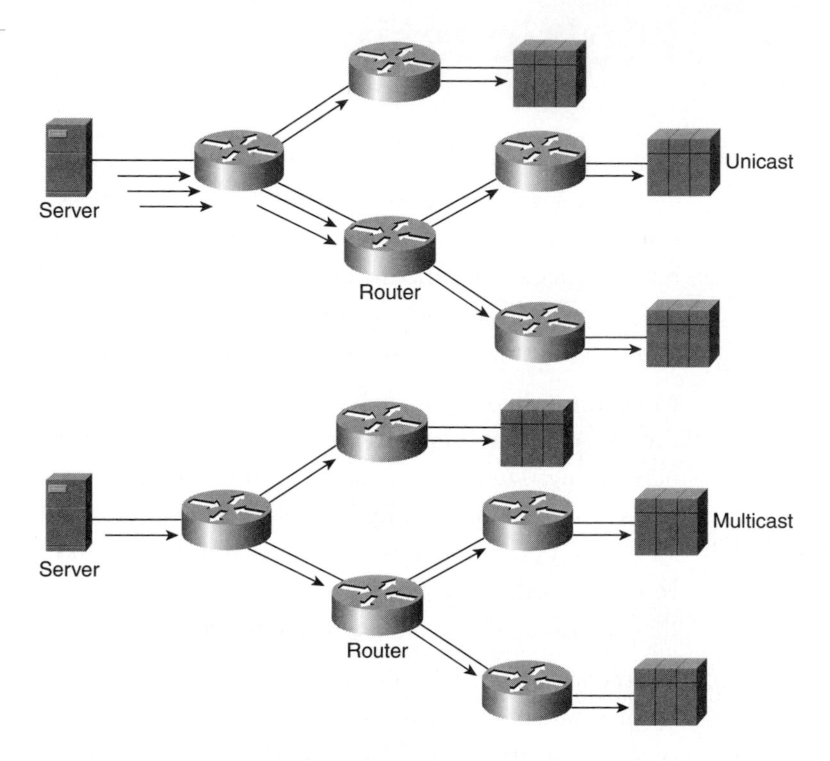

Multicast technologies are generally categorized according to whether they operate between hosts and routers, between switches and routers, or between routers. The multicast technologies operating between hosts and routers enable multicast data streams while simultaneously constraining unnecessary multicast traffic. Multicast technologies between switches and routers are generally designed to constrain unnecessary multicast traffic. And multicast technologies between routers are designed to enable multicast traffic flows while limiting unnecessary or unwanted multicast traffic.

This chapter is organized according to this loose categorization of multicast technologies. The chapter begins by discussing the fundamentals of multicasting, including the details of multicast addressing. Next, the multicast technologies that operate between hosts and routers are described in detail. This is followed by a discussion of multicast technologies operating between switches and routers. Then a survey of intradomain and interdomain multicast technologies operating between routers is presented. Details of how to configure multicast technologies are included in context.

Multicast Addressing

By now, you probably know that the Class D IP address space is used for multicast addressing. However, you might not have thought much about Layer 2 multicast addresses: Hosts on a LAN rely on MAC addresses to process multicast traffic. The following sections describe what each portion of the Class D address space is used for and detail the connection between IP multicast addresses and Layer 2 multicast addresses. Specifically, each of the following multicast addressing concepts are explored:

- Multicast groups
- IP multicast addresses
- Layer 2 multicast addresses
- Mapping between Layer 2 and Layer 3 multicast addresses

This sets a foundation from which we can delve into the various multicast technologies.

Multicast Groups

Multicasting is based on the concept of a group. A *multicast group* is an arbitrary group of receivers that expresses an interest in receiving a particular data stream. This group has no physical or geographical boundaries—the hosts can be located anywhere on the Internet. Hosts that are interested in receiving data flowing to a particular group must join the group using Internet Group Management Protocol (IGMP). Hosts must be a member of the group to receive the data stream. An example of a multicast data stream is video on demand within an enterprise. Multicast groups are identified by IP multicast addresses, which are discussed in the next section.

IP Multicast Addresses

IP multicast addresses specify an arbitrary group of IP hosts that have joined a group and want to receive traffic sent to this group.

The Internet Assigned Numbers Authority (IANA) controls the assignment of IP multicast addresses. IANA assigned the Class D address space to be used for IP multicast. Class D addresses are distinguished by the presence of the high-order bits 1110 in the first octet. All IP multicast group destination addresses fall in the range from 224.0.0.0 through 239.255.255.255. The source IP address for multicast data is always the unicast source address. The following sections describe specially allocated IP multicast address spaces: reserved link local addresses, globally scoped addresses, limited-scope addresses, and GLOP addresses.

Reserved Link Local Addresses

The IANA reserved *link local addresses* in the range from 224.0.0.0 through 224.0.0.255 to be used by network protocols on a local network segment. Packets with these addresses must never be forwarded by a router unless the router is configured to bridge data. These packets remain local on a particular LAN segment and are always sent with a TTL of 1.

Network protocols use these addresses for automatic router discovery and to communicate important routing information. For example, OSPF uses the IP addresses 224.0.0.5 and 224.0.0.6 to exchange OSPF link-state information. Table 9-1 lists some well-known link local IP addresses.

Table 9-1 Well-Known Link Local IP Multicast Addresses

Link Local IP Address	Usage
224.0.0.1	All systems on this subnet
224.0.0.2	All routers on this subnet
224.0.0.5	OSPF routers
224.0.0.6	OSPF designated routers
224.0.0.9	RIP Version 2 routers
224.0.0.10	EIGRP routers
224.0.0.12	DHCP server/relay agent
224.0.0.13	All PIM routers
224.0.0.22	IGMP
224.0.0.25	Router-to-switch (such as RGMP)

You are probably familiar with most of these link local addresses (PIM and IGMP are defined later in this chapter). This list is useful as a ready reference.

Globally Scoped Addresses

Addresses in the range from 224.0.1.0 through 238.255.255.255 are called *globally scoped addresses*. They can multicast data between organizations and across the Internet.

Some of these addresses have been reserved for use by multicast applications through IANA. For example, IP address 224.0.1.1 has been reserved for *Network Time Protocol*

(NTP), used to synchronize clocks on network devices. The IANA also assigned the two group addresses 224.0.1.39 and 224.0.1.40 for Auto-RP, a PIM sparse-mode mechanism (discussed later in this chapter). 224.0.1.39 is used by candidate rendezvous points, and 224.0.1.40 is used by rendezvous point-mapping agents, which are described in the section "Protocol-Independent Multicast."

Limited-Scope Addresses

Addresses in the range from 239.0.0.0 through 239.255.255.255 are called *limited-scope addresses* or *administratively scoped addresses*. These addresses are described in RFC 2365. Limited-scope addresses are constrained to a local group or organization. With limited-scope addresses, the same addresses might be in use at different locations for different multicast sessions (similar to private IP addresses with NAT).

Routers are typically configured with filters to prevent multicast traffic in this address range from flowing outside an autonomous system (AS) or any user-defined domain. Within an AS or domain, the limited-scope address range can be further subdivided so that local multicast boundaries can be defined. This subdivision, called *address scoping*, allows for address reuse between these smaller domains. For example, 239.254.0.0/15 can be used as a site-local scope, whereas 239.192.0.0/14 can be used as an organization-local scope.

GLOP Addresses

First of all, GLOP is not an acronym. The Merriam-Webster online dictionary site, www.m-w.com, defines *glop* as "a thick semiliquid substance (as food) that is usually unattractive in appearance."

RFC 2770 proposes that the 233.0.0.0/8 address range be reserved for statically defined addresses by organizations that already have an AS number reserved. Quoting from the RFC: "This describes an experimental policy for use of the class D address space using 233.0.0.0/8 as the experimental statically assigned subset of the class D address space." This practice is called *GLOP addressing*. The AS number of the domain is embedded in the second and third octets of the 233.0.0.0/8 address range.

For example, AS 62010 is written in hexadecimal format as F23A. Separating the two octets F2 and 3A results in 242 and 58 in decimal format. These values result in a subnet of 233.242.58.0 that would be globally reserved for AS 62010 to use.

RFC 3138, "Extended Assignments in 233/8," a product of the Multicast Deployment Working Group (MBONED), extends the mapping of the GLOP addresses to the private AS space, 64512 to 65535. RFC 3138 states, "While the technique described in RFC 2770 has been successful, the assignments are inefficient in those cases in which a /24 is too small or the user doesn't have its own AS." RFC 3138 proposes the extended GLOP, or EGLOP, address space to be 233.252.0.0 to 233.255.255.255.

NOTE

The 232/8 address range, a subset of the globally scoped addresses, is now allocated by IANA for the multicast technology Source-Specific Multicast (SSM), which is discussed later in this chapter.

Layer 2 Multicast Addresses

Ethernet hardware addresses are 48 bits, expressed as 12 hexadecimal digits. In these 12 hex digits, the first six digits match the vendor of the Ethernet interface. This is called the vendor code or the *Organizational Unique Identifier (OUI)*. The last six digits specify the interface serial number for that interface vendor.

The MAC address is written in many formats. Normally, it is written as a colon-separated list, such as 08:00:20:1B:60:42, but sometimes, it is expressed as 08-00-20-1B-60-42 (Microsoft Windows OS output) or 0800.201B.6042 (Cisco IOS Software output).

NICs on a LAN segment normally receive only packets destined for their burned-in MAC address or the broadcast MAC address. A way had to be devised for multiple hosts to receive the same packet and still be able to differentiate between multicast groups.

The IEEE LAN specifications made provisions for the transmission of broadcast and multicast packets. In the 802.3 standard, a value of 1 for the last bit in the first octet indicates a broadcast or multicast frame. This is why the first byte of every vendor MAC address is even: The MAC address of a NIC or interface is used as a source MAC address and, therefore, cannot be confused with a broadcast or multicast MAC address. Figure 9-2 shows the location of the broadcast/multicast bit in an Ethernet frame.

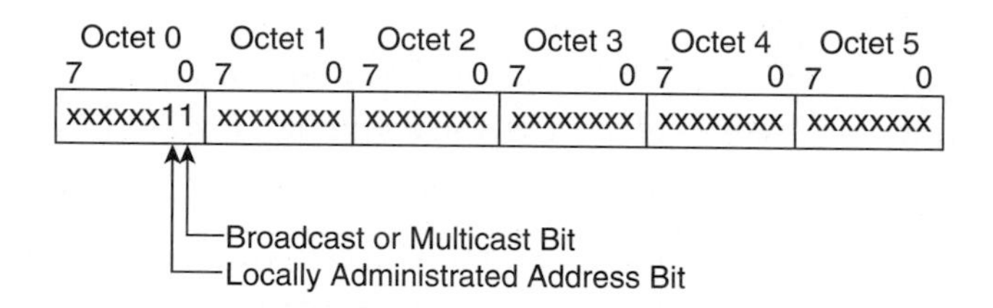

Figure 9-2 Multicast and broadcast Ethernet frames have a 1 in the last bit of the first octet.

When the IEEE assigns an Ethernet address block to a vendor, a block of 2^{25} addresses is reserved: a unicast block of 2^{24} addresses and a multicast block of 2^{24} addresses, differentiated by a 0 or 1 in the last bit of the first octet. Thus, multicast groups for vendor-specific uses tend to match the nonmulticast vendor assignments except for a 1 in the last bit of the first octet.

The following is a partial list of vendor codes owned by Cisco and used by various Cisco networking devices:

```
00000C   01-00-0C-CC-CC-CC is used for CDP, VTP, DISL, DTP, and PAgP
         01-00-0C-DD-DD-DD is used for CGMP (Cisco Group Management Protocol)
00067C
0006C1
00070D   2511 router
001007   Catalyst 1900
00100B
```

```
00100D    Catalyst 2924-XL
001011    Cisco 75xx
00101F    Catalyst 2901
001029    Catalyst 5000
00102F    Catalyst 5000
001079    Catalyst 5500
00107B
0010A6
0010F6
00400B    formerly Crescendo
00500F
005050
005069    formerly PixStream
0050BD
0050E2
006009    Catalyst 5000
00602F
00603E    100Mbps interface
006047
00605C
006070    2524 and 4500 routers
006083    3620/3640 routers
00801C
008024    formerly Kalpana
0090B1
00902B    Ethernet Switches and Light Streams
009086
009092
0090AB
0090B1
0090F2    Ethernet Switches and Light Streams
00A0C9    PIX firewall
00C01D    formerly Grand Junction Networks
00E014
00E01E
00E034
00E04F
00E08F    Catalyst 2900
00E0A3    Catalyst 1924
00E0B0
00E0F7
00E0F9
00E0FE
```

In this list, notice that the second character is always even. Cisco can use any of these addresses with a 1, replacing the second character for whatever Layer 2 multicast purposes Cisco deems appropriate. For each of these vendor codes, Cisco has 2^{24} Layer 2 multicast addresses to work with.

Mapping Between Layer 2 and Layer 3 Multicast Addresses

The IANA owns a block of Ethernet MAC addresses that start with 01:00:5E in hexadecimal format. Half of the 01:00:5E block is allocated for multicast addresses. The range from 0100.5E00.0000 through 0100.5E7F.FFFF is the available range of Ethernet MAC addresses for multicast.

This allocation allows for 23 bits in the Ethernet address to map to an IP multicast group address. The mapping places the lower 23 bits of the IP multicast group address into these available 23 bits in the Ethernet address (see Figure 9-3).

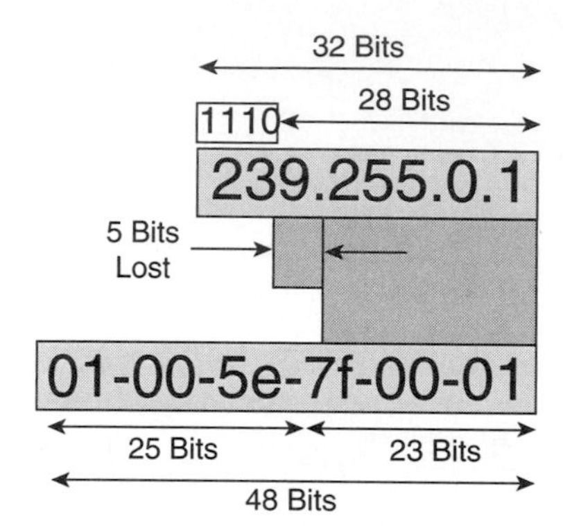

Figure 9-3
28 bits are available for IP multicast, and 23 bits are available for Layer 2 multicast.

When vendors take one of their OUIs and switch the eighth bit of the first octet to a 1 to create a multicast address space, the multicast MAC addresses do not map to any multicast IP addresses. For example, Cisco's 01-00-0C-DD-DD-DD, used for CGMP, comes from the 01-00-0C address space and the 00-00-0C OUI, but it is not part of the 01-00-5E address space and, therefore, does not map to any IP multicast address.

To view the multicast MAC addresses listened to on a CatOS switch, enter the command **show cam system**. For example, when CGMP is enabled on a CatOS switch, the 01-00-0C-DD-DD-DD address is added to the list of listened-to multicast MAC addresses. To view the multicast MAC addresses listened to on a 2900XL or 3500XL switch, enter the command **show mac-address-table self**.

NOTE

Vinton Cerf, a codeveloper of TCP/IP, wrote RFC 2468 as a memoriam to Jon Postel. It is titled "I Remember IANA," and it was published in 1998. It is definitely worth taking some time to read.

IP-to-MAC multicast mapping causes 5 bits in the IP address space to be lost. It turns out that this loss of 5 bits worth of information was not originally intended. Here is the story behind the loss of the 5 bits. When Dr. Steve Deering, the creator of modern multicasting, was doing his seminal research on IP multicast in the late 1980s, he approached his manager, Jon Postel, with the need for 16 OUIs to map all 28 bits worth of Layer 3 IP multicast address into unique Layer 2 MAC addresses. A single OUI provides 24 bits worth of unique MAC addresses to the organization.

Unfortunately, at the time of Deering's request to Postel, the IEEE charged $1000 for each OUI assigned, which meant that Deering was requesting that his manager spend $16,000 so he might continue his research. Due to budget constraints, Postel agreed to purchase a single OUI for Deering. However, Postel also chose to reserve half of the MAC addresses in this OUI for other graduate research projects, so ultimately, he granted Deering only half an OUI. This gave Deering only 23 bits worth of MAC address space to map 28 bits of IP multicast addresses, as shown in Figure 9-4.

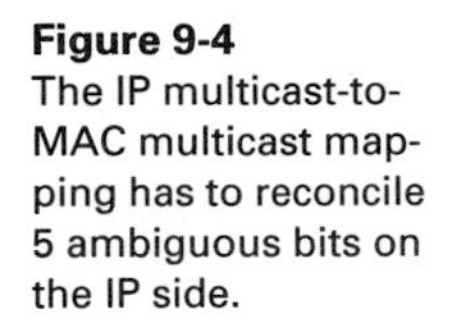

Figure 9-4
The IP multicast-to-MAC multicast mapping has to reconcile 5 ambiguous bits on the IP side.

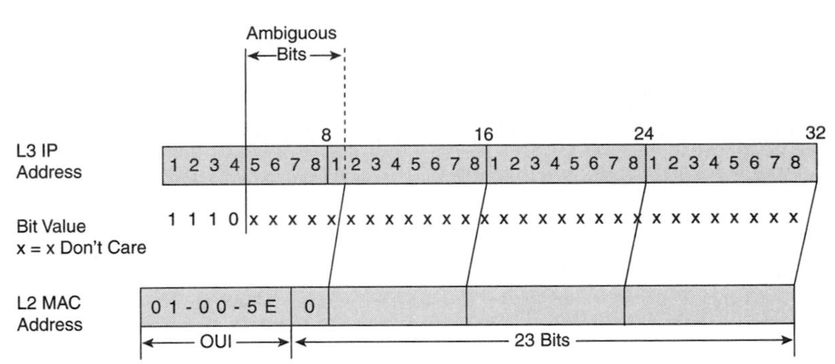

Because the upper 5 bits of the IP multicast address are dropped in this mapping, the resulting address is not unique. In fact, 32 different multicast group IDs map to the same Ethernet address, as shown in Figure 9-5.

Figure 9-5
32 IP multicast addresses map to a single multicast MAC address.

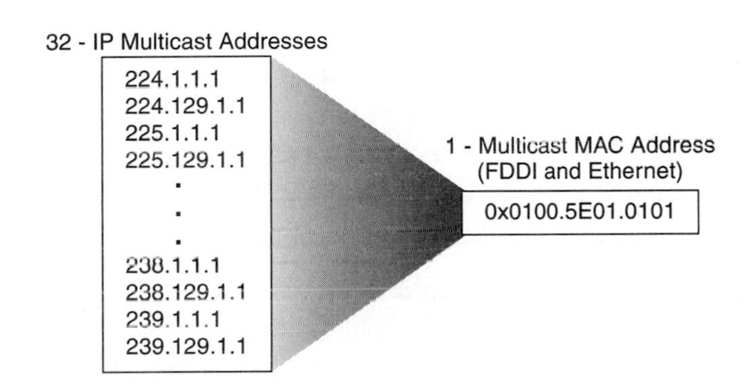

Notice that, if you fix the values of the second, third, and fourth octets, the first octet of an IP multicast address can take on any value between 224 and 239 and still map to the same multicast MAC address. In addition, if you fix the values of the first, third, and fourth octets, the values of the second octet for two IP multicast addresses can differ by 128 and still map to the same multicast MAC address. The preceding two sentences just rephrase the fact that bits 5, 6, 7, 8, and 9 of a multicast IP address are irrelevant with respect to the multicast MAC address that is mapped to. 32 IP multicast addresses map to a single multicast MAC address. This mapping is analogous to what you see in modular arithmetic, used in math and computer science, in which two numbers are congruent modulo *n* if they differ by a multiple of *n*. For example, 3 is congruent to 15 modulo 12, because 15 – 3 = 12 * 1. Or, you could think of it as 3 p.m. is the same as 15:00 in military time.

As an exercise in mapping between Layer 2 and Layer 3 multicast addresses, you might try determining the 32 IP multicast addresses associated with the multicast

MAC address 01-00-5E-00-00-0A. The answers are all associated with the IP link local multicast address used by Enhanced IGRP (EIGRP).

This completes our analysis of Layer 2 and Layer 3 multicast addressing. Next, we consider Internet Group Management Protocol (IGMP), which lets hosts participate in multicast streams while permitting some constraints on the set of hosts receiving the streams.

Internet Group Management Protocol

The primary purpose of *Internet Group Management Protocol (IGMP)* is to permit hosts to communicate their desire to receive multicast traffic to the IP multicast router(s) on the local network segment. This, in turn, permits the IP multicast router(s) to join the specified multicast group and to begin forwarding the multicast traffic onto the network segment. A user initiates IGMP communication indirectly via an application that requests data from a multicast group.

IGMP dynamically registers individual hosts in a multicast group on a particular LAN. Hosts identify group memberships by sending IGMP messages to their local multicast router. With IGMP, routers listen to IGMP messages and periodically send out queries to discover which groups are active or inactive on a particular subnet. Both hosts and routers on a connected segment must support compatible versions of IGMP.

The initial specification for IGMP Version 1 (IGMPv1) is documented in 1989 in RFC 1112, "Host Extensions for IP Multicasting," by Steve Deering. Since that time, many problems and limitations with IGMPv1 have been discovered. This led to the development of the IGMPv2 specification, which was ratified in November 1997 as RFC 2236.

Even before IGMPv2 had been ratified, work on the next generation of the IGMP protocol, IGMPv3, had already begun. IGMPv3 is specified in the Internet draft "Internet Group Management Protocol, Version 3." This document was expanded upon in the November 2001 Internet draft document "IGMPv3 for SSM," which defines the modifications required to the host and router portions of IGMPv3 to support SSM.

IGMP versions are described in the following sections. By default, Cisco routers use IGMP Version 2, but the version you use can be configured.

IGMP Version 1

In *IGMP Version 1 (IGMPv1)*, only the following two types of IGMP messages exist:

- Membership query
- Membership report

Membership queries and membership reports have significance only to the local subnet, so the TTL of these packets is always set to 1. This also prevents the packets from accidentally being forwarded off the local subnet, causing confusion on other subnets.

IGMPv1 has a relatively simple mode of operation. Hosts send out unsolicited *IGMP membership reports* corresponding to a particular multicast group to indicate that they are interested in joining that group, as shown in Figure 9-6. This mechanism lets hosts asynchronously join multicast groups, thus reducing join latency for the end system if no other members are present on the segment. Membership reports also respond to membership queries.

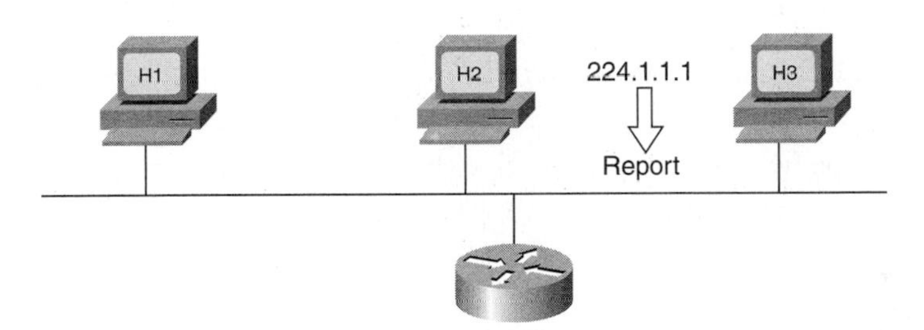

Figure 9-6 IGMPv1 has the simplest packet format, specifying the message type and multicast group address.

One router on the LAN segment is elected to send queries. The election is made by the respective multicast routing protocol (such as PIM), because IGMPv1 has no election mechanism. The router periodically (every 60 seconds) sends out an *IGMP membership query* to the all-hosts multicast address, 224.0.0.1, to verify that at least one host on the subnet is still interested in receiving traffic directed to that group. Such a membership query is also called a *general query* because it is sent to the all-hosts address. Only one member per multicast group responds to a query with a report. This is to save bandwidth on the subnet and CPU processing by the hosts. This process is called *report suppression*. The report-suppression mechanism is accomplished as follows:

- When a host receives the query, it starts a countdown timer for each multicast group of which it is a member. The countdown timers are each initialized to a random count within a given time range of 0 to 10 seconds.
- When a countdown timer reaches 0, the host sends a membership report for the group associated with the countdown timer to notify the router that the group is still active.
- However, if a host receives a membership report before its associated countdown timer reaches 0, it cancels the countdown timer associated with the multicast group, thereby suppressing its own report.

In Figure 9-7, H2's time expires first, so it responds with its membership report. H1 and H3 cancel their timers associated with the group, thereby suppressing their reports.

Figure 9-7
IGMPv1 uses report suppression to save bandwidth and CPU processing on hosts.

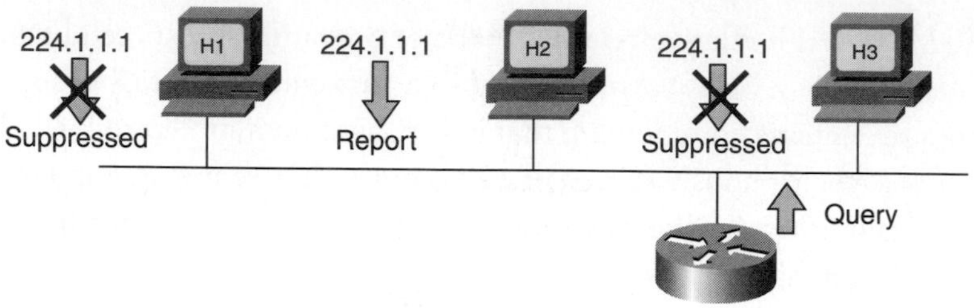

RFC 1112 describes the specification for IGMPv1. Figure 9-8 shows the packet format for an IGMPv1 message. The Version field always has a value of 1 (or 0001 in binary). The Type field is populated with either 1 or 2 (decimal representation), corresponding to membership query or membership report, respectively.

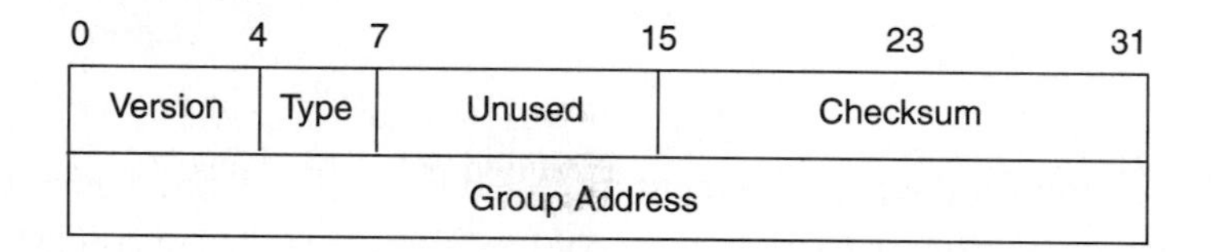

Figure 9-8
IGMPv1 has the simplest packet format, specifying the message type and multicast group address.

0	4	7	15	23 31
Version	Type	Unused	Checksum	
Group Address				

Finally, you need to consider the mechanism IGMPv1 uses for a host to leave a multicast group. The rule in multicast membership is that as long as one member of a multicast group is present on the segment, the group data must be forwarded to that segment. Only one active member of a multicast group is required to keep interest in that group.

For an IGMPv1 host to leave a multicast group, it simply waits for a timeout. The router sends 3 membership queries, 60 seconds apart, on the local segment. If no membership report is heard after 3 queries, the group times out. This is a major weakness of IGMPv1. It can take up to 3 minutes for a host to leave a multicast group, resulting in unwanted multicast traffic during this time.

Next, we look at IGMP Version 2, which adds functionality allowing for a host to proactively leave a group.

IGMP Version 2

IGMP Version 2 (IGMPv2) works basically the same way as Version 1. The join mechanism is exactly the same. Most of the changes between IGMPv1 and IGMPv2 are primarily to address the issues of leave and join latencies and ambiguities in the original protocol specification.

With IGMPv2, hosts can actively communicate to the local multicast routers their intention to leave the group. A host that wants to leave a group sends an *IGMP leave-group message* to the all-routers multicast address, 224.0.0.2. This allows end systems to tell the router they are leaving the group. This reduces the leave latency for the group on the segment when the member leaving is the last member of the group.

After a host sends a leave-group message, the local multicast router sends an *IGMP group-specific query* and determines whether any remaining hosts are interested in receiving the traffic. The group-specific query was added in Version 2 to allow the router to restrict membership queries to a single group instead of all groups. This is an optimized way to quickly find out if any members are left in a group without asking all groups for a report. The difference between the group-specific query and the general query of Version 1 is that a general (membership) query is multicast to the all-hosts address (224.0.0.1), and a group-specific query for Group G is multicast to the Group G multicast address.

If there are no replies to a group-specific query within three seconds, the router times out the group and stops forwarding the traffic. This is pictured in Figure 9-9.

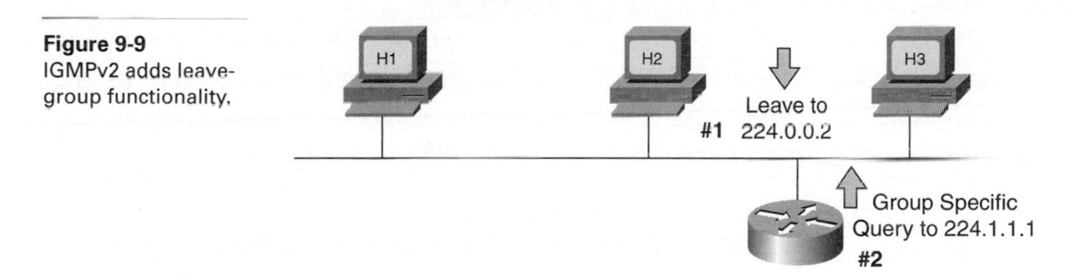

Figure 9-9
IGMPv2 adds leave-group functionality.

Another addition to IGMPv2 is that the process of electing a router querier is built into the protocol instead of relying on a multicast routing protocol. This would often result in multiple queriers on a single multiaccess network. The lowest unicast IP address of the IGMP-speaking routers is elected as the querier. All IGMP-speaking routers initially think that they are the querier, but they relinquish that role if a lower IP address membership query is heard on the same segment. If the currently elected querier fails to issue a query within a specified time limit, a timer in the other IGMPv2 routers times out and causes them to reinitiate the election process.

Finally, IGMPv2 added the query-interval response time, which controls the burstiness of membership reports. The query-interval response time is added in query packets to convey to the membership how much time they have to respond to a query with a report. This value, between 1 and 25 seconds, is used by the IGMPv2 hosts as the upper boundary when randomly choosing the value of their response timers (the default is 10 seconds). You might want to again read about the report-suppression mechanism, detailed in the preceding section, to put this response timer in context.

RFC 2236 describes the specification for IGMPv2. Figure 9-10 shows the packet format for an IGMPv2 message.

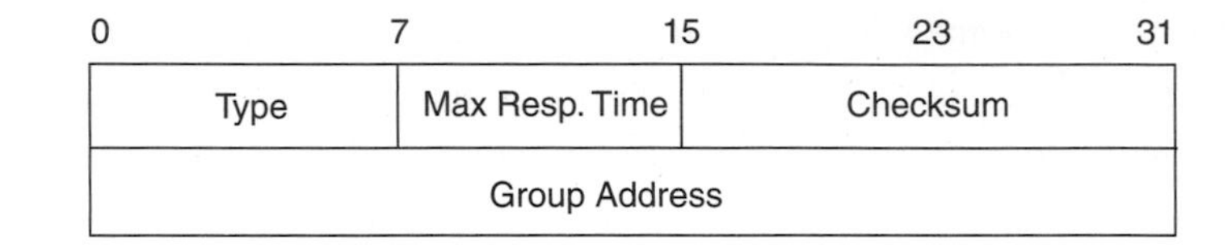

Figure 9-10 IGMPv2 packets merge the 4-bit Version and Type fields in IGMPv1 into a single 8-bit Type field.

In Version 2, the following four types of IGMP messages exist:

- Membership query (type 0x11)
- Version 1 membership report (type 0x12)
- Version 2 membership report (type 0x16)
- Leave group (type 0x17)

In IGMPv2, the old IGMPv1 4-bit Version field is merged with the old 4-bit Type field to create a new 8-bit Type field. By assigning IGMPv2 type codes 0x11 and 0x12 as the membership query (Versions 1 and 2) and the v1 membership report, respectively, backward compatibility with the IGMPv1 and IGMPv2 packet formats is maintained. Think about it!

The Group address field is identical to the IGMPv1 version of this field, except that it is set to 0.0.0.0 for general queries.

The addition of the leave-group message in IGMP Version 2 greatly reduces the leave latency compared to IGMP Version 1. Unwanted and unnecessary traffic can be stopped much sooner. IGMPv2 is the default setting for Cisco routers.

IGMP Version 3

IGMP Version 3 (IGMPv3) is the next step in the evolution of IGMP. IGMPv3 adds support for *source filtering*, which lets a multicast receiver (host) tell a router which groups it wants to receive multicast traffic from, and from which sources this traffic is expected. This membership information lets Cisco IOS Software forward traffic from only sources from which receivers requested the traffic. Among other things, this added feature can prevent rogue multicast servers from monopolizing the bandwidth on a segment; receivers must specifically request the servers from which multicast traffic is desired. IGMPv3 helps conserve bandwidth by allowing hosts to specify the sources of multicast traffic.

IGMPv3 has separate packet formats for membership queries and membership reports. Figure 9-11 shows the query packet format for an IGMPv3 message.

Figure 9-11 IGMPv3 membership query packets include the number of multicast sources and the source IP addresses.

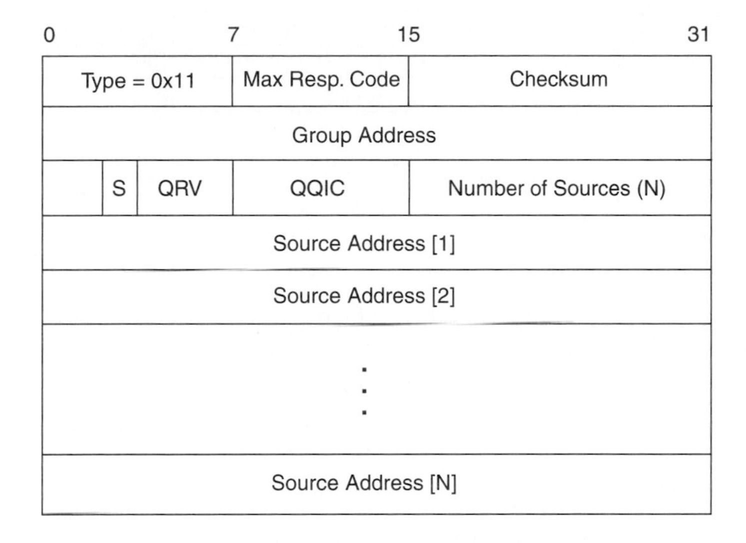

Table 9-2 describes the significant fields in an IGMPv3 query message.

Table 9-2 IGMPv3 Query Packet Field Descriptors

Query Message Field	Description
Type = 0x11	IGMP query.
Max resp. code	Maximum response time (in seconds).
Group address	Multicast group address. This address is 0.0.0.0 for general queries.
S	S flag. This flag indicates that processing by routers is being suppressed. When set to 1, the S flag indicates to any receiving multicast routers that they are to suppress the normal timer updates they perform upon hearing a query.
QRV	Querier Robustness Value. This value affects timers and the number of retries. Routers adopt the QRV value from the most recently received query as their own.
QQIC	Querier's Query Interval Code. The query interval value (in seconds) used by the querier. Multicast routers that are not the current querier adopt the QQIC value from the most recently received query as their own.
Number of sources [N]	The number of sources present in the query. This number is nonzero for a group-and-source query.
Source address [1...N]	The source's address.

Figure 9-12 shows the report packet format for an IGMPv3 message.

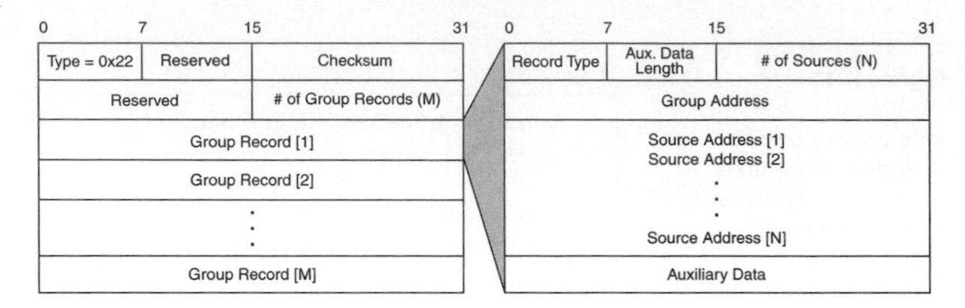

Figure 9-12 IGMPv3 membership report packets include records detailing which multicast groups the host wants to join and the sources from which to receive the multicast traffic.

Table 9-3 describes the significant fields in an IGMPv3 report message. The Auxiliary Data field, if present, contains additional information pertaining to the group record. According to the January 2002 IETF draft for IGMPv3, implementations of IGMPv3 must not include any auxiliary data (that is, the Aux Data Len field must be set to 0) in any transmitted group record and must ignore any auxiliary data present in any received group record.

Table 9-3 **IGMPv3 Report Packet Field Descriptors**

Report Message Field	Description
# of group records [M]	The number of group records present in the report.
Group record [1...M]	A block of fields containing information regarding the sender's membership in a single multicast group on the interface from which the report was sent.
Record type	The group record type. The possible values are as follows: MODE_IS_INCLUDE MODE_IS_EXCLUDE CHANGE_TO_INCLUDE_MODE CHANGE_TO_EXCLUDE_MODE ALLOW_NEW_SOURCES BLOCK_OLD_SOURCES
# of sources [N]	The number of sources present in the record.
Source address [1...N]	The source's address.

In IGMPv3, the following types of IGMP messages exist:

- Version 3 membership query (Type 0x11)
- Version 3 membership report (Type 0x22)

An implementation of IGMPv3 must also support the following three message types, for interoperation with previous versions of IGMP:

- Type 0x12 Version 1 Membership Report (RFC 1112)
- Type 0x16 Version 2 Membership Report (RFC-2236)
- Type 0x17 Version 2 Leave Group (RFC-2236)

The IETF draft states that, in order to take full advantage of IGMPv3's capabilities, a system's IP application programming interface (API) must support the following operation:

IPMulticastListen(socket,interface,multicast-address,filter-mode,source-list)

where:

- socket is an implementation-specific parameter used to distinguish among different requesting entities (such as programs or processes) within the system.
- interface is a local identifier of the network interface on which reception of the specified multicast address is to be enabled or disabled. Interfaces can be physical (such as an Ethernet interface) or virtual (such as the endpoint of a Frame Relay virtual circuit).
- multicast-address is the IP multicast address to which the request pertains.
- filter-mode can be either INCLUDE or EXCLUDE. In INCLUDE mode, reception of packets sent to the specified multicast address is requested only from IP source addresses listed in the source-list parameter. In EXCLUDE mode, reception of packets sent to the given multicast address is requested from all IP source addresses except those listed in the source-list parameter.
- source-list is an unordered list of zero or more IP unicast addresses from which multicast reception is desired or not desired, depending on the filter mode.

For example, the IGMP join operation is equivalent to IPMulticastListen(socket,interface,multicast-address,EXCLUDE,{}), and the IGMP leave operation is equivalent to IPMulticastListen(socket,interface,multicast-address,INCLUDE,{}) where {} is an empty source list.

IGMPv3 supports applications that explicitly signal sources from which they want to receive traffic. With IGMPv3, receivers signal membership to a multicast group in the following two modes:

- **INCLUDE mode**—In this mode, the receiver announces membership to a multicast group and provides a list of source addresses (the INCLUDE list) from which it wants to receive traffic.
- **EXCLUDE mode**—In this mode, the receiver announces membership to a multicast group and provides a list of source addresses (the EXCLUDE list) from

which it does not want to receive traffic. The host receives traffic only from sources whose IP addresses are not listed in the EXCLUDE list. In order to receive traffic from all sources, the old behavior of IGMPv2, a host uses EXCLUDE mode membership with an empty EXCLUDE list.

In Figure 9-13, host H1 has joined group 224.1.1.1 but wants to receive traffic from only Source 1.1.1.1. Host H1 informs the PIM designated router (DR) R3 that it is only interested in multicast traffic from Source 1.1.1.1 for group 224.1.1.1. Router R3 can then prune the traffic sourced by 2.2.2.2. PIM DRs are described in the section "Protocol-Independent Multicast." A PIM DR is elected on each multiaccess segment.

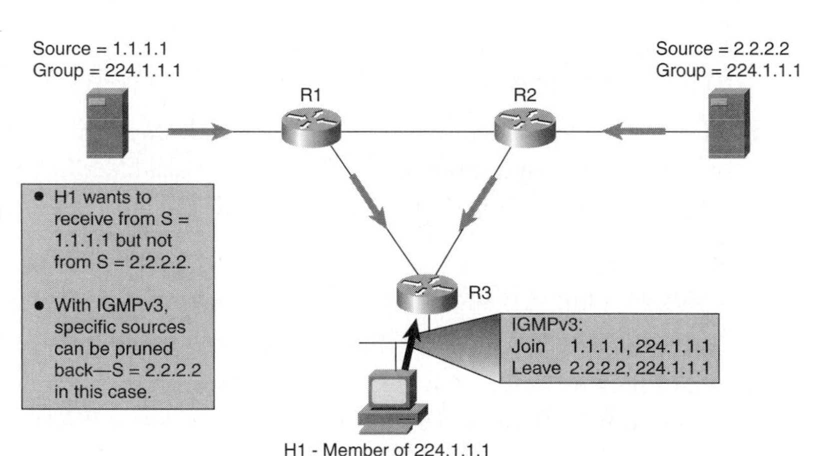

Figure 9-13 IGMPv3 membership report packets include records detailing the multicast groups the host wants to join and the sources from which to receive the multicast traffic.

One of the major benefits of IGMPv3 is its integration with SSM to provide enhanced multicast service. IGMPv3 is the industry-designated standard protocol for hosts to signal channel subscriptions in SSM. SSM was introduced in Cisco IOS Software Release 12.1(3)T. SSM support for IGMPv3 was introduced in 12.1(5)T. For SSM to rely on IGMPv3, IGMPv3 must be available in last-hop routers and host operating system network stacks and must be used by the applications running on those hosts.

IGMPv3 benefits include the following:

- **Enables new multicast services**—SSM. See the section "Source-Specific Multicast" later in this chapter.
- **Optimized bandwidth utilization**—A receiver might request to receive traffic only from explicitly known sources.
- **Improved security**—No denial-of-service attacks from unknown sources.

We do not explore IGMPv3 any further. However, we want to emphasize that IGMPv3 will likely begin to play a greater role in multicast deployments because of the recent growth in popularity of SSM.

To complete our look at IGMP, we next consider IGMP configuration on Cisco routers.

Configuring IGMP

If, by default, no user is registered to a specific group in a subnet, the router does not forward multicast traffic for that group into that subnet. That means that a router needs to receive an IGMP report for a multicast group in order to add it to the multicast routing table and to start forwarding traffic for that group. To enable IGMP on an interface so that hosts can begin joining multicast groups, you need to globally enable multicast routing with the **ip multicast routing** command and configure a multicast routing protocol on the appropriate interface(s). Later in this chapter, we discuss configuring multicast routing protocols, but an example of this is configuring PIM dense mode on the appropriate interface with the command **ip pim dense-mode**.

To test multicast configurations (as soon as your network is set up to support multicast), it is useful to configure a Cisco router to be a member of the multicast group that you are testing. As soon as you have configured the router to be a member of the multicast group, you can ping the multicast address from devices in the multicast network to test functionality for that multicast group. The router responds to ICMP echo request packets addressed to a group of which it is a member. To configure the router to join a multicast group, use the interface command **ip igmp join-group** *group-address*. Enabling a multicast routing protocol on an interface (without the global command **ip multicast routing**) and entering the command **ip igmp join-group** on an interface are both methods of enabling IGMP on the interface, but it won't do much good if the router can't forward multicast traffic!

You might also want to limit the multicast groups supported on a given LAN segment. Multicast routers send IGMP host query messages to determine which multicast groups have members on the router's attached local networks. The routers then forward to these group members all packets addressed to the multicast group. You can place a filter on each interface that restricts the multicast groups that hosts on the subnet serviced by the interface can join. To filter multicast groups allowed on an interface, use the interface command **ip igmp access-group** *access-list*.

If you want to tweak the interval that an IGMP router uses to send host query messages (the default is 60 seconds), use the interface command **ip igmp query-interval** *seconds*. To change the interval that an IGMPv2 router uses for the query-interval response time (which tells the member hosts how much time they have to respond to a query with a report), use the interface command **ip igmp query-max-response-time** *seconds* (the default is 10 seconds). Example 9-1 demonstrates how to verify that these two settings have been entered correctly, using the **show ip igmp interface** command.

Example 9-1 show ip igmp interface Displays IGMP Timers

```
Router(config-if)#ip igmp query-interval 120
Router(config-if)#ip igmp query-max-response-time 20
Router(config-if)#end
Router#show ip igmp interface e0
Ethernet0 is up, line protocol is up
Internet address is 10.1.3.1, subnet mask is 255.255.255.0
IGMP is enabled on interface
Current IGMP version is 2
CGMP is disabled on interface
IGMP query interval is 120 seconds
IGMP querier timeout is 240 seconds
IGMP max query response time is 20 seconds
Inbound IGMP access group is not set
Multicast routing is enabled on interface
Multicast TTL threshold is 0
Multicast designated router (DR) is 10.1.3.1 (this system)
IGMP querying router is 10.1.3.1 (this system)
Multicast groups joined: 224.0.1.40 224.2.127.254
```

To determine whether a group is active on an interface, use the command **show ip igmp group**, as shown in Example 9-2. This command is also useful for confirming that hosts are successfully leaving multicast membership.

Example 9-2 show ip igmp group Displays Active Multicast Groups

```
Router#show ip igmp group
IGMP Connected Group Membership
Group Address Interface Uptime Expires Last Reporter
224.1.1.1 Ethernet0 6d17h 00:02:31 1.1.1.11
```

The output shows that group 224.1.1.1 is active on Ethernet 0 and has been for 6 days and 17 hours. The group will expire and be deleted in 2 minutes and 31 seconds if an IGMP membership report for this group is not heard in that time. The last host to report membership was 1.1.1.11.

Also, by default, the router uses IGMP Version 2. All systems on the subnet must support the same version. You need to configure the router for Version 1 if your hosts do not support Version 2. To select the IGMP version that the router uses, enter the interface command **ip igmp version** {**3** | **2** | **1**}.

Finally, to determine which router is the IGMPv2 querier on a multiaccess network, use the **show ip igmp interface** command, as demonstrated in Example 9-3.

Example 9-3 Use the show ip igmp interface Command to Determine the IGMP Querier

```
Router(config-if)#ip igmp version 2
Router(config-if)#end
Router#show ip igmp interface e0
Ethernet0 is up, line protocol is up
Internet address is 1.1.1.1, subnet mask is 255.255.255.0
IGMP is enabled on interface
Current IGMP version is 2
CGMP is disabled on interface
IGMP query interval is 60 seconds
IGMP querier timeout is 120 seconds
IGMP max query response time is 10 seconds
Inbound IGMP access group is not set
Multicast routing is enabled on interface
Multicast TTL threshold is 0
Multicast designated router (DR) is 1.1.1.1 (this system)
IGMP querying router is 1.1.1.1 (this system)
Multicast groups joined: 224.0.1.40 224.2.127.254
```

The designated router is a different function (determined by PIM) and is listed separately in the output. The output also shows that IGMP Version 2 is enabled on the interface.

This completes our investigation of the Internet Group Management Protocol. In the remainder of this chapter, we explore multicast technologies that govern the behavior of multicast traffic between either switches and routers or between routers. Next, we discuss multicast technologies designed to constrain unnecessary traffic on switch ports.

Constraining Layer 2 Multicast Traffic

The default behavior for a Layer 2 switch is to forward all multicast traffic to every port that belongs to the destination LAN on the switch, as pictured in Figure 9-14. This behavior defeats the purpose of the switch, which is to limit traffic to the ports that need to receive the data.

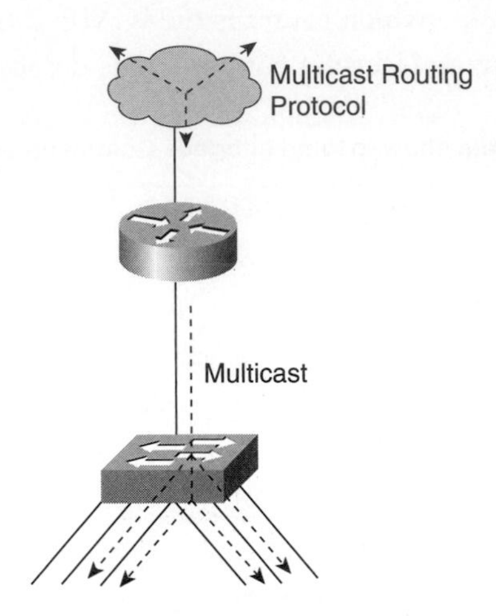

Figure 9-14 A switch's default behavior is to forward multicast traffic out all ports, just as it does with broadcasts and unknown unicasts.

There are three ways to constrain IP multicast in a Layer 2 switching environment:

- Cisco Group Management Protocol (CGMP)
- IGMP snooping
- Router-Port Group Management Protocol (RGMP)

CGMP and IGMP snooping are used on subnets that include end users or receiver clients. RGMP is used on routed segments that contain only routers, such as in a collapsed backbone (where you have switches connected only to routers). Each of these is discussed in the following sections.

Cisco Group Management Protocol

Cisco Group Management Protocol (CGMP) is a Cisco-developed protocol that allows Catalyst switches to leverage IGMP information on Cisco routers (or route processors) to make Layer 2 forwarding decisions. CGMP must be configured on both the multicast routers (CGMP servers) and the Layer 2 switches (CGMP clients). The result is that, with CGMP, IP multicast traffic is delivered only to Catalyst switch ports that are attached to interested receivers. All other ports that have not explicitly requested the traffic do not receive it unless these ports are connected to a multicast router. Multicast router ports must receive every IP multicast data packet.

The following sections detail CGMP operation and configuration on Catalyst switches and Cisco routers.

CGMP Operation

The basic operation of CGMP is shown in Figure 9-15. When the host joins multicast group 224.1.2.3 (part A in the figure), it multicasts an unsolicited IGMP membership report message with its *Unicast Source Address (USA)*, 0080.C7A2.1093, to the target

Group Destination Address (GDA), 0100.5E01.0203. (The IP multicast group 224.1.2.3 maps to the GDA.) The IGMP report is passed through the switch to the router for normal IGMP processing. The router (which must have CGMP enabled on this interface) receives the IGMP report and processes it as it normally would, but it also creates a CGMP join message and sends it to the switch (part B in Figure 9-15). The router, or route processor, is considered the CGMP server, and the Catalyst Layer 2 switch is the CGMP client.

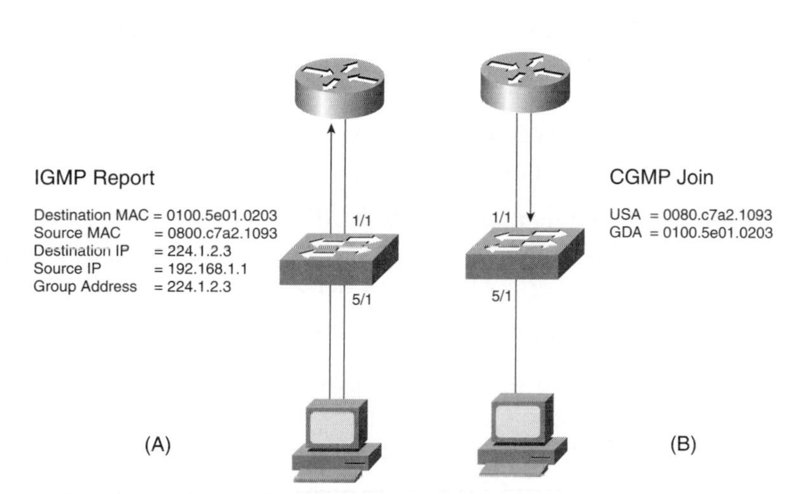

Figure 9-15 CGMP communicates between Cisco routers and Catalyst switches to allow the switches to forward multicast data out only ports that have interested receivers.

The switch receives this CGMP join message and adds the port to its Content-Addressable Memory (CAM) table for that multicast group. All subsequent traffic directed to this multicast group is forwarded out the port for that host. The router port also is added to the entry for the multicast group. Multicast routers must listen to all multicast traffic for every group because the IGMP control messages also are sent as multicast traffic. With CGMP, the switch must listen only to CGMP join and CGMP leave messages from the router. The rest of the multicast traffic is forwarded using the CAM table exactly how the switch was designed.

CGMP frames are Ethernet frames with the destination MAC address 01-00-0C-DD-DD-DD and a SNAP header with the value 0x2001. The CGMP frames contain the following fields:

- **Version**—1 or 2.
- **Message Type**—Join or Leave.
- **Count**—The number of multicast/unicast address pairs in the message. The value reflects the number of GDA and USA entries.
- **GDA**—The MAC address associated with the multicast group.

- **USA**—The MAC unicast addresses of the devices that want to join the multicast group.

The switch needs to be aware of all router ports so that they are automatically be added to any newly created multicast entries. The switch learns of multicast router ports when it receives a CGMP join to GDA 0000.0000.0000 from a router interface USA. These messages are generated on all the router's interfaces that are configured to run CGMP. You can also manually specify multicast router ports on a Catalyst switch. For example, on a CatOS switch, you use the command **set multicast router** *mod/port*.

The precise method that a host uses to join a group with CGMP is as follows:

Step 1 A new host asks to receive traffic for a GDA, so the host sends an IGMP membership report.

Step 2 The router receives the IGMP report, processes it, and sends a CGMP message to the switch. The router copies the multicast group's destination MAC address into the CGMP join's GDA field and copies the host's source MAC address into the CGMP join's USA field.

Step 3 A switch with CGMP enabled needs to listen for 0100.0CDD.DDDD addresses (CGMP messages use this multicast destination address). The switch's processor looks into the USA's CAM table. After the USA is seen in the CAM table, the switch knows on which port the USA is located. Then it either creates a new *static* entry for the GDA and links the USA port to it along with all router ports or simply adds the USA port to the list of the ports for this GDA (if the static entry already exists).

Now we describe how CGMP allows switch ports to be deleted for a particular group (CGMP leave messages enable this process). When IGMPv1 is running on the hosts, the hosts do not send IGMP leave messages. In this case, the router can send CGMP leave messages, but only if it does not receive a reply to three consecutive IGMP queries (which can take up to 3 minutes). This means that, with IGMPv1 on the hosts, *no* port will be deleted from a group if any users are still interested in that group.

CGMP has a mechanism called *CGMP fast-leave* (also known as CGMP fast-leave processing and CGMP leave) that lets the switch quickly remove a port from its list of ports receiving multicast traffic for a particular multicast group. This functionality requires the hosts to use IGMPv2, because CGMP fast-leave relies on IGMPv2 leave group messages.

CGMP fast-leave processing allows the switch to detect IGMPv2 leave messages sent by hosts to the all-router multicast address (224.0.0.2). When the switch receives a leave message, it starts a query-response timer and sends a message on the port on which that leave was received to determine if there is still a host willing to receive this multicast group on that port. If this timer expires before a CGMP join message is received, the port is pruned from the multicast tree for the multicast group specified in the original leave message. If it is the last port in the multicast group, it forwards the IGMP leave message to all router ports. The router then starts the normal deletion process by sending a group-specific query. Because no responses are received, the router removes this group from the multicast routing table for that interface. It also sends a CGMP leave message to the switch that will erase the group from the static table. Fast-leave processing ensures optimal bandwidth management for all hosts on a switched network, even when multiple multicast groups are in use simultaneously.

Table 9-4 summarizes CGMP messages. The second message in the list requires CGMP fast-leave. Recall that the switch recognizes router ports when it receives CGMP join messages with GDA 0000.0000.0000.

Table 9-4 CGMP Messages Between Routers and Switches

GDA	USA	Join/Leave	Message
Multicast MAC	Host MAC	Join	Add port to group.
Multicast MAC	Host MAC	Leave	Delete port from group.
0000.0000.0000	Router MAC	Join	Assign router port.
0000.0000.0000	Router MAC	Leave	Unassign router port.
Multicast MAC	00-00-00-00-00-00	Leave	Delete group.
0000.0000.0000	0000.0000.0000	Leave	Delete all groups.

When CGMP fast-leave processing is enabled, two entries are added to the system CAM, 0100.5E00.0001 and 0100.5E00.0002, because IGMP leave uses 224.0.0.2 and IGMP query uses 224.0.0.1. Also, CGMP does not prune multicast traffic for any IP multicast address that maps to the MAC address range of 0100.5E00.0000 to 0100.5E00.00FF. The reserved link local IP multicast addresses in the range 224.0.0.0 to 224.0.0.255 are used solely for IP multicast traffic on a LAN segment.

Finally, a source-only network is a segment with only a source multicast and no real client. Therefore, there is a chance that no IGMP reports will be generated on that segment. However, in this case, CGMP is designed to restrict the flooding of this source to all hosts on the switch. If a router detects multicast traffic on one interface with no IGMP report, it is identified as a multicast source-only network. The router generates a CGMP join message for itself, and the switch simply adds this group (with only the router port).

Due to a conflict with Hot Standby Router Protocol (HSRP), CGMP leave processing is disabled by default on Catalyst switches. HSRP uses MAC address 0100.5E00.0002, which is the same one used by IGMP leave messages. With CGMP fast-leave, all HSRP packets go to the switch CPU. Because HSRP is not an IGMP packet, the switch regenerates the packets and sends them to all router ports. Therefore, you must not enable CGMP leave processing between HSRP peers when troubleshooting HSRP problems. Doing so can lead to excessive CPU load on the switch, which in turn affects the switch's ability to regenerate HSRP packets.

With this solid introduction to CGMP operation, you are now ready to configure and verify CGMP. Actually, you could configure it without understanding how it works, but you might have difficulty troubleshooting if problems are associated with CGMP.

CGMP Configuration

To configure CGMP, you first need to verify that your Cisco equipment supports it. The trend appears to be away from CGMP with newer Catalyst switches and toward IGMP snooping. CGMP clients are supported on the following platforms:

- **Catalyst 5000 series**—Supported for all software versions above 2.3.
- **Catalyst 2901, 2902, 2926T, 2926F, and 2926G**—Supported for all software versions above 2.3.
- **Catalyst 4000 series (without Supervisor Engine III), Catalyst 2948G, Catalyst 2980G, and Catalyst 4912**—Supported for all software versions.
- **Catalyst 2900XL and Catalyst 3500XL**—Supported for all software versions above 11.2(8)SA3.
- **Catalyst 1900 and 2820**—Supported since software version 6.x.

CGMP is not supported on the Catalyst 6000 or 2950 families of switches. Also, Catalyst 3550 multilayer switches and Catalyst 4006 switches with a Supervisor Engine III can act as CGMP servers but not as CGMP clients. The 6000 family, the 2950, the 3550, and the 4006 with Supervisor Engine III all support IGMP snooping.

Next, we detail the configurations for CGMP on CatOS switches and IOS-based switches. Then we show you how to configure Cisco routers for CGMP.

CatOS Switch To configure CGMP (as a client) on a CatOS switch, use the command **set cgmp enable.** (Disable it with **set cgmp disable.**) Verify the CGMP configuration with the command **show cgmp statistics** [*vlan_number*]. These commands are shown in Example 9-4.

Example 9-4 Configuring CGMP on Catalyst 4000 and 5000 Supervisor Modules

```
Console> (enable) set cgmp enable

CGMP support for IP multicast enabled.
Console> (enable) show cgmp statistics 1

CGMP enabled

CGMP statistics for vlan 1:
valid rx pkts received           211915
invalid rx pkts received         0
valid cgmp joins received        211729
valid cgmp leaves received       186
valid igmp leaves received       0
valid igmp queries received      3122
igmp gs queries transmitted      0
igmp leaves transmitted          0
failures to add GDA to EARL      0
topology notifications received  80
number of CGMP packets dropped   2032227
```

To enable CGMP fast-leave, use the command **set cgmp leave enable**. (Disable it with **set cgmp leave disable.**) Verify the fast-leave configuration with the command **show cgmp leave**. These commands are demonstrated in Example 9-5.

Example 9-5 Configuring CGMP Fast-Leave on Catalyst 4000 and 5000 Supervisor Modules

```
Console> (enable) set cgmp leave enable

CGMP leave processing enabled.
Console> (enable)
Console> (enable) show cgmp leave

CGMP:        enabled
CGMP leave: enabled
```

To display information about multicast router ports learned dynamically using CGMP, use the command **show multicast router cgmp** [*mod/port*] [*vlan_id*]. Example 9-6 demonstrates this command.

Example 9-6 show multicast router cgmp Lists CGMP-Discovered Router Ports

```
Console> (enable) show multicast router cgmp

CGMP enabled
IGMP disabled

Port       Vlan
---------  ----------------
 2/1       99
 2/2       255
 7/9       2,99

Total Number of Entries = 3
'*' - Configured
```

To display information about multicast groups on the switch, use the command **show multicast group**, as shown in Example 9-7.

Example 9-7 show multicast group Displays Multicast Groups Cached by the Switch

```
Console> (enable) show multicast group

CGMP enabled
IGMP disabled

VLAN  Dest MAC/Route Des  Destination Ports or VCs / [Protocol Type]
----  ------------------  ----------------------------------------------------
1     01-00-11-22-33-44*  2/6-12
1     01-11-22-33-44-55*  2/6-12
1     01-22-33-44-55-66*  2/6-12
1     01-33-44-55-66-77*  2/6-12

Total Number of Entries = 4
```

The * indicates that the entry was manually configured in the CAM table. (For example, the first entry is set with the command **set cam static 01-00-11-22-33-44 2/6-12.**)

The multicast MAC addresses in the output are vendor-specific addresses that do not map to IP multicast addresses.

IOS-Based Switch On IOS-based switches supporting CGMP client configuration, CGMP is enabled by default. To enable fast-leave processing, use the global configuration command **cgmp leave-processing**.

Use the command **show cgmp** to display the state of the CGMP-learned multicast groups and routers, as shown in Example 9-8.

Example 9-8 show cgmp IOS-Based Switch Command Displays Information for a CGMP Client Switch

```
Switch#show cgmp

CGMP is running.
CGMP Fast Leave is not running.
CGMP Allow reserved address to join GDA.
Default router timeout is 300 sec.

vLAN     IGMP MAC Address   Interfaces
------  ----------------   -----------
    1    0100.5e01.0203      Fa0/8
    1    0100.5e00.0128      Fa0/8

vLAN     IGMP Router        Expire   Interface
------  ----------------  --------  ----------
    1    0060.5cf3.d1b3     197 sec   Fa0/8
```

To delete CGMP information learned by the switch, use the command **clear cgmp** [**vlan** *vlan-id*] | [**group** [*address*] | **router** [*address*]]. Example 9-9 illustrates some variations on this command, first showing you how to delete all groups and all routers in all VLANs, and then how to delete all groups and routers on VLAN 2, and then how to delete a router address on VLAN 2.

Example 9-9 clear cgmp Allows You to Prescriptively Remove Multicast Entries Learned Via CGMP

```
Switch#clear cgmp group

Switch#clear cgmp vlan 2

Switch#clear cgmp vlan 2 router 0012.1234.1234
```

Notice the difference in the MAC address command input format used by CatOS and IOS-based switches. Configuring CGMP on Cisco routers is discussed next.

Configuring CGMP on Cisco Routers and Route Processors CGMP should be enabled only on routers connected to Catalyst switches. Enabling CGMP triggers a CGMP join message to initiate CGMP operation with a directly connected Catalyst switch.

To enable CGMP for IP multicast on a LAN, use the command **ip cgmp** [**proxy**] in interface configuration mode.

When the **proxy** keyword is specified, the CGMP proxy function is enabled: Any router on the segment that is not CGMP-capable is advertised by the proxy router. The proxy router advertises the existence of other non-CGMP-capable routers by sending a CGMP join message with the MAC address of the non-CGMP-capable router and a GDA of 0000.0000.0000.

Finally, to clear all group entries from the caches of Catalyst switches, use the **clear ip cgmp** privileged-mode command. This command sends a CGMP leave message with a group address of 0000.0000.0000 and a unicast address of 0000.0000.0000 (see Table 9-4). This message instructs the switch to clear all group entries it has cached.

This completes our look at CGMP. Next, we consider IGMP snooping, an alternative to CGMP that serves the same purpose. IGMP snooping is replacing CGMP on newer devices.

IGMP Snooping

IGMP snooping constrains the flooding of multicast traffic by dynamically configuring the interfaces so that multicast traffic is forwarded only to interfaces associated with IP multicast devices. The switch listens to IGMP traffic between the host and the router (see Figure 9-16) and keeps track of multicast groups and member ports.

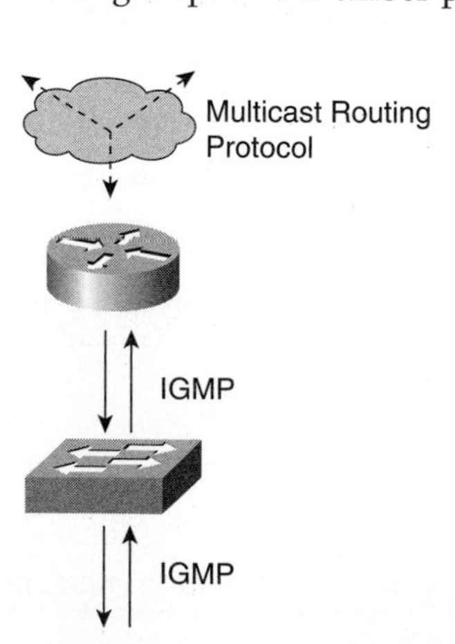

Figure 9-16
IGMP snooping lets a switch watch IGMP join and leave messages between a host and a router in order to manage multicast traffic.

When the switch receives an IGMP join report from a host for a particular multicast group, the switch adds the host port number to the associated multicast forwarding table entry. When it receives an IGMP leave group message from a host, it removes the host port from the table entry. After it relays the IGMP queries from the multicast router, it deletes entries periodically if it does not receive any IGMP membership reports from the multicast clients.

IGMP Snooping Operation

When a host wants to join an IP multicast group, it sends an IGMP join specifying the IP multicast group it wants to join (for example, group 224.1.2.3). The switch hardware recognizes that the packet is an IGMP report and redirects it to the switch CPU. The switch installs a new multicast forwarding table entry for 0100.5E01.0203 and adds the host port and the router port to that entry (note the IP-to-MAC multicast address mapping between 224.1.2.3 and 0100.5E01.0203). The switch then relays the join from the host to all multicast router ports. This process is pictured in Figure 9-17. The segment's designated multicast router adds the outgoing interface to the outgoing interface list for the group and begins forwarding multicast traffic for 224.1.2.3 to this segment.

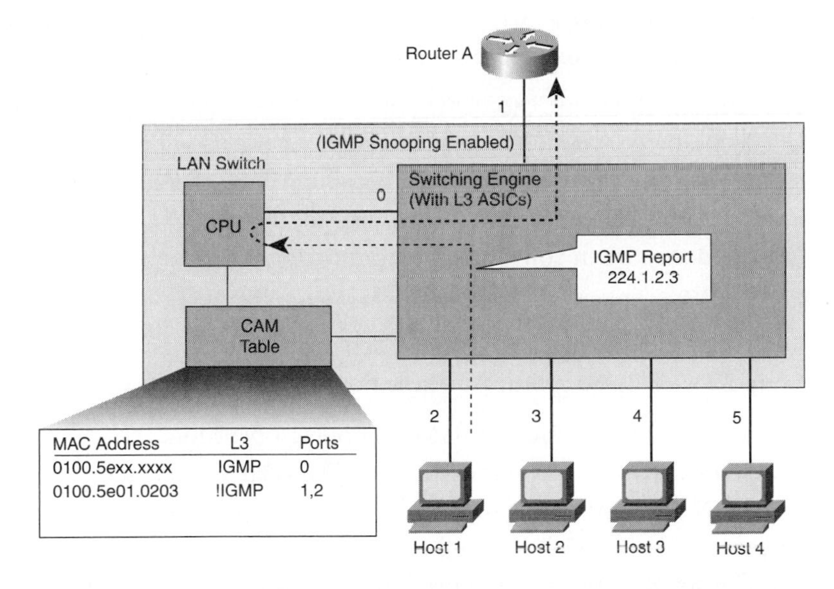

Figure 9-17 Host 1 sends an IGMP join message to 224.1.2.3.

When a second host in this VLAN wants to join group 224.1.2.3, it sends out an IGMP join for this group (see Figure 9-18). The switch hardware recognizes that this is an IGMP control packet and redirects it to the switch CPU. Because the switch already has a multicast forwarding table entry for 0100.5E01.0203 in this VLAN, it just adds the second host port to the entry. Because this is not the first host joining the group, the switch suppresses the report (does not send it to the router).

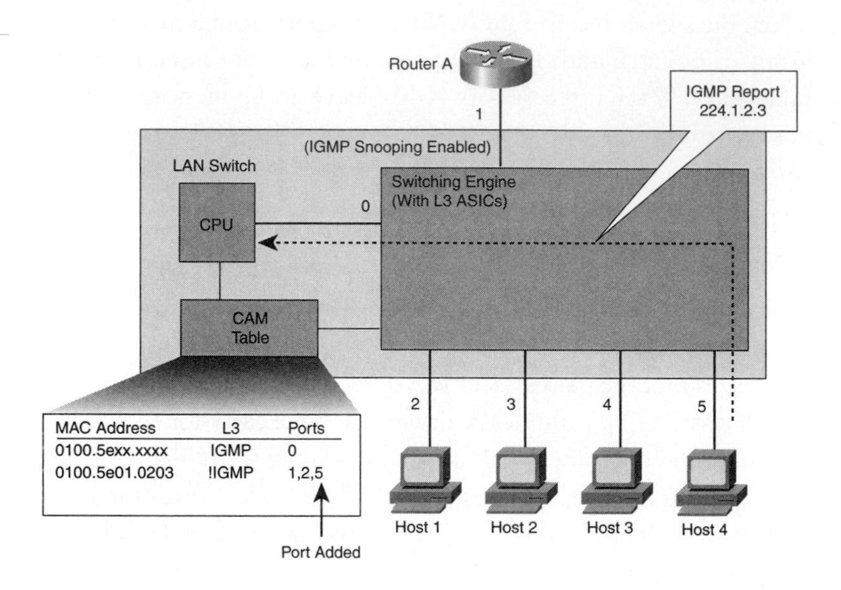

Figure 9-18
Host 5 sends an IGMP join message for group 224.1.2.3.

The entry in the multicast forwarding table tells the switching engine to send frames addressed to 0100.5E01.0203 that are not IGMP packets (!IGMP) to the router and to the hosts that have joined the group.

The switch architecture can distinguish IGMP information packets from other packets for the multicast group. The switch recognizes the IGMP packets through its filter engine. This prevents the CPU from becoming overloaded with multicast data frames, which are not redirected to the CPU. Instead, the multicast traffic hits the MAC group entry, and the switch constrains the traffic to only ports that have been added to that group entry.

At this point, if Host 1 is a multicast server and Host 5 is a multicast receiver for the group 224.1.2.3, the multicast traffic flows as pictured in Figure 9-19.

Continuing our discussion of IGMP snooping operation, the router sends IGMP general queries every 60 seconds by default. The switch floods these queries on all ports in the VLAN. Hosts that are interested in a multicast group respond with an IGMP join for each group in which they are interested. (Contrast this with the earlier discussion, where hosts proactively announced their intention to join a multicast group.) The switch intercepts these IGMP joins, and only the first join per VLAN and per IP multicast group is forwarded on the multicast router ports. Subsequent reports for the same VLAN and group are suppressed (are not sent to the router).

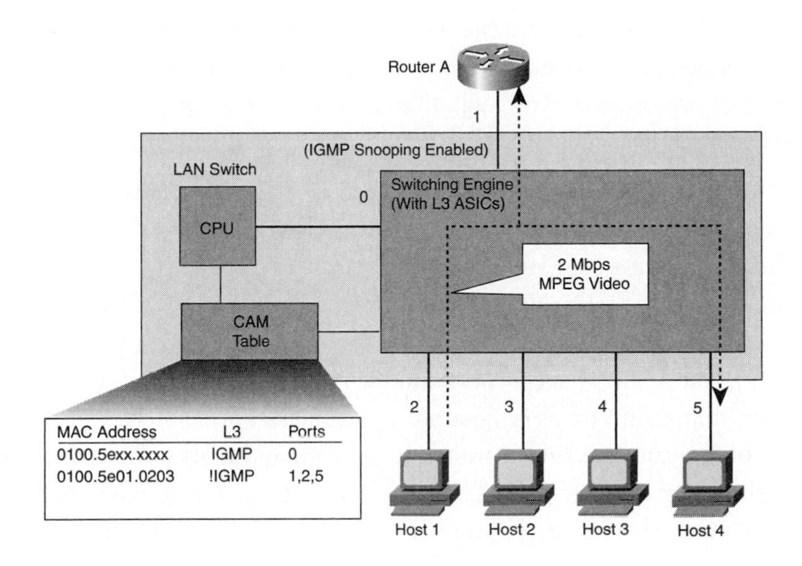

Figure 9-19
IGMP snooping has effectively constrained the multicast traffic to a minimal set of ports, and the CPU does not have to intercept multicast data traffic.

The designated multicast router for a segment continues forwarding multicast traffic to that VLAN as long as at least one host in the VLAN wants to receive multicast traffic. The switch forwards IP multicast group traffic to only hosts that are listed in the forwarding table for the respective IP multicast group.

When hosts want to leave a multicast group, they can either ignore the periodic general queries sent by the multicast router (IGMPv1 host behavior) or send an IGMP leave (IGMPv2 host behavior). As with CGMP, when the switch receives a leave message, it sends out a MAC-based general query on the port on which it received the leave message to determine whether any devices connected to this port are interested in traffic for the specific multicast group. If this port is the last port in the VLAN, the switch sends a MAC-based general query to all ports in the VLAN. MAC-based general queries are addressed to the Layer 2 GDA for which the IGMP leave message was received. At Layer 3, MAC-based general queries are addressed to 244.0.0.1 (all hosts), and in the IGMP header, the group address field is set to 0.0.0.0.

If no IGMP join is received for any of the IP multicast groups that map to the MAC multicast group address, the port is removed from the multicast forwarding entry. If the port is not the last nonmulticast router port in the entry, the switch suppresses the IGMP leave (does not send it to the router). If the port *is* the last nonmulticast router port in the entry, the IGMP leave is forwarded to the multicast router ports, and the MAC group forwarding entry is removed.

NOTE

If there is a spanning-tree state change on a port, an EtherChannel change on a port, or a VLAN ID change on a port, the IGMP snooping-learned multicast groups for the respective port on the VLAN are deleted on the Catalyst switch.

When the router receives the IGMP leave, it sends several IGMP group-specific queries. If no join messages are received in response to the queries, and no downstream routers are connected through that interface, the router removes the interface from the outgoing interface list for that IP multicast group entry in the multicast routing table.

Another consideration you always need to make is whether you have an IP multicast source-only environment. In this case, the switch learns the IP multicast group from the IP multicast data stream and forwards traffic to only the multicast router ports. If the segment contains only one multicast server (multicast source) and no client, you might end up with a situation in which you do not have any IGMP packets in that segment, but you do have a lot of multicast traffic. The switch then forwards the traffic from that group to every host on the segment. Fortunately, a switch running IGMP snooping can detect these multicast streams and adds a multicast entry for that group with only the router port. These entries are flagged internally and are aged periodically. After this aging, the multicast address is relearned after a few seconds if the traffic continues.

NOTE

Do not use fast-leave processing if more than one host is connected to each port. If fast-leave is enabled when more than one host is connected to a port, some hosts might be dropped inadvertently. Fast-leave is supported with IGMPv2 hosts only (not IGMPv1 hosts).

Finally, *IGMP snooping fast-leave processing* (also known as immediate-leave processing) allows the switch processor to remove an interface from a forwarding-table's port list entry without first sending out a MAC-based general query on the port. With IGMP snooping fast-leave enabled, when an IGMP leave is received on a port, the port is immediately removed from the multicast forwarding entry (or the entire entry is removed). Fast-leave processing ensures optimal bandwidth management for all hosts on a switched network, even when multiple multicast groups are in use simultaneously.

Next, we detail the configuration of IGMP snooping on Catalyst switches.

Configuring IGMP Snooping

You can run IGMP snooping with the following hardware and software:

- Catalyst 5000 family—With Supervisor IIG or IIIG, or Supervisor III with NFFC or NFFC II, and software version 4.3, minimum
- Catalyst 6000 family—Supported without restriction
- Catalyst 4006 with Supervisor Engine III
- Catalyst 3550 series
- Catalyst 2950 series

When IGMP snooping is enabled, the switch responds to periodic multicast router queries to all VLANs with only one join request per MAC multicast group, and the switch creates one entry per VLAN in the Layer 2 forwarding table for each MAC multicast group for which it receives an IGMP join request. All hosts interested in this multicast traffic send join requests and are added to the forwarding table entry.

Layer 2 multicast groups learned through IGMP snooping are dynamic. If you specify group membership for a multicast group address statically, your setting supersedes any automatic manipulation by IGMP snooping. Multicast group membership lists can consist of both user-defined and IGMP snooping-learned settings.

Configuring IGMP snooping on Catalyst switches is similar to configuring CGMP clients, except that there is no accompanying router configuration.

CatOS Switch IGMP snooping is enabled by default on the CatOS switches just listed, except for the Catalyst 5000 family (which is disabled by default). In any case, to enable IGMP snooping, use the command **set igmp enable** (and disable with **set igmp disable**). On a Catalyst 5000 family switch, enabling IGMP snooping disables CGMP.

To enable IGMP snooping fast-leave, use the command **set igmp fastleave enable** (and disable with **set igmp fastleave disable**). To verify the IGMP snooping configuration (including fast-leave), enter the command **show igmp statistics**, as shown in Example 9-10.

Example 9-10 Verify IGMP Snooping on a CatOS Switch with the Command show igmp statistics

```
Console> (enable) show igmp statistics
IGMP enabled
IGMP fastleave enabled

IGMP statistics for vlan 1:
Total valid pkts rcvd:            18951
Total invalid pkts recvd          0
General Queries recvd             377
Group Specific Queries recvd      0
MAC-Based General Queries recvd   0
Leaves recvd                      14
Reports recvd                     16741
Other Pkts recvd                  0
Queries  Xmitted                  0
GS Queries Xmitted                16
Reports Xmitted                   0
Leaves Xmitted                    0
Failures to add GDA to EARL       0
Topology Notifications rcvd       10
```

To display the multicast router ports learned dynamically through IGMP, use the command **show multicast router igmp**, as shown in Example 9-11.

Example 9-11 show multicast router igmp Lists the IGMP-Discovered Multicast Router Ports

```
Console> (enable) show multicast router igmp

IGMP enabled

Port       Vlan
---------  ----------------
 1/1       1
 2/1       2,99,255

Total Number of Entries = 2
'*' - Configured
```

To statically configure MAC multicast groups on a CatOS switch, use the command **set cam static** {*multicast_mac*} *mod/ports* [*vlan*]. As in Example 9-7, use the **show multicast group** command to display information about the multicast groups on the switch.

IOS-Based Switch By default, IGMP snooping is globally enabled. When it is globally enabled or disabled, it is also enabled or disabled in all existing VLAN interfaces. By default, IGMP snooping is enabled on all VLANs, but it can be enabled and disabled on a per-VLAN basis.

If IGMP global snooping is disabled, you cannot enable VLAN snooping. If global IGMP snooping is enabled, you can either enable or disable snooping on a VLAN basis.

To globally enable IGMP snooping, enter the global configuration command **ip igmp snooping**. The global command **no ip igmp snooping vlan** *vlan-id* disables IGMP snooping on a particular VLAN.

When you enable IGMP immediate-leave processing, the switch immediately removes a port from the IP multicast group when it detects an IGMPv2 leave message on that port. Immediate-leave processing allows the switch to remove an interface from the forwarding table after receiving a leave message without first sending out group-specific queries from the interface. You should use the immediate-leave feature only when a single receiver is present on every port in the VLAN. Immediate-leave is disabled by default. Use the **ip igmp snooping vlan** *vlan-id* **immediate-leave** global configuration command to enable IGMP immediate-leave processing on a VLAN interface.

To verify the IGMP snooping configuration, use the command **show ip igmp snooping** [**vlan** *vlan-id*], as shown in Example 9-12. You can specify a particular VLAN to limit the output.

Example 9-12 show ip igmp snooping Displays the IGMP Snooping Configuration on an IOS-Based Switch

```
Switch(config)#ip igmp snooping
Switch(config)#ip igmp snooping vlan 1 immediate-leave
Switch(config)#ip igmp snooping vlan 2 immediate-leave
Switch(config)#exit
Switch#show ip igmp snooping

vlan 1
----------
  IGMP snooping is globally enabled
  IGMP snooping is enabled on this Vlan
  IGMP snooping immediate-leave is enabled on this Vlan
  IGMP snooping mrouter learn mode is pim-dvmrp on this Vlan
vlan 2
----------
  IGMP snooping is globally enabled
  IGMP snooping is enabled on this Vlan
  IGMP snooping immediate-leave is enabled on this Vlan
  IGMP snooping mrouter learn mode is cgmp on this Vlan
vlan 3
----------
  IGMP snooping is globally enabled
  IGMP snooping is enabled on this Vlan
  IGMP snooping immediate-leave is disabled on this Vlan
  IGMP snooping mrouter learn mode is cgmp on this Vlan
<output omitted>
```

Notice that the learn mode varies in the output per VLAN. Multicast-capable router ports are added to the forwarding table for every IP multicast entry. The switch learns of such ports through one of these methods:

- Snooping on PIM and DVMRP packets
- Listening to CGMP self-join packets from other routers

- Statically connecting to a multicast router port with the **ip igmp snooping mrouter interface** *interface-id* global configuration command

You can configure the switch either to snoop on Protocol-Independent Multicast/Distance Vector Multicast Routing Protocol (PIM/DVMRP) packets or to listen to CGMP self-join packets. By default, the switch snoops on PIM/DVMRP packets on all VLANs.

To learn of multicast router ports through only CGMP self-join packets, use the **ip igmp snooping vlan** *vlan-id* **mrouter learn cgmp** global configuration command. When this command is used, the router listens only to CGMP self-join packets and no other CGMP packets.

To learn of multicast router ports through only PIM/DVMRP packets, use the **ip igmp snooping vlan** *vlan-id* **mrouter learn pim-dvmrp** global configuration command.

Example 9-13 shows how to specify the multicast router statically and via the CGMP learning method. Two methods are demonstrated for configuring an IOS-based switch to discover multicast routers; the available discovery methods include CGMP, PIM, DVMRP, and static configuration.

Example 9-13 Configuring IOS-Based Switches to Discover Multicast Routers

```
Switch(config)#ip igmp snooping vlan 1 mrouter interface fastethernet0/6
Switch(config)#ip igmp snooping vlan 2 mrouter learn cgmp
```

To view the Layer 2 multicast entries for the switch or for a VLAN, use the command **show mac-address-table multicast** [**vlan** *vlan-id*] [**user** | **igmp-snooping**] [**count**], as shown in Example 9-14. The **user** option displays only the user-configured multicast entries.

Example 9-14 There Are Numerous Options for Viewing GDAs and GDA Counts on the Switch

```
Switch#show mac-address-table multicast vlan 1

Vlan    Mac Address     Type    Ports
----    -----------     ----    -----
   1    0100.5e00.0128  IGMP    Fa0/11
   1    0100.5e01.1111  USER    Fa0/5, Fa0/6, Fa0/7, Fa0/11

Switch#show mac-address-table multicast count

Multicast Mac Entries for all vlans: 10
```

Example 9-14 There Are Numerous Options for Viewing GDAs and GDA Counts on the Switch (Continued)

```
Switch#show mac-address-table multicast vlan 1 count

Multicast Mac Entries for vlan 1: 2

Switch# show mac-address-table multicast vlan 1 user

vlan   mac address       type        ports
-----+---------------+-------+--------------------
1     0100.5e02.0203    user      Fa0/1,Fa0/2,Fa0/4

Switch# show mac-address-table multicast vlan 1 igmp-snooping count

Number of igmp-snooping programmed entries : 1
```

Finally, to statically configure MAC multicast groups on an IOS-based switch, use the global command **ip igmp snooping vlan** *vlan-id* **static** *mac-address* **interface** *interface-id*. Example 9-15 demonstrates this command on a Catalyst 2950.

Example 9-15 Multicast Groups Can Be Configured Statically for a Given Interface and a Specific VLAN

```
Switch(config)#ip igmp snooping vlan 1 static 0100.5e02.0203 interface
  fastethernet0/6
```

Next, we show you how to configure IGMP snooping with the CatIOS, a term used to describe a single Cisco IOS Software image available for a Catalyst 6000 family switch with an MSFC or a Catalyst 4006 switch with a Supervisor Engine III. The image enables a single command line for configuring both routing and switching for the device.

Catalyst 4006 with Supervisor Engine III The Catalyst 4006 with Supervisor Engine III has only one option for an OS image—the integrated Cisco IOS Software. The first release is 12.1(8a)EW. The 4006 with Supervisor Engine III supports a wide range of modules, such as a 48-port 10/100/1000BaseT Gigabit Ethernet switching module.

Because the Layer 2 and Layer 3 functionality for the 4006 with Supervisor Engine III is integrated into a single operating system, you need to pay close attention to whether a given technology or configuration applies to a Layer 2 port or a Layer 3 interface.

To use IGMP snooping, first you need to configure a Layer 3 interface in the subnet for multicast routing, such as PIM (PIM is discussed later in this chapter). Enabling PIM on an interface also enables IGMP operation on that interface. Before you can enable PIM on an interface, you first have to globally enable multicast routing with the global configuration command **ip multicast-routing**. When IP multicast routing is disabled, IP multicast traffic data packets are not forwarded by the Catalyst 4006 switch with Supervisor Engine III. However, IP multicast control traffic continues to be processed and forwarded.

IGMP snooping is globally enabled by default, but it won't function unless you first proceed as detailed in the preceding paragraph. If IGMP snooping is disabled, use the global configuration command **ip igmp snooping** to enable it. You verify the configuration with the command **show ip igmp snooping**, as demonstrated in Example 9-16. The **include globally** option causes only output lines that contain the string "globally" to be displayed.

Example 9-16 Enabling IGMP Snooping on a Catalyst 4006 with Supervisor Engine III

```
4006S3(config)#ip igmp snooping
4006S3(config)#end
4006S3#show ip igmp snooping | include globally
  IGMP snooping is globally enabled
```

To enable IGMP snooping in a VLAN, enter the global command **ip igmp snooping vlan** *vlan_ID* and verify with the command **show ip igmp snooping vlan** *vlan_ID*. This is demonstrated in Example 9-17.

Example 9-17 Enabling and Verifying IGMP Snooping for a Particular VLAN

```
4006S3(config)#ip igmp snooping vlan 10
4006S3(config)#end
4006S3#show ip igmp snooping vlan 10
vlan 10
IGMP snooping is globally enabled
IGMP snooping is enabled on this Vlan
IGMP snooping intermediate-leave is disabled on this Vlan
IGMP snooping mrouter learn mode is pim-dvmrp on this Vlan
IGMP snooping is running in IGMP_ONLY mode on this Vlan
```

To enable IGMP fast-leave processing on an interface, enter the global command **ip igmp snooping vlan** *vlan_ID* **immediate-leave**, as shown in Example 9-18.

Example 9-18 Enabling and Verifying IGMP Snooping Fast-Leave Processing

```
4006S3(config)#ip igmp snooping vlan 10 immediate-leave
Configuring immediate leave on vlan 10
4006S3(config)#end
4006S3#show ip igmp interface vlan 10
IGMP snooping is globally enabled
IGMP snooping is enabled on this Vlan
IGMP snooping immediate-leave is enabled on this Vlan
IGMP snooping mrouter learn mode is pim-dvmrp on this Vlan
IGMP snooping is running in IGMP-ONLY mode on this Vlan
```

To configure the learning method for IGMP snooping, use the global command **ip igmp snooping vlan** *vlan_ID* **mrouter learn** [**cgmp** | **pim-dvmrp**].

To configure a static connection to a multicast router, enter the **ip igmp snooping mrouter vlan** *vlan-id* **interface** *interface-id* command, as demonstrated in Example 9-19.

Example 9-19 Configuring the IGMP Learning Method and a Static Connection to a Multicast Router

```
4006S3(config)#ip igmp snooping vlan 1 mrouter learn cgmp
4006S3(config)#ip igmp snooping vlan 10 mrouter interface fastethernet 3/2
4006S3(config)#end
4006S3#show ip igmp snooping mrouter vlan 10
vlan            ports
-----+----------------------------------------
 10   Fa3/2
```

To determine the multicast router interfaces on the switch learned via IGMP snooping, enter the command **show ip igmp snooping mrouter vlan** *vlan_ID*, as shown in Example 9-20.

Example 9-20 Displaying Multicast Router Interfaces

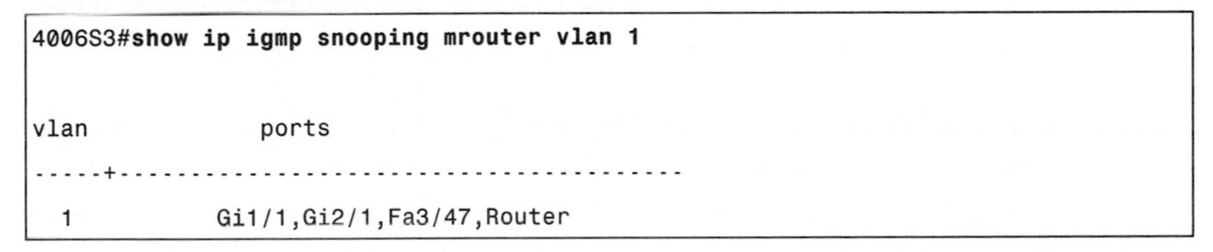

```
4006S3#show ip igmp snooping mrouter vlan 1

vlan            ports
-----+----------------------------------------
  1          Gi1/1,Gi2/1,Fa3/47,Router
```

To display MAC address multicast entries for a VLAN, enter the command **show mac-address-table multicast vlan** *vlan_ID* [**count**], as shown in Example 9-21.

Example 9-21 Displaying GDAs for a VLAN

```
4006S3#show mac-address-table multicast vlan 1

Multicast Entries
 vlan    mac address     type    ports
-------+---------------+-------+-----------------------------------------
   1    0100.5e01.0101     igmp Switch,Gi6/1
   1    0100.5e01.0102     igmp Switch,Gi6/1
   1    0100.5e01.0103     igmp Switch,Gi6/1
   1    0100.5e01.0104     igmp Switch,Gi6/1
   1    0100.5e01.0105     igmp Switch,Gi6/1
   1    0100.5e01.0106     igmp Switch,Gi6/1
4006S3#show mac-address-table multicast vlan 1 count

Multicast MAC Entries for vlan 1:    6
```

Finally, to statically configure MAC multicast groups, use the interface command **ip igmp snooping vlan** *vlan_ID* **static** *mac_address* **interface** *interface_num*, as demonstrated in Example 9-22.

Example 9-22 Configuring a Static GDA on an Interface

```
4006S3(config)#ip igmp snooping vlan 10 static 0100.5e02.0203 interface
  fastethernet 3/7
```

You can see many similarities between the command syntax used by IOS-based switches in the preceding section and the commands in this section. This is the idea Cisco had in mind with the 4006 Supervisor Engine III IOS and the Native IOS on the Catalyst 6000. The command structures of the various platforms are becoming more similar over time.

Catalyst 6000 Family Switch Running Native IOS The Catalyst 6000 family switches with Multilayer Switch Feature Cards (MSFCs) have two options for software images: the Hybrid image (CatOS on the Supervisor Engine and an IOS image on the MSFC) and Native IOS (a single IOS image for both the Supervisor and MSFC).

Just as with the 4006 Supervisor Engine III, IGMP snooping is enabled by default. But for it to be functional, you have to configure **ip multicast-routing** in global configuration mode and enable a multicast routing protocol on an interface. However, one difference is that, with the exception of globally enabling IGMP snooping, *all IGMP snooping commands are supported only on VLAN interfaces.*

If IGMP snooping is disabled, you can enable it with the global configuration command **ip igmp snooping** and verify the configuration with the command **show ip igmp interface vlan** *vlan_ID*, as shown in Example 9-23.

Example 9-23 Verifying the IGMP Snooping Configuration on Catalyst 6000 Native IOS

```
Router#show ip igmp interface vlan 200 | include globally
  IGMP snooping is globally enabled
```

To configure IGMP snooping for a VLAN, enter the command **ip igmp snooping** in interface VLAN configuration mode. Verify the configuration with the command **show ip igmp interface vlan** *vlan_ID*, as demonstrated in Example 9-24.

Example 9-24 Verifying the IGMP Snooping Configuration on a VLAN Interface

```
Router#interface vlan 200
Router(config-if)#ip igmp snooping
Router(config-if)#end
Router#show ip igmp interface vlan 200
Vlan200 is up, line protocol is up
  Internet address is 172.20.52.94/27
  IGMP is enabled on interface
  Current IGMP version is 2
  CGMP is disabled on interface
  IGMP query interval is 60 seconds
  IGMP querier timeout is 120 seconds
  IGMP max query response time is 10 seconds
  Last member query response interval is 1000 ms
  Inbound IGMP access group is not set
  IGMP activity: 0 joins, 0 leaves
  Multicast routing is enabled on interface
  Multicast TTL threshold is 0
  Multicast designated router (DR) is 172.20.52.94 (this system)
  IGMP querying router is 172.20.52.94 (this system)
```

continues

Example 9-24 Verifying the IGMP Snooping Configuration on a VLAN Interface (Continued)

```
No multicast groups joined
IGMP snooping is globally enabled
IGMP snooping is enabled on this interface
IGMP snooping fast-leave is disabled on this interface
IGMP snooping querier is disabled on this interface
```

To enable IGMP fast-leave processing in a VLAN, enter the VLAN interface command **ip igmp snooping fast-leave**, as shown in Example 9-25.

Example 9-25 Configuring IGMP Fast-Leave

```
Router#interface vlan 200
Router(config-if)#ip igmp snooping fast-leave
Configuring fast leave on vlan 200
Router(config-if)#end
Router#show ip igmp interface vlan 200 | include fast-leave
IGMP snooping fast-leave is enabled on this interface
```

To specify the learning method for IGMP snooping, use the interface VLAN command **ip igmp snooping mrouter learn** {**cgmp** | **pim-dvmrp**}.

To configure a static connection to a multicast router, enter the VLAN interface command **ip igmp snooping mrouter interface** {*type slot/port*} or **ip igmp snooping mrouter interface vlan** *vlan_ID*, as demonstrated in Example 9-26. The *type* is **ethernet**, **fastethernet**, **gigabitethernet**, or **tengigabitethernet**. The interface to the router must be in the VLAN where you enter the command. Verify the configuration with the **show ip igmp snooping mrouter interface** *vlan_ID* command.

Example 9-26 Configuring a Static Connection to a Multicast Router

```
Router(config)#interface vlan 200
Router(config-if)#ip igmp snooping mrouter interface vlan 200
Router(config-if)#end
Router#show ip igmp snooping mrouter interface vlan 200

vlan            ports
-----+----------------------------------------
 200  Fa5/8
```

To display multicast router interfaces, use the command **show ip igmp snooping mrouter interface** *vlan_ID*. It is demonstrated in Example 9-27.

Example 9-27 Displaying Multicast Router Interfaces

```
Router#show ip igmp snooping mrouter interface vlan 1

vlan            ports
-----+----------------------------------------
  1          Gi1/1,Gi2/1,Fa3/48,Router
```

To display MAC multicast addresses for a VLAN, use the command **show mac-address-table multicast** *vlan_ID* [**count**], as shown in Example 9-28.

Example 9-28 Displaying Multicast MAC Address Entries and Counts for a VLAN

```
Router#show mac-address-table multicast vlan 1

vlan   mac address     type    qos             ports
-----+---------------+--------+---+--------------------------------
  1  0100.5e02.0203  static   --  Gi1/1,Gi2/1,Fa3/48,Router
  1  0100.5e00.0127  static   --  Gi1/1,Gi2/1,Fa3/48,Router
  1  0100.5e00.0128  static   --  Gi1/1,Gi2/1,Fa3/48,Router
  1  0100.5e00.0001  static   --  Gi1/1,Gi2/1,Fa3/48,Router,Switch
Router#show mac-address-table multicast 1 count

Multicast MAC Entries for vlan 1:    4
```

Finally, to statically configure MAC multicast groups for a Layer 2 port, use the VLAN interface command **ip igmp snooping static** *mac_address* **interface** {*type slot/port*}, as shown in Example 9-29.

Example 9-29 Configuring a Static GDA for a Port

```
Router#interface vlan 200
Router(config-if)#ip igmp snooping static 0100.5e02.0203 interface fastethernet
  5/11

Configuring port FastEthernet5/11 on group 0100.5e02.0203 vlan 200
```

This completes our discussion of configuring IGMP snooping for Catalyst switches.

TECH NOTE: CGMP VERSUS IGMP SNOOPING

IGMP snooping, like CGMP, is not a standard. IGMP snooping is a methodology for switches to intercept IGMP host reports. If it is implemented in hardware, it has no impact on performance. However, if it is implemented in software, the switch's performance can be expected to decrease. Vendors other than Cisco do support IGMP snooping; for example, the Nortel Baystack 420 supports IGMP snooping in hardware.

Because IGMP control messages are transmitted as multicast packets, they are indistinguishable from multicast data at Layer 2. A switch running IGMP snooping must examine every multicast data packet to check if it contains any pertinent IGMP control information. If IGMP snooping is implemented on a low-end switch with a slow CPU, this could have a severe performance impact when data is transmitted at high rates. The solution is to implement IGMP snooping on high-end switches with special ASICs that can perform the IGMP checks in hardware. CGMP is ideal for low-end switches without special hardware.

If there are CGMP switches in the network, join and leave suppression does not occur as it does in an IGMP snooping-only environment. In a network that has both IGMP and CGMP switches, all join and leave messages are forwarded to the multicast routers so that the router can generate CGMP join and leave messages.

Cisco makes the decision of whether to configure CGMP or IGMP snooping easy because, with the exception of the Catalyst 5000 series, the set of devices supporting a CGMP client and the set of devices supporting IGMP snooping are mutually exclusive, and the switches that support IGMP snooping support it in hardware.

Because a CGMP client (switch) is needed for CGMP, and because CGMP is not supported on the client side on virtually all new Catalyst devices (such as the Catalyst 2950, Catalyst 3550, Catalyst 4000 Supervisor Engine III, and Catalyst 6000), and because all the new devices support IGMP snooping in hardware, the natural inference to make is that CGMP is being replaced by IGMP snooping.

In some networks, due to hardware limitations, you might not be able to run IGMP snooping on all switches. In this case, you might need to run CGMP on some switches in the same network. You would likely enable CGMP on only older wiring closet switches such as the Catalyst 1900, 2820, 2900XL, and 2500XL. Note that this is a special case. The switch running IGMP snooping detects CGMP messages and detects that some switches in the network are running CGMP. Therefore, it moves to a special IGMP-CGMP mode and disables the proxy reporting (so that the switch will not act as a proxy for sending IGMP reports from connected hosts to connected routers). This is absolutely necessary for the proper operation of CGMP, because routers use the IGMP report's source MAC address to create a CGMP join. Routers running CGMP need to see all host IGMP reports, so proxy reporting must be disabled in this case. Any reports sent to the router should be only those strictly needed for IGMP snooping.

The last Layer 2 multicast constraint mechanism we need to explore is RGMP. This is followed by the final multicast topic for this chapter—multicast routing protocols.

Router-Port Group Management Protocol

Router-Port Group Management Protocol (RGMP) lets a router communicate to a switch the IP multicast group(s) for which the router would like to receive or forward traffic. RGMP is a Cisco-proprietary protocol designed for switched Ethernet backbone networks running PIM sparse mode, sparse-dense mode, source-specific mode, or bidirectional mode (these modes are discussed later in this chapter).

RGMP-enabled switches and router interfaces in a switched network support directly connected, multicast-enabled hosts that *receive* multicast traffic, but they do not support directly connected, multicast-enabled hosts that *source* multicast traffic. A multicast-enabled host can be a workstation or a multicast application running in a router.

In Figure 9-20, the sources for the two different multicast groups (the source for group A and the source for group B) send traffic into the same switched network. Without RGMP, traffic from source A is unnecessarily flooded from switch A to switch B and then to router B and router D. Also, traffic from source B is unnecessarily flooded from switch B to switch A and then to router A and router C. With RGMP enabled on all routers and switches in this network, traffic from source A would not flood router B and router D. Also, traffic from source B would not flood router A and router C. Traffic from both sources would still flood the link between switch A and switch B. Flooding over this link would still occur because RGMP does not restrict traffic on links toward other RGMP-enabled switches with routers behind them.

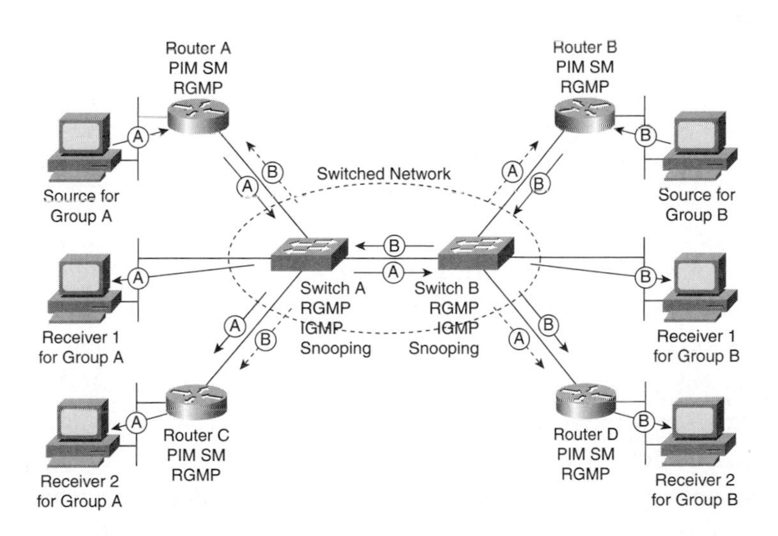

Figure 9-20 RGMP constrains unnecessary multicast traffic between routers in the campus network backbone.

By restricting unwanted multicast traffic in a switched network, RGMP increases the available bandwidth for all other multicast traffic in the network and saves the routers' processing resources.

Figure 9-21 shows the RGMP messages sent between an RGMP-enabled router and an RGMP-enabled switch. The router sends simultaneous PIM hello and RGMP hello messages to the switch. The PIM hello message is used to locate neighboring PIM routers. The RGMP hello message instructs the switch to restrict all multicast traffic on the interface from which the switch received the RGMP hello message. RGMP messages are sent to the multicast address 224.0.0.25, which is the link local multicast address reserved by the IANA for sending IP multicast traffic from routers to switches. If RGMP is not enabled on both the router and the switch, the switch automatically forwards all multicast traffic out the interface from which the switch received the PIM hello message.

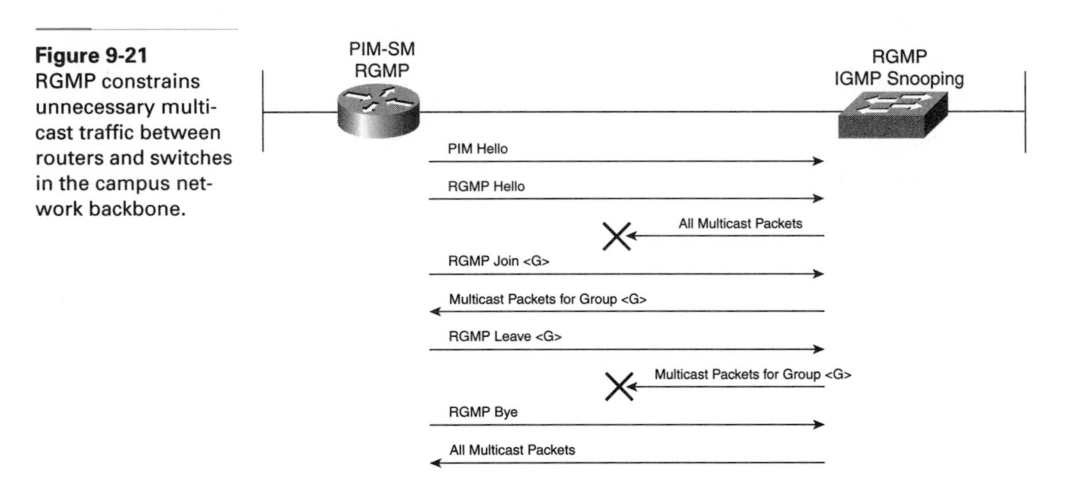

Figure 9-21 RGMP constrains unnecessary multicast traffic between routers and switches in the campus network backbone.

The router sends the switch an RGMP join <G> message (where G is the multicast group address) when the router wants to receive traffic for a specific multicast group. The RGMP join message instructs the switch to forward multicast traffic for group <G> out the interface from which the switch received the RGMP hello message. The router sends the switch an RGMP join <G> message for a multicast group even if the router is only forwarding traffic for the multicast group into a switched network. By joining a specific multicast group, the router can determine whether another router is also forwarding traffic for the multicast group into the same switched network. If two routers are forwarding traffic for a specific multicast group into the same switched network, the two routers use a PIM mechanism to determine which router should continue forwarding the multicast traffic into the network.

The router sends the switch an RGMP leave <G> message when the router wants to stop receiving traffic for a specific multicast group. The RGMP leave message instructs the switch to stop forwarding the multicast traffic on the port on which the switch received the PIM and RGMP hello messages. An RGMP-enabled router cannot send an RGMP leave <G> message until the router does not receive or forward traffic from any source for a specific multicast group. The router sends the switch an RGMP bye message when RGMP is disabled on the router. The RGMP bye message instructs the switch to forward to the router all IP multicast traffic on the port from which the switch received the PIM and RGMP hello messages, as long as the switch continues to receive PIM hello messages on the port.

Configuring RGMP

RGMP is presently supported on only the following switch platforms:

- Catalyst 6000 with software version 5.4 or later
- Catalyst 5000 with software version 5.4 or later

RGMP is supported on the following Cisco router IOS releases:

- 12.2 Mainline release (Major release)
- 12.1E
- 12.1T (beginning with version 12.1(5)T1)
- 12.0S (beginning with version 12.0(10)S)
- 12.0ST (beginning with version 12.0(11)ST)

Here are some configuration restrictions to keep in mind with RGMP:

- You need to run RGMP on both the routers and the switches.
- You need to enable IGMP snooping on the switches.
- CGMP and RGMP cannot interoperate on the same switched network. If RGMP is enabled on a switch or router interface, CGMP is automatically disabled on that switch or router interface; if CGMP is enabled on a switch or router interface, RGMP is automatically disabled on that switch or router interface.

Before you enable RGMP on a router, ensure that the following features are enabled:

- IP routing
- IP multicast
- PIM in sparse mode, sparse-dense mode, source-specific mode, or bidirectional mode

To configure RGMP on a router, enter the Cisco IOS Software command **ip rgmp** in interface configuration mode. RGMP is disabled by default. Verify the configuration with the command **show ip igmp interface**, as shown in Example 9-30.

Example 9-30 Verifying RGMP IOS Configuration on a Router

```
Router#show ip igmp interface

Ethernet1/0 is up, line protocol is up
  Internet address is 10.0.0.0/24
    IGMP is enabled on interface
  Current IGMP version is 2
  RGMP is enabled
  IGMP query interval is 60 seconds
  IGMP querier timeout is 120 seconds
  IGMP max query response time is 10 seconds
  Last member query response interval is 1000 ms
  Inbound IGMP access group is not set
  IGMP activity: 1 joins, 0 leaves
  Multicast routing is enabled on interface
  Multicast TTL threshold is 0
  Multicast designated router (DR) is 10.0.0.0 (this system)
  IGMP querying router is 10.0.0.0 (this system)
  Multicast groups joined (number of users):
      224.0.1.40(1)
```

If RGMP is not enabled on an interface, no RGMP information is displayed in the **show ip igmp interface** command output for that interface. Next, we discuss RGMP configuration on CatOS switches.

RGMP is disabled by default on a Catalyst switch. Recall that RGMP is supported only on the Catalyst 5000 and 6000 families of switches. To enable RGMP on a CatOS switch, use the command **set rgmp enable**. Verify the RGMP configuration with the command **show rgmp group** [*mac_addr*] [*vlan_id*] or the command **show rgmp group count** [*vlan_id*]. These commands are demonstrated in Example 9-31.

Example 9-31 RGMP Configuration and Verification on a CatOS Switch

```
Console> (enable) set rgmp enable

RGMP enabled.
```

Example 9-31 RGMP Configuration and Verification on a CatOS Switch (Continued)

```
Console> show rgmp group

Vlan     Dest MAC/Route Des     RGMP Joined Router Ports
--------------------------------------------------------------------------------
-------
1        01-00-5e-00-01-28      5/1,5/15
1        01-00-5e-01-01-01      5/1
2        01-00-5e-27-23-70*     3/1, 5/1
Total Number of Entries = 3
'*' - Configured
Console> show rgmp group count 1

Total Number of Entries = 2
```

You can view RGMP statistics with the command **show rgmp statistics** [*vlan*]. The statistics are cleared with the command **clear rgmp statistics**. These commands are shown in Example 9-32.

Example 9-32 Displaying and Clearing RGMP Statistics

```
Console> show rgmp statistics 23

RGMP enabled
RGMP Statistics for vlan <23>:
Receive:
Valid pkts:    20
Hellos:    10
Joins:    5
Leaves:    5
Byes:    0
Discarded:    0
Transmit:
Total Pkts:    10
Failures:    0
Hellos:    10
Joins:    0
Leaves:    0
Byes:    0
Console> (enable) clear rgmp statistics
```

The command **show multicast router** displays detected RGMP-capable routers. A plus next to the router port indicates that it is an RGMP-capable router, as shown in Example 9-33.

Example 9-33 Displaying RGMP-Capable Routers

```
Console> show multicast router

Port     Vlan
------   ------
5/1 +    1
5/14 +   2
5/15     1

Total Number of Entries = 3
'*' - Configured
'+' - RGMP-capable
```

Finally, a useful command is available to display Layer 2 multicast protocols configured on the switch. The command **show multicast protocols status** displays the status of the Layer 2 multicast protocols on the switch, as shown in Example 9-34.

Example 9-34 Displaying All Layer 2 Multicast Protocols for the Switch

```
Console> show multicast protocols status

IGMP disabled
IGMP fastleave enabled
RGMP enabled
GMRP disabled
```

In Example 9-34, you see a reference to GMRP. GMRP is one of the Layer 2 multicast protocols that we have not discussed. We present an overview of GMRP in the following Tech Note. GMRP and MVR are two multicast protocols that are supported on newer Cisco IOS Software and CatOS releases. They are being deployed more frequently in campus networks.

This completes the discussion of RGMP and the technologies used to constrain Layer 2 multicast traffic.

TECH NOTE: GMRP AND MVR

Generic Attribute Registration Protocol (GARP), part of the IEEE 802.1p specification, implements a service that provides a generic Layer 2 transport mechanism to propagate information throughout a switching domain. The information is then propagated in the form of attributes, types, values, and semantics, which are defined by the application employing the service, such as the *GARP VLAN Registration Protocol (GVRP)*. GVRP is a GARP application that provides IEEE 802.1Q-compliant VLAN pruning and dynamic VLAN creation on 802.1Q trunk ports (similar to VTP, but not Cisco-proprietary). The beauty of GARP applications is that when a vendor implements these technologies, they interoperate with other vendors' devices.

GARP Multicast Registration Protocol (GMRP) is a GARP application that provides a constrained multicast flooding facility similar to IGMP snooping. GMRP software components run on both the switch and the host. On the host, in an IP multicast environment, you must use IGMP with GMRP; the host GMRP software spawns Layer 2 GMRP versions of the host's Layer 3 IGMP control packets. The switch receives both the Layer 2 GMRP and the Layer 3 IGMP traffic from the host. The switch forwards the Layer 3 IGMP control packets to the router and uses the received GMRP traffic to constrain multicasts at Layer 2 in the host's VLAN. GMRP is supported on Catalyst 29xxG, Catalyst 4000 series, Catalyst 5000 family, and Catalyst 6000 family switches.

Another recent Cisco multicast technology that is proving useful in campus networks is *Multicast VLAN Registration (MVR) Protocol*. MVR provides the ability to have a single multicast stream traverse the network regardless of the number of users or VLANs that want to have access to it. MVR is designed for applications using wide-scale deployment of multicast traffic across a service provider network (for example, the broadcast of multiple television channels over a service provider network).

Some of the advantages of MVR are as follows:

- MVR reduces each multicast group to a single data stream on the backbone regardless of the number of users or VLANs.
- MVR scales with the number of multicast groups provided, not with the number of subscribers.
- MVR speeds up joins and leaves, because they are done locally.
- MVR can limit the number of multicast sessions per port.
- MVR relies on a single multicast VLAN to be shared in the network while subscribers remain in separate VLANs.
- MVR provides the ability to continuously send multicast streams in the multicast VLAN but to isolate the streams from the subscriber VLANs for bandwidth and security reasons.

MVR is supported on Catalyst 2900XL, 3500XL, 2950 series, and 3550 series switches.

The remainder of this chapter focuses on multicast technologies that are designed to operate strictly among routers. These multicast technologies allow routers to communicate information among themselves for the purpose of building multicast data paths between multicast sources and receivers.

IP Multicast Routing

The subject area encompassing IP multicast routing, as alluded to in this chapter's introduction, includes a vast and changing array of multicast technologies that push the envelope of modern campus network design. The definitive text for IP multicast routing is Beau Williamson's book *Developing IP Multicast Networks*, Volume I, published by Cisco Press. Every network engineer needs to own this book. Multicast aficionados are all looking forward to the publication of Volume II. In his book, Williamson carefully details the operation of IP multicast routing protocols, arguably one of the most difficult concepts to master in networking.

Suffice it to say that IP multicast routing is a field in and of itself, and the subject matter cannot be afforded much justice in the context of a single chapter in a book. IP multicast routing protocols operate strictly within the domain of routers; this book focuses on switching technologies. We deliberately present a superficial look at IP multicast routing to give you an idea of its role in campus networks and some guidance on how to configure some of the protocols. Further study is recommended outside the scope of this book if you are interested in mastering IP multicast routing design and deployment techniques.

The IP multicast routing protocols we touch on are Distance Vector Multicast Routing Protocol (DVMRP), Core-Based Trees (CBT), Multicast OSPF (MOSPF), Protocol-Independent Multicast (PIM), Source-Specific Multicast (SSM), Multicast Source Discovery Protocol (MSDP), and Multiprotocol BGP (MBGP). The configuration focuses on the various implementations of PIM. DVMRP, CBT, and PIM depend on the concept of distribution trees, which is discussed next.

Distribution Trees

The campus network is composed of a collection of subnetworks connected by routers. When the source of a multicast data stream is located on one subnet and the receivers are located on different subnets, there needs to be a way of determining how to connect the destinations (multicast receivers) to the source (multicast server). For unicast transmissions, this is the handled by IP routing protocols.

Routing multicast traffic is a more complex problem. A multicast address identifies a particular transmission session rather than a specific physical destination. An individual host can join an ongoing multicast session by using IGMP to communicate this desire to an appropriately configured router on the subnet.

Because the number of receivers for a multicast session can potentially be large, the source should not need to know all the relevant addresses. Instead, the routers must somehow be able to translate multicast addresses into host addresses. The basic principle involved in multicast routing is that routers interact with each other to exchange information about multicast groups and neighboring multicast routers.

To deliver multicast packets to all receivers, designated routers construct a tree that connects all members of an IP multicast group. A *distribution tree* specifies a unique forwarding path between the subnet of the source and each subnet containing members of the multicast group.

A distribution tree has just enough connectivity so that there is only one loop-free path between every pair of routers. Because each router knows which of its lines belong to the tree, the router can copy an incoming multicast datagram onto all the outgoing branches. This action generates the minimum needed number of datagram copies. Because messages are replicated only when the tree branches, the number of copies of the messages transmitted through the network is minimized.

Because multicast groups are dynamic, with members joining or leaving a group at any time, the distribution tree must be dynamically updated. Branches that contain new members must be added. Branches in which no listeners exist must be discarded, or pruned.

There are two basic distribution tree construction techniques: source-specific trees and shared trees. Each of these is discussed in the following sections.

Source Distribution Trees (Shortest-Path Trees)

The simplest form of a multicast distribution tree is a *source tree*. Its root is the source of the multicast traffic, and its branches form a spanning tree through the network to the receivers. Because this tree uses the shortest path through the network, it also is often called a shortest-path tree (SPT). For every source of a multicast group, there is a corresponding SPT.

Figure 9-22 shows an example of an SPT for group 224.1.1.1 rooted at Source 1 and connecting two receivers, Receiver 1 and Receiver 2. The special notation of (S,G), pronounced "S comma G," denotes an SPT where S is the source's IP address and G is the multicast group address.

Notice that this notation implies that a separate SPT exists for every individual source that originates a stream for a multicast group. Therefore, if Source 2 is also sending traffic to group G to Receiver 1 and Receiver 2, a separate (S,G) SPT would look as shown in Figure 9-23.

The dense-mode option of Protocol-Independent Multicast (PIM-DM), discussed later in this chapter, utilizes source distribution trees. DVMRP also uses SPTs. In general, SPTs require more memory on routers and greater bandwidth on router links, but you get optimal paths from a multicast source to all receivers, thus minimizing delay.

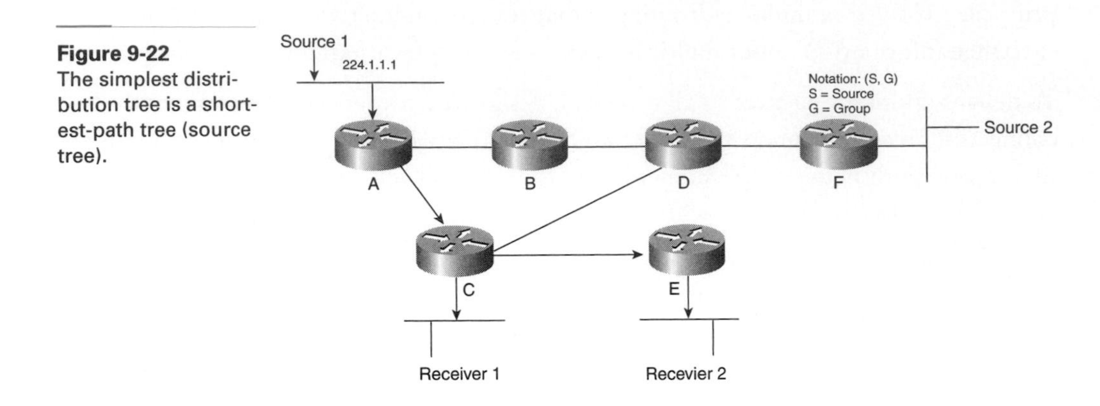

Figure 9-22
The simplest distribution tree is a shortest-path tree (source tree).

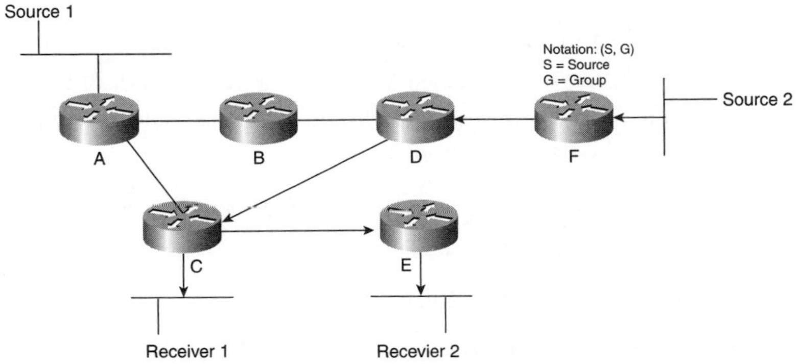

Figure 9-23
A unique SPT is built for each source.

Shared Distribution Tree

Unlike source trees, which have their roots at the source, *shared trees* use a single common root placed at a chosen point in the network. Depending on the multicast routing protocol, this root is called a rendezvous point (RP) or core. Protocol-Independent Multicast sparse mode (PIM-SM) utilizes shared distribution trees. The RP terminology is used in the context of PIM-SM. This chapter uses the RP terminology. PIM-SM and CBT both use the shared-tree approach to multicast routing.

Figure 9-24 shows a shared tree with the RP located at Router D. When using a shared tree, sources must send their traffic to the root in order for the traffic to reach the receivers. In this example, multicast group traffic from Source 1 and Source 2 travels to the rendezvous point (Router D) and then down the shared tree to Receiver 1 and Receiver 2. Because all sources in the multicast group use a common shared tree, a wildcard notation represents the tree: (*,G), pronounced "star comma G." In this case, the * (asterisk) means any source, and the G represents the multicast group.

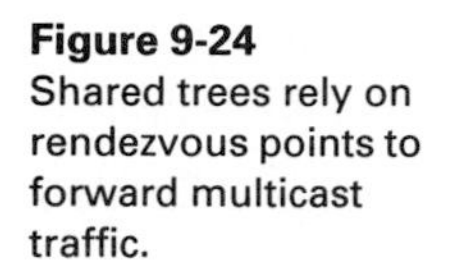

Figure 9-24
Shared trees rely on rendezvous points to forward multicast traffic.

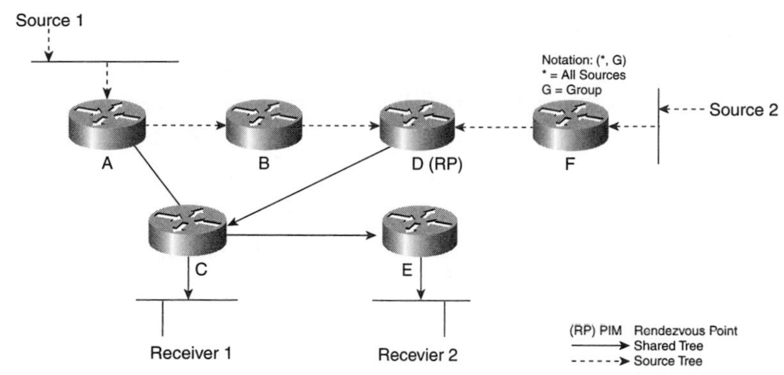

In general, shared trees require less router memory but might result in suboptimal paths from a multicast source to the receivers. This might introduce additional delay.

Now that you have a good idea of what source trees and shared trees are about, we next take a brief look at IP multicast routing protocols used to propagate multicast traffic within a routing domain: DVMRP, CBT, PIM dense mode, PIM sparse mode, PIM sparse-dense mode, and MOSPF. (MOSPF does not rely on source trees or shared trees.)

Distance Vector Multicast Routing Protocol

Distance Vector Multicast Routing Protocol (DVMRP) was codeveloped by Steve Deering while he was working toward his dissertation at Stanford. DVMRP was first publicized in RFC 1075, "Distance Vector Multicast Routing Protocol," in November 1988. Deering's dissertation was titled *Multicast Routing in a Datagram Network* and was published in December 1991. DVMRP, the oldest multicast routing protocol, is the precursor to modern multicast routing protocols.

DVMRP Version 2 (of 3) was used to build the *MBONE*, a multicast backbone across the public Internet, by building tunnels between DVMRP-capable Sun workstations running the *mroute* daemon (process). At one time, the MBONE was used widely in the research community to transmit the proceedings of various conferences and to permit desktop conferencing. The vision of the MBONE was to improve communication, just as radio, television, and the Internet have done over the last century. One of the ultimate goals is to make high-quality, real-time, many-to-many videoconferencing over the Internet as common as using a telephone or home computer. "At that point, the MBONE technology will be so deeply integrated in our everyday lives that the MBONE will cease to exist as a separate entity from the Internet itself." This is a quote from a great Web site, www.savetz.com/mbone, detailing the history of the MBONE. This quote encapsulates what many multicast developers envision for the not-too-distant future of the Internet.

DVMRP was derived from RIP and uses a technique known as Reverse Path Forwarding (RPF) on source distribution trees. When a router receives a packet, it floods the packet out all paths except the one that leads back to the packet's source. RPF allows this flooding to occur only if the packet passes the RPF check: It must have been received on the interface chosen by DVMRP to be the one leading back to the source. This flooding process allows a data stream to reach all LANs (possibly multiple times). If a router is attached to a set of LANs that do not want to receive a particular multicast group, the router can send a prune message back up the source distribution tree to stop subsequent packets from traveling where there are no members. This is where the phrase "flood-and-prune" originated. Flood-and-prune behavior is common to multicast routing protocols that rely solely on source trees. This is illustrated in Figure 9-25.

Figure 9-25
DVMRP relies on the flood-and-prune method.

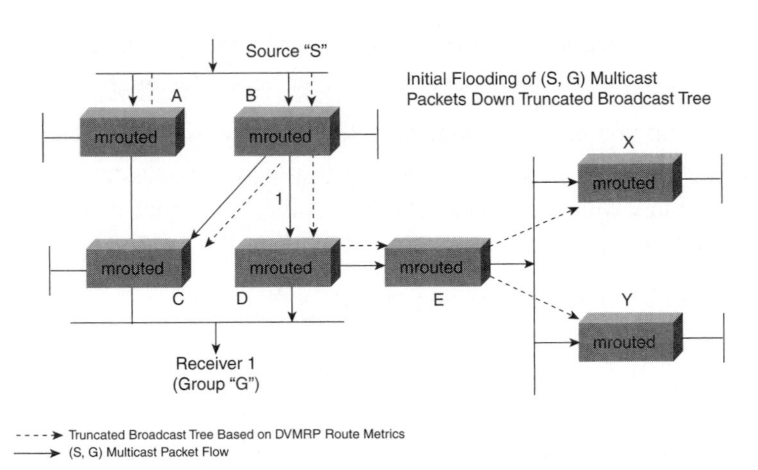

DVMRP periodically refloods in order to reach any new hosts that want to receive a particular group. There is a direct relationship between the time it takes for a new receiver to get the data stream and the frequency of flooding.

DVMRP implements its own unicast routing protocol in order to determine which interface leads back to the source of the data stream. This unicast routing protocol is similar to RIP and is based purely on hop counts (with a maximum hop count of 32). As a result, the path that the multicast traffic follows might not be the same as the path that the unicast traffic follows.

In general, DVMRP deployment is appropriate only for a large number of densely distributed receivers located in close proximity to the source.

DVMRP has significant scaling problems because of the necessity to flood frequently. This limitation is exacerbated by the fact that early implementations of DVMRP did not implement pruning.

DVMRP is not directly supported by Cisco devices, but multicast routing protocols, such as PIM, can be configured on Cisco devices to interoperate with DVMRP.

Core-Based Trees

Core-Based Trees (CBT) is a multicast routing protocol introduced in September 1997 in RFC 2201, "Core Based Trees (CBT) Multicast Routing Architecture." Quoting from the RFC, "CBT is a multicast routing architecture that builds a single delivery tree per group which is shared by all the group's senders and receivers. Most multicast algorithms build one multicast tree per sender (the tree is identical for senders on the same subnetwork), the tree being rooted at the sender's subnetwork. The primary advantage of the shared tree approach is that it typically offers more favorable scaling characteristics than all other multicast algorithms."

CBT is protocol-independent in that it uses information in the IP unicast routing table to calculate the next-hop router in the direction of the core of the shared tree.

When a receiver joins a multicast group via IGMP, its local CBT router looks up the multicast address and obtains the address of the *core* router for the group. The router then sends a join message for the group toward the core. At each router on the way to the core, forwarding state is instantiated for the group, and an acknowledgment is sent back to the previous router. In this way, a multicast tree is built, as shown in Figure 9-26.

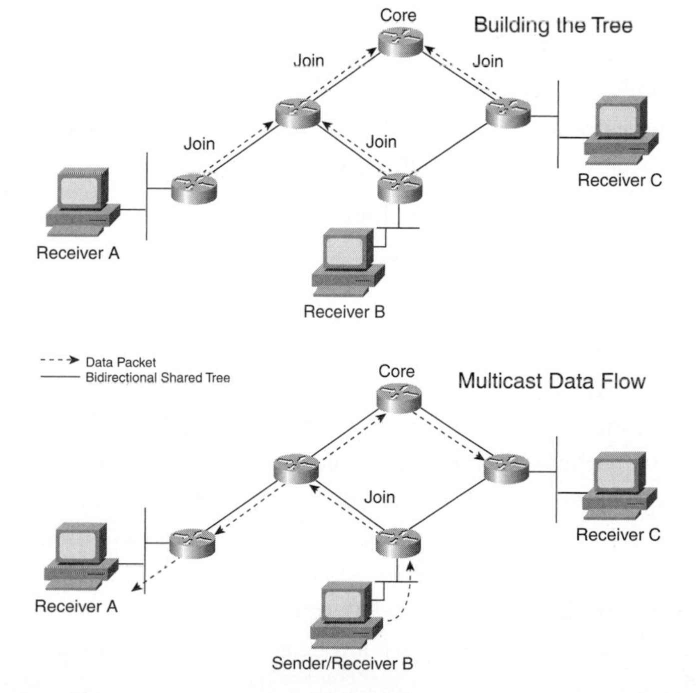

Figure 9-26
CBT builds a bidirectional shared tree based on the tree's core.

If a multicast group member sends data to the group, the packets reach the local router, which forwards them to any of its neighbors that are on the multicast tree. Each router that receives a packet forwards it out all its interfaces that are on the tree except the one that the packet came from. The style of tree that CBT builds is called a *bidirectional shared tree* because the routing state is bidirectional. Packets can flow both up the tree toward the core and down the tree away from the core, depending on the source's location. This scenario is in contrast to the *unidirectional shared trees* that are built by PIM-SM, as you will see later in this chapter.

IP multicast does not require senders to a group to be members of the group, so it is possible that a sender's local router is not on the tree. In this case, the packet is forwarded to the next hop toward the core. Eventually the packet either reaches a router that is on the tree, or it reaches the core, where it is then distributed along the multicast tree.

CBT also allows multiple core routers to be specified, adding a little redundancy in case the core becomes unreachable. CBT Version 2 never properly solved the problem of how to map a group address to a core's address. In addition, good core placement is a difficult problem. Without good core placement, CBT trees can be inefficient, so CBT is unlikely to be used as a global multicast routing protocol.

However, within a limited routing domain, CBT is efficient in terms of the amount of state (active multicast groups and associated forwarding information) that routers need to keep. Only routers on the distribution tree for a group keep forwarding state for that group, and no router needs to keep information about any source. Thus, CBT scales much better than flood-and-prune protocols, especially for sparse groups in which only a small proportion of subnetworks have members. In general, multicast routing protocols employing shared trees must maintain less multicast state than multicast routing protocols employing source trees.

Although CBT Version 3 development is progressing, despite CBT's advantages, it is apparent that CBT remains an experimental protocol with no actual network deployments to speak of. CBT is not currently supported on Cisco routers.

Protocol-Independent Multicast

Protocol-Independent Multicast (PIM) gets its name from the fact that it is IP routing protocol-independent. That is, regardless of which unicast routing protocols are used to populate the unicast routing table (including static routes), PIM uses this information to perform multicast forwarding; hence, it is protocol-independent. Although we refer to PIM as a multicast routing protocol, it actually uses the existing unicast routing table to perform the Reverse Path Forwarding (RPF) check instead of maintaining a separate multicast route table. Because PIM doesn't have to maintain its own routing table, it doesn't send or receive multicast route updates like other protocols, such as

MOSPF or DVMRP. Because multicast route updates don't have to be sent, the PIM overhead is significantly reduced in comparison to MOSPF and DVMRP.

Each multicast-enabled router is configured explicitly to select which interfaces are to use PIM. These interfaces connect either to neighboring PIM routers in the network or to host systems on a LAN. If there are multiple PIM routers on a LAN, they are considered neighbors.

PIM sends hello messages every 30 seconds to discover the existence of other PIM routers on the network and to elect the Designated Router (DR) on multiaccess networks (not the same as the OSPF DR). The DR is responsible for sending IGMP host-query messages to all hosts on a LAN segment. The DR is also responsible for sending joins to the RP for members on the multiaccess network and for sending registers to the RP for a source on the multiaccess network (informing the RP of an active source). For multiaccess networks, such as Ethernet, the PIM hello messages are multicast to the all-PIM-routers multicast group address, 224.0.0.13.

To elect the DR, each PIM node on a multiaccess network examines the received PIM hello messages from its neighbors and compares the IP address of its interface with the IP address of its PIM neighbors. The PIM neighbor with the highest IP address is elected the DR. If no PIM hellos have been received from the elected DR after a certain period (this is configurable), the DR election mechanism is run again to elect a new DR. This process is illustrated in Figure 9-27.

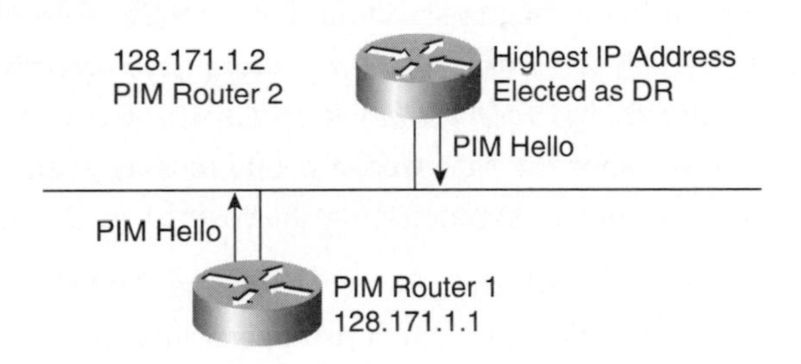

Figure 9-27 PIM hello messages are used on multi-access segments to elect a Designated Router (DR).

The next several sections describe PIM sparse mode, PIM dense mode, and bidirectional PIM. Sparse mode and dense mode have been supported by the Cisco IOS Software since Release 10.0. Bidirectional PIM support began with Cisco IOS Software Release 12.1(2)T.

Protocol-Independent Multicast Sparse Mode

Protocol-Independent Multicast sparse mode (PIM-SM) was developed as a result of the ongoing development efforts with CBT. The work on CBT encouraged development to improve on its limitations while keeping the beneficial properties of shared trees. The equivalent of a CBT core is called a rendezvous point (RP) in PIM-SM, but it largely serves the same purpose. PIM-SM and CBT are sometimes collectively called center-based tree protocols; *center* refers to PIM-SM's RP or CBT's core.

With PIM-SM, when a sender starts sending, whether it is a member or not, its local router receives the packets and maps the group address to the RP's address. It then encapsulates each packet in another IP packet and sends it as a unicast packet directly to the RP.

For multiaccess segments in PIM sparse-mode networks, a DR is elected. The DR is responsible for sending joins to the RP for members on the multiaccess network and for sending register messages to the RP for sources on the multiaccess network.

When a receiver joins a group via IGMP, its local router initiates a join message that travels hop-by-hop to the RP instantiating forwarding state for the group. However, this state is unidirectional. It can be used only by packets flowing from the RP toward the receiver, not for packets flowing back up the tree toward the RP. Data from senders is de-encapsulated at the RP and flows down the shared tree to all the receivers.

PIM-SM is an improvement over CBT in that discovery of senders and tree building from senders to receivers are separate functions. PIM-SM unidirectional trees are not necessarily ideal distribution trees, but they do start data flowing to the receivers. As soon as this data is flowing, *a receiver's local router can then initiate a transfer from the shared tree to a shortest-path tree* (source tree) by sending a source-specific join message toward the source, as shown in Figure 9-28. Restated, *the* default *behavior of PIM-SM in Cisco IOS Software is that routers with directly connected members join the shortest-path tree as soon as they detect a new multicast source.* When data starts arriving along the shortest-path tree, a prune message can be sent back up the shared tree toward the source to avoid getting the traffic twice.

Unlike other shortest-path tree protocols such as DVMRP and PIM dense mode (discussed in the next section), in which prune state exists everywhere that there are no receivers, with PIM-SM, source-specific state exists only on the shortest-path tree. Also, low-bandwidth sources such as those sending Real-Time Control Protocol (RTCP) receiver reports do not trigger the transfer to a shortest-path tree, a scenario that further helps scaling by eliminating unnecessary source-specific state.

Because PIM-SM can optimize its distribution trees after formation, it is less critically dependent on the RP location than CBT is on the core location. Hence, the primary requirement for choosing an RP is load balancing.

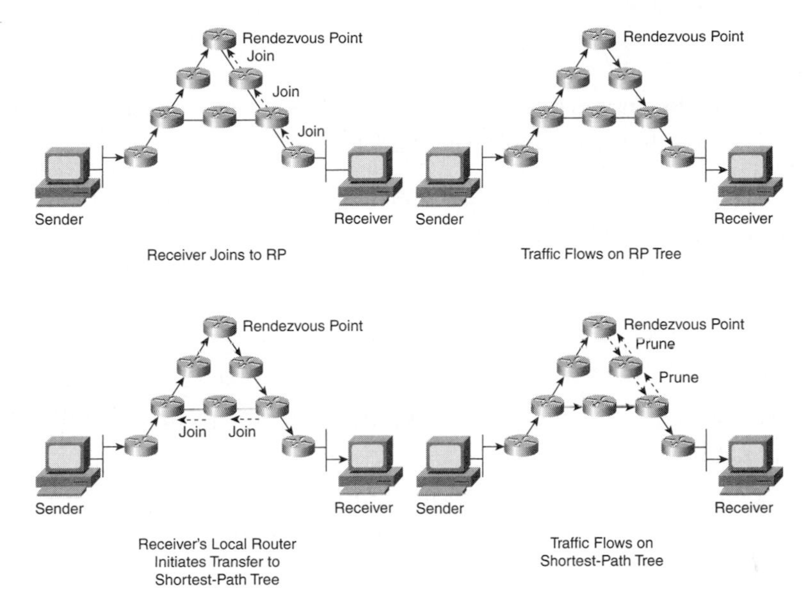

Figure 9-28 Unlike CBT, PIM-SM includes a mechanism for converting from a shared tree to a source tree.

To perform multicast-group-to-RP mapping, one common PIM-SM implementation method, Auto-RP, redistributes a list of candidates to be RPs to all routers. When a router needs to perform this mapping, it uses a special hash function to hash (use an algorithm to generate a number from a string of characters) the group address into the list of candidate RPs to decide which RP to join. Except in rare failure circumstances, all the routers within the domain perform the same hash and come up with the same choice of RP for the same multicast group. The RP might or might not be in an optimal location, but this situation is offset by the ability to switch to a shortest-path tree. However, the dependence on this hash function and the requirement to achieve convergence on a list of candidate RPs does limit PIM-SM's scaling. As a result, it is also best deployed within a routing domain, although the size of such a domain might be large.

One of the design goals of sparse mode is to optimize multicast traffic flow among sparsely populated senders and receivers. However, sparse mode can in fact be used where there is a dense distribution of multicast receivers because of its capability to switch over to a source tree after the initial shared tree is built.

A moment ago, we touched on the Auto-RP method that PIM-SM can use for multicast-group-to-RP mappings. For an RP to be useful, PIM routers need to know how to find the RP and how to determine which RP is appropriate for a given multicast group. There are four ways to locate an RP and distribute group-to-RP mappings:

- Auto-RP
- PIMv2 Bootstrap Router

- Static RP
- Anycast RP

We'll briefly discuss the first three options. Anycast RP is an application of MSDP and is discussed in the section "Anycast RP."

www.cisco.com/univercd/cc/td/doc/cisintwk/intsolns/mcst_sol/rps.htm is a useful Web site that lists four Cisco-recommended methods for configuring a rendezvous point. The methods detailed on the Web site use a *combination* of the four methods just listed. Auto-RP, PIMv2 Bootstrap Router, and Static RP are discussed in the following sections.

Auto-RP *Auto-RP* is a feature introduced in Cisco IOS Software Release 11.1 that automates the distribution of group-to-RP mappings in a PIM network. This feature has the following benefits:

- It's easy to use multiple RPs within a network to serve different group ranges.
- It allows load balancing among different RPs and arrangement of RPs according to the location of group participants.
- It avoids inconsistent, manual RP configurations that can cause connectivity problems.

Multiple RPs can be used to serve different group ranges or serve as backups of each other. To make Auto-RP work, a router must be designated as an *RP mapping agent*, which receives the RP-announcement messages from the RPs and arbitrates conflicts. The RP mapping agent then sends the consistent group-to-RP mappings to all other routers. Thus, all routers automatically discover which RP to use for the groups they support.

Multicast is used to distribute group-to-RP mapping information via two special IANA-assigned multicast groups:

- **Cisco-Announce Group**—224.0.1.39
- **Cisco-Discovery Group**—224.0.1.40

Because multicast is used to distribute this information, a catch-22 can occur if these groups operate in sparse mode: Routers would have to know an RP's address before they can learn it via Auto-RP messages. Therefore, it is recommended that these particular two multicast groups always run in dense mode so that this information is flooded throughout the network.

Multiple *candidate RPs* can be defined via configuration so that, in the case of an RP failure, the other candidate RP can assume RP responsibility. The candidate RPs begin multicasting their candidacy to be the RP via RP-Announce messages sent via the Cisco-Announce group, 224.0.1.39.

The RP mapping agents multicast the contents of their group-to-RP mapping cache to the Cisco-Discovery group, 224.0.1.40.

PIMv2 Bootstrap Router Cisco developed Auto-RP prior to the publication of PIMv2 by an IETF multicast working group in RFC 2117. PIMv2 specifies the bootstrap router method for multicast group-to-RP mappings.

A *bootstrap router (BSR)* provides a fault-tolerant automated RP discovery and distribution mechanism. With BSR, routers dynamically learn the group-to-RP mappings. PIMv2, which introduced BSR, was introduced in Cisco IOS Software Release 11.3T.

If all routers in the network are running PIMv2, you can configure a BSR instead of Auto-RP—they are very similar. With BSR configuration, you configure RP candidates (as in Auto-RP) and BSRs (similar to Auto-RP mapping agents).

Auto-RP supports administratively scoped zones. This can be important when you're trying to prevent high-rate group traffic from leaving a campus and consuming too much bandwidth on WAN links. BSR does not support administrative scoping. This is one of the main differences between the two approaches and is one reason that Auto-RP is generally preferred in a network consisting exclusively of Cisco routers.

PIM Version 1, together with the Auto-RP feature, can perform the same tasks as the PIM Version 2 BSR. However, Auto-RP is a standalone protocol, separate from PIM Version 1, and is Cisco-proprietary. PIM Version 2 is a standards track protocol from the IETF. Cisco recommends that you use PIM Version 2.

Either the BSR or Auto-RP should be chosen for a given range of multicast groups. If there are PIM Version 1 routers in the network, you do not use the BSR.

PIM uses the BSR to discover and announce RP information for each group prefix to all the routers in a PIM domain. This is the same function accomplished by Auto-RP, but the BSR is part of the PIM Version 2 specification. The BSR mechanism interoperates with Auto-RP on Cisco PIMv2 routers.

To avoid a single point of failure, you can configure several candidate BSRs in a PIM domain. A BSR is elected from among the candidate BSRs automatically; they use bootstrap messages to discover which BSR has the highest priority. This router then announces to all PIM routers in the PIM domain that it is the BSR.

Routers that are configured as candidate RPs then unicast to the BSR the group range for which they are responsible. The BSR includes this information in its bootstrap messages and disseminates it to all PIM routers in the domain. Based on this information, all routers can map multicast groups to specific RPs. As long as a router is receiving the bootstrap message, it has a current RP map.

NOTE

If router interfaces are configured in sparse mode, Auto-RP can still be used if all routers are configured with a static RP address for the Auto-RP groups: 224.0.1.39 and 224.0.1.40. The default is for Auto-RP learned information to take precedence.

Static RP Static RP requires hard-coded RP addresses to be configured on every multicast router except the RP itself. All routers must use the same RP address for the same multicast groups. With this option, RP failover is not possible unless Anycast RPs are also in use (see the section "Multicast Source Discovery Protocol").

If you configure PIM to operate in sparse mode, you must choose one or more routers to be RPs. *No command explicitly configures a router to be an RP. It learns to become an RP based on the static RP configuration on the other multicast routers.* Senders to a multicast group use RPs to announce their existence, and receivers of multicast packets use RPs to learn about new senders. The Cisco IOS Software can be configured so that packets for a single multicast group can use one or more RPs.

When the static RP method is used, you can specify the multicast groups for which an individual RP is responsible. The default setting is for a statically configured RP to be responsible for all multicast groups. Multiple RPs can be configured using the static RP method.

The RP address specified in a static RP configuration is used by first-hop routers to send PIM register messages on behalf of a host sending a packet to the group. The RP address is also used by last-hop routers to send PIM join and prune messages to the RP to inform it of group membership. You must configure the RP address on all routers except the RP router.

A PIM router can be an RP for more than one group. Only one RP address for a multicast group can be used at a time within a PIM domain. The conditions specified by the access list in the static RP configuration determine for which groups the router is an RP.

We have discussed three methods of locating an RP and distributing group-to-RP mappings: Auto-RP, PIMv2 Bootstrap Router, and Static RP. The fourth method, Anycast RP, is detailed in the section "Multicast Source Discovery Protocol."

Protocol-Independent Multicast Dense Mode

Protocol-Independent Multicast dense mode (PIM-DM) is similar to DVMRP and can interoperate with DVMRP. PIM-DM protocol works best when numerous members belong to each multicast group. PIM floods the multimedia packet out to all routers in the network and then prunes routers that do not support members of that particular multicast group.

With dense mode, unlike sparse mode, the DR on a multiaccess segment has no function, except in the case that IGMPv1 is in use. In this case, the DR functions as the IGMP Querier for the multiaccess network.

PIM dense mode is most useful when

- Senders and receivers are in close proximity to one another.
- There are few senders and many receivers.
- The volume of multicast traffic is high.
- The stream of multicast traffic is constant.

PIM-DM uses the underlying unicast routing table to make RPF checks and flood the network with multicast data and then prune paths based on uninterested receivers, as shown in Figure 9-29.

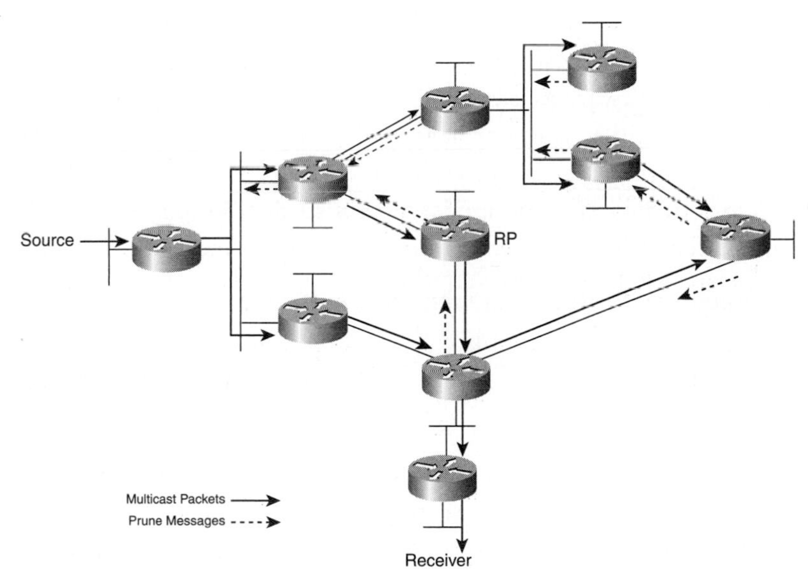

Figure 9-29
PIM-DM, like DVMRP, uses the flood-and-prune method to build a source tree.

PIM-DM uses a push model to enable multicast traffic flow, whereas PIM-SM uses a pull model. PIM-SM uses an explicit join mechanism, and traffic is sent only where it is requested.

Some advantages of PIM-DM are as follows:

- A minimal number of commands are required for configuration.
- It has a simple mechanism for reaching all possible receivers and eliminating distribution to uninterested receivers.
- Simple behavior is easier to understand and therefore easier to debug.
- It does not require a separate multicast routing protocol.

A disadvantage of PIM-DM is that the flood-and-prune process repeats every 3 minutes. PIM-DM is most effective in a network that has a large number of densely distributed receivers located in close proximity to the source.

One last PIM topic relating to dense mode that has become common in recent years is *PIM sparse-dense mode*. Cisco has implemented an alternative to choosing just dense mode or just sparse mode on a router interface. This was necessitated by a change in the paradigm for forwarding multicast traffic via PIM that became apparent during its development. It turned out that it was more efficient to choose sparse or dense on a per-group basis rather than on a per-router interface basis. Sparse-dense mode facilitates this ability. Network engineers can configure sparse-dense mode, introduced in Cisco IOS Software Release 11.1. This configuration option allows individual groups to be run in either sparse or dense mode, depending on whether RP information is available for that group. If the router learns RP information for a particular group, the group is treated as sparse mode; otherwise, the group is treated as dense mode. In general, Cisco now recommends using sparse-dense mode when configuring PIM.

Bidirectional Protocol-Independent Multicast

Bidirectional PIM (bidir-PIM) is a variant of the PIM suite of multicast routing protocols for IP multicast and is an extension of PIM-SM. Bidir-PIM resolves some limitations of PIM-SM for groups with a large number of sources.

Bidir-PIM is based on the draft-kouvelas-pim-bidir-new-00.txt IETF protocol specification. This draft and other drafts referenced in it can be found at ftp://ftpeng.cisco.com/ipmulticast/drafts. Cisco router support for bidir-PIM began with Cisco IOS Software Release 12.1(2)T.

A router can simultaneously support PIM-DM, PIM-SM, and bidir-PIM or any combination of them for different multicast groups. In bidirectional mode, traffic is routed only along a bidirectional shared tree that is rooted at the group's RP. In bidir-PIM, the RP's IP address acts as the key to having all routers establish a loop-free spanning-tree topology rooted in that IP address.

Membership in a bidirectional group is signaled via explicit join messages. Traffic from sources is unconditionally sent up the shared tree toward the RP and is passed down the tree toward the receivers on each branch of the tree.

Bidir-PIM is designed to be used for many-to-many applications within individual PIM domains. Multicast groups in bidirectional mode can scale to an arbitrary number of sources without incurring overhead due to the number of sources.

Bidir-PIM is derived from the mechanisms of PIM-SM and shares many SPT operations as well. Bidir-PIM also has unconditional forwarding of source traffic toward the RP upstream on the shared tree, but no registering process for sources as in PIM-SM. These modifications are necessary and are sufficient to allow forwarding of traffic in all routers solely based on the (*,G) multicast routing entries. This feature eliminates any source-specific state and allows scaling capability for an arbitrary number of sources. Figure 9-30 shows the state created per router for a unidirectional shared tree and a source tree.

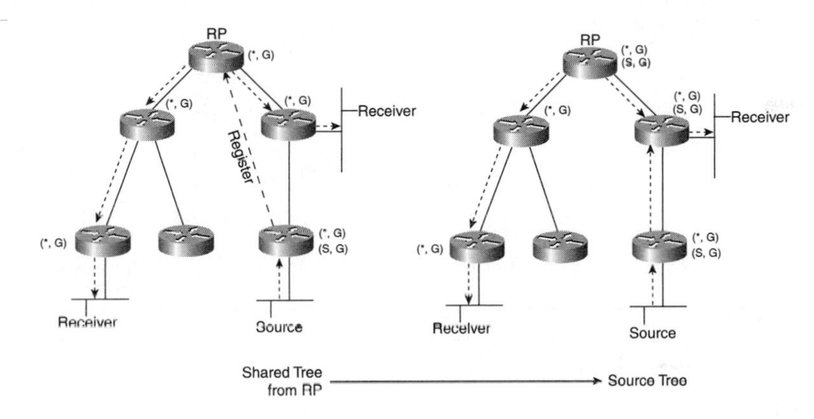

Figure 9-30 The multicast states differ for a PIM-SM shared tree and a PIM-DM source tree in a PIM domain.

Figure 9-31 shows the state created per router for a bidirectional shared tree.

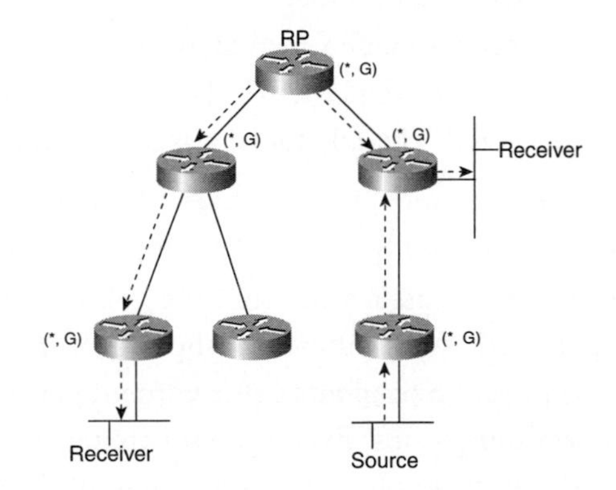

Figure 9-31 (S,G) state is avoided altogether in a bidir-PIM shared tree.

When packets are forwarded downstream from the RP toward receivers, there are no fundamental differences between bidir-PIM and PIM-SM. Bidir-PIM deviates substantially from PIM-SM when passing traffic from sources upstream toward the RP.

PIM-SM cannot forward traffic in the upstream direction of a tree, because it accepts traffic from only one RPF interface. This interface (for the shared tree) points toward the RP, thereby allowing only downstream traffic flow. In this case, upstream traffic is first encapsulated into unicast register messages, which are passed from the source's DR toward the RP. In a second step, the RP joins an SPT that is rooted at the source. Therefore, in PIM-SM, traffic from sources traveling toward the RP does not flow upstream in the shared tree, but downstream along the source's SPT until it reaches the RP. From the RP, traffic flows along the shared tree toward all receivers, as shown in Figure 9-32.

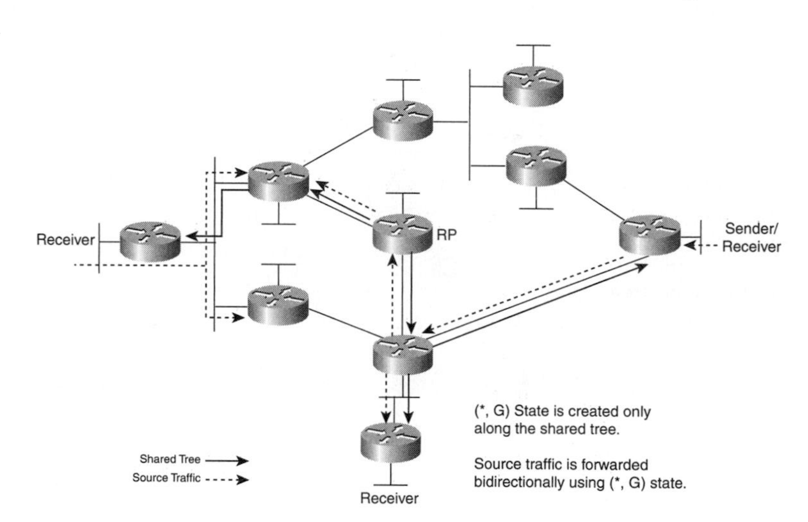

Figure 9-32 Bidir-PIM allows multicast traffic to flow up the shared tree toward the RP.

In bidir-PIM, the packet-forwarding rules have been improved over PIM-SM, allowing traffic to be passed up the shared tree toward the RP. To avoid multicast packet looping, bidir-PIM introduces a new mechanism called *designated forwarder (DF)* election, which establishes a loop-free SPT rooted at the RP. On each link, the router with the best path to the RP is elected as the DF. The DF is responsible for forwarding traffic upstream toward the RP. The DF concept is key to bidir-PIM operation.

The implementation of bidir-PIM, in which all the forwarding on a link is centered around the DF, ensures highly robust PIM-SM multicast networks and eliminates possible loops. All the multicast traffic from the link toward the RP and in the opposite direction passes the DF. With bidir-PIM, the role of the DR (in PIM-SM) is replaced by the DF. All (*,G) joins are originated (forwarded) via the DF, which eliminates the possibilities for forwarding loops. Even if downstream routers on the link use customized unicast routes, the election of a DF ensures that all those routers know who the DF is and use it for forwarding (*,G) joins. This again eliminates the multicast forwarding loops that were possible in regular PIM-SM due to misconfiguration.

So bidir-PIM is a big improvement over PIM-SM:

- It eliminates all (S,G) state in the network.
- SPTs between sources and RPs are eliminated.
- Source traffic flows both up and down the shared tree.

In practical terms, bidir-PIM allows many-to-many applications to scale and permits a virtually unlimited number of sources. The only downside is that the configuration is slightly more complicated than PIM-SM.

Multicast OSPF

The last intradomain multicast routing protocol we consider is *Multicast Open Shortest Path First (MOSPF)*. MOSPF, specified in RFC 1584 ("Multicast Extensions to OSPF," published in March 1994), is the multicast extension to OSPF. OSPF is a unicast link-state routing protocol. MOSPF requires OSPF as the underlying routing protocol. John Moy created OSPF and MOSPF and authored the RFCs for both OSPF and MOSPF. RFC 1131, "The OSPF Specification," published in October 1989, is the original OSPF RFC.

NOTE

Because MOSPF is an extension of OSPF, MOSPF builds very efficient distribution trees. This is one of its strengths. MOSPF also has the potential for interdomain functionality as a result of OSPF's capability to handle protocol redistribution between routing domains. However, for practical purposes, MOSPF is an intradomain multicast routing protocol.

Link-state routing protocols work by having each router periodically send a routing message listing its neighbors and how far away they are (as measured by the particular routing protocol metric). These routing messages are flooded throughout the entire network, so every router can build a map of the network. This map is then used to build forwarding tables (using the Dijkstra algorithm) so that the router can decide quickly which is the correct next hop for a particular packet.

This concept is extended to multicast simply by having each router also list in a routing message the groups for which it has local receivers. Thus, given the map and the locations of the receivers, a router can also build a multicast forwarding table for each group. A sample MOSPF network is pictured in Figure 9-33.

MOSPF has some disadvantages that have prevented it from being adopted on a wide-scale basis. MOSPF suffers from a poor ability to scale. With flood-and-prune protocols, data traffic is an *implicit* message about where there are senders, so routers need to store unwanted state where there are no receivers. With MOSPF, there are *explicit* messages about where all the receivers are, so the routers need to store unwanted state where there are no senders.

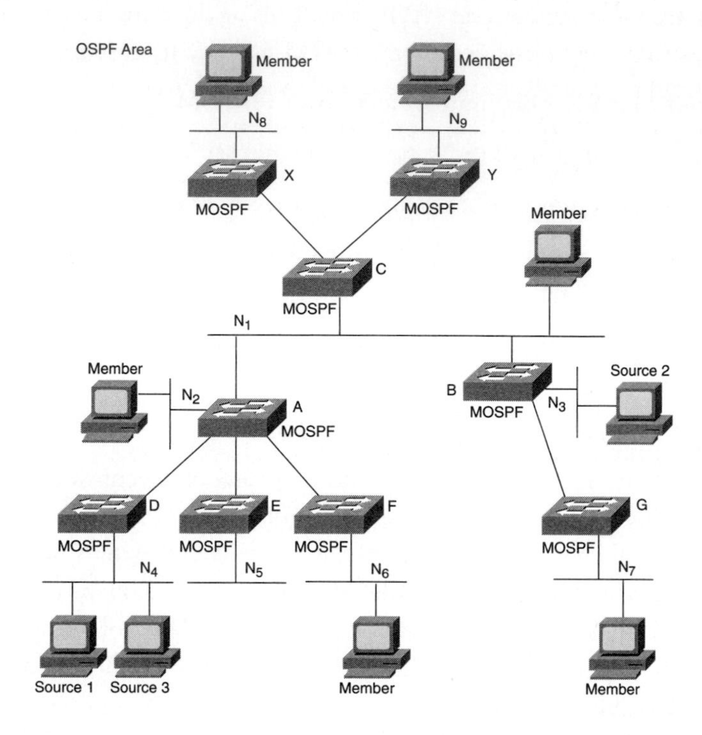

Figure 9-33 MOSPF is an extension to OSPF, incorporating Type 6 Link State Advertisements.

Also, MOSPF requires frequent flooding of link-state membership, which hinders performance as CPU demands grow rapidly to keep track of current network topology (source-group pairs). The Dijkstra algorithm must be run for every single multicast source. This can be a big problem on a network with unstable links or in a network-with a large number of simultaneously active source-group pairs (routers have to maintain too much information relating to the entire network topology). Finally, multicast applications permit any user in the network to create a new source-group pair. There is no way for a network administrator to control the number of source-group pairs in the network. A network administrator has little control over MOSPF's bringing down the network as multicast applications become more popular with users.

Cisco routers do not support MOSPF. MOSPF incorporates a Type 6 LSA that is not recognized by Cisco routers. You can configure a Cisco router to ignore Type 6 LSAs to prevent the router from generating syslog messages resulting from the dropped Type 6 LSAs.

This completes the discussion of intradomain multicast routing protocols. Now we look at some interdomain multicast routing protocols: Multicast Source Discovery Protocol (MSDP), Multiprotocol BGP (MBGP), and Source-Specific Multicast (SSM).

Multicast Source Discovery Protocol

Multicast Source Discovery Protocol (MSDP) was developed for peering between ISPs. ISPs did not want to rely on an RP maintained by a competing ISP to provide service to their customers. MSDP allows each ISP to have its own local RP and still forward and receive multicast traffic to and from the Internet.

MSDP is a mechanism to connect multiple PIM-SM domains. MSDP allows multicast sources for a group to be known to all RPs in different domains. Each PIM-SM domain uses its own RPs and need not depend on RPs in other domains. An RP runs MSDP over TCP to discover multicast sources in other domains.

An RP in a PIM-SM domain has MSDP peering relationships with MSDP-enabled routers in other domains. The peering relationship occurs over a TCP connection, where primarily a list of sources sending to multicast groups is exchanged. The TCP connections between RPs are achieved by the underlying routing system. The receiving RP uses the source lists to establish a source path.

The purpose of this topology is to have domains discover multicast sources in other domains. If the multicast sources are of interest to a domain that has receivers, multicast data is delivered over the normal source-tree building mechanism in PIM-SM.

MSDP depends heavily on BGP or MBGP for interdomain operation. Cisco recommends that you run MSDP in RPs in your domain that are RPs for sources sending to global groups to be announced to the Internet.

Figure 9-34 illustrates MSDP operating between two MSDP peers. PIM uses MSDP as the standard mechanism to register a source with a domain's RP.

Figure 9-34
MSDP shares SA messages via TCP to reflect interest in multicast group membership between rendezvous points in different routing domains.

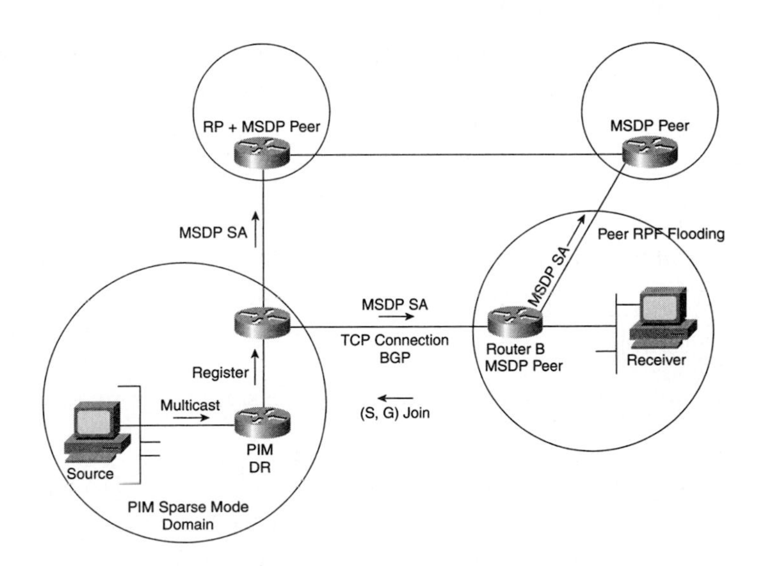

When MSDP is configured, the following sequence occurs. When a source's first data packet is registered by the first-hop router, that same data packet is decapsulated by the RP and is forwarded down the shared tree. That packet is also re-encapsulated in a *Source Active (SA)* message that is immediately forwarded to all MSDP peers. SA messages are the key to MSDP operation. The SA message identifies the source, the group the source is sending to, and the address or the originator ID of the RP, if configured. If the peer is an RP and has a member of that multicast group, the data packet is decapsulated and forwarded down the shared tree in the remote domain.

Each MSDP peer receives and forwards the SA message away from the originating RP to achieve *peer-RPF flooding*. The concept of peer-RPF flooding is with respect to forwarding SA messages. The router examines the BGP or MBGP routing table to determine which peer is the next hop toward the SA message's originating RP. Such a peer is called an *RPF peer*. The router forwards the message to all MSDP peers other than the RPF peer.

If the MSDP peer receives the same SA message from a *non-RPF* peer toward the originating RP, it drops the message. Otherwise, it forwards the message to all its MSDP peers.

When an RP for a domain receives an SA message from an MSDP peer, it determines whether it has any group members interested in the group that the SA message describes. If the (*,G) entry exists with a nonempty outgoing interface list, the domain is interested in the group, and the RP triggers an (S,G) join toward the source, as pictured in Figure 9-34.

NOTE

MSDP support began with Cisco IOS Software Release 12.0(7)T.

MSDP has the following benefits:

- It breaks up the shared multicast distribution tree. You can make the shared tree local to your domain. Your local members join the local tree, and join messages for the shared tree never need to leave your domain.
- PIM-SM domains can rely on their own RPs only, thus decreasing reliance on RPs in another domain. This increases security, because you can prevent your sources from being known outside your domain.
- Domains with only receivers can receive data without globally advertising group membership.
- Global source multicast routing table state is not required, thus saving on memory.

Now that you have a rudimentary understanding of MSDP, we can now describe the fourth way for multicast routers to locate a rendezvous point and distribute group-to-RP mappings within a routing domain: Anycast RP.

Anycast RP

Anycast RP is a useful application of MSDP. Although MSDP is designed for interdomain multicast applications, some smart person figured out that MSDP could be used within a routing domain to provide redundancy and load-sharing capabilities. Enterprise customers typically use Anycast RP to configure a PIM-SM network to meet fault tolerance requirements within a single multicast domain.

In Anycast RP, two or more RPs are configured with the same IP address on loopback interfaces. The Anycast RP loopback address should be configured with a 32-bit mask, making it a host address. All the downstream routers should be configured to know that the Anycast RP loopback address is the IP address of their local RP. *IP routing automatically selects the topologically closest RP for each source and receiver.* Assuming that the sources are evenly spaced around the network, an equal number of sources register with each RP. That is, the process of registering the sources is shared equally by all the RPs in the network.

Because a source might register with one RP and receivers might join to a different RP, a method is needed for the RPs to exchange information about active sources. This information exchange is done with MSDP, as shown in Figure 9-35.

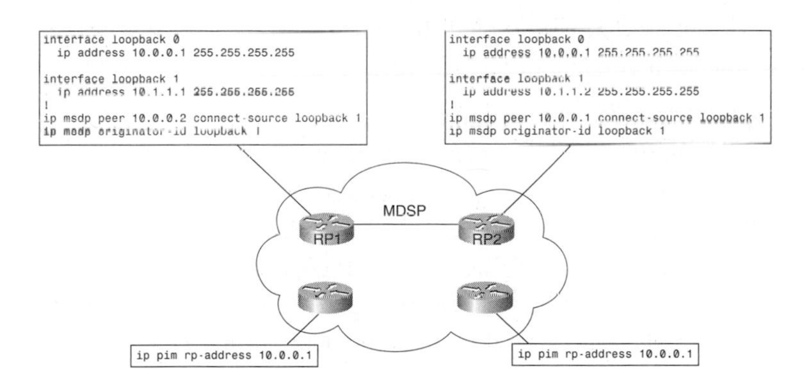

Figure 9-35 Anycast RP leverages MSDP in a PIM-SM domain to provide optimal multicast group-to-RP mapping.

In Anycast RP, all the RPs are configured to be MSDP peers of each other (full mesh). When a source registers with one RP, an SA message is sent to the other RPs, informing them that there is an active source for a particular multicast group. The result is that each RP knows about the active sources in the area of the other RPs. If any of the RPs fails, IP routing converges, and one of the RPs becomes the active RP in more than one area. New sources register with the backup RP. Receivers join toward the new RP, and connectivity is maintained.

The RP is normally needed only to start new sessions with sources and receivers. The RP facilitates the shared tree so that sources and receivers can directly establish a multicast data flow. If a multicast data flow is already directly established between a source and the receiver, an RP failure won't affect that session. Anycast RP ensures that new sessions with sources and receivers can begin at any time.

Multiprotocol BGP

Multiprotocol BGP (MBGP) adds capabilities to BGP to enable multicast routing policy throughout the Internet and to connect multicast topologies within and between BGP autonomous systems. That is, MBGP is an enhanced BGP that carries IP multicast routes. BGP carries two sets of routes—one set for unicast routing and one set for multicast routing. PIM uses the routes associated with multicast routing to build data distribution trees.

It is now possible to configure BGP peers that exchange both unicast and multicast network layer reachability information (NLRI).

MBGP is useful when you want a link dedicated to multicast traffic, perhaps to limit which resources are used for which traffic. Perhaps you want all multicast traffic exchanged at one network access point (NAP). MBGP allows you to have a unicast routing topology that is different from a multicast routing topology. Thus, you have more control over your network and resources.

Prior to MBGP, the only way to do interdomain multicast routing was to use the BGP infrastructure that was in place for unicast routing. If those routers were not multicast-capable, or if you had differing policies where you wanted multicast traffic to flow, you could not support it.

Figure 9-36 illustrates a simple example of unicast and multicast topologies that are incongruent and therefore are not possible without MBGP. Autonomous systems 100, 200, and 300 are each connected to two NAPs that are FDDI rings. One is used for unicast peering and therefore the exchanging of unicast traffic. The Multicast-Friendly Interconnect (MFI) ring pictured in Figure 9-36 is used for multicast peering and therefore the exchanging of multicast traffic. Each router is unicast- and multicast-capable.

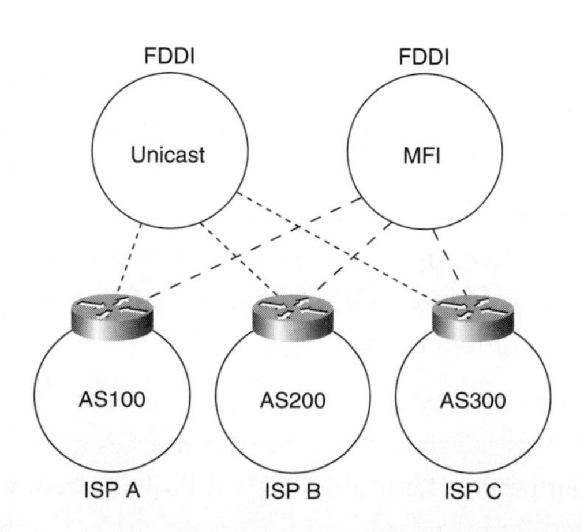

Figure 9-36 Facilitating interdomain multicast traffic with incongruent unicast and multicast topologies was difficult or impossible prior to MBGP.

Figure 9-37 is a topology of unicast-only routers and multicast-only routers. The two routers on the left are unicast-only routers (that is, they don't support or are not configured to do multicast routing). The two routers on the right are multicast-only routers. Routers A and B support both unicast and multicast routing. The four routers in the middle are connected to a single NAP.

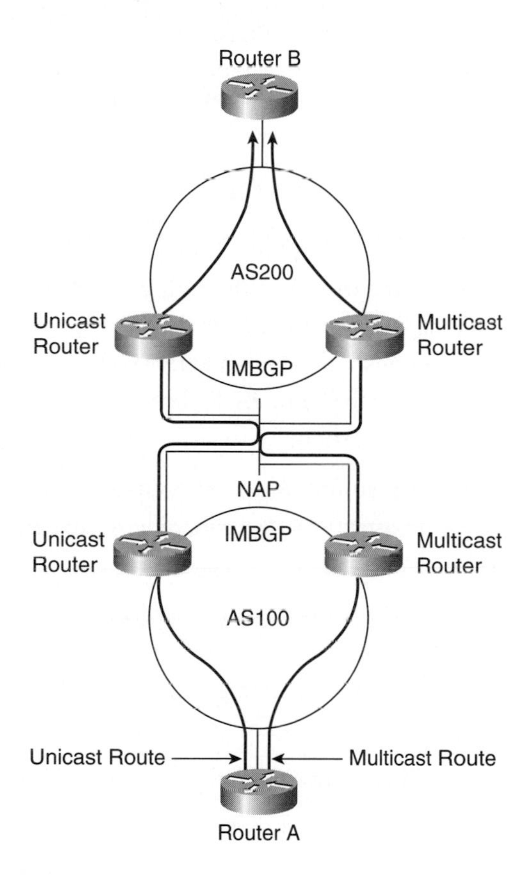

Figure 9-37
MBGP allows both unicast and multicast routing policies between autonomous systems.

Only unicast traffic can travel from Router A to the unicast routers to Router B and back. Multicast traffic could not flow on that path, so another routing table is required. Multicast traffic uses the path from Router A to the multicast routers to Router B and back.

Figure 9-37 illustrates an MBGP environment with a separate unicast route and multicast route from Router A to Router B. MBGP allows these routes to be noncongruent. Both the autonomous systems must be configured for Internal MBGP (IMBGP).

A multicast routing protocol, such as PIM, uses the MBGP routing table to perform RPF lookups for multicast-capable sources. Thus, packets can be sent and accepted on the multicast topology but not on the unicast topology.

MBGP offers the following benefits:

- A set of autonomous systems can support incongruent unicast and multicast topologies.
- A set of autonomous systems can support congruent unicast and multicast topologies that have different policies (BGP filtering configurations).
- All of BGP's routing policy capabilities can be applied to MBGP.
- All the BGP commands can be used with MBGP.

MBGP solves part of the interdomain multicast problem by allowing autonomous systems to exchange multicast RPF information in the form of MBGP Multicast NLRI. Because this is accomplished using an extension to the BGP protocol to make it support multiple protocols, the same BGP configuration knobs are available for both unicast and multicast information.

The separation of unicast and multicast prefixes into separate unicast and multicast routing information bases (RIB) permits unicast and multicast traffic to follow different paths if desired.

MBGP is only one piece of the overall interdomain multicast solution. PIM must still be used to build the multicast distribution trees, perform RPF checks, and forward multicast traffic. PIM-SM is recommended because it permits the use of MSDP, which solves most of the remaining issues, as described in the preceding section.

MBGP support on Cisco routers began with Cisco IOS Software Release 11.1(20)CC.

Source-Specific Multicast

The *Source-Specific Multicast (SSM)* feature is an extension of IP multicast in which datagram traffic is forwarded to receivers from only multicast sources to which the receivers have explicitly joined. For multicast groups configured for SSM, only source-specific multicast distribution trees (no shared trees) are created.

The first meeting of the SSM IETF working group was in August 2000. SSM support on Cisco routers began with Cisco IOS Software Release 12.1(3)T. SSM normally works in concert with IGMPv3; support for IGMPv3 began with Cisco IOS Software Release 12.1(5)T.

SSM is a datagram delivery model that best supports one-to-many applications, also known as broadcast applications. SSM is a core networking technology for the Cisco implementation of IP multicast solutions targeted for audio and video broadcast application environments.

Protocol-Independent Multicast source-specific mode (PIM-SSM) is the routing protocol that supports the implementation of SSM. It is derived from PIM-SM. IGMPv3 supports source filtering, which is required for SSM. To run SSM with IGMPv3, SSM must be supported in the Cisco IOS Software router, the host where the application is running, and the application itself. IGMP v3lite and URL Rendezvous Directory (URD) are two Cisco-developed transition solutions that enable the immediate development and deployment of SSM services without the need to wait for the availability of full IGMPv3 support in host operating systems and SSM receiver applications. IGMP v3lite is a solution for application developers that allows immediate development of SSM receiver applications switching to IGMPv3 as soon as it becomes available. URD is a solution for content providers and content aggregators that lets them deploy receiver applications that are not yet SSM-enabled (through support for IGMPv3). IGMPv3, IGMP v3lite, and URD interoperate with each other, so both IGMP v3lite and URD can easily be used as transitional solutions toward full IGMPv3 support in hosts.

NOTE

IGMPv3 uses new membership report messages that might not be recognized correctly by older IGMP snooping switches. In this case, hosts will not properly receive traffic. This situation is not an issue if URD or IGMP v3lite is used with hosts in which the operating system is not upgraded for IGMPv3, because IGMP v3lite and URD rely only on IGMPv1 or IGMPv2 membership reports.

An established network, in which IP multicast service is based on PIM-SM, can support SSM services. SSM can also be deployed alone in a network without the full range of protocols that are required for interdomain PIM-SM (for example, MSDP, MBGP, Auto-RP, or BSR) if only SSM service is needed.

If SSM is deployed in a network already configured for PIM-SM, only the last-hop routers must be upgraded to a Cisco IOS Software image that supports SSM. Routers that are not directly connected to receivers do not have to upgrade to a Cisco IOS Software image that supports SSM. In general, these non-last-hop routers must only run PIM-SM in the SSM range (232.0.0.0/8) and might need additional access control configuration to suppress MSDP signaling, registering, or PIM-SM shared-tree operations from occurring within the SSM range.

Benefits of SSM

First, with other forms of IP multicasting, applications must acquire a unique IP multicast group address, because traffic distribution is based only on the IP multicast group address used. If two applications with different sources and receivers use the same IP multicast group address, receivers of both applications receive traffic from the senders of both applications. Even though the receivers, if programmed appropriately, can filter out the unwanted traffic, this situation would cause generally unacceptable levels of unwanted traffic. In SSM, traffic from each source is forwarded between routers in the network independent of traffic from other sources. Thus, different sources can reuse multicast group addresses in the SSM range, 232.0.0.0/8.

Second, in SSM, multicast traffic from each individual source is transported across the network only if it is requested from a receiver. In contrast, other forms of IP multicasting forward traffic from any active source sending to a multicast group to all receivers requesting that multicast group. In Internet broadcast applications, this behavior is highly undesirable, because it allows unwanted sources to easily disturb the actual Internet broadcast source by simply sending traffic to the same multicast group. This situation depletes bandwidth at the receiver side with unwanted traffic and thus disrupts the undisturbed reception of the Internet broadcast. In SSM, this type of denial of service (DoS) attack cannot be made by simply sending traffic to a multicast group.

Third, SSM is easy to install and provision in a network, because it does not require the network to maintain which active sources are sending to multicast groups, as with other forms of IP multicasting (such as PIM-SM and MSDP). RP management in PIM-SM and MSDP is required for the network to learn about active sources. This management is not necessary in SSM. Thus, SSM is easier to install and manage and therefore is easier to operationally scale in deployment. Another factor that contributes to SSM's ease of installation is that it can leverage preexisting PIM-SM networks and requires only the upgrade of last-hop routers to support IGMPv3, IGMP v3lite, or URD.

The three benefits just described make SSM ideal for Internet broadcast-style broadcast applications. These benefits can be summarized as follows:

- SSM solves multicast address allocation problems. It allows content providers to use the same group ranges, because flows are differentiated by both source and group.
- SSM improves security. It helps prevent certain DoS attacks originated by bogus sources of multicast traffic.
- SSM is easy to install and provision in a network.

This completes our exploration of IP multicast routing protocols. We finish the chapter with a brief look at configuring PIM in a campus network. We forego the configuration of BSR, bidir-PIM, SSM, MSDP, and MBGP. Also, DVMRP, CBT, and MOSPF are not supported on Cisco routers. Our configurations are restricted to PIM-DM, PIM-SM, PIM sparse-dense mode, Auto-RP, Static RP, and Anycast RP.

Configuring IP Multicast Routing with PIM

When PIM is enabled on the interface of a router running Cisco IOS Software Release 11.3(2)T or later, PIM's default operational mode is Version 2.

There are two approaches to using PIM Version 2. You can use Version 2 exclusively in your network, or you can migrate to Version 2 by employing a mixed PIM version environment. The options available depend on whether your network includes all Cisco routers:

- If your network contains all Cisco routers, you can use either Auto-RP or BSR.
- If you have routers other than Cisco in your network, you need to use BSR.
- If you have PIM Version 1 and PIM Version 2 Cisco routers and routers from other vendors, you must use both Auto-RP and BSR.

For each of the PIM options we consider, you must first globally enable IP multicast routing. The global configuration command to enable IP multicast routing is **ip multicast-routing**. This enables the Cisco IOS Software to forward IP multicast packets.

A second requirement is to configure PIM on the individual interfaces participating in multicast routing. To cut to the chase, some experts now recommend configuring PIM sparse-dense mode on all interfaces on a router that will participate in IP multicast routing.

Enabling PIM on an interface also enables IGMP operation on that interface. An interface can be configured to be in dense mode, sparse mode, or sparse-dense mode. The mode determines how the router populates its multicast routing table and how the router forwards multicast packets it receives from its directly connected LANs. You must enable PIM in one of these modes for an interface to perform IP multicast routing. There is no default mode setting. By default, multicast routing is disabled on an interface.

In populating the multicast routing table, dense-mode interfaces are always added to the table. Sparse-mode interfaces are added to the table only when periodic join messages are received from downstream routers or when a directly connected member is on the interface. When forwarding from a LAN, sparse-mode operation occurs if an RP is known for the group. If it is, the packets are encapsulated and sent toward the RP. When no RP is known, the packet is flooded in a dense-mode fashion. If the multicast traffic from a specific source is sufficient, the receiver's first-hop router might send join messages toward the source to build a source-based distribution tree.

Before describing how multicast routes are learned dynamically via PIM-DM, we begin by demonstrating how to configure static multicast routes.

Static Mroutes

Just as you occasionally need to configure a static route to prescribe next-hop behavior for a particular unicast route, there might be cases in which you need to configure a *static multicast route* (static mroute). The global configuration command to configure a static mroute is **ip mroute** *source-address mask* [*protocol as-number*] {*rpf-address* | *type number*} [*distance*].

This command allows you to statically configure where multicast sources are located (even though the unicast routing table might show something different).

The *source-address mask* pair specifies the multicast source's IP address. *protocol* specifies which unicast routing protocol you are using. *as-number* specifies the autonomous system number of the routing protocol you are using (if applicable).

rpf-address specifies the mroute's incoming interface. *rpf-address* can be a host IP address of a directly connected system or a network/subnet number. When it is a route, a recursive lookup is done from the unicast routing table to find a directly connected system. If the *rpf-address* argument is not specified, the interface *type number* value is used as the incoming interface.

The *distance* determines whether a unicast route, a DVMRP route, or a static mroute should be used for the RPF lookup. Lower distances are preferred. If the static mroute has the same distance as the other two RPF sources, the static mroute takes precedence. The default is 0.

Example 9-35 configures all multicast sources via a single interface, serial0.

Example 9-35 Example 9-35Static Mroutes Can Be Configured Using Almost the Same Syntax Used for Static Routes

```
Router(config)#ip mroute 0.0.0.0 255.255.255.255 serial0 200
```

Because the distance is set at 200, the static mroute takes effect only if the dynamically discovered unicast routes for a given destination become unavailable.

Configuring PIM-DM

Initially, a dense-mode interface forwards multicast packets until a prune message is received from a downstream router. Then, the dense-mode interface periodically forwards multicast packets out the interface until the same conditions occur. Dense mode assumes that multicast group members are present. Dense mode routers never send a join message. They do send prune messages as soon as they determine they have no members or downstream PIM routers. A dense-mode interface is subject to multicast flooding by default.

To configure PIM on an interface to be in dense mode, use the interface configuration command **ip pim dense-mode**.

Configuring PIM-SM

A sparse-mode interface is used for multicast forwarding only if a join message is received from a downstream router or if group members are directly connected to the interface. Sparse mode assumes that no other multicast group members are present.

When sparse-mode routers want to join the shared path, they periodically send join messages toward the RP. When sparse-mode routers want to join the source path, they periodically send join messages toward the source. They also send periodic prune messages toward the RP to prune the shared path.

To configure PIM on an interface to be in sparse mode, use the interface configuration command **ip pim sparse-mode**.

PIM-SM THRESHOLDS AND LEAF ROUTERS

Recall that the *default* behavior of PIM-SM is that routers with directly connected multicast group members join the shortest-path tree as soon as they detect a new multicast source. An optional PIM-SM configuration that might be appropriate for your network is to change this default behavior by configuring SPT thresholds (to control when to switch over to the SPT). SPT thresholds are specified in Kbps and can be used with access lists to specify to which groups the threshold applies. The default SPT threshold is 0 Kbps. This means that any and all sources are immediately switched to the SPT. If an SPT threshold of "infinity" is specified for a group, the sources are not switched to the SPT and remain on the shared tree.

To configure when a PIM leaf router should join the shortest-path source tree for the specified group, use the global configuration command **ip pim spt-threshold** {*kbps* | **infinity**} [**group-list** *access-list*].

If a source sends at a rate greater than or equal to the *kbps* value, a PIM join message is triggered toward the source to construct a source tree. If the **infinity** keyword is specified, all sources for the specified group use the shared tree. Specifying a group list access list indicates what groups the threshold applies to.

If the traffic rate from the source drops below the threshold *kbps* value, after a certain amount of time the leaf router switches back to the shared tree and sends a prune message toward the source.

To complete the PIM-SM configuration, you need to configure the router to learn group-to-RP mappings. This requires static RP, Auto-RP, or Anycast RP configuration (we forego the BSR option). Static RP, Auto-RP, and Anycast RP configurations are discussed after the next section.

Configuring Sparse-Dense Mode

If you configure either the **ip pim sparse-mode** or **ip pim dense-mode** interface configuration command, sparseness or denseness is applied to the interface as a whole. However, some environments might require PIM to run in a single region in sparse mode for some groups and in dense mode for other groups.

An alternative to enabling only dense mode or only sparse mode is to enable sparse-dense mode. In this case, the interface is treated as dense mode if the group is in dense mode; the interface is treated as sparse mode if the group is in sparse mode. The router must know about an RP if the interface is in sparse-dense mode and you want to treat the group as a sparse group.

If you configure sparse-dense mode, the idea of sparseness or denseness is applied to the group on the router, and the network manager should apply the same concept throughout the network.

Another benefit of sparse-dense mode is that Auto-RP information can be distributed in a dense-mode manner, yet multicast groups for user groups can be used in a sparse-mode manner. Thus, there is no need to configure a default RP on the routers (as in the case of static RP).

When an interface is treated in dense mode, it is populated in the outgoing interface list of the multicast routing table when any of the PIM neighbors on the interface have not pruned for the group.

When an interface is treated in sparse mode, it is populated in the multicast routing table's outgoing interface list when a PIM neighbor on the interface has received an explicit join message.

To enable PIM to operate in the same mode as the group, use the interface configuration command **ip pim sparse-dense-mode**. Cisco recommends entering this command on all multicast interfaces in a new PIM installation (with the exception of multipoint Frame Relay interfaces).

Configuring Static RP

Static RP requires a hard-coded RP address to be configured on every router in the network. All routers must have the same RP address. With this option, RP failover is not possible unless Anycast RPs are also in use. Refer to the section "Static RP" for a theoretical backdrop of the static RP configuration.

To configure the RP's address, use the global configuration command **ip pim rp-address** *rp-address* [*access-list*] [**override**].

The *access-list* option allows you to specify which multicast groups the RP handles. The default is 224.0.0.0/4, which includes 224.0.1.39 and 224.0.1.40. It is used by Auto-RP. The implication is that this router attempts to operate these groups in sparse mode. This is normally undesirable. It can often lead to problems in which some routers in the network are trying to run these groups in dense mode (the usual method) while others are trying to use sparse mode. This results in some routers in the network being deprived of Auto-RP information. This in turn can result in members of some groups not receiving multicast traffic.

Static definitions for the group and RP address of the **ip pim rp-address** command can be used together with dynamically learned group and RP address mapping through Auto-RP or BSR. Group and RP address mappings learned through Auto-RP and BSR take precedence over mappings statically defined by the **ip pim rp-address** command without the **override** keyword. Commands with the **override** keyword take precedence over dynamically learned mappings.

Configuring Auto-RP

In a pure PIM-SM environment, a *default RP* is an RP to which other PIM routers point via a static configuration (using the **ip pim rp-address** command). If you are setting up Auto-RP in a new internetwork, *you do not need a default RP, because you configure all the interfaces for sparse-dense mode.* Cisco recommends using sparse-dense mode in new PIM deployments.

If all interfaces are configured for sparse-dense mode, to successfully implement Auto-RP and prevent any groups other than 224.0.1.39 and 224.0.1.40 from operating in dense mode, Cisco recommends configuring a sink RP (also known as an "RP of last resort"). A ***sink RP*** is a statically configured RP that *might or might not actually exist in the network.* Configuring a sink RP does not interfere with Auto-RP operation because, by default, Auto-RP messages supersede static RP configurations. Cisco recommends configuring a sink RP for all possible multicast groups in your network, because it is possible for an unknown or unexpected source to become active, possibly due to a user with malicious intent or due to a carelessly configured multicast server. If no RP is configured to limit source registration, the group might revert to dense-mode operation and be flooded with data.

If you are adding Auto-RP to an *existing sparse-mode cloud,* a default RP is required. In this case, the assumption is that a static RP configuration is in place. To minimize disruption of the existing multicast infrastructure, use the default RP for global groups (224.*x*.*x*.*x*), such as 224.0.1.1 (used by NTP) and the Auto-RP groups (224.0.1.39 and 224.0.1.40). At the least, a standard configuration is to use a default RP strictly for the two Auto-RP groups. This kick-starts the Auto-RP process. Again, it is likely that you already have the default RP configured, because we are assuming that an existing sparse-mode configuration is in place. Remember that, by default, RPs discovered dynamically through Auto-RP take precedence over statically configured RPs. You should work under the assumption that it is desirable to use a second RP for local multicast groups to be handled by Auto-RP.

Find another router to serve as the RP for the local groups. The RP mapping agent can double as an RP. Assign the whole range of 239.*x*.*x*.*x* to that RP, or assign a subrange of it (for example, 239.2.*x*.*x*). These addresses are described in RFC 2365 as being constrained to a local group or organization.

To designate a router as a candidate RP, use the global configuration command **ip pim send-rp-announce** *type number* **scope** *ttl-value* [**group-list** *access-list*] [**interval** *seconds*].

This command causes the router to announce its candidacy for being an RP via the 224.0.1.39 Cisco-Announce group. The **interval** keyword, introduced in Cisco IOS

Software Release 12.1(2)T, specifies the interval between RP announcements in seconds (the default is 60).

Example 9-36 shows how to use the **ip pim send-rp-announce** command. This command results in the sending of RP announcements out all PIM-enabled interfaces for a maximum of 31 hops. The IP address by which the RP is to be identified is the IP address associated with Ethernet interface 0. Access list 5 describes the groups for which this router serves as RP (the administratively scoped groups).

Example 9-36 Configuring Candidate RPs for Auto-RP

```
Router(config)#ip pim send-rp-announce ethernet0 scope 31 group-list 5
Router(config)#access-list 5 permit 239.0.0.0 0.255.255.255
```

Next, you need to configure one or more RP mapping agents for the PIM-SM domain. The RP mapping agent is the router that sends the authoritative discovery packets telling other routers which group-to-RP mapping to use. Such a role is necessary in the event of conflicts (such as overlapping group-to-RP ranges). Find a router whose connectivity is unlikely to be interrupted, and assign it the role of RP mapping agent. All routers within TTL hops from the source router receive Cisco-Discovery group messages via 224.0.1.40. To assign the role of RP mapping agent for a router, use the global configuration mode command **ip pim send-rp-discovery** [*type number*] **scope** *ttl-value*.

Configure this command on the router designated as an RP mapping agent. Specify a TTL large enough to cover your PIM domain.

When Auto-RP is used, the following steps occur:

Step 1 The RP mapping agent listens on the well-known group address CISCO-RP-ANNOUNCE (224.0.1.39), which candidate RPs send to.

Step 2 The RP mapping agent sends RP-to-group mappings in an Auto-RP RP discovery message to the well-known group CISCO-RP-DISCOVERY (224.0.1.40). The TTL value limits how many hops the message can take.

Step 3 PIM designated routers listen to this group and use the RPs they learn about from the discovery message.

This completes the two major steps for Auto-RP: configuring candidate RPs (**ip pim send-rp-announce**) and configuring mapping agents (**ip pim send-rp-discovery**). These commands are used whether the multicast router interfaces are configured with **ip pim sparse-mode** or **ip pim sparse-dense-mode**). If you were to enter the commands on the routers in your network as described so far in this section, Auto-RP would be functional on your network at this point. The Auto-RP groups might have been communicated via a default RP (if the interfaces are configured with **ip pim sparse-mode** in an existing sparse-mode cloud) or without a default RP (if the interfaces are configured with **ip pim sparse-dense-mode** in a new internetwork).

The multicast network can now be tested with a multicast application, such as IP/TV.

The only remaining consideration is dealing with filters associated with group-to-RP mappings. This is the subject of the next section.

PIM-SM Filtering If you are adding Auto-RP to an existing sparse-mode cloud, you might have some filters already in place. In this case, you need to check for the presence of the **ip pim accept-rp** global configuration command configured throughout the network. This command instructs routers to accept join or prune messages destined for a specified RP.

If the **ip pim accept-rp** command is not configured on any router, there is no further configuration involving this command. In routers that are already configured with the **ip pim accept-rp** command, you must specify the command again to accept the newly advertised RP (via Auto-RP). To accept all RPs advertised with Auto-RP and reject all other RPs by default, use the **ip pim accept-rp auto-rp** command on the PIM routers.

If all interfaces are in sparse mode, a default RP is needed to support the two well-known groups 224.0.1.39 and 224.0.1.40. Auto-RP relies on these two well-known groups to collect and distribute RP mapping information. When this is the case and the **ip pim accept-rp auto-rp** command is configured, another **ip pim accept-rp** command accepting the default RP must be configured, as shown in Example 9-37.

Example 9-37 Default RP Must Be Able to Handle the Auto-RP Groups

```
Router(config)#ip pim accept-rp 128.171.1.1 2
Router(config)#access-list 2 permit 224.0.1.39
Router(config)#access-list 2 permit 224.0.1.40
```

Finally, to filter incoming Auto-RP announcement messages coming from the RP, use the global configuration command **ip pim rp-announce-filter rp-list** *access-list* **group-list** *access-list*. This command is used on RP mapping agents to prescribe the candidate RPs for particular multicast groups. Cisco recommends that, if you use more than one RP mapping agent, you make the filters among them consistent so that there are no conflicts in mapping state when the announcing agent goes down. Example 9-38 illustrates how to use this command.

Example 9-38 Filtering Candidate RPs to Be Used for Particular Multicast Groups

```
Router(config)#ip pim rp-announce-filter rp-list 1 group-list 2
Router(config)#access-list 1 permit 10.0.0.1
Router(config)#access-list 1 permit 10.0.0.2
Router(config)#access-list 2 permit 224.0.0.0 15.255.255.255
```

These commands configure the router to accept RP announcements from RPs in access list 1 for group ranges described in access list 2.

Configuring Anycast RP

Anycast RPs provide for RP redundancy without your having to use Auto-RP or BSR. RP failover occurs at roughly the same speed as the unicast routing protocol converges. Refer back to the section "Anycast RP" for a review of the Anycast RP mechanism.

Anycast RP requirements are as follows:

- All Anycast RPs are configured to use the same IP address.
- All Anycast RPs advertise this IP address as a host route (/32). This causes the DRs in the network to see only the closest RP. If there is a metric tie with the unicast routing protocol, the normal RPF mechanism selects only one path back to the RP. The path selected is the one that has the highest next-hop address.
- All Anycast RPs are tied together via MSDP peering sessions.
- The **ip msdp originator-id** command controls the IP address that is sent in any SA messages that are originated by an RP. This is done to clarify which RP originated the SA message. If this were not done, all RPs would originate SA messages using the same IP address.

Figure 9-38 illustrates the Anycast RP method. In this figure, the top routers are Anycast RPs. Because two routers are configured with the same RP address, there is no single point of failure. At any given time, only one RP address is active (either per group or, as in this case, for all groups). Anycast RP provides excellent redundancy and load balancing. The load balancing is achieved through MSDP, which allows RPs to exchange information about active sources. For this type of scenario, when an RP changes, convergence is normally on the order of seconds.

Because the top routers are Anycast RPs, they are configured with the same IP address, 10.1.1.1, on interface Loopback 1. The network 10.1.1.1/32 is advertised by the unicast routing protocol on each router.

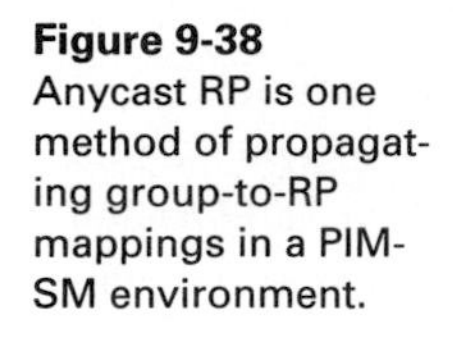

Figure 9-38
Anycast RP is one method of propagating group-to-RP mappings in a PIM-SM environment.

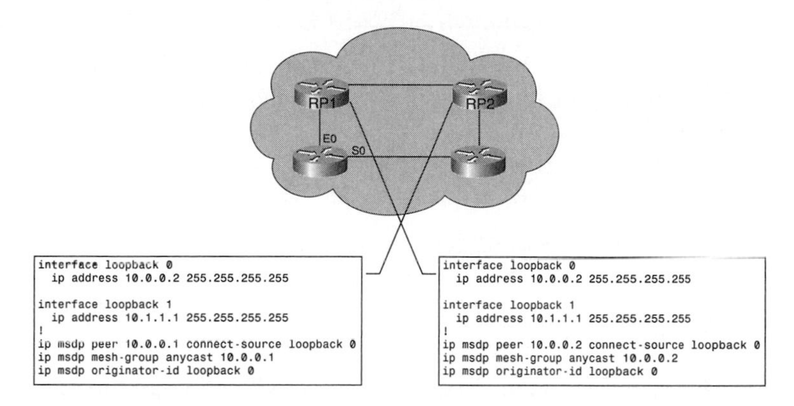

The MSDP peering is between the IP addresses 10.0.0.2 and 10.0.0.1, configured on Loopback 0. The MSDP peering addresses must be unique, unlike the address used for the RPs. Similar to BGP syntax for establishing a neighbor relationship, you also configure the MSDP command **ip msdp peer** *peer-address* **connect-source loopback 0** on each MSDP peer. As with BGP, Loopback 0's IP address becomes the source IP address for the TCP-based MSDP connection.

An MSDP mesh group is a group of MSDP-configured routers that have fully meshed MSDP connectivity among themselves. SA messages received from a peer in a mesh group are not forwarded to other peers in the same mesh group. Mesh groups are used to reduce SA message flooding. For this particular example, mesh groups are not necessary, because there are only two Anycast RPs (we include mesh groups for illustration purposes). However, if more than two Anycast RPs exist, you need to configure MSDP mesh groups to prevent looping of SA messages. To configure the MSDP mesh group, use the command **ip msdp mesh-group** *mesh-name peer-address*. We use the name **anycast** in our example to emphasize that we are using MSDP for the purposes of Anycast RP. *peer-address* is the IP address of the MSDP peer. This must be the same peer address that's used with the **ip msdp peer** command.

Earlier, we noted that the **ip msdp originator-id** command controls the IP address that is sent in any SA messages originated by the RP. This command is used on each router. **originator-id** references Loopback 0 on each router (not the loopback used for identifying the RP).

Finally, to make this work, you must ensure that **ip pim sparse-mode** is configured on each of the router interfaces and that a static RP assignment is made on all four multicast routers pointing to the Anycast RP's IP address: **ip pim rp-address 10.1.1.1.**

AVOIDING ANYCAST RP/ROUTER ID CONFLICTS

The most difficult part of Anycast RP configuration is the use of loopback IP addresses. It is common to use the Anycast RP solution in an OSPF and/or BGP environment. Both OSPF and BGP use their own router IDs in their respective operation. You need to be sure to avoid Anycast RP/router ID conflicts, because a conflict will break your OSPF or BGP operation. To ensure that a loopback address used for Anycast RP is not accidentally used as a router ID, you have three options:

- Configure the Anycast RP as the lowest IP address.
- Use a secondary IP address on the loopback for the Anycast IP address.
- Use the respective router ID commands in OSPF and BGP to statically configure the router ID.

This completes the configuration of Anycast RP. Although this section presented a static Anycast RP configuration, you can also configure Anycast RP with Auto-RP. (In fact, many people consider this option the ideal way to configure PIM for a routing domain.)

The remaining consideration for PIM configuration is how to configure PIM on Frame Relay hub-and-spoke topologies. Although Frame Relay is off the beaten path for a multilayer switching book, we describe PIM over Frame Relay here for the sake of completion: This topic closes the circle on the range of PIM configuration options.

PIM and Multipoint Frame Relay Interfaces

You configure PIM with point-to-point subinterfaces no differently than we have described for PIM up to this point. However, if you are using a physical interface as a Frame Relay interface, it acts as a multipoint interface. You to make special considerations when configuring PIM for physical Frame Relay interfaces or multipoint Frame Relay subinterfaces.

To provide efficient IP multicast support in Frame Relay networks, the underlying Frame Relay network architecture should be designed in a hub-and-spoke topology. In the hub-and-spoke topology, each remote router can also act as a hub, and each connection to another remote site can act as a spoke (in a hierarchical fashion). In a multiple-hub topology, the load associated with sending broadcast and multicast data can be distributed across multiple central hub sites rather than concentrated at a single central site. Thus, even though data might require extra hops to get to a particular location, data delivery is more efficient in a hub-and-spoke network than in other network topologies. This design also provides a scalable, hierarchical network that greatly reduces the central router's resource requirements, allowing the Frame Relay network to utilize the advantages of IP multicast applications.

Multicast Operation in a Frame Relay Network Typically, a Frame Relay network consists of a physical Layer 2 point-to-point network that is partially meshed, appearing to the router's Layer 3 functions as a logical LAN. In this situation, a remote-site router expects all routers in the network to receive any broadcast or multicast packet sent. Figure 9-39 shows the network topology of the logical Layer 3 perspective of a router in a Frame Relay network.

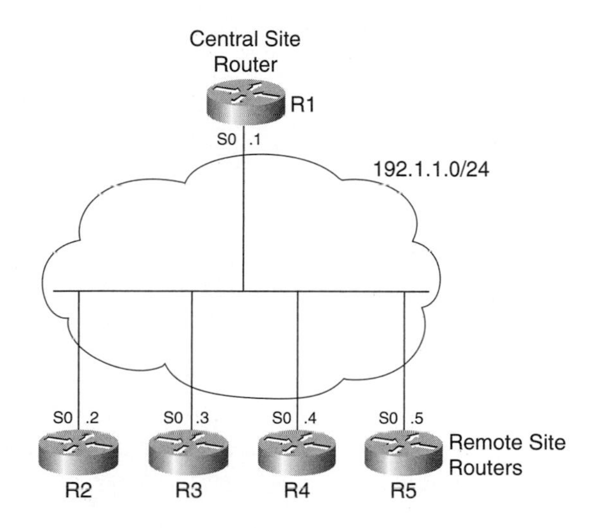

Figure 9-39 The Layer 2 Frame Relay topology appears to the router's Layer 3 functions as a logical LAN.

In reality, however, broadcast or multicast traffic sent by a remote-site router is received only by the central-site router. Other remote-site routers do not receive the broadcast or multicast traffic, because each remote-site router is connected point-to-point to the central-site router, as shown in Figure 9-40.

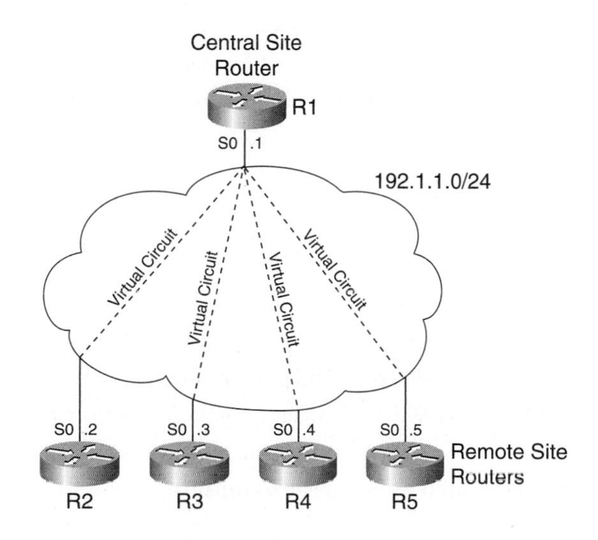

Figure 9-40 A multipoint Frame Relay (sub)interface does not propagate broadcast or multicast traffic sent by a spoke router.

If a remote-site router sends a PIM prune message, only the central-site router receives the prune message, as shown in Figure 9-41. Consequently, other remote-site routers cannot override this prune message. This can prevent members of multicast groups from receiving multicast traffic that they want.

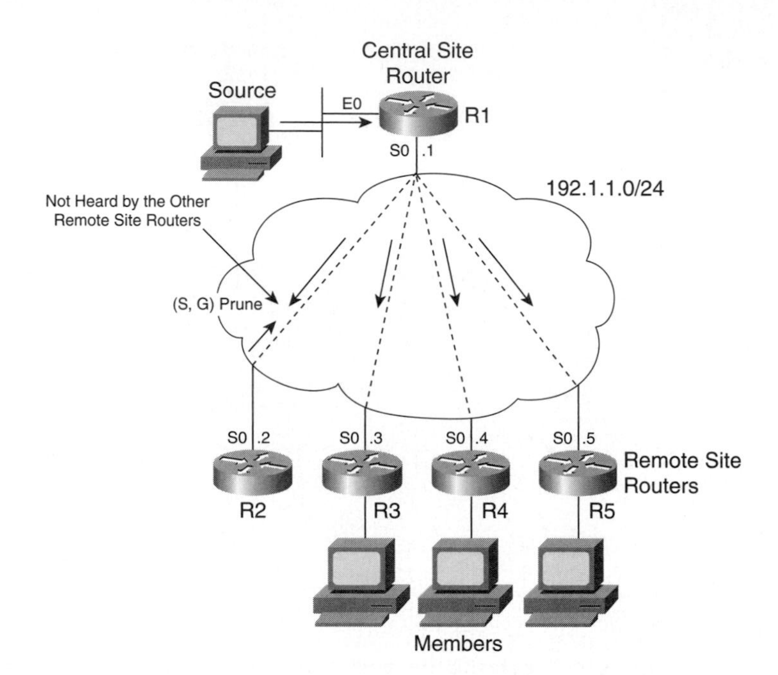

Figure 9-41
The spoke routers cannot override a prune message in this topology.

When nonbroadcast multiaccess (NBMA) networks, such as Frame Relay networks, need to send broadcast or multicast data packets on a main interface, routers perform a *pseudobroadcast*. This process uses the broadcast queue, which operates independently of the normal interface queue, as illustrated in Figure 9-42. The broadcast queue is commonly configured to prevent routing updates from being dropped from the queue.

A pseudobroadcast is the router's way of mimicking a broadcast multiaccess network. To understand a pseudobroadcast, imagine seven WAN locations that are all running OSPF. When one router sends an OSPF hello packet to an IP multicast group address, the router's data link layer replicates the hello packet, sending one copy to each WAN neighbor. In this example, six copies of the hello packet are created and sent over the link to the multiaccess WAN. Consequently, the router that sent the initial OSPF hello packet was required to use six times the resources that a router connected to true broadcast medium such as Ethernet would use.

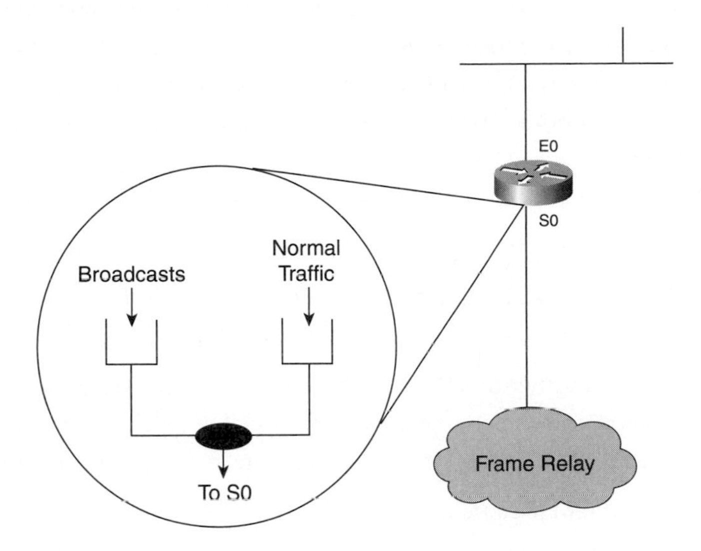

Figure 9-42
The serial interface uses a broadcast queue, independent of the normal interface queue.

The limitations of pseudobroadcasts in an NBMA network such as Frame Relay are as follows:

- The pseudobroadcast solution is not scalable for IP multicast applications (which might consist of high traffic rates) because pseudobroadcasts are process-switched.
- Pseudobroadcast data packets are treated as broadcast traffic and are sent to all neighbors on the WAN regardless of their need to receive these packets. This situation can quickly lead to packets, including control data, being dropped from the broadcast queue due to oversubscription.

Cisco created a solution to these problems—PIM NMBA mode.

PIM NBMA Mode The NBMA mode software feature is one solution to configuring IP multicast within a Frame Relay network. (This feature can also be used in ATM networks.) This PIM feature allows you to configure a router to send packets to only neighbors that want to receive them. A router in PIM NBMA mode treats each remote PIM neighbor as if it were connected to the router through a point-to-point link.

In a Frame Relay network that uses IP multicast, NBMA mode improves router performance for the following reasons:

- Traffic is fast-switched rather than process-switched.
- Routers receive traffic only for the multicast groups to which they are joined.

To enable NBMA mode on an interface, use the **ip pim nbma-mode** command in interface configuration mode. This command allows the router to track each neighbor's IP address when a PIM join message is received from that neighbor. The router can also track the neighbor's interface in the outgoing interface list for the multicast groups that the neighbor joins. This information allows the router to forward data destined for a particular multicast group to only neighbors that have joined that particular group.

When using the **ip pim nbma-mode** command, note the following usage guidelines:

- This command applies to only PIM sparse-mode configurations, because its functionality is dependent on the PIM sparse-mode join message.
- As the number of PIM neighbors increases, the outgoing interface list increases. Each interface entry requires additional resources from the NBMA mode-enabled router, thereby increasing data replication time and memory utilization.

Regarding the first guideline, the Cisco IOS Software lets you (with a warning) configure dense mode or sparse-dense mode on a multipoint Frame Relay interface configured with **ip pim nbma-mode**, but you should always use this command in conjunction with the **ip pim sparse-mode** command.

Auto-RP and NBMA Mode The NBMA mode feature does not support PIM dense mode. Auto-RP relies on dense-mode flooding of data for the Auto-RP groups, 224.0.1.39 and 224.0.1.40. Without dense-mode flooding capability, multicast routers in a Frame Relay network using Auto-RP might have problems receiving RP mapping information unless the mapping agent is placed in the appropriate location within the network or a more costly full-mesh architecture is created.

Use the following guidelines when placing the mapping agent (MA) in your network:

- All candidate RPs must be connected to the mapping agent.
- All mapping agents must be connected to all PIM routers.

As shown in Figure 9-43, if the mapping agent were placed below router R2, only the central-site router, R1, would receive the Auto-RP messages containing RP mapping information. In this situation, the central-site router would not resend the Auto-RP messages to routers R3, R4, and R5.

However, if the mapping agent is placed above the central-site router, R1, the central-site router, which has direct connections to the remote-site routers, forwards the RP mapping information to routers R2, R3, R4, and R5. This is illustrated in Figure 9-44.

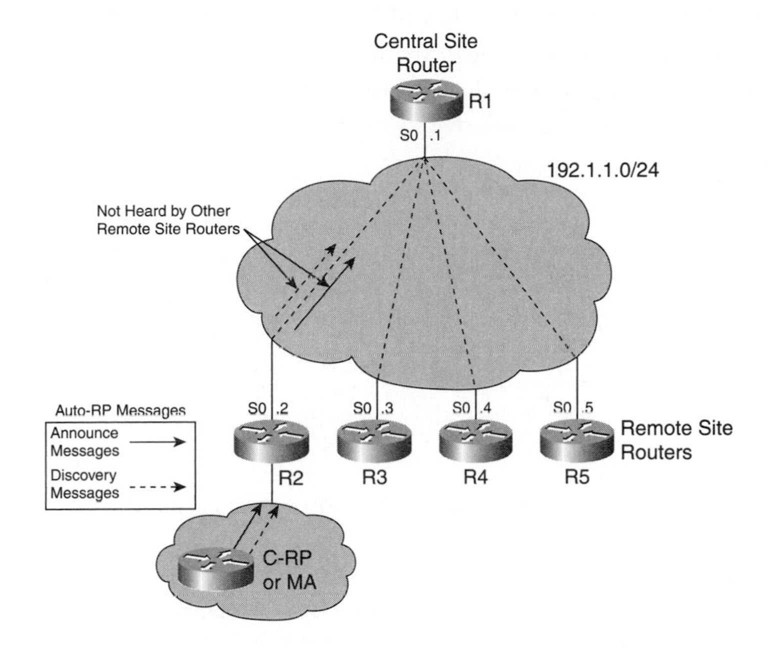

Figure 9-43
The PIM mapping agent should not be placed at a remote site in a Frame Relay network.

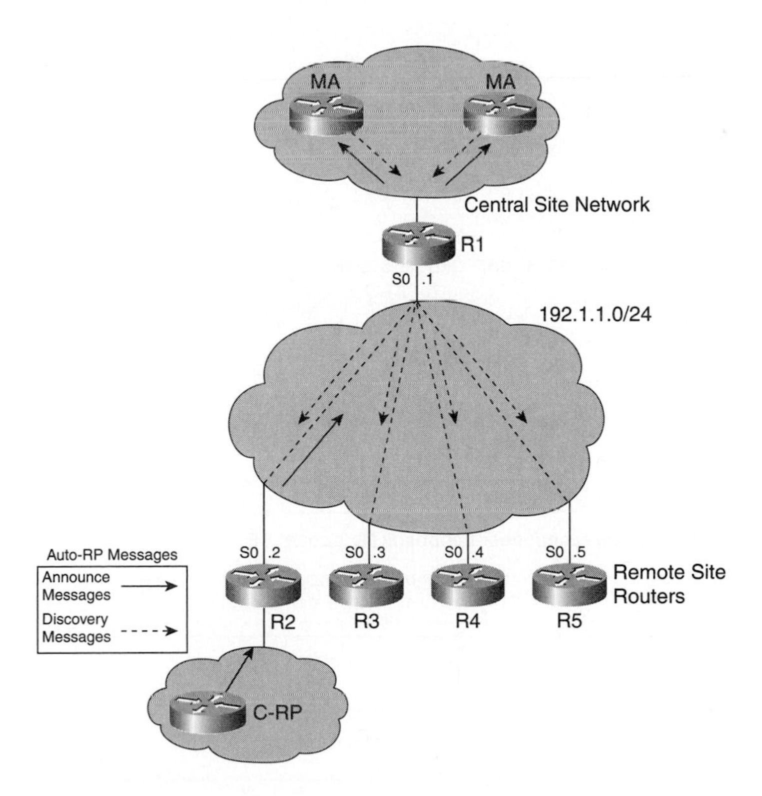

Figure 9-44
The PIM mapping agents should be placed at or "above" the hub site.

PIM Example with a Multipoint Frame Relay Interface To configure the network pictured in Figure 9-45, the PIM commands are the same as before, except for one additional command: **ip pim nbma-mode** on the multipoint Frame Relay (sub)interface.

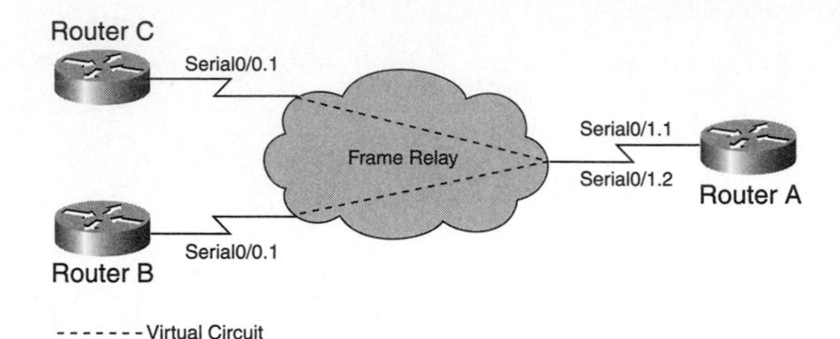

Figure 9-45 Configuring PIM on a multipoint Frame Relay (sub)interface requires only one additional command on the hub router's interface.

Example 9-39 shows the relevant portion of the configuration for the hub router (RouterA).

Example 9-39 Configuring PIM NBMA Mode for a Multipoint Frame Relay (Sub)interface

```
ip multicast-routing
!
interface Loopback10
ip address 10.1.1.1 255.255.255.255
ip pim sparse-dense-mode
!
interface Serial0/1
ip address 10.1.48.1 255.255.255.248
ip pim nbma-mode
ip pim sparse-mode
encapsulation frame-relay
frame-relay interface-dlci 100
frame-relay interface-dlci 200
!
ip pim send-rp-announce Loopback10 scope 16
ip pim send-rp-discovery Loopback10 scope 16
!
```

RouterA is configured to be both a candidate RP and a mapping agent in this case. This is common with Auto-RP configuration.

We complete this chapter with a discussion of IP PIM verification commands.

Verifying PIM Configuration

Several **show** commands are available for viewing details related to your PIM configuration. Some of the more commonly used ones are

- **show ip mroute**
- **show ip pim interface**
- **show ip pim neighbor**
- **show ip pim rp**

All the options discussed for these commands are available in Cisco IOS Software Release 12.1(5)T and later.

Each of these commands is used to verify or troubleshoot a PIM configuration. As you might have deduced, the hard part of PIM is understanding what is going on behind the scenes. In fact, we have almost completely avoided any careful scrutiny as to how multicast routing protocols such as PIM operate behind the scenes. To be able to troubleshoot PIM, you must understand these behind-the-scenes operations. Troubleshooting PIM and other multicast routing protocols is beyond the scope of this book.

PIM, like the other multicast routing protocols, has an intricate set of mechanisms for dealing with every particular topological configuration. For a carefully presented exposition detailing the algorithms used by IP multicast routing protocols, we refer you to Beau Williamson's consummate book *Developing IP Multicast Networks*, Volume I, published by Cisco Press.

Each of the IP PIM **show** commands are discussed in the next sections.

show ip mroute

Use the **show ip mroute** command to display the contents of the IP multicast routing table. This command displays all groups and sources.

The Cisco IOS Software populates the multicast routing table by creating (S,G) entries from (*,G) entries. In creating (S,G) entries, the software uses the best path to that destination group found in the unicast routing table via RPF.

Example 9-40 displays sample output from the **show ip mroute** command for a router operating in PIM sparse mode.

Example 9-40 Use show ip mroute to Display the Contents of the IP Multicast Routing Table

```
Router#show ip mroute
IP Multicast Routing Table
Flags: D - Dense, S - Sparse, C - Connected, L - Local, P - Pruned
       R - RP-bit set, F - Register flag, T - SPT-bit set
Timers: Uptime/Expires
Interface state: Interface, Next-Hop, State/Mode

(*, 224.0.255.3), uptime 5:29:15, RP is 198.92.37.2, flags: SC
  Incoming interface: Tunnel0, RPF neighbor 10.3.35.1, Dvmrp
  Outgoing interface list:
    Ethernet0, Forward/Sparse, 5:29:15/0:02:57

(198.92.46.0/24, 224.0.255.3), uptime 5:29:15, expires 0:02:59, flags: C
  Incoming interface: Tunnel0, RPF neighbor 10.3.35.1
  Outgoing interface list:
    Ethernet0, Forward/Sparse, 5:29:15/0:02:57
```

This command is the primary tool used to troubleshoot PIM. Just as it takes time to learn how to interpret the entries in a unicast routing table, it takes much effort to decipher the contents of the multicast routing table. The key is to understand what each flag means for each entry. The S and C flags appear in Example 9-40. The S flag indicates that the entry is operating in sparse mode. The C flag means that a member of the multicast group is present on the directly connected interface. Everything appears to be working OK with this output.

The RPF neighbor indicates where the multicast traffic will come from for each group.

The Dvmrp output in the first entry indicates that RPF information is learned from the DVMRP routing table. Recall that the Cisco IOS Software does not support DVMRP directly. It supports interoperation with routers that support DVMRP.

show ip pim interface

To display information about PIM interfaces, use the command **show ip pim interface** [*type number*][*rp-address*] [**detail**].

Example 9-41 displays sample output for this command.

Example 9-41 show ip pim interface Displays Multicast Information Specific to Each Interface

```
Router#show ip pim interface

Address          Interface          Mode     Neighbor  Query     DR
                                             Count     Interval
198.92.37.6      Ethernet0          Dense    2         30        198.92.37.33
198.92.36.129    Ethernet1          Dense    2         30        198.92.36.131
10.1.37.2        Tunnel0            Dense    1         30        0.0.0.0
```

The **detail** keyword provides considerably more information, as shown in Example 9-42.

Example 9-42 detail Option of show ip pim interface Provides Extensive Information for an Interface

```
Router#show ip pim interface fastethernet 0/1 detail
FastEthernet0/1 is up, line protocol is up
  Internet address is 172.16.8.1/24
  Multicast switching:process
  Multicast packets in/out:0/0
  Multicast boundary:not set
  Multicast TTL threshold:0
  PIM:enabled
    PIM version:2, mode:dense
    PIM DR:172.16.8.1 (this system)
    PIM neighbor count:0
    PIM Hello/Query interval:30 seconds
        PIM State-Refresh processing:enabled
        PIM State-Refresh origination:enabled, interval:60 seconds
    PIM NBMA mode:disabled
    PIM ATM multipoint signalling:disabled
    PIM domain border:disabled
  Multicast Tagswitching:disabled
```

Among other things, this output shows the PIM mode and which router is the PIM DR for the connected segment.

show ip pim neighbor

To list the PIM neighbors discovered by the router, use the command **show ip pim neighbor** [*type number*].

Example 9-43 shows sample output for this command.

Example 9-43 You Can Easily Find All Your PIM Neighbors

```
Router#show ip pim neighbor



PIM Neighbor Table
Neighbor Address  Interface          Uptime    Expires   Mode
171.68.0.70       FastEthernet0      2w1d      00:01:24  Sparse
171.68.0.91       FastEthernet0      2w6d      00:01:01  Sparse (DR)
171.68.0.82       FastEthernet0      7w0d      00:01:14  Sparse
171.68.0.86       FastEthernet0      7w0d      00:01:13  Sparse
171.68.0.80       FastEthernet0      7w0d      00:01:02  Sparse
171.68.28.70      Serial2.31         22:47:11  00:01:16  Sparse
171.68.28.50      Serial2.33         22:47:22  00:01:08  Sparse
171.68.27.74      Serial2.36         22:47:07  00:01:21  Sparse
171.68.28.170     Serial0.70         1d04h     00:01:06  Sparse
171.68.27.2       Serial1.51         1w4d      00:01:25  Sparse
171.68.28.110     Serial3.56         1d04h     00:01:20  Sparse
171.68.28.58      Serial3.102        12:53:25  00:01:03  Sparse
```

This output shows the neighbors that are off of a total of five physical interfaces. You can see that only sparse mode is being used on this router. The uptime for each neighbor is also displayed.

show ip pim rp

To display active RPs that are cached with associated multicast routing entries, use the command **show ip pim rp** [**mapping** | **metric**] [*rp-address*].

The **mapping** keyword displays all group-to-RP mappings of which the router is aware. The **metric** keyword displays the unicast routing metric to the RPs, either configured statically or learned via Auto-RP or BSR.

This command's output is fairly simple if the keywords are omitted, as shown in Example 9-44.

Example 9-44 Output of the show ip pim rp Command Displays the RPs

```
Router#show ip pim rp

Group:227.7.7.7, RP:10.10.0.2, v2, v1, next RP-reachable in 00:00:48
```

Adding the **mapping** keyword gives you the group-to-RP mappings for each RP, as shown in Example 9-45.

Example 9-45 Output of the show ip pim rp mapping Command Displays All the Group-to-RP Mappings

```
Router#show ip pim rp mapping

PIM Group-to-RP Mappings
This system is an RP (Auto-RP)
This system is an RP-mapping agent

Group(s) 227.0.0.0/8
  RP 10.10.0.2 (?), v2v1, bidir
    Info source:10.10.0.2 (?), via Auto-RP
         Uptime:00:01:42, expires:00:00:32
Group(s) 228.0.0.0/8
  RP 10.10.0.3 (?), v2v1, bidir
    Info source:10.10.0.3 (?), via Auto-RP
         Uptime:00:01:26, expires:00:00:34
Group(s) 229.0.0.0/8
  RP 10.10.0.5 (mcast1.cisco.com), v2v1, bidir
    Info source:10.10.0.5 (mcast1.cisco.com), via Auto-RP
         Uptime:00:00:52, expires:00:00:37
Group(s) (-)230.0.0.0/8
  RP 10.10.0.5 (mcast1.cisco.com), v2v1, bidir
    Info source:10.10.0.5 (mcast1.cisco.com), via Auto-RP
         Uptime:00:00:52, expires:00:00:37
```

Useful output here includes the fact that this router is an RP and an RP mapping agent. All the group-to-RP mappings are listed as well.

Summary

Multicasting is a broad subject area that encompasses a wide range of technologies. This subject area is also one of the fastest-changing areas in networking. Multicasting allows one-to-many or many-to-many communication on campus networks or over the Internet. The catch is that all the routers between the source of the multicast traffic and the receivers of the traffic must be configured to support IP multicasting. The future Internet will be fully multicast-enabled. At that point, many exciting things will happen, including distance education opportunities that are acceptable to students and instructors.

The set of multicast technologies can be roughly subdivided into host-to-router, switch-to-router, and router-to-router technologies. We discussed IGMP, which is the host-to-router technology relied on for multicasting. We explored CGMP, IGMP snooping, and RGMP, which are switch-to-router technologies designed to constrain multicast traffic. We also surveyed the numerous router-to-router multicast technologies, which were categorized according to their functionality within or between routing domains. These technologies include PIM, BSR, CBT, DVMRP, MOSPF, SSM, MSDP, and MBGP.

Finally, you learned how to configure PIM dense mode, sparse mode, and sparse-dense mode in the context of static RP, Auto-RP, and Anycast RP. Configuring PIM over Frame Relay multipoint interfaces requires the addition of the PIM NBMA feature.

Chapter 10 briefly explores AVVID, security, and performance in campus networks.

Check Your Understanding

Test your understanding of the concepts covered in this chapter by answering these review questions. Answers are listed in Appendix A, "Check Your Understanding Answer Key."

1. What is the command to configure an interface on a Cisco router to participate in a particular multicast group?
   A. **ip pim sparse-mode**
   B. **ip igmp**
   C. **ip igmp join-group**
   D. **ip pim dense-mode**
2. What is the default interval IGMP uses to send host query messages?
   A. 10 seconds
   B. 20 seconds
   C. 30 seconds
   D. 60 seconds
3. Which multicast routing protocol uses source active messages?
   A. PIM-SM
   B. SSM
   C. MSDP
   D. MBGP
4. Which protocol uses the MAC multicast address 01-00-5E-00-00-16?
   A. OSPF
   B. IGMP
   C. PIM
   D. Auto-RP
5. How many multicast IP addresses map to a single multicast MAC address?
   A. 1
   B. 4
   C. 16
   D. 32

6. What is the range for administratively scoped multicast IP addresses?

   A. 224.0.0.0/4

   B. 232.0.0.0/8

   C. 239.0.0.0/8

   D. 252.0.0.0/8

7. Which multicast routing protocols use a shared distribution tree?

   A. MOSPF

   B. PIM-DM

   C. PIM-SM

   D. CBT

8. Which multicast routing protocol eliminates all (S,G) state in the network?

   A. PIM-DM

   B. PIM-SM

   C. Bidir-PIM

   D. DVMRP

9. Which version of IGMP adds support for source filtering and is relied on for a full implementation of SSM?

   A. 1

   B. 2

   C. 3

   D. 4

10. How often are PIM hello messages sent, and to what multicast address?

    A. 10; 224.0.0.1

    B. 30; 224.0.0.13

    C. 15; 224.0.0.2

    D. 60; 224.0.0.2

Key Terms

Anycast RP An application of MSDP that serves as a method to design a PIM-SM network to meet fault-tolerance requirements within a routing domain. In Anycast RP, two or more RPs are configured with the same IP address on loopback interfaces.

Auto-RP A PIM feature that automates the distribution of group-to-RP mappings.

Bidir-PIM (Bidirectional PIM) A variant of the PIM suite of multicast routing protocols for IP multicast and an extension of PIM-SM. Bidir-PIM resolves some limitations of PIM-SM for groups that have a large number of sources. Bidir-PIM eliminates all (S,G) state in the network.

BSR (bootstrap router) A fault-tolerant, automated RP discovery and distribution mechanism. With BSR, routers dynamically learn the group-to-RP mappings. BSR was introduced with PIMv2, specified in RFC 2117.

candidate RP A router that announces its candidacy to be an RP for a PIM-SM network via the Cisco-Announce group, 224.0.1.39.

CBT (Core-Based Trees) A multicast routing protocol introduced in September 1997 in RFC 2201, CBT builds a shared tree like PIM-SM. CBT uses information in the IP unicast routing table to calculate the next-hop router in the direction of the core.

CGMP (Cisco Group Management Protocol) A Cisco-developed protocol that runs between Cisco routers and Catalyst switches to leverage IGMP information on Cisco routers to make Layer 2 forwarding decisions on Catalyst switches. With CGMP, IP multicast traffic is delivered only to Catalyst switch ports that are attached to interested receivers.

CGMP fast-leave Also known as CGMP fast-leave processing, CGMP immediate-leave processing, CGMP immediate leave, and CGMP leave. A CGMP feature requiring IGMPv2 on the hosts that allows the switch to quickly remove a port from its list of ports receiving multicast traffic for a particular multicast group.

core In CBT, the router that serves as the root of the shared tree. Similar to PIM-SM's RP.

default RP RP in a pure PIM-SM domain to which other PIM routers point via the **ip pim rp-address** command. It is normally used as a means of providing the group-to-RP mapping for the Auto-RP groups, but it can also be used as a fallback mechanism for PIM.

DF (Designated Forwarder) A PIM construct that establishes a loop-free SPT rooted at the RP. On each link, the router with the best path to the RP is elected as the DF. The DF is responsible for forwarding traffic upstream toward the RP.

distribution tree A unique forwarding path for a multicast group between the source and each subnet containing members of the multicast group.

DVMRP (Distance Vector Multicast Routing Protocol) The first multicast routing protocol, publicized in RFC 1075 in 1998. Unlike PIM, DVMRP builds a multicast routing table separate from the unicast routing table.

GARP (Generic Attribute Registration Protocol) Part of the IEEE 802.1p specification, GARP implements a service that provides a generic Layer 2 transport mechanism to propagate information throughout a switching domain. The information is then propagated in the form of attributes, types, values, and semantics, which are defined by the application employing the service.

GDA (Group Destination Address) The notation used with CGMP to denote a Layer 2 multicast group address.

GLOP addressing RFC 2770 proposes that the 233.0.0.0/8 address range be reserved for statically defined addresses by organizations that already have an AS number reserved. The second and third octets are populated with the decimal equivalents of the first and second bytes of the hexadecimal representation of the AS number.

globally scoped addresses A multicast address space in the range from 224.0.1.0 through 238.255.255.255. These addresses can be used to multicast data between organizations and across the Internet.

GMRP (GARP Multicast Registration Protocol) An application that provides a constrained multicast flooding facility similar to IGMP snooping.

GVRP (GARP VLAN Registration Protocol) A GARP application that provides IEEE 802.1Q-compliant VLAN pruning and dynamic VLAN creation on 802.1Q trunk ports (similar to VTP).

IGMP snooping fast-leave An IGMP feature that allows the switch processor to remove an interface from the port list of a forwarding-table entry without first sending out a MAC-based general query on the port. When an IGMP leave is received on a port, the port is immediately removed from the multicast forwarding entry.

IGMP (Internet Group Management Protocol) The protocol used to allow hosts to communicate to local multicast routers their desire to receive multicast traffic.

IGMP leave-group message A message used by IGMPv2 to allow a host to announce its intention to leave a multicast group. The leave message is sent to the all-routers multicast address, 224.0.0.2.

IGMP membership query A message used by IGMP and sent to the all-hosts multicast address, 224.0.0.1, to verify that at least one host on the subnet is still interested in receiving traffic directed to that group. Also called a general query.

IGMP membership report A message used by IGMP to indicate a host's interest in receiving traffic for a particular multicast group.

IGMP snooping A Layer 2 multicast-constraining mechanism (like CGMP) used by switches. It allows multicast traffic to be forwarded to interfaces associated with IP multicast devices. The switch snoops on the IGMP traffic between the host and the router and keeps track of multicast groups and member ports.

IGMP Version 1 The version of IGMP that relies solely on membership query messages and membership report messages.

IGMP Version 2 The version of IGMP that adds the ability of a host to proactively leave a multicast group and the ability of a router to send a group-specific query (as opposed to a general query).

IGMP Version 3 The version of IGMP that adds support for source filtering.

limited-scope address Also called an administratively scoped address. These addresses fall in the range 239.0.0.0 through 239.255.255.255 and are limited for use by a local group or organization.

link local address The IANA-reserved multicast address space in the range from 224.0.0.0 through 224.0.0.255 to be used by network protocols on a local network segment. These packets remain local to a particular LAN segment and are always sent with a TTL of 1.

MBGP (Multiprotocol BGP) An extension to BGP that enables multicast routing policy throughout the Internet. MBGP allows the interconnection of multicast topologies within and between BGP autonomous systems. With MBGP you can configure BGP peers that exchange both unicast and multicast network layer reachability information.

MBONE A multicast backbone across the public Internet, built with tunnels between DVMRP-capable Sun workstations running the *mroute* daemon (process).

MOSPF (Multicast OSPF) The multicast routing protocol specified in RFC 1584. MOSPF is a set of extensions to OSPF that enable multicast forwarding decisions.

MSDP (Multicast Source Discovery Protocol) A TCP-based mechanism to connect multiple PIM-SM domains. MSDP allows multicast sources for a group to be learned via source active messages by RPs in different domains.

multicast group An arbitrary group of hosts that expresses an interest in receiving a particular data stream via multicast. This group has no physical or geographical boundaries. Hosts that are interested in receiving data flowing to a particular group must join the group using IGMP.

MVR (Multicast VLAN Registration) Provides the ability to have a single multicast stream traverse the network via a dedicated multicast VLAN, regardless of the number of receivers (all residing on VLANs separate from the multicast VLAN). Designed for applications using wide-scale deployment of multicast traffic, such as television channels, across a service provider network.

NTP (Network Time Protocol) Used to synchronize clocks on network devices. The multicast address 224.0.1.1 is reserved for NTP operation.

PIM (Protocol Independent Multicast) The IP multicast routing protocol that derives its information using RPF checks based on the unicast routing table. PIM has several variations, including dense mode, sparse mode, sparse-dense mode, source-specific mode, and bidirectional mode.

PIM-DM (Protocol Independent Multicast Dense Mode) A PIM variation that builds a source tree for each multicast source and uses flood-and-prune behavior.

PIM-SM (Protocol Independent Multicast Sparse Mode) A PIM variation that builds shared trees. The root of the tree is called the rendezvous point, similar to CBT's core.

PIM sparse-dense mode An alternative to pure dense mode or pure sparse mode for a router interface. Sparse-dense mode allows individual groups to be run in either sparse or dense mode, depending on whether RP information is available for that group. If the router learns RP information for a particular group, it is treated as sparse mode; otherwise, it is treated as dense mode.

pseudobroadcast A process used by routers in a multipoint Frame Relay environment that uses a broadcast queue operating independently of the normal interface queue. Pseudobroadcast allows the router to simulate a broadcast environment in a nonbroadcast multiaccess network.

RGMP (Router-Port Group Managment Protocol) Allows a router to communicate to a switch the IP multicast group for which the router wants to receive or forward traffic. RGMP is designed for switched Ethernet backbone networks running PIM.

RP mapping agent A router that receives RP-announcement messages from the candidate RPs and arbitrates conflicts. The RP mapping agent sends group-to-RP mappings to all multicast-enabled routers via the Cisco-Discovery group, 224.0.1.40.

SA (Source Active) A message used between MSDP peers that identifies the sources and associated multicast groups.

shared tree A multicast distribution tree that uses a root placed at a chosen point in the network. PIM-SM and CBT both use shared trees. When using a shared tree, sources must send their traffic to the root for the traffic to reach the receivers.

sink RP Also known as the RP of last resort. A statically configured RP for all multicast groups in a PIM network with all interfaces configured for sparse-dense mode. It serves as a mechanism to successfully implement Auto-RP and prevent any groups other than 224.0.1.39 and 224.0.1.40 from operating in dense mode. Auto-RP-discovered group-to-RP mappings take precedence over those configured via a default RP.

source filtering A capability added to IGMPv3 that lets a multicast host signal to a router the groups for which it wants to receive multicast traffic and the sources from which it wants to receive the traffic.

source tree A multicast distribution tree with its root at the source of the multicast traffic and whose branches form a spanning tree through the network to the receivers. Also called the shortest-path tree (SPT). Every source of a multicast group has a corresponding SPT.

SSM (Source Specific Multicast) An extension of IP multicast in which datagram traffic is forwarded to receivers from only multicast sources to which the receivers have explicitly joined. For multicast groups configured for SSM, only source-specific multicast distribution trees (no shared trees) are created. SSM relies on IGMPv3.

After completing this chapter, you will be able to perform tasks related to the following:

- Elements of a campus network security policy
- Types of physical security controls
- Types of logical security controls
- Types of logical access controls
- Local and centralized authentication
- IP permit lists on CatOS switches
- Privilege levels on IOS-based switches
- Port security, SPAN, and syslog

Chapter 10

Security

The President and CEO of Cisco Systems, John Chambers, has often been heard saying, "The Internet is changing the way we work, live, play, and learn." These changes occur in how we experience e-commerce, real-time information access, e-learning, and expanded communication options. It will not be long before it is normal for enterprises to use the Internet to provide free telephone service. Also, in the not-too-distant future, it will be commonplace to take a class via the Internet, where both teacher and students can see and hear the other students with quality video and audio, while sharing applications and utilizing whiteboards. As a society, we are just beginning to unlock the Internet's potential. But with the Internet's unparalleled growth comes unprecedented exposure of personal data, critical enterprise resources, government secrets, and so forth. Every day, hackers pose an increasing threat to these entities with several different types of attacks. These attacks have become both more prolific and easier to implement. The two primary reasons for this problem are the ubiquity of the Internet and the pervasiveness of easy-to-use operating systems and development environments.

A campus network consists of various devices and applications that should be secured to create an environment in which standard hacker tools cannot be used to compromise the network. Routers, switches, hosts, applications, and the network itself are possible targets of attacks. This chapter focuses on securing Catalyst switches in a campus network, and thus provides one piece of the security puzzle. A great reference for general network security is *Managing Cisco Network Security* (Cisco Press). It describes how to secure Cisco network devices other than Catalyst switches.

This chapter begins with an overview of formulating a security policy for a campus network. This is followed by an analysis of the common security options and associated configurations for Catalyst switches.

Campus Network Security Policy

Up to this point, most of our discussions have been motivated primarily by facilitating connectivity and providing access. This is, in some sense, completely antithetical to the goal of securing a campus network. In fact, normally the sequence of events in network implementations is to first enable connectivity and then to implement a security policy. Security is typically manifested as a layer on top of the base network architecture. Security restricts access and limits connectivity. This is the dichotomy of networking that is now standard operating procedure: Enable high-speed access and then prescriptively restrict access with security and rate-limiting features.

How a security policy is designed and implemented is much less formulaic than the network design and implementation processes discussed to this point. You could argue that implementing security is more of an art form than a science. A security implementation can consist entirely of UNIX-based or Microsoft-based software tools throughout the network, or it can consist of a heavy dose of dedicated hardware security devices. The trend is definitely moving toward greater reliance on dedicated hardware security devices within campus networks. Host-based personal firewalls and virus-prevention software provide additional security.

The design and implementation of a security policy are site-specific. After you identify the critical assets and analyze the risks to the network, you design the security policy by defining the guidelines and procedures to be followed by campus personnel. The policy defines access rights and privileges for the network users, provides guidelines for connecting external networks and connecting devices to a network, and adding new software to systems. Most security policies outline network device management issues, user access to the network, traffic-flow policies, and route filtering. To be effective, the procedures detailed in the security policy should be concise and to the point. You want a document that most people will actually read. A document consisting of no more than 10 pages should suffice to begin with. Technical implementation details should not be included because they change over time. The policy's design takes careful planning to ensure that all security issues are adequately addressed.

To address all security-related issues in the security policy, consider each of the following areas:

- **Defining the physical security controls**—These controls pertain to the physical infrastructure, physical device security, and physical access to network devices.
- **Defining the logical security controls**—These controls restrict logical access to equipment and create boundaries between different cable segments. When traffic is logically filtered between networks, logical access controls provide security.

Logical security controls also include logging and reporting of unauthorized access or attempted access.

- **Ensuring system and data integrity**—You want to ensure that any traffic on the network is valid traffic, such as supported services, unspoofed traffic, and data that has not been altered (nonrepudiated traffic).
- **Ensuring data confidentiality**—This pertains to encryption. You have to decide which data to encrypt and which data to leave as clear text.
- **Developing policies and procedures for the staff that is responsible for the campus network**—The people responsible for maintaining and upgrading the network should have specific guidelines to aid them in carrying out their tasks in accordance with the campus security policies.
- **Developing appropriate security awareness training for users of the campus network**—Employees should be provided with adequate training to educate them about the many problems and ramifications of security-related issues.

A campus security policy is carefully constructed based on these considerations. This chapter explores only the first two considerations: physical and logical access controls (for Catalyst switches). The remaining controls are covered in a standard network security course.

Physical Security Controls

Securing physical devices and media is often the most overlooked aspect of network security. *Physical security controls* deal with the physical infrastructure, physical device security, and physical access.

Physical network infrastructure involves media selection and the path of the physical cabling. You want to minimize the possibility of an intruder's eavesdropping on the data traversing the network.

Optical fiber is more secure than twisted-pair cable because it does not radiate an appreciable amount of energy. Wiretaps on twisted pair or fiber can be detected using a time domain reflectometer (TDR) or optical time domain reflectometer (OTDR), respectively. These devices measure signal attenuation and the length of installed cables; however, they can also be used to detect illegal wiretaps.

You need to design the network cable infrastructure to minimize inappropriate access to the network. For example, underground cables between buildings should be at least 40 inches underground and placed within sealed conduits.

You need to secure physical devices to prevent intruders from taking the most blatant approach to compromising the network. Physical device security includes the following considerations:

- **Physical location**—Network infrastructure equipment should be physically located in areas with restricted access. Switches, firewalls, modems, intrusion detection systems, and routers should be secured in a lockable wiring closet or network operation center that does not provide access through the ceiling. You also need to physically secure Simple Network Management Protocol (SNMP), Domain Name System (DNS), Network Time Protocol (NTP), Network File System (NFS), Hypertext Transfer Protocol (HTTP), File Transfer Protocol (FTP), Terminal Access Controller Access Control System (TACACS+), Remote Authentication Dial-In User Service (RADIUS), Kerberos, system logging (syslog), and network auditing servers.
- **Physical access**—Physical access to wiring closets and network operation centers should be restricted. Access should be permitted only if a person is specifically authorized or requires access to perform his or her job.
- **Environmental safeguards**—Adequate safeguards should be implemented to protect critical network resources. This includes fire prevention, detection, suppression, and protection; water hazard prevention, detection, and correction; electric power supply protection; temperature control; humidity control; natural disaster protection; protection from magnetic fields; and procedures to protect against dust and dirt. A policy limiting the use of food, drink, and smoking materials in sensitive areas should be set out.

Physical security controls are often overlooked, but they play just as important a role in network security as the usual precautions, such as authentication and encryption.

Logical Security Controls

Logical security controls create boundaries between networks and provide logical access controls. The boundaries between network segments control traffic flow between VLANs in particular. A security policy can incorporate detailed routing policies, in which routes for separate networks and subnets are announced and accepted as needed. Routing policies are normally incorporated at the distribution layer. Filtering routes is a useful means to exert control over who can source traffic and to what destination; this type of filtering is normally done using distribution lists and route maps on Layer 3 devices.

IP spoofing is the process of faking the source identity of network traffic. IP spoofing is standard operating procedure for hackers. Route filtering does not protect from spoofing attacks, but it can make them harder to carry out. Firewalls are commonly used to provide protection against spoofing attacks.

The remainder of this chapter focuses on logical access controls. Logical access controls include prevention controls and detection controls. *Prevention controls* uniquely identify authorized devices and users and deny access to unauthorized devices and users. *Detection controls* log and report the activities of authorized users and devices and to log and report unauthorized access or attempted access to systems, programs, and data.

Logical Access Controls

The methods for providing logical access controls on Catalyst switches primarily consist of authentication mechanisms and port security. The next sections discuss the following configuration options for logical access control:

- Basic logical access controls—Local authentication, IP permit lists, HTTP access
- Server-based authentication mechanisms—AAA with TACACS+
- Privilege levels
- Banner messages
- Port security

Authentication mechanisms vary from simple password protection to the use of dedicated servers with software employing TACACS+, RADIUS, Kerberos, or IEEE 802.1x protocols. Authentication can be configured for out-of-band interfaces and in-band interfaces. Additional security options include timeouts, privilege levels, and banner messages. Port security is a security option commonly configured on access-layer switches, where users enter the network.

Basic Logical Access Controls

Catalyst switches can be accessed in several different ways. Each method of accessing the device should have a password applied to prevent unauthorized access.

Out-of-band management options include console port and auxiliary port access. In-band management options include Trivial File Transfer Protocol (TFTP) servers, network management software such as CiscoWorks 2000, Hypertext Transfer Protocol (HTTP), and virtual terminal (vty) ports that are used for terminal access. Each Cisco device has five vty ports by default. You can create more virtual terminal ports on some devices, such as access servers (for example, the Cisco 2511 router), if you need to have more than five simultaneous connections to a device.

The basic configuration of login and privileged-mode passwords, as well as timeouts for Catalyst switches, are described in the "Passwords" and "Remote Access" sections of Chapter 3. We proceed to describe some variations on the basic access control options for CatOS and IOS-based switches.

CatOS Switch This section describes the configuration of the following features:

- Maximum login attempts and lockout times
- IP permit lists
- HTTP access

First, you can limit the number of login attempts for normal mode and privileged mode on CatOS switches. You can also configure the lockout period following the specified number of failed login attempts.

The command **set authentication login attempt** {*count*} limits the number of normal-mode login attempts for console and Telnet access before a lockout is enforced. The command **set authentication login lockout** {*time*} specifies how long in seconds a lockout is enforced following the specified number of failed console or Telnet normal-mode login attempts.

The command **set authentication enable attempt** {*count*} limits the number of privileged-mode login attempts for console and Telnet access before a lockout is enforced. The command **set authentication login lockout** {*time*} specifies how long in seconds a lockout is enforced following the specified number of failed console or Telnet privileged-mode login attempts.

Second, you can limit access to the switch using IP permit lists. This feature is analogous to the access class feature used to limit Telnet access to Cisco routers. An IP permit list prevents inbound Telnet and SNMP access to the switch from unauthorized source IP addresses. All other TCP/IP services (such as IP traceroute and IP ping) continue to work normally when you enable the IP permit list. Outbound Telnet, TFTP, and other IP-based services are unaffected by the IP permit list. Telnet attempts from unauthorized source IP addresses are denied a connection. SNMP requests from unauthorized IP addresses receive no response; the request times out. You can configure up to 100 entries in the permit list. Each entry consists of an IP address and subnet mask pair in dotted-decimal format and information on whether the IP address is part of the SNMP permit list, Telnet permit list, or both lists. The bits set to 1 in the mask are checked for a match with the source IP address of incoming packets, and the bits set to 0 are not checked. This process allows sets of IP addresses to be conveniently specified. If you do not specify the mask for an IP permit list entry, or if you enter a host name instead of an IP address, the mask has an implicit value of all bits set to 1, which matches only the IP address of that host. If you do not specify SNMP or Telnet for the type of permit list for the IP address, the IP address is added to both the SNMP and Telnet permit lists.

To specify addresses to add to the IP permit list, use the command **set ip permit** *ip_address* [*mask*] [**all** | **snmp** | **telnet** | **ssh**]. Secure Shell Version 1 (SSH) is an alternative to Telnet with the added benefit of encrypting the traffic associated with remotely configuring the switch; SSH uses port 22 and requires SSH client software, such as SecureCRT. Use the command **show ip permit** to verify the IP permit list configuration.

To apply the permit list, you need to enable the IP permit list using the command **set ip permit enable** [**ssh** | **snmp** | **telnet**]. If you do not specify the **ssh**, **snmp**, or **telnet** keyword, all three of the permit lists are enabled. Example 10-1 illustrates the use of IP permit lists.

Example 10-1 Configuring an IP Permit List

```
Console> (enable) set ip permit 172.16.0.0 255.255.0.0 telnet
172.16.0.0 with mask 255.255.0.0 added to Telnet permit list.
Console> (enable) set ip permit 172.20.52.32 255.255.0.0 snmp
172.20.52.32 with mask 255.255.0.0 added to Snmp permit list.
Console> (enable) set ip permit 172.20.52.3 all
172.20.52.3 added to IP permit list.
Console> (enable) set ip permit 172.20.52.31 255.255.255.224 ssh
172.20.52.31 with mask 255.255.255.224 added to Ssh permit list.
Console> (enable) set ip permit enable

Telnet, Snmp and Ssh permit list enabled
Console> (enable) show ip permit

   Telnet permit list enabled.
   Ssh permit list enabled.
   Snmp permit list enabled.
Permit List        Mask               Access-Type
----------------   ----------------   -------------
172.16.0.0         255.255.0.0        telnet
172.20.0.0         255.255.0.0        snmp
172.20.52.0        255.255.255.224   ssh
172.20.52.3                           telnet ssh snmp
```

continues

Example 10-1 Configuring an IP Permit List (Continued)

```
Denied IP Address Last Accessed Time Type
---------------- ------------------ ------
Denied IP Address   Last Accessed Time Type    Telnet Count   SNMP Count
-----------------   ------------------ ------  ------------   ----------
172.100.101.104     01/20/97,07:45:20  SNMP              14         1430
172.187.206.222     01/21/97,14:23:05  Telnet             7          236
```

Use the command **clear ip permit** {*ip_address* [*mask*] | **all**} [**ssh** | **snmp** | **telnet**] to clear an IP permit list entry. Also, use the command **set ip permit disable** [**ssh** | **snmp** | **telnet**] to disable the IP permit list.

Finally, to configure the switch as an HTTP server so that you can configure the switch through a web browser, proceed as follows:

Step 1 Assign an IP address to the switch, if necessary, using the command **set interface sc0** [*ip_addr/netmask*].

Step 2 Enable the HTTP server on the switch with the command **set ip http server enable.**

Step 3 Configure the HTTP port (the default is 80; perform this step only if you need to change the default). Use the command **set ip http port** *port_number* **default.**

Step 4 Verify the HTTP server configuration with the command **show ip http.**

After the switch is set up as an HTTP server, you can use the Catalyst Web Interface (CWI) to configure the Catalyst 6000, 5000, and 4000 family switches. It is a browser-based tool that consists of a graphical user interface (GUI) that runs on the client, Catalyst CV 5.0 (the Catalyst version of CiscoView 5.0), and an HTTP server that runs on the switch. Complete instructions for using CWI and CV 5.0 can be found at www.cisco.com/univercd/cc/td/doc/product/lan/cat6000/cfgnotes/78_10697.htm.

A GUI alternative to the CLI and Simple Network Management Protocol (SNMP) interfaces, the CWI provides a real-time graphical representation of the switch and detailed information such as port status, module status, type of chassis, and modules. The CWI uses HTTP to download Catalyst CV from the server to the client.

IOS-Based Switch This section describes the configuration of the following features:

- The **login local** option in line-configuration mode
- Using **access-class** to limit IP sources for login
- HTTP access

First, the **login** option in line-configuration mode on an IOS-based switch indicates where to find the login information. If **login** is specified without a keyword, the system uses the line as the login (a username is not required). The user is prompted for the password of the line itself. The other **login** options indicate that the specific user must log in. The keyword after login indicates where to find the user information. The **login local** statement indicates that the information is to be found locally in one of the **username** *username* **password** *password* lines in the configuration file.

It is recommended that users log in to the system with a username and password rather than having everyone use the line's password. Having users log in to the device makes it easier to track down who is accessing the device and what changes they have made. By default, passwords are stored in clear-text format in the router configuration. The only exception to this is the enable secret password, which is automatically encrypted. Password encryption can be compromised, so it should be used in combination with other methods of security. To encrypt all passwords appearing in the configuration file, use the **service password-encryption** command in global configuration mode.

Second, the **access-class** *access-list-number* {**in** | **out**} command applies an access list to a console or vty line. The access list is a standard or extended access list; the list specifies what source and destination IP addresses are permitted for Telnet access. The **in** | **out** condition at the end of the **access-class** statement restricts incoming or outgoing connections between a particular Cisco device and the addresses in the access list.

Finally, Cisco IOS Software allows you to use a Web browser to issue IOS commands to your network device. This allows for an alternative method of configuring IOS-based switches, but this is not recommended, in general, because of the security problems associated with opening port 80. By default, HTTP access is disabled. To enable HTTP access on a Cisco IOS command-based switch or router, enter the command **ip http server**. You can then use an access list to filter the access to the network device's HTTP management by applying the access list with the **ip http access-class** *access-list-number* command. Password security for web access is similar to console and virtual terminal access. Use the command **ip http authentication** [**aaa** | **enable** | **local** | **tacacs**] to specify where the authentication information is contained. These authentication options are explored more in the next section.

AAA and TACACS+

Although usernames and passwords can be configured directly on a network device, this type of configuration does not scale well. It is generally recommended that security be handled at a centralized location rather than on individual devices. This is done using authentication, authorization, and accounting (AAA), which allows user security to be defined on a central server. The TACACS+ protocol provides detailed accounting information and administrative control over the authentication and authorization process. RADIUS and Kerberos options are also available, but we restrict our discussion to TACACS+. Server software is required for an AAA/TACACS+ implementation. Cisco Secure Access Control Server software is commonly used to provide both AAA and TACACS+ services for network devices.

Next, we demonstrate how to configure AAA and TACACS+ for authentication on CatOS and IOS-based switches. Authorization and accounting options are also available but are not described here. A common practice is to use TACACS+ as the primary authentication method, with local authentication as a backup method in case the TACACS+ server is unreachable.

CatOS Switch To configure basic authentication with TACACS+ on a CatOS switch for console and vty access, perform the following steps:

Step 1 Enable TACACS authentication for accessing normal mode by issuing the command **set authentication login tacacs enable.**

Step 2 Enable TACACS authentication for accessing privileged mode by issuing the command **set authentication enable tacacs enable.**

Step 3 Define the server by issuing the command **set tacacs server** *ip_addr* [**primary**]. The **primary** keyword forces this server to be the primary TACACS+ server.

Step 4 Define the TACACS+ server key (this is optional with TACACS+). This causes the switch-to-server data to be encrypted. If this is used, the key must agree with the server. Use the command **set tacacs key** *key.*

Verify the TACACS+ configuration with the commands **show authentication** and **show tacacs.** Example 10-2 illustrates this process.

Example 10-2 Configuring Authentication with TACACS+ on a CatOS Switch

```
Console> (enable) set authentication login tacacs enable

tacacs login authentication set to enable for console and telnet session.
Console> (enable) set authentication enable tacacs enable

tacacs enable authentication set to enable for console and telnet session.
Console> (enable) set tacacs server 172.20.52.3

```

Example 10-2 Configuring Authentication with TACACS+ on a CatOS Switch (Continued)

```
172.20.52.3 added to TACACS server table as primary server.
Console> (enable) set tacacs server 172.20.52.2 primary

172.20.52.2 added to TACACS server table as primary server.
Console> (enable) set tacacs server 172.20.52.10

172.20.52.10 added to TACACS server table as backup server.
 Console> (enable) set tacacs key Secret_TACACS_key

The tacacs key has been set to Secret_TACACS_key.

Console> (enable) show authentication



Login Authentication:  Console Session   Telnet Session
---------------------  ---------------   ---------------
tacacs                 enabled(primary)  enabled(primary)
radius                 disabled          disabled
local                  enabled           enabled
Enable Authentication: Console Session   Telnet Session
---------------------- ----------------- ---------------
tacacs                 enabled(primary)  enabled(primary)
radius                 disabled          disabled
local                  enabled           enabled
Console> (enable) show tacacs

Tacacs key: Secret_TACACS_key
Tacacs login attempts: 5
Tacacs timeout: 30 seconds
Tacacs direct request: enabled

Tacacs-Server                              Status
----------------------------------------   -------
172.20.52.3
172.20.52.2                                primary
172.20.52.10
```

You can also configure the TACACS+ timeout interval and the number of login attempts. Use the **set tacacs timeout** *seconds* command to set the TACACS+ timeout interval. Use the command **set tacacs attempts** *number* to configure the number of allowed login attempts.

IOS-Based Switch To configure basic authentication with TACACS+ on an IOS-based switch for console and vty access, perform the following steps:

Step 1 In global configuration mode, enable AAA/TACACS+ with the command **aaa new-model**.

Step 2 Enable authentication at login, and create one or more lists of authentication methods. Use the command **aaa authentication login** {**default** | *list-name*} **group** *method1* [*method2*...]. The default keyword creates a default list that is used when a named list is not specified in the **login authentication** command (Step 4). The default method list is automatically applied to all lines.

Step 3 Enable authentication for privileged mode, and create one or more lists of authentication methods. Use the command **aaa authentication enable default group** *method1* [*method2*...]. If more than one method is given, a subsequent method is used only if the preceding method returns an error (as opposed to a login failure).

Step 4 In line configuration mode, apply the authentication list using the command **login authentication** {**default** | *list-name*}. If the list-name is used, it overrides the **default** list (if configured in Step 2). If configured in Step 2, the default list is used if no **login authentication** command is entered for the line.

Step 5 In global configuration mode, define a TACACS+ server with the command **tacacs-server host** *name*, where name is either a hostname or an IP address for the server.

Step 6 Define an encryption key for communication between the switch and the TACACS+ daemon. Use the command **tacacs-server key** *key*.

You verify the TACACS+ configuration with the command **show tacacs**. This process parallels the configuration of AAA and TACACS+ on Cisco routers. Next, we consider the options available for configuring privilege levels, which selectively assign access to commands according to the specified privilege levels.

Privilege Levels

It might be appropriate for various members of the campus network staff to have different levels of access privileges for the networking devices comprising the network.

AAA's authorization option can be used to require users to supply a valid username and password pair to execute certain commands, as verified by an authorization server (for example, a TACACS+ server).

Another method of restricting access to commands on Catalyst switches is the use of privilege levels. This option is available on IOS-based switches, the Catalyst 6000 Native IOS, and the Catalyst 4000 Supervisor Engine III IOS.

By default, the Cisco IOS Software has two modes of password security: user EXEC mode and privileged EXEC mode. You can configure up to 16 hierarchical levels of commands for each mode. By configuring multiple passwords, you can allow different sets of users to have access to specified commands. For example, if you want many users to have access to the **clear line** command, you can assign it level 2 security and distribute the level 2 password fairly widely. If you want more restricted access to the **configure** command, you can assign it level 3 security and distribute that password to a more-restricted group of users.

To set a command's privilege level, use the global configuration command **privilege** *mode* **level** *level command*, where mode specifies the mode, such as **configure** for global configuration mode, **exec** for user mode, and **line** for line configuration mode.

To specify the enable password for a particular privilege level, use the global configuration command **enable password level** *level password*. To log into a specified privilege level, a user enters the command **enable** *level*. The **show privilege** command shows the privilege level configuration.

The commands in Example 10-3 set the **configure** command to privilege level 14 and establish SecretPswd14 as the password users must enter to use level 14 commands.

Example 10-3 Configuring Privilege Levels

```
Switch(config)#privilege exec level 14 configure
Switch(config)#enable secret level 14 SecretPswd14
```

The privilege-level options for IOS-based switches are extremely useful for allowing selective access to IOS commands. Care must be taken to ensure that you don't inadvertently allow access to more commands than you had planned. The next section explores the use of banners as an added means of security for Catalyst switches.

Banners

Banners can be used as a deterrent to inappropriate access to campus network devices. Use the banner to state specifically the actions that will be taken for unauthorized access. Do not advertise the site name or network data that might provide information to unauthorized users. These banners provide recourse in case a device is compromised and the perpetrator is caught. In many companies, the legal department must approve the banner messages.

On a CatOS switch, use the message-of-the-day (MOTD) command **set banner motd** *c* [*text*] *c*, where *c* is the delimiting character marking the beginning and end of the message, as shown in Example 10-4.

Example 10-4 Configuring a Login Banner

```
Console> (enable) set banner motd %
*** Unauthorized Access Prohibited ***
***  All transactions are logged   ***
***  Violations reported to the Godfather  ***
------------ Notice Board ------------
----Contact Stanford at 1-go-stanford for access problems----
%
MOTD banner set
```

On an IOS-based switch, the banner is configured with the global configuration command **banner motd** *d message d*, where d is the delimiter for the message. The message of the day is the text that appears onscreen when you open a Telnet session or console port connection to the switch. The banner can be up to 255 characters long.

The next section discusses a logical access control typically used on access-layer switches: port security.

Port Security

The access layer is the entry point for users to access the network. Cable connections are generally pulled from an access-layer switch to offices, labs, and cubicles in a campus network. For this reason, the network devices at the access layer are the most physically vulnerable. Anyone can plug a station into an unsecured access-layer switch. You should take two precautions at the access layer:

- **Port security**—Limit the MAC addresses that are permitted on switch ports to prevent unauthorized users from gaining access to the network.
- **VLAN management**—VLAN 1 is the default VLAN of all ports. VLAN 1 is traditionally the management VLAN, meaning that users entering the network on ports that were not configured for a VLAN would be assigned to the switch block's management VLAN. It is recommended that the management VLAN be moved to another VLAN to prevent users from entering the management VLAN on an unconfigured port.

You can use port security to block input to an Ethernet, Fast Ethernet, or Gigabit Ethernet port when the MAC address of the station attempting to access the port is different from any of the MAC addresses specified for that port. If a security violation occurs, you can configure the port to go into either shutdown mode or restrictive mode. The shutdown mode option allows you to specify whether the port is permanently disabled or disabled for only a specified time. The default is for the port to shut down permanently. Restrictive mode allows you to configure the port to remain enabled during a security violation and to drop only packets that come in from insecure hosts.

When a secure port receives a packet, the packet's source MAC address is compared to the list of secure source addresses that were manually configured or autoconfigured (sticky-learned) on the port. If a MAC address of a device attached to the port differs from the list of secure addresses, the port either shuts down permanently, shuts down for the time you have specified, or drops incoming packets from the insecure host. A port's behavior depends on how you configure it to respond to a security violation. If a security violation occurs, the link LED for that port turns orange.

As soon as a MAC address has been associated with a given port, that MAC address cannot be "moved" to another port with security enabled, or the port will go into either shutdown mode or restrictive mode (according to your configuration). You have to use the **clear cam dynamic** command before "moving" the MAC address to a secured port.

The next two sections describe how to configure port security on CatOS and IOS-based switches.

CatOS Switch To enable port security on a CatOS switch, use the command **set port security** *mod_num/port_num* **enable**. You can specify the number of MAC addresses to secure on a port with the command **set port security** *mod_num/port_num* **maximum** *num_of_mac*. After you allocate the maximum number of MAC addresses on a port, you can either specify the secure MAC address for the port manually or have the port dynamically configure the MAC address of the connected devices. Out of a maximum allocated number of MAC addresses on a port, you can manually configure all of them, allow all of them to be autoconfigured, or configure some manually and allow the rest to be autoconfigured. When you manually change the maximum number of MAC addresses associated with a port to be greater than the default value (1) and then manually enter some of the authorized MAC addresses, any remaining MAC addresses configure automatically (are sticky-learned). For example, if you configure a port's port security to have a maximum of ten MAC addresses but you add only two MAC addresses, the next eight new source MAC addresses received on that port are added to the port's secured MAC address list.

To manually add MAC addresses to the list of secure addresses, use the command **set port security** *mod_num/port_num mac_addr.*

Should a host connect to a port and violate the configured port security, either shutdown or restrict mode ensues, with the following characteristics:

- **Shutdown**—This shuts down the port permanently or for a specified time. Permanent shutdown is the default mode.
- **Restrict**—This drops all packets from insecure hosts but remains enabled.

To specify the security violation action to be taken, use the command **set port security** *mod_num/port_num* **violation** {**shutdown** | **restrict**}. To specify how long a port is to remain disabled in the event of a security violation, enter the command **set port security** *mod_num/port_num* **shutdown** *time*, where *time* is in minutes. The default is for the port to be permanently shut down.

To disable a port's port security, use the command **set port security** *mod_num/port_num* **disable**. Verify the port security configuration with the command **show port security** [*mod_num/port_num*].

Example 10-5 summarizes the configuration commands for port security.

Example 10-5 Port Security Configuration on a CatOS Switch

```
Console> (enable) set port security 2/1 enable <--- using the learned MAC address

Port 2/1 port security enabled with the learned mac address.
Trunking disabled for Port 2/1 due to Security Mode
Console> (enable) show port 2/1

Port  Name               Status     Vlan       Level  Duplex Speed Type
----- ------------------ ---------- ---------- ------ ------ ----- ------------
 2/1                     connected  522        normal   half   100 100BaseTX

Port  Security Secure-Src-Addr   Last-Src-Addr     Shutdown Trap     IfIndex
----- -------- ----------------- ----------------- -------- -------- -------
 2/1  enabled  00-90-2b-03-34-08 00-90-2b-03-34-08 No       disabled 1081
<output omitted>
Console> (enable) set port security 2/1 enable 00-90-2b-03-34-08 <-- manually
  specify secure MAC address
Port 2/1 port security enabled with 00-90-2b-03-34-08 as the secure mac address
Trunking disabled for Port 2/1 due to Security Mode <----- port security on access
  ports only
```

Example 10-5 Port Security Configuration on a CatOS Switch (Continued)

```
Console> (enable) set port security 2/1 disable

Port 2/1 port security disabled.
Console> (enable) set port security 4/7 maximum 20 <--- maximum number of 20 MAC
  addresses to be secured

Maximum number of secure addresses  set to 20 for port 4/7.
Console> (enable) set port security 4/7 violation restrict

Port security violation on port 4/7 will cause insecure packets to be dropped.
Console> (enable) set port security 4/7 shutdown 600
Console> (enable) show port security 4/7

Port  Security Violation Shutdown-Time Age-Time Max-Addr Trap     IfIndex
----- -------- --------- ------------- -------- -------- -------- -------
 3/24  enabled  shutdown           600       60       20 disabled     921

Port  Num-Addr Secure-Src-Addr   Age-Left Last-Src-Addr     Shutdown/Time-Left
----- -------- ----------------- -------- ----------------- ------------------
 4/7         4 00-e0-4f-ac-b4-00       60 00-e0-4f-ac-b4-00       no          -
               00-11-22-33-44-55        0
               00-11-22-33-44-66        0
               00-11-22-33-44-77        0
```

Port security is a useful tool for preventing unwanted hosts from accessing the network.

IOS-Based Switch To enable port security on an IOS-based switch, enter the interface configuration command **switchport port-security.** (Use the **no** form of this command to disable port security.) Use the command **switchport port-security maximum** *max_addrs* to set the maximum number of MAC addresses allowed on the interface.

You can also set an interface's security violation mode; the default is **shutdown**. The three available modes are as follows:

- **shutdown**—The interface is shut down immediately following a security violation.
- **restrict**—A security violation sends a trap to the network management station.
- **protect**—When the port secure addresses reach the allowed maximum on the port, all packets with unknown source addresses are dropped.

To configure security violation mode, enter the command **switchport port-security violation** {**shutdown** | **restrict** | **protect**}.

The privileged-mode command **show port security** [**interface** *interface-id* | **address**] verifies the port security configuration.

This completes our discussion of prevention logical access controls for campus networks. We complete the chapter with a look at two detection logical access controls: Switched Port Analyzer (SPAN) and syslog.

Switched Port Analyzer

A *Switched Port Analyzer (SPAN)* session is an association of a destination port with a set of source ports, configured with parameters that enable the monitoring of network traffic. You can configure multiple SPAN sessions in a switched network. SPAN sessions do not interfere with the switches' normal operation. You can enable or disable SPAN sessions with CLI or SNMP commands. When enabled, a SPAN session might become active or inactive based on various events or actions that are indicated by a syslog message (syslog is discussed in the next section). All network traffic, including multicast and bridge protocol data unit (BPDU) packets, can be monitored using SPAN.

After the system is on, a SPAN destination session remains inactive until the destination port is operational. A destination port (also called a monitor port) is a switch port in which SPAN sends packets for analysis. (A protocol analyzer or remote monitoring [RMON] probe is connected to this port for traffic analysis.) After a port becomes an active destination port, it does not forward any traffic except that required for the SPAN session. By default, an active destination port disables incoming traffic (from the network to the switching bus) unless you specifically enable the port. If incoming traffic is enabled for the destination port, it is switched in the destination port's native VLAN. The destination port does not participate in spanning tree while the SPAN session is active.

Only one destination port is allowed per SPAN session, and the same port cannot be a destination port for multiple SPAN sessions. A switch port configured as a destination port cannot be configured as a source port. Not all ports can be destination ports. For example, EtherChannel ports cannot be SPAN destination ports. For these destination ports, sometimes the SPAN commands are accepted, even though SPAN does not work.

If a SPAN destination port's trunking mode is **on** or **nonegotiate** during SPAN session configuration, the SPAN packets forwarded by the destination port have the encapsulation specified by the trunk type; however, the destination port stops trunking.

A *source port* is a switch port monitored for network traffic analysis. The traffic through the source ports is categorized as ingress, egress, or both. You can monitor one or more source ports in a single SPAN session with user-specified traffic types (ingress, egress, or both). Ingress SPAN copies network traffic received by the source ports for analysis at the destination port. Egress SPAN copies network traffic transmitted from the source ports for analysis at the destination port. Figure 10-1 illustrates the use of source and destination SPAN ports with a switch probe.

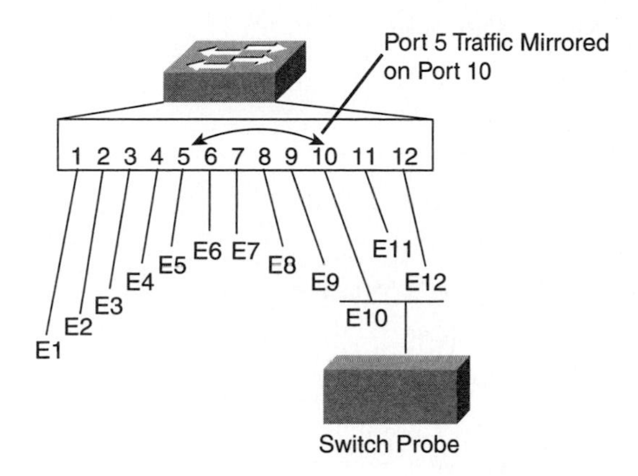

Figure 10-1
Using SPAN.

You can configure source ports in any VLAN. You can configure VLANs as source ports (*src_vlans*), which means that all ports in the specified VLANs are source ports for the SPAN session. You can also configure trunk ports as source ports and mix them with nontrunk source ports.

Source ports are administrative (*Admin Source*) or operational (*Oper Source*) or both. Administrative source ports are the source ports or source VLANs specified during SPAN session configuration. Operational source ports are the source ports monitored by the destination port. For example, when source VLANs are used as the administrative source, the operational source is all the active ports in all the specified VLANs. The operational sources are always active ports. If a port is not in the spanning tree, it is not an operational source.

On a CatOS switch, use the command **set span** {*src_mod/src_ports* | *src_vlan*} *dest_mod/dest_port* [**rx** | **tx** | **both**] to configure SPAN. **rx** refers to traffic received at the source. **tx** refers to traffic transmitted from the source. When monitoring a VLAN on a switch, you must monitor both transmit and receive traffic (**both**). You cannot monitor only transmit (**tx**) traffic or only receive (**rx**) traffic. The command **show span** verifies the SPAN configuration.

Example 10-6 illustrates SPAN configuration on a CatOS switch, first with a port and then with a VLAN serving as the source.

Example 10-6 SPAN Configuration on a CatOS Switch

```
Console> (enable) set span 2/4 3/6

Overwrote Port 3/6 to monitor transmit/receive traffic of Port 2/4
Incoming Packets disabled. Learning enabled.
Console> (enable) show span



Destination      : Port 3/6
Admin Source     : Port 2/4
Oper Source      : None
Direction        : transmit/receive
Incoming Packets: disabled
Learning         : enabled
Filter           : -
Status           : active

----------------------------------------------
Total local span sessions:  1
Console> (enable)
Console> (enable) set span 522 2/1

Overwrote Port 2/1 to monitor transmit/receive traffic of VLAN 522
Incoming Packets disabled. Learning enabled.
Console> (enable) show span

Destination      : Port 2/1
Admin Source     : VLAN 522
Oper Source      : Port 2/1-2
Direction        : transmit/receive
Incoming Packets: disabled
Learning         : enabled
Filter           : -
Status           : active

----------------------------------------------
Total local span sessions:  1
```

Use the command **set span disable** [*dest_mod/dest_port* | **all**] to disable SPAN on a CatOS switch.

On an IOS-based switch, use the global configuration commands **monitor session** *session_number* **source interface** *interface-id* [**both** | **rx** | **tx**] and **monitor session** *session_number* **destination interface** *interface-id* to set the source and destination ports, respectively. Use the **no** forms of these commands to remove SPAN source and destination ports. Verify the SPAN configuration with the **show monitor** command.

The last section of this chapter explores syslog configuration on Catalyst switches.

Syslog

The system message-logging software on Catalyst switches can save messages in a log file or direct them to other devices. The system message-logging facility has these features:

- It provides you with logging information for monitoring and troubleshooting.
- It allows you to select the types of logging information captured.
- It allows you to select the destination of captured logging information.

By default, the switch logs normal but significant system messages to its internal buffer and sends these messages to the system console. You can specify which system messages should be saved based on the type of facility (such as CDP, DTP, IP, SNMP, Telnet, or VTP) and the severity level (see Table 10-1). Messages are time-stamped to enhance real-time debugging and management; for this to be useful, the switch clock needs to be set, or you need to configure NTP on the switch.

Table 10-1 Severity Levels

Severity Level	Keyword	Description
0	**emergencies**	System unusable
1	**alerts**	Immediate action required
2	**critical**	Critical conditions
3	**errors**	Error conditions
4	**warnings**	Warning conditions
5	**notifications**	Normal but significant conditions
6	**informational**	Informational messages
7	**debugging**	Debugging messages

You can access logged system messages using the switch CLI or by saving them to a syslog (system logging) server. The switch software saves syslog messages in an internal buffer. You can monitor system messages remotely by accessing the switch through Telnet or the console port, or by viewing the logs on a syslog server.

The system log message format for a CatOS switch is *mm/dd/yyy:hh/mm/ss:facility-severity-MNEMONIC:description*. The system log message elements are specified in Table 10-2.

Table 10-2 System Log Message Elements

Element	Description
mm/dd/yyy:hh/mm/ss	Date and time of the error or event. This information appears only if configured using the **set logging timestamp enable** command.
facility	Indicates the facility to which the message refers (for example, SNMP, SYS, and so on).
severity	Single-digit code from 0 to 7 that indicates the message's severity.
MNEMONIC	Text string that uniquely describes the error message.
description	Text string containing detailed information about the event being reported.

The system log message for an IOS-based switch is structured as *%FACILITY-SEVERITY-MNEMONIC*: *Message-text*. *FACILITY* is a code consisting of two or more uppercase letters that indicate the facility to which the message refers. A facility can be a hardware device, a protocol, or a module of the system software. *SEVERITY*, as shown in Table 10-1, is a single-digit code from 0 to 7 that reflects the condition's severity; the lower the number, the more serious the situation. *MNEMONIC* is a code that uniquely identifies the message.

The next section describes syslog configuration for CatOS and IOS-based switches.

CatOS Switch You can change the severity level for each logging facility using the **set logging level** *facility severity* privileged-mode command. The **set logging server facility** command described in Step 2 in the following list is an alternative means of setting the logging level.

You can verify the system logging information with the **show logging** command. To enable or disable the logging time stamp, use the privileged-mode command **set logging timestamp** {**enable** | **disable**}.

To configure the switch to log messages to a syslog server, perform the following tasks in privileged mode:

Step 1 Specify the IP address of one or more syslog servers with the command **set logging server** *ip_addr*.

Step 2 Set the facility and severity levels for syslog server messages with the commands **set logging server facility** *server_facility_parameter* and **set logging server severity** *server_severity_level*. The first command specifies the type of system messages to capture. *server_facility_parameter* specifies the logging facility of the syslog server. Valid values include **local0**, **local1**, **local2**, **local3**, **local4**, **local5**, **local6**, and **local7** (there are other options). These values are just tags to allow you to differentiate between logs from different sources. They are not intrinsically linked to severity levels. *server_severity_level* specifies the severity level of system messages to capture.

Step 3 Enable system message logging to configured syslog servers with the command **set logging server enable**.

Step 4 Verify the configuration with the **show logging** command.

Example 10-7 illustrates configuring syslog support. It shows how to specify a syslog server, set the facility and severity levels, and enable logging to the server.

Example 10-7 Configuring Syslog on CatOS Switches

```
Console> (enable) set logging server 10.10.10.100

10.10.10.100 added to System logging server table.
Console> (enable) set logging server facility local5

System logging server facility set to <local5>
Console> (enable) set logging server severity 5

System logging server severity set to <5>
Console> (enable) set logging server enable

System logging messages will be sent to the configured syslog servers.
Console> (enable) show logging


Logging buffer size:          200
        timestamp option:     disabled
```

continues

Example 10-7 Configuring Syslog on CatOS Switches (Continued)

```
Logging history size:          1
Logging console:               enabled
Logging server:                enabled
{syslog.bigcorp.com}
        server facility:       LOCAL5
        server severity:       notifications(5)

Facility             Default Severity          Current Session Severity
-------------        -----------------------   -----------------------
cdp                  3                         3
drip                 2                         5
dtp                  5                         5
dvlan                2                         5
earl                 2                         5
fddi                 2                         5
filesys              2                         5
gvrp                 2                         5
ip                   2                         5
kernel               2                         5
mcast                2                         5
mgmt                 5                         5
mls                  5                         5
pagp                 5                         5
protfilt             2                         5
pruning              2                         5
radius               2                         5
security             2                         5
snmp                 2                         5
spantree             2                         5
sys                  5                         5
tac                  2                         5
tcp                  2                         5
telnet               2                         5
tftp                 2                         5
udld                 4                         5
vmps                 2                         5
vtp                  2                         5
```

Example 10-7 Configuring Syslog on CatOS Switches (Continued)

```
0(emergencies)        1(alerts)             2(critical)
3(errors)             4(warnings)           5(notifications)
6(information)        7(debugging)
Console> (enable)
```

The severity levels are conveniently listed at the end of the output.

IOS-Based Switches Message logging is enabled by default on IOS-based switches. However, to send messages to any destination other than the console, use the global configuration command **logging** *host*. To build a list of syslog servers that receive logging messages, enter this command more than once.

By default, system messages are not time-stamped. To enable time stamps on log messages showing the time since the system was rebooted, use the global configuration command **service timestamps log uptime**. To enable time stamps on log messages showing the date and time, use the command **service timestamps log datetime** [**msec**] [**localtime**] [**show-timezone**]. Depending on the options selected, the time stamp can include the date, the time in milliseconds relative to the local time zone, and the time zone name.

You can limit messages displayed to the selected device by specifying the messages' severity level.

To limit messages logged to the console, use the command **logging console** *level*. The default is for the console to receive all messages (up through severity level 7).

When connected via Telnet or SSH to a Catalyst switch, you need to enter the privileged command **terminal monitor** to display messages you would normally see via a console connection. The default is for terminal lines to receive all messages (assuming that **terminal monitor** has been entered). To disable displaying logging messages altogether, use the privileged configuration command **terminal no monitor**. To limit messages logged to the terminal lines, use the command **logging monitor** *level*.

To limit messages logged to a syslog server, use the command **logging trap** *level*. By default, syslog servers receive informational messages and numerically lower severity levels. To configure the syslog server's logging facility, use the command **logging facility** *facility-type*. The default facility is **local7**.

The syslog server can be configured on a UNIX machine (syslogd is the built-in syslog daemon). See www.cisco.com/univercd/cc/td/doc/product/lan/c3550/1218ea1/3550scg/swlog.htm#xtocid5 for details on configuring syslog on UNIX hosts.

To display the addresses and levels associated with the current logging setup and any other logging statistics, use the **show logging** command.

Summary

Creating and diligently implementing a security policy is a fundamental responsibility of the campus networking staff. The network security policy addresses physical security controls, logical security controls, system and data integrity, data confidentiality (encryption), policies and procedures for campus networking staff, and security awareness training for users of the campus network. This chapter discussed physical and logical access controls.

Physical security controls deal with the physical infrastructure, physical device security, and physical access. To ensure physical device security, attention should be focused on the physical location of equipment, physical access to equipment, and environmental safeguards.

Logical security controls create boundaries between networks and provide logical access controls. The boundaries between network segments control traffic flow between VLANs. Routing policy is configured primarily at the distribution layer. Logical access controls can be grouped into prevention controls and detection controls. The prevention controls we explored were local authentication, centralized authentication with AAA and TACACS+, IP permit lists, privilege levels, banners, and port security. The detection controls we looked at were SPAN and syslog.

Check Your Understanding

Test your understanding of the concepts covered in this chapter by answering these review questions. Answers are listed in Appendix A, "Check Your Understanding Answer Key."

1. Which of the following are addressed by a network security policy?
   - A. Encryption
   - B. Data integrity
   - C. Security awareness training
   - D. Mobile IP
2. Physical security controls include which of the following?
   - A. Physical infrastructure
   - B. Physical addresses
   - C. Device security
   - D. Physical access
3. Physical device security takes into account which of the following?
   - A. Physical location of equipment
   - B. Port security
   - C. Environmental safeguards
   - D. IP telephony architecture
4. What are two functions of logical security controls?
   - A. Physical location of equipment
   - B. Create boundaries between networks
   - C. Environmental safeguards
   - D. Provide logical access controls
5. What are the two types of logical access controls?
   - A. Prevention
   - B. Imperative
   - C. Causative
   - D. Detection

6. Which of the following are prevention logical access controls?
   A. Port security
   B. Mobile IP
   C. Authentication
   D. IP spoofing
7. At which layer of the hierarchical campus network model do routing policy and route filtering take place to help maintain logical security controls?
   A. Remote
   B. Distribution
   C. Access
   D. Core
8. Which of the following are detection logical access controls?
   A. Physical location of equipment
   B. SPAN
   C. Environmental safeguards
   D. Syslog
9. Which privileged-mode command is required to view debug messages from a vty session on an IOS-based switch?
   A. **logging monitor**
   B. **access-class**
   C. **terminal monitor**
   D. **logging console**
10. Which line-configuration command on an IOS-based switch calls an authentication procedure referred to by the authentication list name auth1?
    A. **login authentication default**
    B. **login auth1**
    C. **login authentication auth1**
    D. **login local**

Key Terms

detection control A logical access control used to log and report the activities of authorized users and devices and to log and report unauthorized access or attempted access to systems, programs, and data.

logical security control A security control used to create boundaries between networks and to provide logical access controls.

physical security control A security control dealing with physical infrastructure, physical device security, and physical access.

prevention control A logical access control used to uniquely identify authorized devices and users and deny access to unauthorized devices and users.

SPAN (Switched Port Analyzer) A tool used with Catalyst switches to enable the capture of traffic. A SPAN session is an association of a destination port with a set of source ports, configured with parameters that enable the monitoring of network traffic.

Appendix A

Check Your Understanding Answer Key

Chapter 1

1. C
2. B
3. False
4. C
5. D
6. B
7. B
8. C
9. B
10. C

Chapter 2

1. D
2. D
3. B
4. C
5. C
6. C
7. C
8. C
9. 2900, 4000, 5000, 6000
10. HSRP

Chapter 3

1. D
2. C
3. A
4. D
5. A
6. C
7. A
8. B
9. D
10. B

Chapter 4

1. A, B
2. C
3. A
4. D
5. C
6. D
7. A, D
8. B
9. C
10. C

Chapter 5

1. C
2. C
3. B
4. B
5. C

6. A
7. A, B, C, D
8. D
9. C
10. D

Chapter 6

1. B
2. A, C
3. B, C, D
4. A, D
5. A, C
6. C
7. A
8. A
9. D
10. A

Chapter 7

1. A, B, D
2. A
3. C
4. D
5. C
6. C
7. B
8. D
9. A, C, D
10. A, B, C

Chapter 8

1. A
2. C
3. D
4. D
5. B, D
6. C
7. C
8. B
9. B
10. A, B, D

Chapter 9

1. C
2. D
3. C
4. B
5. D
6. C
7. C, D
8. C
9. C
10. B

Chapter 10

1. A, B, C
2. A, C, D
3. A, C
4. B, D
5. A, D

6. A, C
7. B
8. B, D
9. C
10. C

Appendix B

Transparent Bridging with Routers

Cisco routers are versatile devices. Although their main use is to route traffic, they also support bridging and several other technologies. There are cases in which you might want to route some protocols and bridge others. Nonroutable protocols such as Local Area Transport (LAT) and NetBEUI must be bridged.

This type of configuration, in which routing and bridging are both utilized on a single device, is ideal for the Catalyst 8500 series switch routers, which have Layer 3 switching speeds approximating Layer 2 switching speeds. Although most Cisco routers can bridge traffic, bridging is normally processed in software, which can lead to a serious bottleneck in your network.

With Integrated Routing and Bridging (IRB), Cisco routers can simultaneously bridge and route traffic. The key to making this work is the *Bridged Virtual Interface (BVI)*. A BVI creates a link between bridged domains and routed domains. Layer 3 protocols can be configured on a BVI, but bridging commands are not permitted.

Generally, IRB is used only for spot fixes within a network because of the negative implications of bridging between VLANs. Inter-VLAN bridging collapses per-VLAN spanning trees into a single large spanning tree for the bridged VLANs. This opens the door to spanning-tree scaling issues, increasing the likelihood of bridging loops. Spanning-tree load balancing is also precluded if you configure inter-VLAN bridging.

This appendix presents two examples of inter-VLAN bridging utilizing IRB and BVIs. Keep in mind that these scenarios are meant only to demonstrate the technology and are not necessarily recommended as a campus LAN design solution. These examples give you some guidance in implementing IRB, which is not an intuitive process.

IRB: Four VLANs, a 4006 Switch with an L3 Module, and a 2513 Router

Figure B-1 illustrates the scenario. The Cisco 2513 router has one 4/16 Token Ring interface (To0) and one 10 Mbps Ethernet interface (E0).

Figure B-1
Inter-VLAN bridging allows VLANs to span a router.

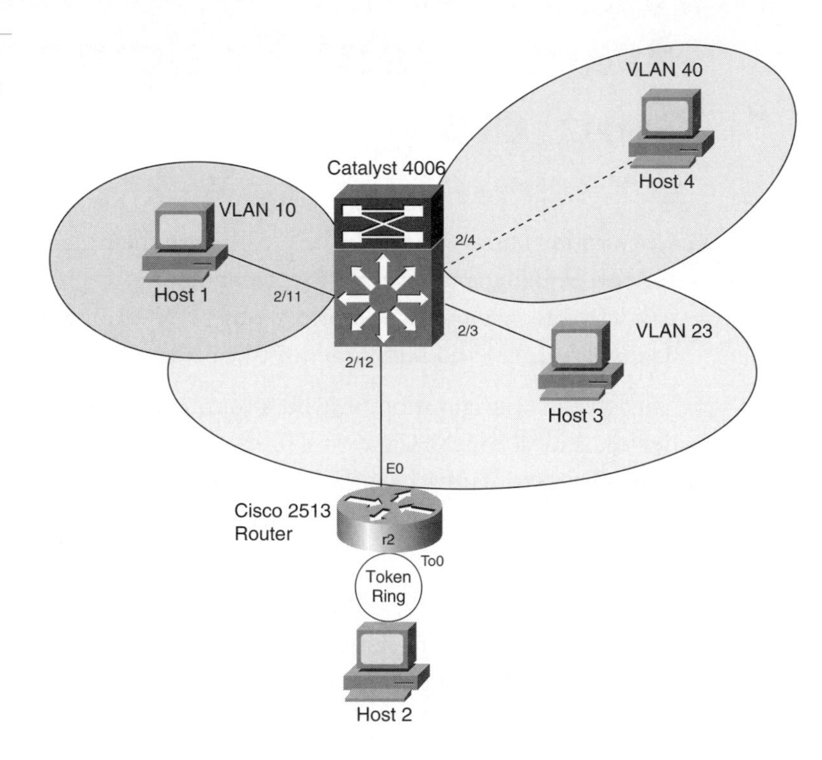

The following data get this scenario under way:

> **NOTE**
> The same default gateway is used for VLAN 10 and VLAN 40. (This is explained later.)

There are four VLANs:

- VLAN 1 (default gateway: 1.1.1.1/24)
- VLAN 10 (default gateway: 1.1.14.1/24)
- VLAN 23 (default gateway: 1.1.23.1/24)
- VLAN 40 (default gateway: 1.1.14.1/24)
- A Gigabit EtherChannel (port channel) is formed between the internal ports 2/1 and 2/2 on the 4006 backplane and the internal interfaces G3 and G4 on the L3 module.
- An 802.1Q trunk is created over the Gigabit EtherChannel (port channel).
- One port-channel subinterface is created for each VLAN on the L3 module:
  - — Port channel 1.1

IP: 1.1.1.1/24

IPX: 1

VLAN 1 (dot1Q—native)

— Port channel 1.10

IP: None

IPX: 10

VLAN 10 (dot1Q)

— Port channel 1.23

IP: 1.1.23.1/24

IPX: 23

VLAN 23 (dot1Q)

— Port channel 1.40

IP: None

IPX: 40

VLAN 40 (dot1Q)

- RIP is configured on the L3 module for network 1.0.0.0.
- The sc0 (management) interface on the Supervisor module is configured for VLAN 1 with IP 1.1.1.2/24.
- A default gateway pointing to 1.1.1.1 is created on the Supervisor module.
- Router 2 (r2), interface E0, connects to port 2/12 on the 4006:
  - — E0 IP address: 1.1.23.22/24
  - — E0 IPX address: 23
- r2, interface To0, connects to Token Ring Host 2:
  - — To0 IP address: 10.1.0.1/24
  - — To0 IPX address: 2
  - — Only transparent bridging is used (no source-route bridging, as is commonly used in Token Ring networks)
- RIP is configured on r2 for networks 1.0.0.0 and 10.0.0.0 (no NAT).
- Host 1 is on VLAN 10 connected to 4006 port 2/11:
  - — Host 1 IP: 1.1.14.11/24
  - — Host 1 IPX: 10
- Host 2 is connected to r2's To0 interface:
  - — Host 2 IP: 10.1.0.2/24
  - — Host 2 IPX: 2

NOTE

IP addresses are not defined for port-channel interfaces 1.10 and 1.40. (This is explained later.)

- Host 3 is on VLAN 23 connected to 4006 port 2/3:
  — Host 3 IP: 1.1.23.33/24
  — Host 3 IPX: 23
- Host 4 is on VLAN 40 connected to 4006 port 2/4:
  — Host 4 IP: 1.1.14.44/24
  — Host 4 IPX: 40

The following configuration tasks must be performed to get the basic setup configured:

Step 1 Configure the Catalyst 4006 as a VTP server, and assign a VTP domain name.

Step 2 Create the VLANs on the Catalyst 4006.

Step 3 Configure the sc0 interface and the default gateway on the 4006.

Step 4 Configure the Gigabit EtherChannel IEEE 802.1Q trunk.

Step 5 Create the port-channel subinterfaces on the L3 module.

Step 6 Assign the ports 2/3, 2/4, 2/11, and 2/12 to the appropriate VLANs.

Step 7 Configure all IP and IPX addresses as specified in the earlier bulleted list.

Step 8 Configure RIP on r2 and the L3 module.

At this point, after completing Steps 1 through 8, the network will be broken. You want subnet 1.1.14.0/24 to span both VLAN 10 and VLAN 40. In addition, you want IP and IPX connectivity between all hosts in the network.

You need to answer the following questions to understand how routing and bridging work in this scenario:

- Why leave off the IP addresses on port channel interfaces 1.10 and 1.40?
- The default gateway for r2 and Host 3 is 1.1.23.1. What is the default gateway for Host 1 and Host 4?
- How does Host 1 on VLAN 10 communicate with Host 4 on VLAN 40 without routing?

The answers to these questions are interrelated. No IP address was configured on port channel 1.10 or 1.40 because you have to bridge the traffic between VLANs 10 and 40. If you had configured IP addresses on port channel interfaces 1.10 and 1.40, routing would have been in effect by default.

However, you want to route traffic among all the VLANs, so you need a mechanism in addition to the present configuration. This is where the BVI comes into the picture. You create a BVI interface to act as a single routed interface on behalf of all the bridged interfaces corresponding to the BVI number. When interfaces are configured for bridging, they are configured with a bridge-group number; the bridge-group number must match the BVI number that you want to associate with these interfaces.

Thus, you need to configure a bridge group on port-channel interfaces 1.10 and 1.40 and configure a BVI that serves these two interfaces (allowing for routing traffic between VLANs 10 and 40 and VLANs 1 and 23).

BRIDGE GROUPS

When you use bridge groups on interfaces, all protocols configured on an interface with a Layer 3 address are routed, and all other protocols are bridged. For example, port channel 1.40 has no IP address configured, but it does have IPX configured, so IPX is routed, but IP (and all other protocols) is bridged within that bridge group.

Also, a BVI cannot be created before IRB is enabled on the router. First, enable IRB with the global configuration command **bridge irb** and then create the BVI with the global command **interface bvi** *bvi_interface_number.*

Follow these steps to configure a bridge group on port-channel interfaces 1.10 and 1.40 and then configure a BVI that serves those two interfaces:

Step 1 Enter the command **bridge-group 1** on port-channel interfaces 1.10 and 1.40 (the bridge-group number is independent of the VLAN numbers). Also, you must define the Spanning-Tree Protocol with the global command **bridge 1 protocol** ieee.

Step 2 Next, as previously mentioned, configure IRB with the global command **bridge irb** and then create the BVI with the command **interface bvi 1**. In **interface bvi 1** configuration mode, you need to configure all the Layer 3 protocols that you left out of the port-channel interface configurations (basically, all the Layer 3 protocols that you are bridging within the bridge group—in our example, this is only IP). Use the default gateway address 1.1.14.1/24 on BVI 1: **ip address 1.1.14.1 255.255.255.0.** It would not make sense to put this command on port-channel interface 1.10 or 1.40 because they share the same IP subnet.

Step 3 Finally, you must tell the router explicitly what protocols you want routed and what protocols you want bridged. If you do not specify that you want the traffic routed, traffic will not be routable between distinct Layer 3 networks for that protocol. Thus, you finish with two global configuration commands: **bridge 1 route ip** and **bridge 1 route ipx.**

NOTE

The other options for this command are **dec**, **ibm**, and **vlan-bridge**. **ieee** is the most common option for switching.

If you leave out the command **bridge 1 route ip,** Host 1 can talk to Host 4 via IP, but not to Host 2 or Host 3. If you leave out the command **bridge 1 route ipx,** Host 1 cannot communicate with any of the other hosts via IPX (the same is true for Host 4), but Host 2 and Host 3 can still communicate via IPX.

IP and IPX traffic between Host 2 and Host 3 is Layer 2 switched through the 4006 (between ports 2/12 and 2/3) because r2's E0 interface and Host 4 are both in VLAN 23.

By creating the BVI, you are forced to configure the command **bridge 1 route ipx**. If the BVI and IRB are removed from the configuration on the L3 module, IPX works throughout this network because IPX networks are configured on each port-channel subinterface. (Remember: When you use bridge groups on interfaces, all protocols configured on an interface with a Layer 3 address are routed.) Assuming that IRB is removed and the remaining bridging commands are retained (**bridge-group 1** on port-channel interfaces 1.10 and 1.40 and the global command **bridge 1 protocol ieee**), you still have IP connectivity between Host 1 and Host 4 and between Host 2 and Host 3, but that's it.

In conclusion, the requirement to have a single IP subnet span both VLANs 10 and 40 through the L3 module forces you to configure IRB to get full IP and IPX connectivity.

Examples B-1, B-2, and B-3 provide the working configurations of the Supervisor, the L3 module, and r2, respectively.

Example B-1 The Complete Output of the Supervisor show config Command

```
4006> (enable) show config
This command shows non-default configurations only.
Use 'show config all' to show both default and non-default configurations.
...........
.................
...................






..


begin
!
# ***** NON-DEFAULT CONFIGURATION *****
!
!
#time: Mon Feb 18 2002, 16:11:56
```

Example B-1 The Complete Output of the Supervisor show config Command (Continued)

```
!
#version 6.2(1)
!
!
#system web interface version(s)
!
#test
!
#system
set system name  4006
!
#frame distribution method
set port channel all distribution mac both
!
#vtp
set vtp domain cisco
set vlan 1 name default type ethernet mtu 1500 said 100001 state active
set vlan 1002 name fddi-default type fddi mtu 1500 said 101002 state active
set vlan 1004 name fddinet-default type fddinet mtu 1500 said 101004 state active
  stp ieee
set vlan 1005 name trnet-default type trbrf mtu 1500 said 101005 state active stp
  ibm
set vlan 10,23,40
set vlan 1003 name token-ring-default type trcrf mtu 1500 said 101003 state active
  mode srb aremaxhop 7 stemaxhop 7 backupcrf off
!
#ip
set interface sc0 1 1.1.1.2/255.255.255.0 1.1.1.255


set interface sl0 down
set interface me1 down
set ip route 0.0.0.0/0.0.0.0         1.1.1.1
!
#spantree
#vlan 1
set spantree priority 8192   1
#vlan 1003
```

continues

Example B-1 The Complete Output of the Supervisor show config Command (Continued)

```
set spantree fwddelay 15     1003
set spantree maxage   20     1003
set spantree priority 8192   1003
#vlan 1005
set spantree fwddelay 15     1005
set spantree maxage   20     1005
set spantree priority 8192   1005
!
#syslog
set logging level cops 2 default
!
#set boot command
set boot config-register 0x2
set boot system flash bootflash:cat4000.6-2-1.bin
set boot system flash bootflash:cat4000.5-5-1.bin
!
#mls
set mls nde disable
!
#port channel
set port channel 2/1-2 11
!
#module 1 : 2-port 1000BaseX Supervisor
!
#module 2 : 34-port Router Switch Card
set vlan 10   2/11
set vlan 23   2/3,2/12
set vlan 40   2/4
set port speed      2/4,2/11  100
set port duplex     2/4,2/11  full
set trunk 2/1  nonegotiate dot1q 1-1005
set trunk 2/2  nonegotiate dot1q 1-1005
set spantree portfast    2/3-4,2/11-12 enable
set port channel 2/1-2 mode on
!
#module 3 empty
```

Example B-1 The Complete Output of the Supervisor show config Command (Continued)

```
!
#module 4 empty
!
#module 5 empty
!
#module 6 empty
end
Console> (enable)
```

Example B-2 The Complete Output of the L3 Module show run Command

```
L3#show run
Building configuration...

Current configuration:
!
version 12.0
no service pad
service timestamps debug uptime
service timestamps log uptime
no service password-encryption
!
hostname L3
!
!
ip subnet-zero
no ip domain-lookup
ipx routing 3030.3030.3030
bridge irb
!
!
!
interface Port-channel1
 no ip address
 no ip directed-broadcast
 hold-queue 300 in
!
```

continues

Example B-2 The Complete Output of the L3 Module show run Command (Continued)

```
interface Port-channel1.1
 encapsulation dot1Q 1 native
 ip address 1.1.1.1 255.255.255.0
 no ip redirects
 no ip directed-broadcast
 ipx encapsulation NOVELL-ETHER
 ipx network 1
!
interface Port-channel1.10
 encapsulation dot1Q 10
 no ip redirects
 no ip directed-broadcast
 ipx encapsulation NOVELL-ETHER
 ipx network 10
 bridge-group 1
!
interface Port-channel1.23
 encapsulation dot1Q 23
 ip address 1.1.23.1 255.255.255.0
 no ip redirects
 no ip directed-broadcast
 ipx encapsulation NOVELL-ETHER
 ipx network 23
!
interface Port-channel1.40
 encapsulation dot1Q 40
 no ip redirects
 no ip directed-broadcast
 ipx encapsulation NOVELL-ETHER
 ipx network 40
 bridge-group 1
!
interface FastEthernet1
 no ip address
 no ip directed-broadcast
 shutdown
!
```

Example B-2 The Complete Output of the L3 Module show run Command (Continued)

```
interface GigabitEthernet1
 no ip address
 no ip directed-broadcast
 shutdown
!
interface GigabitEthernet2
 no ip address
 no ip directed-broadcast
 shutdown
!
interface GigabitEthernet3
 no ip address
 no ip directed-broadcast
 no negotiation auto
 channel-group 1
!
interface GigabitEthernet4
 no ip address
 no ip directed-broadcast
 no negotiation auto
 channel-group 1
!
interface BVI1
 ip address 1.1.14.1 255.255.255.0
 no ip directed-broadcast
!
router rip
 passive-interface FastEthernet1
 network 1.0.0.0
!
ip classless
!
!
!
!
bridge 1 protocol ieee
```

continues

Example B-2 The Complete Output of the L3 Module show run Command (Continued)

```
 bridge 1 route ip
 bridge 1 route ipx
bridge 1 priority 16000
alias exec sr show run
alias exec ct config t
!
line con 0
 transport input none
line aux 0
line vty 0 4
 privilege level 15
 no login
!
end

L3#
```

Example B-3 The Complete Output of r2's show run Command

```
r2#show run
Building configuration...

Current configuration:
!
version 12.0
service timestamps debug uptime
service timestamps log uptime
no service password-encryption
!
hostname r2
!
!
!
!
!
!
```

Example B-3 The Complete Output of r2's show run Command (Continued)

```
ip subnet-zero
no ip domain-lookup
!
ipx routing 2222.2222.2222
!
cns event-service server
!
!
!
!
process-max-time 200
!
interface Ethernet0
 ip address 1.1.23.22 255.255.255.0
 no ip directed-broadcast
 ipx network 23
!
interface Serial0
 no ip address
 no ip directed-broadcast
 no ip mroute-cache
 shutdown
 no fair-queue
!
interface Serial1
 no ip address
 no ip directed-broadcast
 shutdown
!
interface TokenRing0
 ip address 10.1.0.1 255.255.255.0
 no ip directed-broadcast
 ipx network 2
 ring-speed 16
```

continues

Example B-3 The Complete Output of r2's show run Command (Continued)

```
!
!
!
router rip
 network 1.0.0.0
 network 10.0.0.0
!
ip classless
no ip http server
!
!
!
!
!
!
!
!
line con 0
 transport input none
line aux 0
 exec-timeout 0 0
line vty 0 4
 login
!
end

r2#
```

IRB Between Three Ethernet Interfaces on a Router

Next, we look at an example utilizing IRB on a lone Cisco router (no switches are involved). Figure B-2 shows a Cisco router with Ethernet interfaces E1, E2, and E3. In this IRB example, **bridge-group 1** and **BVI 1** do the following:

- Route IP traffic
- Bridge IPX traffic
- Bridge and route AppleTalk traffic

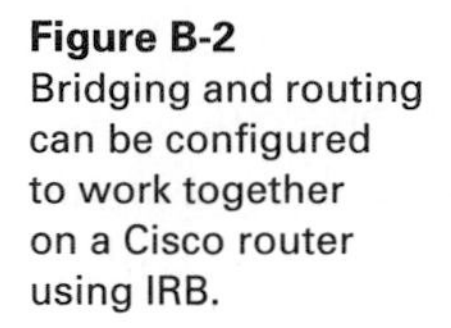

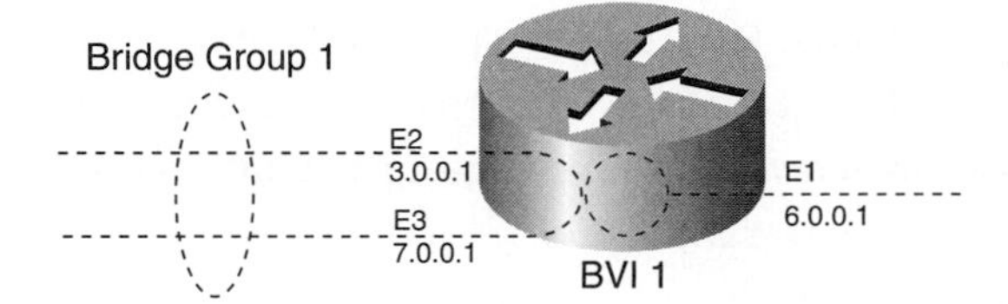

Figure B-2
Bridging and routing can be configured to work together on a Cisco router using IRB.

No VLANs are involved here because no switch is in sight. Similarly, no trunks or EtherChannels are used in this example. You simply have a router that you want to use to route and bridge traffic as specified. Example B-4 shows the relevant portions of the router configuration.

Example B-4 The Commands Needed to Configure IRB with IP, IPX, and AppleTalk

```
appletalk routing
!
interface Ethernet 1
 ip address 5.0.0.1 255.0.0.0
 appletalk cable-range 35-35 35.1
 appletalk zone Engineering
!
interface Ethernet 2
 ip address 3.0.0.1 255.0.0.0
 bridge-group 1
!
interface Ethernet 3
 ip address 7.0.0.1 255.0.0.0
 bridge-group 1
!
interface BVI 1
 no ip address
 appletalk cable-range 33-33 33.1
 appletalk zone Accounting
!
bridge irb
bridge 1 protocol ieee
 bridge 1 route appletalk
 bridge 1 route ip
 no bridge 1 bridge ip
```

The last command disables the possibility of bridging IP in bridge group 1. **bridge 1 route appletalk** enables routing AppleTalk between interface BVI1 and interface E1 (AppleTalk is bridged between interfaces E2 and E3). All IPX traffic is bridged. (Remember: When you use bridge groups on interfaces, all protocols configured on an interface with a Layer 3 address are routed, and all other protocols are bridged within that bridge group.)

8500 SWITCH ROUTERS

Using IRB as demonstrated in this example is most common on 8500 series switch routers. The 8500 series switch routers switch at Layer 3 essentially at the same rate as they switch at Layer 2. However, most Cisco routers require software bridging, so this configuration is not recommended in general. The 8500 series routers permit a design model wherein VLANs are terminated by the router, with Layer 3 switching between the terminated VLANs. The IRB option, however, allows you to configure campus-wide VLANs on 8500s without the latency of other Cisco routers. IRB is the only way to configure a router to allow VLANs to span the router, preserving VLAN tags as they transit the router.

Appendix C

Miscellaneous Campus LAN Switch Technologies

A number of new and emerging technologies are available for campus networks. Some of these technologies originate in the IEEE working groups and are open-standard alternatives to previously implemented vendor-specific solutions. Others are Cisco-specific technologies that provide additional automation, redundancy, scalability, or acceleration. There are also technologies that have been around for a while but are relegated to this appendix because they are less-common configuration options.

We look at a sampling of these technologies:

- Unidirectional link detection (UDLD)
- Switch Virtual Interface (SVI)
- Class of service (CoS), type of service (ToS), Differentiated Services Code Point (DSCP)
- Access Control Entity (ACE) and Access Control Parameter (ACP)
- Remote Switched Port Analyzer (RSPAN)
- Server Load Balancing (SLB)
- IEEE 802.1x

This appendix presents a brief introduction to each of these campus network technologies.

Unidirectional Link Detection

A *unidirectional* link is one in which only the transmit or receive portion of the link is functional. The unidirectional link detection (UDLD) protocol allows devices connected through fiber-optic or twisted-pair copper cabling to monitor the cables' physical configuration and detect when a unidirectional link exists. When a unidirectional link is detected, UDLD shuts down the affected port and alerts the user. Unidirectional links can cause a variety of problems, including spanning-tree topology loops.

UDLD is a Layer 2 protocol that works with Layer 1 mechanisms, such as autonegotiation, to determine a link's physical status. At Layer 1, autonegotiation handles physical signaling and fault detection. UDLD also performs tasks that autonegotiation cannot, such as detecting the identities of neighbors and shutting down misconnected ports. When both autonegotiation and UDLD are enabled, Layer 1 and Layer 2 detection features can work together to prevent physical and logical unidirectional connections and malfunctioning of other protocols.

A unidirectional link occurs whenever traffic transmitted by the local device over a link is received by the neighbor, but traffic transmitted from the neighbor is not received by the local device. For example, if one of the fiber strands in a pair is disconnected, as long as autonegotiation is active, the link does not stay up. In this situation, the logical link is undetermined, and UDLD does not take any action. If both fibers are working normally at Layer 1, UDLD at Layer 2 determines whether those fibers are connected correctly and whether traffic is flowing bidirectionally between the correct neighbors. This check cannot be performed by autonegotiation because autonegotiation is a Layer 1 feature.

The switch periodically transmits UDLD messages (packets) to neighbor devices on ports that have UDLD enabled. If the messages are echoed to the sender within a specific time and they lack a specific acknowledgment, the link is flagged as unidirectional, and the port is shut down. Devices on both ends of the link must support UDLD in order for the protocol to successfully identify and disable unidirectional links.

UDLD has been available on CatOS switches since CatOS 5.1. It is also available on 2900 XL, 3500 XL, 2950, and 3550 Catalyst switches.

CatOS Releases 5.4(3) and later support Aggressive UDLD (AUDLD) mode, which is disabled by default. With aggressive mode enabled, when a port on a bidirectional link stops receiving UDLD packets, UDLD tries to reestablish the connection with the neighbor. After eight failed retries, UDLD puts the port into *errdisable* state. A port is in errdisable state if it is enabled in NVRAM but is disabled at runtime by any process. Because the NVRAM configuration for the port is enabled (you have not disabled the port), the port status is shown as errdisable.

Enabling UDLD aggressive mode provides additional benefits in the following cases:

- One side of a link has a port stuck (both tx and rx).
- One side of a link remains up while the other side of the link has gone down.

In these cases, AUDLD mode disables one of the ports on the link and stops the loss of traffic.

Next, we describe how to configure UDLD on Catalyst switches. First of all, UDLD is disabled by default on Catalyst switches. On a CatOS switch, you must enable UDLD globally before any port can use UDLD. To globally enable UDLD on a CatOS switch, use the command **set udld enable.** To enable UDLD on a port, use the command **set udld enable** *mod_num/port_num*. To enable AUDLD on a port, use the command **set udld aggressive-mode enable** *mod_num/port_num*. You can verify the configuration with the command **show udld port** [*mod_num*[*/port_num*]]. Example C-1 demonstrates configuring UDLD on a Catalyst 4000 switch.

Example C-1 Configuring UDLD on a CatOS Switch

```
Console> (enable) set udld enable

UDLD enabled globally
Console> (enable) show udld

UDLD      : enabled
Cons*ole> (enable)
Console> (enable) set udld enable 4/1

UDLD enabled on port 4/1
Console> (enable) show udld port 4/1

UDLD      : enabled
Message Interval: 15 seconds
Port      Admin Status  Aggressive Mode Link State
--------  ------------  --------------- ---------
 4/1      enabled       disabled        bidirectional
Console> (enable)
Console> (enable) set udld aggressive-mode enable 4/1

Aggressive UDLD enabled on port 4/1.
Console> (enable)
Console> (enable) show udld port 4/1

UDLD      : enabled
Message Interval: 10 seconds
Port      Admin Status  Aggressive Mode Link State
--------  ------------  --------------- ---------
 4/1      enabled       enabled         bidirectional
Console> (enable)
```

On an IOS-based switch, you can configure UDLD on the entire switch (use the global configuration command **udld enable**) or on an individual port (use the interface command **udld enable**). Use the **udld reset** privileged-mode command to reset all ports that have been shut down by UDLD.

Switch Virtual Interfaces

A *switch virtual interface (SVI)* is a virtual interface in a Catalyst switch that represents a VLAN of switch ports as one interface to the routing or bridging function in the system. Figure C-1 illustrates two SVIs and the basic commands that apply to each type of interface logically associated with the SVIs: routed (Layer 3), SVI, and switchport (Layer 2). Only one SVI can be associated with a VLAN, but you need to configure an SVI for a VLAN only when you want to route between VLANs or to bridge nonroutable protocols between VLANs. By default, an SVI is created for the default VLAN (VLAN 1) to permit remote switch administration. Additional SVIs must be explicitly configured.

Figure C-1
SVIs are integral to the logical forwarding model of a Catalyst 4006 Supervisor Engine III.

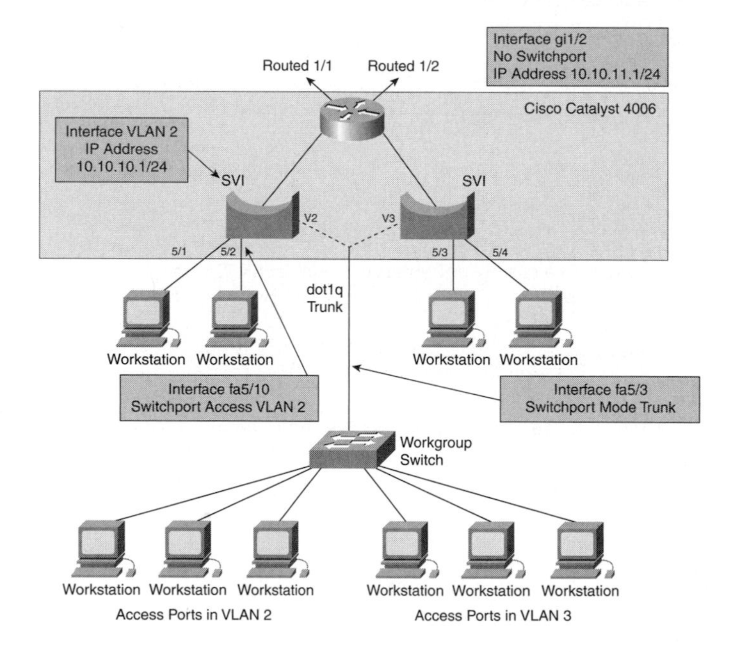

SVIs are supported on the Catalyst 4006 Supervisor Engine III, Catalyst 6000 switches running Native IOS, Catalyst 6000 switches with an MSFC, Catalyst 3550 switches running the enhanced multilayer software image, and Cisco 2600/3600/3700 routers with 16- or 36-port Ethernet switch network modules running Cisco IOS Software

Release 12.2(8)T or later. On all of these Cisco devices, SVIs are created the first time you enter the **interface vlan** *vlan* global configuration command. The VLAN corresponds to the VLAN tag associated with data frames on an ISL or 802.1Q encapsulated trunk or the VLAN ID configured for an access port. You configure a VLAN interface for each VLAN for which you want to route traffic.

Quality of Service

Quality of service (QoS) refers to a network's ability to provide improved service to selected network traffic over various underlying technologies, including Frame Relay, ATM, Ethernet, SONET, and IP-routed networks. In particular, QoS features offer improved and more predictable network service by providing the following services:

- Supporting dedicated bandwidth
- Improving loss characteristics
- Avoiding and managing network congestion
- Shaping network traffic
- Setting traffic priorities across the network

QoS relies on prioritization values as a means of implementing measurable effects on network traffic. QoS prioritization values are carried in Layer 3 packets and Layer 2 frames:

- Layer 2 class of service (CoS) values range between 0 for low priority and 7 for high priority:
  - Layer 2 Inter-Switch Link (ISL) frame headers carry an IEEE 802.1p CoS value. Layer 2 802.1Q frame headers also carry a CoS value, as shown in Figure C-2.
  - On LAN ports configured as Layer 2 ISL trunks, all traffic consists of ISL frames. On LAN ports configured as Layer 2 802.1Q trunks, all traffic consists of 802.1Q frames except for traffic in the native VLAN.
- Layer 3 IP precedence values—The IP version 4 specification defines the 3 most-significant bits of the 1-byte type of service (ToS) field as IP precedence. IP precedence values range between 0 for low priority and 7 for high priority.
- Layer 3 Differentiated Services Code Point (DSCP) values—The IETF defines the 6 most-significant bits of the 1-byte IP ToS field as the DSCP. The per-hop behavior represented by a particular DSCP value can be configured. DSCP values range between 0 and 63. Layer 3 IP packets can carry either an IP precedence value or a DSCP value. DSCP values are backward-compatible with IP precedence values.

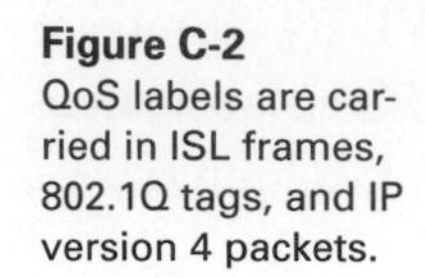

Figure C-2 QoS labels are carried in ISL frames, 802.1Q tags, and IP version 4 packets.

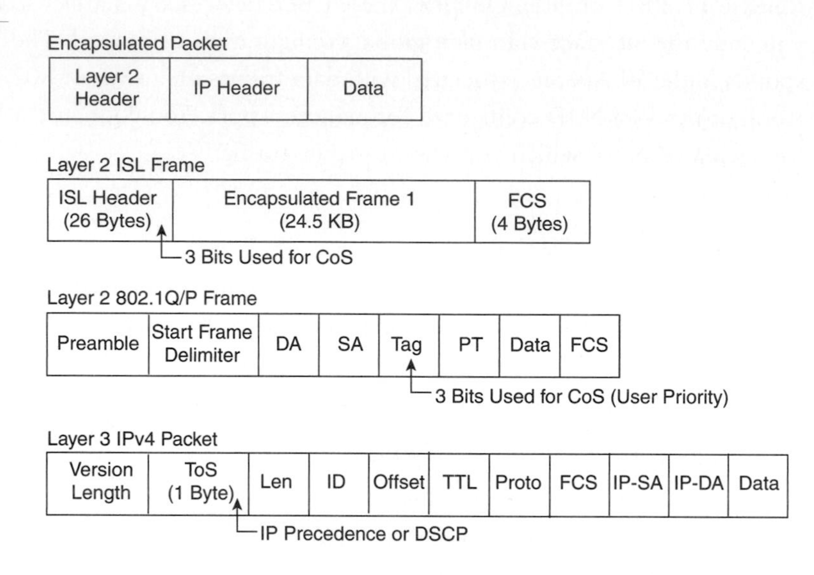

Breaking down the most-significant 6 bits in the ToS field and comparing them to the most-significant 3 bits in the same field, you see that the DSCP values map to IP precedence values as follows: 0 to 7 maps to 0, 8 to 15 maps to 1, 16 to 23 maps to 2, 24 to 31 maps to 3, 32 to 39 maps to 4, 40 to 47 maps to 5, 48 to 55 maps to 6, and 56 to 63 maps to 7.

Standard QoS terminology includes labels, classification, marking, scheduling, congestion avoidance, and policing. Virtually all QoS implementations can be characterized by a combination of these terms:

- *Labels* are prioritization values carried in Layer 3 packets and Layer 2 frames, as just listed.
- *Classification* is the selection of traffic to be marked.
- *Marking* is the process of setting a Layer 3 DSCP value in a packet; sometimes, this is extended to include setting Layer 2 CoS values.
- *Scheduling* is the assignment of Layer 2 frames to a queue. QoS involves assigning frames to a queue based on Layer 2 CoS values.
- *Congestion avoidance* is the process by which QoS reserves ingress and egress port capacity for Layer 2 frames with high-priority Layer 2 CoS values. QoS implements congestion avoidance with Layer 2 CoS value-based drop thresholds. A *drop threshold* is the percentage of queue buffer utilization above which frames with a specified Layer 2 CoS value are dropped, leaving the buffer available for frames with higher-priority Layer 2 CoS values.

- *Policing* is limiting bandwidth used by a flow of traffic. Policing can mark or drop traffic. A *policer* is a traffic-policing mechanism.

Describing QoS theory and application in detail is beyond the scope of this book. Common Catalyst QoS configurations entail classifying traffic using ACLs or class maps; policing and marking traffic using policy maps; specifying CoS-to-DSCP, IP precedence-to-DSCP, or DSCP-to-CoS mappings; and configuring Weighted Random Early Detection (WRED) for congestion avoidance.

Class maps and policy maps are components of Cisco's Modular quality of service command-line interface (Modular QoS CLI). The Modular QoS CLI provides a rich hierarchical model for configuring QoS: linking the various constructs, such as class maps, map classes, policy maps, and service policies. See www.cisco.com/univercd/cc/td/doc/product/software/ios122/122cgcr/fqos_c/fqcprt8/index.htm for more information. In the next section, we restrict our attention to QoS on Catalyst 2950 switches.

QoS on Catalyst 2950 Switches

The Catalyst 2950 offers a much more comprehensive approach to QoS than that of the existing Catalyst 2900 XL and 3500 XL switches. Catalyst 2950 switches support the rapid evolution of integrated voice and video traffic and the proliferation of bandwidth-intensive applications. Market requirements have increased for more granularity of QoS capabilities in Layer 2 and Layer 3 devices, providing an end-to-end QoS implementation. Because the Catalyst 2950 (enhanced image models) and the Catalyst 3550 switches are based on Native Cisco IOS Software, the QoS implementation is similar to that of the Catalyst 6000 family of switches. Weighted-priority scheduling avoids the starvation of packets in the lower-priority queues by providing bandwidth allocation to all queues of transmission. The packet-transmission process for each egress port starts from the highest CoS queue, sends the maximum preconfigured number of packets, and moves on to the next-lower CoS queue. By using this scheme, low-priority queues have the opportunity to transmit packets even though the high-priority queues are not empty. The Catalyst 2950 also offers twice the number of queues—four CoS queues per interface versus two queues on the XL switches. This allows for better classification of traffic and the ability to configure more QoS granularity.

The next section explores the specific methodology employed for QoS on Catalyst 2950 switches.

ACLs, ACPs, and ACE

The Catalyst 2950 series switches incorporate the configuration of ACPs, or QoS masks, to process QoS data. Before configuring ACLs on the Catalyst 2950 switches, users must have a thorough understanding of ACPs, which serve as QoS masks in the switch CLI commands and output. Packets can be classified based on Layer 2 through Layer 4 fields on the Catalyst 2950 switches. A mask can be a combination of either multiple Layer 3 and Layer 4 fields or multiple Layer 2 fields. Layer 2 fields cannot be combined with Layer 3 or Layer 4 fields.

Each ACE has a mask and a rule. The classification field or mask is the field of interest on which users want to perform an action. The specific values associated with a given mask are called rules. The Catalyst 2950 switches allow for the configuration of ACLs, which include ACPs and ACEs, to administer QoS actions configured on the switch. Within an ACL, both ACP and ACE parameters exist. ACPs are configured to filter data. These parameters are source and destination MAC addresses, IP addresses, or TCP/UDP ports. ACEs dictate the specific parameters within an ACP, such as a policy per interface.

Network engineers can configure ACPs globally, setting parameters for traffic that match the configured ACP (such as an IP subnet). As soon as the ACP has been configured, the engineer can create and set parameters for class-map and policy-map ACEs, which include configuring policies and specifying actions that the switch will take (such as setting a DSCP value or dropping traffic that exceeds a predefined bandwidth).

From this point on, any traffic traveling through the interface to which the ACE was applied is compared with the ACE to determine if it is conforming or nonconforming. As soon as the packet is classified and policed/marked, it is queued according to the CoS or DSCP value set and the CoS-DSCP or DSCP-CoS mapping configured by the engineer. After it has been placed in the appropriate holding queue for transmission, on egress it is scheduled based on the scheduling algorithm configured. Examples C-2 and C-3 illustrate these concepts.

Example C-2 ACEs and ACPs on 2950 Switches

```
ACL 1
ACE 1 permit ip 171.20.0.0 0.0.255.255 (ACP 1)
ACE 2 permit ip 171.20.10.0 0.0.0.255 (ACP 2)
```

Example C-3 ACE and ACP with Layer 4 Parameter

```
permit tcp any any eq 80 (TCP and 80 are examples of an ACP)
```

Based on a configured ACE, which is applied to an interface or set of interfaces, the Catalyst 2950 switch can now apply class maps and policy maps to these ACEs for QoS granularity. Thus, the Catalyst 2950 series now supports Layer 2 through Layer 4 processing.

There should be no **deny** statements in QoS ACEs. This is because a **permit** in a QoS ACE indicates that a specific QoS action will be taken for that packet, and a **deny** statement indicates that control should be transferred to the next ACE.

The ACLs in Catalyst 2950 switches are essentially restricted versions of the ACLs in the Catalyst 3550 series. The key differences are as follows:

- Number of entries
- Syntax restrictions
- Class-map restrictions
- Policy-map restrictions
- Service-policy restrictions

Hence, an ACL from the Catalyst 2950 imported to the Catalyst 3550 will work, but an ACL from a Catalyst 3550 imported to the Catalyst 2950 will not work.

Remote Switched Port Analyzer

Remote Switched Port Analyzer (RSPAN) has all the features of SPAN plus support for source ports and destination ports distributed across multiple switches, allowing remote monitoring of multiple switches across your network. RSPAN is supported on Catalyst 4000 and 6000 family switches.

The traffic for each RSPAN session is carried over a user-specified RSPAN VLAN that is dedicated for that RSPAN session in all participating switches. A reflector port is the mechanism you use to copy packets onto an RSPAN VLAN. The reflector port forwards only the traffic from the RSPAN source session with which it is affiliated. Any device connected to a port set as a reflector port loses connectivity until the RSPAN source session is disabled.

The SPAN traffic from the sources is copied onto the RSPAN VLAN through the reflector port and is then forwarded over trunk ports carrying the RSPAN VLAN to RSPAN destination ports monitoring the RSPAN VLAN. Traffic sent out through the source port is also sent out on the reflector port. Because the reflector port is an access port in loopback mode, the traffic is switched out with no VLAN tag and is immediately sent back to the switch. In the loopback, the traffic is encoded into the RSPAN VLAN. A switch with an RSPAN destination session receives the traffic, as shown in Figure C-3.

The traffic type for sources (ingress, egress, or both) in an RSPAN session can be different for source switches, but it must be the same for all source ports on a given switch.

To configure RSPAN, you need to create an RSPAN VLAN that does not exist on any of the switches. Use the command **set vlan** *vlan_num* **rspan**. With VTP enabled in the network, you can create the RSPAN VLAN in one switch, and VTP propagates it to the other switches in the VTP domain. You can use VTP pruning to get efficient flow of RSPAN traffic, or manually delete the RSPAN VLAN from all trunks that do not need to carry the RSPAN traffic.

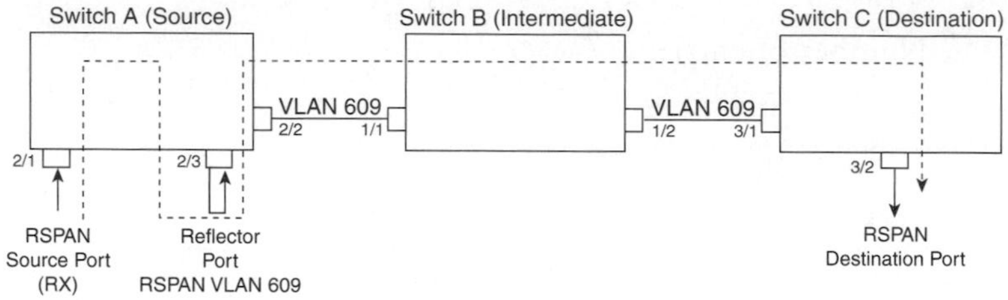

Figure C-3 RSPAN operation relies on reflector ports.

To configure RSPAN source ports or source VLANs, use the command **set rspan source** {*mod/ports...* | *vlans...*} {*rspan_vlan*} **reflector** *mod/port* [**rx** | **tx** | **both**].

To configure RSPAN destination ports, use the command **set rspan destination** {*mod_num/port_num*} {*rspan_vlan*}. Verify the RSPAN configuration with the commands **show vlan** and **show rspan.** Example C-4 illustrates RSPAN configuration and verification.

Example C-4 RSPAN Configuration and Verification

```
Console> (enable) set vlan 500 rspan


vlan 500 configuration successful
Console> (enable) show vlan
<output omitted>
VLAN DynCreated  RSPAN
---- ---------- --------
1    static     disabled
2    static     disabled
3    static     disabled
99   static     disabled
500  static     enabled
Console> (enable) set rspan source 2/3 500 reflector 2/34 rx
Rspan Type       : Source
Destination      : -
Reflector        : Port 2/34
Rspan Vlan       : 500
```

Example C-4 RSPAN Configuration and Verification (Continued)

```
Admin Source     : Port 2/3
Oper Source      : Port 2/3
Direction        : receive
Incoming Packets: -
Learning         : -
Filter           : -
Status           : active

Console> (enable) 2001 May 02 13:22:17 %SYS-5-SPAN_CFGSTATECHG:remote span source
  session
active for remote span vlan 500
Console> (enable) set rspan source 200 500

Rspan Type       : Source
Destination      : -
Rspan Vlan       : 500
Admin Source     : VLAN 200
Oper Source      : None
Direction        : transmit/receive
Incoming Packets: -

Learning         : -
Multicast        : enabled
Filter           : -
Console> (enable) set rspan destination 3/1 500
Rspan Type       : Destination
Destination      : Port 3/1
Rspan Vlan       : 500
Admin Source     : -
Oper Source      : -
Direction        : -
Incoming Packets: disabled
Learning         : enabled
Filter           : -
Status           : active
```

To disable RSPAN source sessions on a switch, use the command **set rspan disable source** [*rspan_vlan* | **all**]. To disable RSPAN destination sessions on a switch, use the command **set rspan disable destination** [*mod_num/port_num* | **all**]. To review, RSPAN configuration requires that a dedicated RSPAN VLAN be created on participating switches, RSPAN source ports and VLANs are specified together with reflector ports, and destination RSPAN ports are specified.

Server Load Balancing

Cisco IOS Software Server Load Balancing (SLB) capabilities, using server cache technology, give network engineers the ability to balance services across multiple servers at speeds of millions of packets per second. IOS SLB is supported on Catalyst 6000 family switches with Supervisor Engine 1 or 2, Catalyst 4840G switches, Cisco 7100 series routers, and Cisco 7200 series routers. SLB functionality was introduced with Cisco IOS Software Release 12.1.

IOS SLB intelligently load-balances TCP/IP traffic across multiple servers, as shown in Figure C-4. It appears as one "virtual" server to the requesting clients. All traffic is directed toward a virtual IP address (virtual server) via DNS. Those requests are then distributed over a series of real IP addresses on servers (real servers). A virtual IP address is an address that is in DNS and most likely has a domain name. A real IP address is physically located on a real server behind IOS SLB.

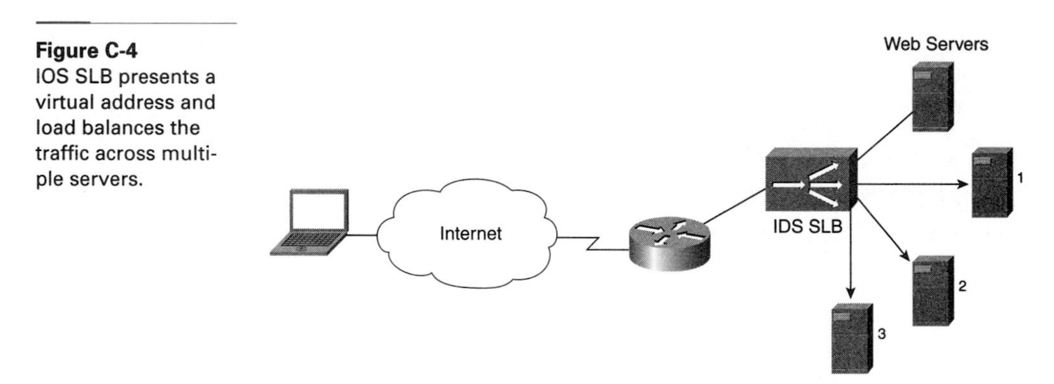

Figure C-4 IOS SLB presents a virtual address and load balances the traffic across multiple servers.

IOS SLB provides the following benefits:

- High performance is achieved by distributing client requests across a cluster of servers.
- Administration of server applications is easier. Clients know only about virtual servers; no administration is required for real server changes.

- Security of the real server is provided, because its address is never announced to the external network. Users are familiar only with the virtual IP address. Additionally, filtering of unwanted traffic can be based on both IP address and IP port numbers.
- Ease of maintenance with no downtime is achieved by allowing physical (real) servers to be transparently placed in or out of service.

IOS SLB allows users to represent a group of network servers (a server farm) as a single server instance, balance the traffic to the servers, and limit traffic to individual servers. The single server instance that represents a server farm is called a virtual server. The servers that comprise the server farm are called real servers.

In this environment, clients are configured to connect to the virtual server's IP address. When a client initiates a connection to the virtual server, the SLB function chooses a real server for the connection based on a configured load-balancing algorithm.

Two algorithm choices can be used for load balancing or determining which server will receive a new connection request:

- **(Weighted) least connections**—The number of sessions assigned to a server is based on the number of current TCP connections.
- **(Weighted) round robin**—The round-robin predictor option directs the network connection to the next server and treats all servers as equals if they are unweighted, regardless of the number of connections or the response time.

The weighted least connections algorithm specifies that the next real server chosen from a server farm for a new connection to the virtual server is the server with the fewest active connections. Each real server is also assigned a weight for this algorithm. When weights are assigned, the server with the fewest connections is based on the number of active connections on each server and on each server's relative capacity. A given real server's capacity is calculated as that server's assigned weight divided by the sum of the assigned weights of all the real servers associated with that virtual server, or $n_1 / (n_1 + n_2 + n_3 \ldots)$.

The weighted round robin algorithm specifies that the real server used for a new connection to the virtual server is chosen from the server farm in a circular fashion. Each real server is assigned a weight, n, that represents its capacity to handle connections, as compared to the other real servers associated with the virtual server. That is, new connections are assigned to a given real server n times before the next real server in the server farm is chosen.

Representing server farms as virtual servers facilitates scalability and availability for the user. The addition of new servers and the removal/failure of existing servers can occur at any time without affecting the virtual server's availability.

IOS SLB can operate in one of two redirection modes: directed or dispatched. In directed mode, the virtual server can be assigned an IP address that is not known to any of the real servers. IOS SLB translates packets exchanged between a client and a real server, translating the virtual server IP address to a real server address via network address translation (NAT). In dispatched mode, the virtual server address is known to the real servers, and IOS SLB redirects packets to the real servers at the media access control (MAC) layer.

IOS SLB relies on a site's firewalls to protect the site from attacks. In general, IOS SLB is no more susceptible to direct attack than is any switch or router.

IOS SLB supports the following protocols:

- Domain Name System (DNS)
- File Transfer Protocol (FTP)
- Hypertext Transfer Protocol (HTTP)
- Hypertext Transfer Protocol over Secure Socket Layer (HTTPS)
- Internet Message Access Protocol (IMAP)
- Mapping of Airline Traffic over IP, Type A (MATIP-A)
- Network News Transfer Protocol (NNTP)
- Post Office Protocol, version 2 (POP2)
- Post Office Protocol, version 3 (POP3)
- RealAudio/RealVideo via HTTP
- Remote Authentication Dial-In User Service (RADIUS)
- Simple Mail Transfer Protocol (SMTP)
- Telnet
- X.25 over TCP (XOT)

SLB is important not only for scaling of web services, but also for servers providing traditional IP services, such as DNS resolution and FTP services. IOS SLB capabilities provide network engineers with an integrated solution to balance services across numerous servers at speeds of millions of packets per second.

IEEE 802.1x

IEEE 802.1x is a client/server-based access control and authentication protocol that restricts unauthorized devices from connecting to a LAN through publicly accessible ports. 802.1x authenticates each user device connected to a switch port before making available any services offered by the switch or the LAN. Until the device is authenticated, 802.1x access control allows only Extensible Authentication Protocol over LAN (EAPOL) traffic through the port to which the device is connected. After authentication is successful, normal traffic can pass through the port.

802.1x controls network access by creating two distinct virtual access points at each port. One access point is an uncontrolled port; the other is a controlled port. All traffic through the single port is available to both access points. Only EAPOL traffic is allowed to pass through the uncontrolled port, which is always open. The controlled port is open only when the device connected to the port has been authenticated by 802.1x. After this authentication takes place, the controlled port opens, allowing normal traffic to pass.

IEEE 802.1x is supported on CatOS switches, Catalyst 2950 switches, and Catalyst 3550 switches. 802.1x is an open-standard alternative to AAA/TACACS+, AAA/RADIUS, or AAA/Kerberos authentication.

Summary

This appendix explored some new and some not-so-new LAN switch technologies. We looked at a sampling of these technologies: UDLD and AUDLD, SVI, CoS, ToS, DSCP, ACE and ACP, RSPAN, SLB, and IEEE 802.1x. These technologies provide enhanced link detection, improved virtual interfaces for Layer 2/3 interplay on newer Catalyst switches and operating systems, a plethora of QoS options, the ability to remotely monitor campus network traffic, increased availability via SLB, and IEEE standards-based authentication.

Glossary

Numerics

1000BaseCX The standard for Gigabit Ethernet transmission over a special balanced 150-ohm cable. Specified in IEEE 802.3z. The distance limitation is 25 m. The connector types are DB-9 or HSSDC.

1000BaseLX The long-wavelength (1300 nm) laser standard for Gigabit Ethernet over fiber-optic cable. Specified in IEEE 802.3z. Supports multimode and single-mode fiber.

1000BaseSX The short-wavelength (850 nm) laser standard for Gigabit Ethernet over fiber-optic cable. Specified in IEEE 802.3z. Supports only multimode fiber.

1000BaseT Provides 1 Gbps bandwidth via four pairs of Category 5 UTP cable (250 Mbps per wire pair).

20/80 rule Only 20 percent of traffic is local to the VLAN, and 80 percent of traffic is destined for other VLANs, a server farm on the network, or locations outside the campus network.

8B/10B encoding The encoding scheme specified in IEEE 802.3z for Gigabit Ethernet. 8B/10B combines two other codes—a 5B/6B code and a 3B/4B code. The mapping represents 8-bit raw data blocks in terms of 10-bit code blocks.

A

access layer The point at which local end users are allowed to attach to the network. Frequently, Layer 2 switches play a significant role at the access layer.

access port A switch port that connects to an end user device or a server.

adjacency table One of three tables used by CEF. The adjacency table is a database of node adjacencies (two nodes are said to be adjacent if they can reach each other via a single Layer 2 hop) and their associated Layer 2 MAC rewrite or next-hop information. Each leaf of the FIB tree offers a pointer to the appropriate next-hop entry in the adjacency table.

aging time Determines the time before an MLS entry is aged out. The default is 256 seconds. You can configure the aging time in the range of 8 to 2032 seconds in 8-second increments.

Anycast RP An application of MSDP that serves as a method to design a PIM-SM network to meet fault-tolerance requirements within a routing domain. In Anycast RP, two or more RPs are configured with the same IP address on loopback interfaces.

ATM LANE (ATM LAN Emulation LANE) A standard defined by the ATM Forum that gives two stations attached via ATM the same capabilities they normally have with Ethernet and Token Ring.

Auto-RP A PIM feature that automates the distribution of group-to-RP mappings.

B

BackboneFast A Catalyst feature that is initiated when a root port or blocked port on a switch receives inferior BPDUs from its designated bridge. Max Age is skipped in order to allow appropriate blocked ports to transition quickly to the Forwarding state in the case of an indirect link failure.

BID (Bridge ID) An 8-byte field consisting of an ordered pair of numbers. The first is a 2-byte decimal number called the bridge priority, and the second is a 6-byte (hexadecimal) MAC address.

Bidir-PIM (Bidirectional PIM) A variant of the PIM suite of multicast routing protocols for IP multicast and an extension of PIM-SM. Bidir-PIM resolves some limitations of PIM-SM for groups that have a large number of sources. Bidir-PIM eliminates all (S,G) states in the network.

Blocking state The STP state in which a bridge listens for BPDUs. This state follows the Disabled state. Forwarding does not occur in the Blocking state.

bootflash: A Flash memory device that stores OS images. It can also store a boot helper image or system configuration information.

BPDU (Bridge Protocol Data Unit) Frame passed between switches to communicate STP information used in the STP decision-making process.

Bridge ID Priority The Bridge Priority and the system ID extension combined.

Bridge Priority The decimal number used to measure the preference of a bridge in the Spanning-Tree Algorithm. The possible values range between 0 and 65,535.

BSR (bootstrap router) A fault-tolerant, automated RP discovery and distribution mechanism. With BSR, routers dynamically learn the group-to-RP mappings. BSR was introduced with PIMv2, specified in RFC 2117.

C

CAM (Content-Addressable Memory) Memory that is accessed based on its contents, not on its memory address. Sometimes called associative memory.

campus network A network that connects devices within and between a collection of buildings. LAN switching and high-speed routing provide connectivity among the buildings. It is common to focus on the role of switching when referring to a campus network.

candidate packet When a source initiates a data transfer to a destination, it sends the first packet to the MLS-RP through the MLS-SE. The MLS-SE recognizes the packet as a candidate packet for Layer 3 switching, because the MLS-SE has learned the MLS-RP's destination MAC addresses and VLANs through MLSP. The MLS-SE learns the candidate packet's Layer 3 flow information (such as the destination address, source address, and protocol port numbers) and forwards the candidate packet to the MLS-RP. A partial MLS entry for this Layer 3 flow is created in the MLS cache.

candidate RP A router that announces its candidacy to be an RP for a PIM-SM network via the Cisco-Announce group, 224.0.1.39.

CatOS The OS used by the Supervisor Engine on chassis-based Catalyst switches (with two exceptions), such as the 4000, 5000, and 6000 series. Two exceptions to this are the Native IOS loaded on a Catalyst 6000 (optional) and the Native IOS loaded on the Catalyst 4000 with Supervisor Engine III. CatOS is distinguished by its reliance on set commands.

CBT (Core-Based Tree) A multicast routing protocol introduced in September 1997 in RFC 2201, CBT builds a shared tree like PIM-SM. CBT uses information in the IP unicast routing table to calculate the next-hop router in the direction of the core.

CEF (Cisco Express Forwarding) A Cisco multilayer-switching technology that allows for increased scalability and performance to meet the requirements for large enterprise networks. CEF has evolved to accommodate the traffic patterns realized by modern networks, characterized by an increasing number of short-duration flows.

central rewrite engine The Catalyst 5000 NFFC contains central rewrite engines (one per bus) to handle PDU header rewrites when inline rewrite is not an option. If central rewrite engines are used, a frame must traverse the bus twice—first with a VLAN tag for the source (on its way to the NFFC), and second with a VLAN tag for the destination (on its way to the egress port).

CGMP (Cisco Group Management Protocol) A Cisco-developed protocol that runs between Cisco routers and Catalyst switches to leverage IGMP information on Cisco routers to make Layer 2 forwarding decisions on Catalyst switches. With CGMP, IP multicast traffic is delivered only to Catalyst switch ports that are attached to interested receivers.

CGMP fast-leave Also known as CGMP fast-leave processing, CGMP immediate-leave processing, CGMP immediate leave, and CGMP leave. A CGMP feature requiring IGMPv2 on the hosts that allows the switch to quickly remove a port from its list of ports receiving multicast traffic for a particular multicast group.

clustering A method of managing a group of switches without having to assign an IP address to every switch.

control plane The portion of hardware and software on a Cisco device that handles Layer 3 traffic forwarding.

core In CBT, the router that serves as the root of the shared tree. Similar to PIM-SM's RP.

core block A switch or set of switches that interconnect multiple switch blocks. A switch block consists of access layer and distribution layer devices in the hierarchical model. The main function of the core block is to pass data with minimum latency between switch blocks.

core layer A high-speed switching backbone in a campus network designed to switch packets with minimum latency.

cost For switches with active 1 Gbps links but no active 10 Gbps links, a link's STP cost is defined according to the following table:

Bandwidth	STP Cost
4 Mbps	250
10 Mbps	100
16 Mbps	62
45 Mbps	39
100 Mbps	19
155 Mbps	14
622 Mbps	6
1 Gbps	4
10 Gbps	2

CST (Common Spanning Tree) The spanning-tree implementation specified in the IEEE 802.1Q standard. CST defines a single instance of spanning tree for all VLANs with BPDUs transmitted over VLAN 1.

D

default Provides a definitive location for IP packets to be sent in case the source device has no IP routing functionality built into it. This is the method most end stations use to access nonlocal networks. The default gateway can be set manually or learned from a DHCP server.

default RP RP in a pure PIM-SM domain to which other PIM routers point via the **ip pim rp-address** command. It is normally used as a means of providing the group-to-RP mapping for the Auto-RP groups, but it can also be used as a fallback mechanism for PIM.

Designated Bridge The bridge containing the designated port for that segment.

Designated Port The bridge port connected to that segment that both sends traffic toward the root bridge and receives traffic from the root bridge over that segment.

detection control A logical access control used to log and report the activities of authorized users and devices and to log and report unauthorized access or attempted access to systems, programs, and data.

device: Hardware that supports a Flash image. This includes bootflash: and PCMCIA cards (Flash PC cards) for CatOS switches.

DF (designated forwarder) A PIM construct that establishes a loop-free SPT rooted at the RP. On each link, the router with the best path to the RP is elected as the DF. The DF is responsible for forwarding traffic upstream toward the RP.

Disabled state The administratively shut-down STP state.

distribution layer The demarcation point between the access and core layers. The distribution layer helps define and differentiate the core. The purpose of this layer is to provide a network boundary definition. This is where packet/frame manipulation takes place.

distribution tree A unique forwarding path for a multicast group between the source and each subnet containing members of the multicast group.

DTP (Dynamic Trunking Protocol) A Cisco-proprietary protocol that autonegotiates trunk formation for either ISL or 802.1Q trunks.

DVMRP (Distance Vector Multicast Routing Protocol) The first multicast routing protocol, publicized in RFC 1075 in 1998. Unlike PIM, DVMRP builds a multicast routing table separate from the unicast routing table.

Dynamic VLAN A VLAN in which end stations are automatically assigned to the appropriate VLAN based on their MAC address. This is made possible via a MAC address-to-VLAN mapping table contained in a VLAN Management Policy Server (VMPS) database.

E

ELAN (emulated LAN) A logical construct, implemented with switches, that provides Layer 2 communication between a set of hosts in a LANE network. See ATM LANE.

enable packet When an MLS-SE receives a packet from the MLS-RP that originated as a candidate packet, it recognizes that the source MAC address belongs to the MLS-RP, that the XTAG matches that of the candidate packet, and that the packet's flow information matches the flow for which the candidate entry was created. The MLS-SE considers this packet an enable packet and completes the MLS entry created by the candidate packet in the MLS cache.

end-to-end VLAN Also known as a campus-wide VLAN. An end-to-end VLAN spans a campus network. It is characterized by the mapping to a group of users carrying out a similar job function (independent of physical location).

EtherChannel A Cisco-proprietary technology that, by aggregating links into a single logical link, provides incremental trunk speeds ranging from 10 Mbps to 160 Gbps.

F

fast aging time The amount of time before the purging of MLS entries that have no more than *pkt_threshold* packets switched within *fastagingtime* seconds after they are created.

FIB (Forwarding Information Base) One of three tables used by CEF. The FIB table contains the minimum information necessary to forward packets; in particular, it does not contain any routing protocol information. This table consists of a four-level hierarchical tree, with 256 branch options per level (reflecting four octets in an IP address).

flow A unidirectional sequence of packets between a particular source and destination that share the same Layer 3 and Layer 4 PDU header information.

flow mask A set of criteria, based on a combination of source IP address, destination IP address, protocol, and protocol ports, that describes a flow's characteristics.

Forward Delay The time that the bridge spends in the Listening and Learning states.

Forwarding state The STP state in which data traffic is both sent and received on a port. It is the "last" STP state, following the Learning state.

G

GARP (Generic Attribute Registration Protocol) Part of the IEEE 802.1p specification, GARP implements a service that provides a generic Layer 2 transport mechanism to propagate information throughout a switching domain. The information is then propagated in the form of attributes, types, values, and semantics, which are defined by the application employing the service.

GBIC (Gigabit Interface Converter) GBICs were developed to allow network engineers to configure gigabit ports on a port-by-port basis for short-wavelength (SX), long-wavelength (LX), and long-haul (LH) interfaces. The converters fit into modular slots on a wide variety of Catalyst switches.

GDA (Group Destination Address) The notation used with CGMP to denote a Layer 2 multicast group address.

globally scoped addresses A multicast address space in the range from 224.0.1.0 through 238.255.255.255. These addresses can be used to multicast data between organizations and across the Internet.

GLOP addressing RFC 2770 proposes that the 233.0.0.0/8 address range be reserved for statically defined addresses by organizations that already have an AS number reserved. The second and third octets are populated with the decimal equivalents of the first and second bytes of the hexadecimal representation of the AS number.

GMRP (GARP Multicast Registration Protocol) An application that provides a constrained multicast flooding facility similar to IGMP snooping.

GVRP (GARP VLAN Registration Protocol) A GARP application that provides IEEE 802.1Q-compliant VLAN pruning and dynamic VLAN creation on 802.1Q trunk ports (similar to VTP).

H

hello interval The amount of time that elapses between the sending of HSRP hello packets. The default is 3 seconds.

hello packet An HSRP protocol data unit used to communicate HSRP information such as the virtual IP address, the hello time, and the hold time. This is also called a hello message.

Hello Time The time interval between the sending of configuration BPDUs.

hold time The amount of time it takes before HSRP routers in an HSRP group declare the active router down. The default is 10 seconds.

HSRP (Hot Standby Router Protocol) The Layer 3 Cisco-proprietary protocol that lets a set of routers on a LAN segment work together to present the appearance of a single virtual router or default gateway to the hosts on the segment. HSRP enables fast rerouting to alternate default gateways should one of them fail.

HSRP group Routers on a subnet, VLAN, or a subset of a subnet participating in an HSRP process. Each group shares a virtual IP address. The group is defined on the HSRP routers.

HSRP state The descriptor for the current HSRP condition for a particular router interface and a particular HSRP group. The possible HSRP states are initial, learn, listen, speak, standby, and active.

I

idle timeout A timer that specifies how long a connection stays active without any keystrokes taking place.

IEEE 802.10 The IEEE standard that provides a method for transporting VLAN information inside the IEEE 802.10 frame (FDDI). The VLAN information is written to the security association identifier (SAID) portion of the 802.10 frame. This allows for transporting VLANs across FDDI backbones.

IEEE 802.1p The IEEE standard that was incorporated into the 1998 version of the IEEE 802.1D standard. It introduces the concept of traffic classes. Layer 2 switches and bridges supporting this standard can be configured to effectively prioritize time-critical traffic, such as voice and video.

IEEE 802.1Q The IEEE standard for identifying VLANs associated with Ethernet frames. IEEE 802.1Q trunking works by inserting a VLAN identifier into the Ethernet frame header.

IEEE 802.3ab The IEEE standard that specifies 1000 Mbps over Category 5 and Category 5e cable (1000BaseT).

IEEE 802.3ae The IEEE standard that specifies 10 Gigabit Ethernet implementations. 10 Gigabit Ethernet is a full-duplex technology targeting the LAN, MAN, and WAN application spaces.

IEEE 802.3z The IEEE standard that specifies 1000 Mbps over fiber-optic cable.

IGMP (Internet Group Management Protocol) The protocol used to allow hosts to communicate to local multicast routers their desire to receive multicast traffic.

IGMP leave-group message A message used by IGMPv2 to allow a host to announce its intention to leave a multicast group. The leave message is sent to the all-routers multicast address, 224.0.0.2.

IGMP membership query A message used by IGMP and sent to the all-hosts multicast address, 224.0.0.1, to verify that at least one host on the subnet is still interested in receiving traffic directed to that group. Also called a general query.

IGMP snooping fast-leave An IGMP feature that allows the switch processor to remove an interface from the port list of a forwarding-table entry without first sending out a MAC-based general query on the port. When an IGMP leave is received on a port, the port is immediately removed from the multicast forwarding entry.

IGMP membership report A message used by IGMP to indicate a host's interest in receiving traffic for a particular multicast group.

IGMP snooping A Layer 2 multicast-constraining mechanism (like CGMP) used by switches. It allows multicast traffic to be forwarded to interfaces associated with IP multicast devices. The switch snoops on the IGMP traffic between the host and the router and keeps track of multicast groups and member ports.

IGMP Version 1 The version of IGMP that relies solely on membership query messages and membership report messages.

IGMP Version 2 The version of IGMP that adds the ability of a host to proactively leave a multicast group and the ability of a router to send a group-specific query (as opposed to a general query).

IGMP Version 3 The version of IGMP that adds support for source filtering.

inline rewrite Some Catalyst 5000 family switching line cards have onboard hardware that performs PDU header rewrites and forwarding, maximizing IP MLS performance. When the line cards perform the PDU header rewrites, this is called inline rewrite. With inline rewrite, frames traverse the switch bus only once.

IOS-based switch A switch running the Cisco IOS. The OS on these switches mirrors that of Cisco routers, except for the addition of the VLAN database configuration mode.

IRDP (ICMP Router Discovery Protocol) Hosts supporting IRDP dynamically discover routers in order to access nonlocal networks. IRDP allows hosts to locate routers (default gateways). Router discovery packets are exchanged between hosts (IRDP servers) and routers (IRDP clients).

ISL (Inter-Switch Link) A Cisco-proprietary encapsulation protocol for creating trunks. ISL prepends a 26-byte header and appends a 4-byte CRC to each data frame.

L

LACP (Link Aggregation Control Protocol) Defined in IEEE 802.3ad. Allows Cisco switches to manage Ethernet channels with non-Cisco devices conforming to the 802.3ad specification.

LANE (LAN Emulation) An ATM Forum standard used to transport VLANs over ATM networks.

Learning state The STP state in which the bridge does not pass user data frames but builds the bridging table and gathers information, such as the source VLANs of data frames. This state follows the Listening state.

limited-scope address Also called an administratively scoped address. These addresses fall in the range 239.0.0.0 through 239.255.255.255 and are limited for use by a local group or organization.

link local address The IANA-reserved multicast address space in the range from 224.0.0.0 through 224.0.0.255 to be used by network protocols on a local network segment. These packets remain local to a particular LAN segment and are always sent with a TTL of 1.

Listening state The STP state in which no user data is passed, but the port sends and receives BPDUs in an effort to determine the active topology. It is during the Listening state that the three initial convergence steps take place—elect a Root Bridge, elect Root Ports, and elect Designated Ports. This state follows the Blocking state.

local VLAN Also known as a geographic VLAN. A local VLAN is defined by a restricted geographic location, such as a wiring closet.

logical security control A security control used to create boundaries between networks and to provide logical access controls.

M

management VLAN The VLAN on the network used for network management.

Max Age An STP timer that controls how long a bridge stores a BPDU before discarding it. The default is 20 seconds.

MBGP (Multiprotocol BGP) An extension to BGP that enables multicast routing policy throughout the Internet. MBGP allows the interconnection of multicast topologies within and between BGP autonomous systems. With MBGP, you can configure BGP peers that exchange both unicast and multicast network layer reachability information.

MBONE A multicast backbone across the public Internet, built with tunnels between DVMRP-capable Sun workstations running the mroute daemon (process).

me1 interface A physical, out-of-band, Ethernet management interface/port on the Catalyst 2948G, 2980G, and 4000 families of switches. This interface is used for network management only and does not support network switching. It acts like a switch port in that it is accessed from a PC's NIC with straight-through cable.

MISTP (Multiple Instances of Spanning Tree) Allows you to group multiple VLANs under a single instance of spanning tree. MISTP combines the Layer 2 load-balancing benefits of PVST+ with the lower CPU load of IEEE 802.1Q.

MLS (multilayer switching) A specific multilayer switching technology employed by various Cisco devices to perform wire-speed PDU header rewrites. The first packet in a flow is routed as normal, and subsequent packets are switched by the MLS-SE based on cached information.

MLSP (Multilayer Switching Protocol) The protocol used to communicate MLS information between the MLS-SE and MLS-RP. In particular, the MLS-SE populates its Layer 2 CAM table with updates received from MLSP packets.

MLS-RP (MLS Route Processor) A Cisco device with a route processor that supports MLS. For example, a Catalyst 3620 router is an MLS-RP.

MLS-RP management interface Sends hello messages, advertises routing changes, and announces VLANs and MAC addresses of interfaces participating in MLS.

MLS-SE (MLS Switching Engine) The set of hardware components on a Catalyst switch, excluding the route processor, that are necessary to support MLS. For example, a Catalyst 5000 with a Supervisor Engine IIIG is an MLS-SE.

MOSPF (Multicast OSPF) The multicast routing protocol specified in RFC 1584. MOSPF is a set of extensions to OSPF that enable multicast forwarding decisions.

MSDP (Multicast Source Discovery Protocol) A TCP-based mechanism to connect multiple PIM-SM domains. MSDP allows multicast sources for a group to be learned via source active messages by RPs in different domains.

MSFC (Multilayer Switch Feature) MSFC1 and MSFC2 are daughter cards to the Catalyst 6000 Supervisor Engine that provide multilayer switching functionality and routing services between VLANs.

MSM (Multilayer Switch Module) A line card for the Catalyst 6000 family of switches that runs the Cisco IOS router software and directly interfaces with the Catalyst 6000 backplane to provide Layer 3 switching.

MST (Mono Spanning Tree) The spanning-tree implementation used by non-Cisco 802.1Q switches. One instance of STP is responsible for all VLAN traffic.

MST (Multiple Spanning Tree) Extends the IEEE 802.1w Rapid Spanning Tree (RST) algorithm to multiple spanning trees, as opposed to the single CST of the original IEEE 802.1Q specification. This extension provides for both rapid convergence and load balancing in a VLAN environment.

multicast group An arbitrary group of hosts that expresses an interest in receiving a particular data stream via multicast. This group has no physical or geographical boundaries. Hosts that are interested in receiving data flowing to a particular group must join the group using IGMP.

multilayer switching Combines Layer 2 switching, Layer 3 routing functionality, and the caching of Layer 4 port information. Multilayer switching provides wire-speed switching enabled by Application-Specific Integrated Circuits (ASICs).

MVR (Multicast VLAN Registration) Provides the ability to have a single multicast stream traverse the network via a dedicated multicast VLAN, regardless of the number of receivers (all residing on VLANs separate from the multicast VLAN). Designed for applications using wide-scale deployment of multicast traffic, such as television channels, across a service provider network.

N

Native IOS A single (optional) IOS image that simultaneously controls the Layer 2, 3, and 4 configuration of a Catalyst 6000 switch. It is also called Catalyst 6000 IOS, Cat IOS, or Supervisor IOS. In this case, the Supervisor module and the Multilayer Switch Feature Card (MSFC) both run a single bundled Cisco IOS image.

native VLAN The VLAN that a trunk port reverts to if trunking is disabled on the port.

NetFlow table One of three tables used by CEF. The NetFlow table provides network accounting data. It is updated in parallel with the CEF-based forwarding mechanism provided by the FIB and adjacency tables.

NFFC (NetFlow Feature Card) The daughter card for a Catalyst 5000 Supervisor module that enables intelligent network services, such as high-performance multilayer switching and accounting and traffic management.

normal mode Also known as Cat IOS and Supervisor IOS. It is the mode on a CatOS switch that allows you to view most switch parameters. Configuration changes are not allowed in this mode.

NTP (Network Time Protocol) Used to synchronize clocks on network devices. The multicast address 224.0.1.1 is reserved for NTP operation.

P

PAgP (Port Aggregation Protocol) A Cisco-proprietary technology that facilitates the automatic creation of EtherChannels by exchanging packets between Ethernet interfaces.

Path Cost An STP measure of how close bridges are to each other. Path Cost is the sum of the costs of the links in a path between two bridges.

physical security control A security control dealing with physical infrastructure, physical device security, and physical access.

PIM (Protocol Independent Multicast) The IP multicast routing protocol that derives its information using RPF checks based on the unicast routing table. PIM has several variations, including dense mode, sparse mode, sparse-dense mode, source-specific mode, and bidirectional mode.

PIM-DM (Protocol Independent Multicast Dense Mode) A PIM variation that builds a source tree for each multicast source and uses flood-and-prune behavior.

PIM-SM (Protocol Independent Multicast Sparse Mode) A PIM variation that builds shared trees. The root of the tree is called the rendezvous point, similar to CBT's core.

PIM sparse-dense mode An alternative to pure dense mode or pure sparse mode for a router interface. Sparse-dense mode allows individual groups to be run in either sparse or dense mode, depending on whether RP information is available for that group. If the router learns RP information for a particular group, it is treated as sparse mode; otherwise, it is treated as dense mode.

port channel A logical interface on a switch or router into which physical interfaces are grouped to form a single logical link.

Port ID A 2-byte STP parameter consisting of an ordered pair of numbers. The first is the port priority, and the second is the port number.

Port Number A numerical identifier used by Catalyst switches to enumerate a port.

Port Priority A configurable STP parameter with values ranging from 0 to 63 on a CatOS switch (the default is 32) and from 0 to 255 on an IOS-based switch (the default is 128). Port Priority influences Root Bridge selection when all other STP parameters are equal.

port-based VLAN Also known as a static VLAN. A port-based VLAN is configured manually on a switch, where ports are mapped, one-by-one, to the configured VLAN. This hard-codes the mapping between ports and VLANs directly on each switch.

PortFast A Catalyst feature that, when enabled, causes an access or trunk port to enter the spanning tree Forwarding state immediately, bypassing the Listening and Learning states.

preempt An HSRP feature that lets the router with the highest priority immediately assume the active role at any time.

prevention control A logical access control used to uniquely identify authorized devices and users and deny access to unauthorized devices and users.

private VLAN A VLAN you configure to have some Layer 2 isolation from other ports within the same private VLAN. Ports belonging to a private VLAN are associated with a common set of supporting VLANs that create the private VLAN structure.

priority An HSRP parameter used to facilitate the election of an active HSRP router for an HSRP group on a LAN segment. The default priority is 100. The router with the greatest priority for each group is elected the active forwarder for that group.

privileged mode The CatOS switch mode used to change the system configuration. This differs from Cisco routers, in which you need to enter global configuration mode to make system changes.

proxy ARP Lets an Ethernet host with no knowledge of routing communicate with hosts on other networks or subnets. Such a host assumes that all hosts are on the same local segment and that it can use ARP to determine their hardware addresses. Routers handle the proxy ARP function.

pseudobroadcast A process used by routers in a multipoint Frame Relay environment that uses a broadcast queue operating independently of the normal interface queue. Pseudobroadcast allows the router to simulate a broadcast environment in a nonbroadcast multiaccess network.

PVST (Per-VLAN Spanning Tree) A Cisco-proprietary spanning tree implementation requiring ISL trunk encapsulation. PVST runs a separate instance of STP for each VLAN.

PVST+ A Cisco-proprietary implementation that allows CST and PVST to exist on the same network.

R

RGMP (Router-Port Group Managment Protocol) Allows a router to communicate to a switch the IP multicast group for which the router wants to receive or forward traffic. RGMP is designed for switched Ethernet backbone networks running PIM.

Root Path Cost The cumulative cost of all links to the Root Bridge.

Root Port The port on a bridge closest to the Root Bridge in terms of Path Cost.

route processor The main system processor in a Layer 3 networking device. Responsible for managing tables and caches and for sending and receiving routing protocol updates.

router discovery The process in which a host determines a default gateway in order to send data beyond the local LAN segment.

router-on-a-stick A method for inter-VLAN routing consisting of an external router with a Fast Ethernet, Gigabit Ethernet, or EtherChannel trunk connecting to a switch, utilizing ISL or 802.1Q. Subinterfaces on the trunk are created to correspond to VLANs in a one-to-one fashion.

RP mapping agent A router that receives RP-announcement messages from the candidate RPs and arbitrates conflicts. The RP mapping agent sends group-to-RP mappings to all multicast-enabled routers via the Cisco-Discovery group, 224.0.1.40.

RSFC (Route Switch Feature Card) A daughter card to the Catalyst 5000/5500 Supervisor Engine IIG and IIIG that provides inter-VLAN routing and multilayer switching functionality. The RSFC runs Cisco IOS router software and directly interfaces with the Catalyst switch backplane.

RSM (Route Switch Module) A line card that interfaces with the Supervisor Engine of a Catalyst 5000/5500 switch to provide inter-VLAN routing functionality. The RSM runs the Cisco IOS.

S

SA (Source Active) A message used between MSDP peers that identifies the sources and associated multicast groups.

sc0 interface The logical, in-band, management interface on a CatOS switch that is associated with the management VLAN. The sc0 interface participates in the functions of a normal switch port, such as spanning tree, CDP, and VLAN membership.

server farm An industry term used to describe a set of enterprise servers housed in a common location, with high-speed links to the intranet, typically deployed in concert with a secure firewall and web caching servers. The services provided by server farms can include database access, e-mail, and DNS.

shared tree A multicast distribution tree that uses a root placed at a chosen point in the network. PIM-SM and CBT both use shared trees. When using a shared tree, sources must send their traffic to the root for the traffic to reach the receivers.

sink RP Also known as the RP of last resort. A statically configured RP for all multicast groups in a PIM network with all interfaces configured for sparse-dense mode. It serves as a mechanism to successfully implement Auto-RP and prevent any groups other than 224.0.1.39 and 224.0.1.40 from operating in dense mode. Auto-RP-discovered group-to-RP mappings take precedence over those configured via a default RP.

sl0 interface The out-of-band management port that relies on SLIP. This interface is found on the Catalyst 2926G, 2948G, 2980G, 4000, 5000, and 6000 families of switches. By connecting a modem to the console port, a network engineer can remotely dial up the switch using SLIP.

source filtering A capability added to IGMPv3 that lets a multicast host signal to a router the groups for which it wants to receive multicast traffic and the sources from which it wants to receive the traffic.

source tree A multicast distribution tree with its root at the source of the multicast traffic and whose branches form a spanning tree through the network to the receivers. Also called the shortest-path tree (SPT). Every source of a multicast group has a corresponding SPT.

SPAN (Switched Port Analyzer) A tool used with Catalyst switches to enable the capture of traffic. A SPAN session is an association of a destination port with a set of source ports, configured with parameters that enable the monitoring of network traffic.

SSM (Source Specific Multicast) An extension of IP multicast in which datagram traffic is forwarded to receivers from only multicast sources to which the receivers have explicitly joined. For multicast groups configured for SSM, only source-specific multicast distribution trees (no shared trees) are created. SSM relies on IGMPv3.

static multicast route Also called an mroute. A manually configured specification of where multicast traffic for a particular group should originate relative to the configured router.

STP (Spanning-Tree Protocol) A Layer 2 protocol that utilizes a special-purpose algorithm to discover physical loops in a network and effect a logical loop-free topology.

Supervisor Engine Also called Supervisor module or Supervisor. Can be thought of as the brains of a chassis-based Catalyst switch. All other modules in the chassis can be accessed or configured from the Supervisor module. The Catalyst OS runs on the Supervisor module.

switch block A set of logically grouped switches and associated network devices that provide a balance of Layer 2 and Layer 3 services. A switch block is a relatively self-contained, independent collection of devices, typically binding a collection of Layer 3 networks. Switch blocks are used as a tool to facilitate communication about switched network design.

switch fabric The hardware architecture in a chassis-based Catalyst switch that enables high-speed point-to-point connections to each line card. The switch fabric provides a mechanism to simultaneously forward packets on all point-to-point connections between the chassis slots.

System ID Extension The 12-bit number of the VLAN or the MISTP instance on Catalyst 6000 family switches that support 4096 VLANs.

T

tracking An HSRP feature that allows you to specify other interfaces on the router for the HSRP process to monitor. If the tracked interface goes down, the HSRP standby router takes over as the active router. This process is facilitated by a decrement to the HSRP priority resulting from the tracked interface line protocol going down.

trunk A point-to-point link connecting a switch to another switch, a router, or a server. A trunk carries traffic for multiple VLANs over the same link. The VLANs are multiplexed over the link with a trunking protocol.

tunnel A mechanism that enables the propagation of packets or frames through a series of devices in such a way that the contents of the packets or frames are not changed in transit.

U

UplinkFast A Catalyst feature that accelerates the choice of a new Root Port when a link or switch fails.

USA (Unicast Source Address) A notation used with CGMP that denotes the Layer 2 address of the source of multicast traffic for a particular group.

V

VACL (VLAN access control list) A generalized access control list applied on a Catalyst 6000 switch that permits filtering of both intra-VLAN and inter-VLAN packets.

VLAN (Virtual LAN) A group of end stations with a common set of requirements, independent of their physical location, that communicate as if they were attached to the same wire. A VLAN has the same attributes as a physical LAN but allows you to group end stations even if they are not located physically on the same LAN segment.

VLAN configuration mode The mode on an IOS-based switch that is accessed from privileged mode with the **vlan database** command. VLAN configuration mode is used to make changes to the VLAN configuration, such as adding and removing VLANs.

VMPS (VLAN Management Policy Server) A Cisco-proprietary solution for enabling dynamic VLAN assignments to switch ports within a VTP domain.

VTP (Virtual Trunking Protocol) A Cisco-proprietary protocol used to communicate information about VLANs between Catalyst switches.

VTP domain Also called a VLAN management domain. A network's VTP domain is the set of all contiguously trunked switches with the same VTP domain name.

VTP pruning A switch feature used to dynamically eliminate, or prune, unnecessary VLAN traffic.

X

XTAG A 1-byte value that the MLS-SE attaches to each VLAN for all MAC addresses learned from the same MLS-RP via MLSP.

Index

Numerics

A

B

C

D

E

F

G

J-L

M

P

Q

R

S

T

U

V